3rd International Conference on Mathematics, Science, and Education (ICMSE 2016)

Journal of Physics: Conference Series Volume 824

Semarang, Indonesia
3 – 4 September 2016

ISBN: 978-1-5108-3963-2
ISSN: 1742-6588

Printed from e-media with permission by:

Curran Associates, Inc.
57 Morehouse Lane
Red Hook, NY 12571

Some format issues inherent in the e-media version may also appear in this print version.

Printed by Curran Associates, Inc. (2017)

Institute of Physics
Dirac House, Temple Back
Bristol BS1 6BE UK

Phone: 44 1 17 929 7481
Fax: 44 1 17 920 0979

techtracking@iop.org

Additional copies of this publication are available from:

Curran Associates, Inc.
57 Morehouse Lane
Red Hook, NY 12571 USA
Phone: 845-758-0400
Fax: 845-758-2633
Email: curran@proceedings.com
Web: www.proceedings.com

Table of contents

Volume 824

The 3rd International Conference on Mathematics, Science and Education 2016

3–4 September 2016, Semarang, Indonesia

Accepted papers received: 13 March 2017
Published online: 18 April 2017

Preface

Papers

Preface

The 3rd International Conference of Mathematics, Science, and Education (ICMSE) 2016 on Semarang, 3-4 September 2016 organized by Faculty of Mathematics and Natural Science, Semarang State University. ICMSE2016 provides a platform to the researchers, experts and practitioners from academia, governments, NGOs, research institutes, and industries to meet and share cutting-edge progress in the fields of mathematics and natural science. It is reflected in this year theme "Contribution of Mathematics and Science Research for Sustainable Life in Facing Global Challenge". The scope of this conference are Mathematics, Biology, Chemistry, and Physics,

We thank to the keynote speakers and all authors of the contributed papers, for the cooperation rendered to us in the publication of the conference proceedings. In particular, we would like to place on record our thanks to the expert reviewers who have spared their time reviewing the papers. We also highly appreciate the assistance offered by many volunteers in the preparation of the conference proceedings, and of course to the sponsors assisting in funding this conference, especially Research, Technology and Higher Education Ministry of Indonesia for supporting this conference.

The committee selected 71 papers from 129 papers presented in this forum to be published in **Journal of Physics: Conference Series** (Institute of Physics Publisher) indexed by Scopus. We hope that this program will further stimulate research in Mathematics, Science, and Education; share research interest and information; and create a forum of collaboration and build trust relationship. We feel honored and privileged to serve the best recent developments in the field of Mathematics and Science Education to you through this exciting program.

Chairperson,

Dr. Margareta Rahayuningsih

IOP Conf. Series: Journal of Physics: Conf. Series **824** (2017) 011001 doi:10.1088/1742-6596/824/1/011001

COMMITTEE

International Scientific Advisory Board

Edy Cahyono (*Chemistry Department, State University of Semarang*)
Rahim Sahar (*Department of Physics, Universiti Teknologi Malaysia*)
Uda Hashim (*Institute of Nano Electronic Engineering, Universiti Malaysia Perlis*)
Sutikno (*Physics Department, State University of Semarang*)
Masturi (*Physics Department, State University of Semarang*)
Putut Marwoto (*Physics Department, State University of Semarang*)
Yu-Cheng Hsu (*Department of Natural Resources and Environmental Studies, National Dong Hwa University*)
Takeshi Sakurai (*Department of Molecular Neuroscience and Integrative Physiology, Kanazawa University*)
Ibnu Maryanto (*Zoology Division Research Centre for Biology, Indonesian Institute of Sciences*)
Hans Dieter Barke (*Westfälische Wilhelms-Universität Münster, Institut für Didaktik der Chemie*)
Roberta Hunter (*Math Education, Institute of Education, Massey University*)
Anuradha Mathrani (*Computer Science and Computer Technology, Massey University*)
Arramel (*Chemical Physics, Department of Physics, National University of Singapore*)

Organizing Committee

Chairperson:
Margareta Rahayuningsih

Members:
Zaenuri
Hardi Suyitno
Wiyanto
Amin Retnoningsih
Wardono
Cepi Kurniawan
Isti Hidayah
Sudarmin
Endang Sugiharti
Aji Purwinarko
Dante Alighiri
Stephani Diah Pamelasari
Muhammad Abdullah

Symposium Documentation

Dr. Margareta Rahayuningsih S.Si, M.Si

IOP Conf. Series: Journal of Physics: Conf. Series **824** (2017) 011001 doi:10.1088/1742-6596/824/1/011001

Prof. Intan Ahmad, Ph.D

Robby Gunawan

IOP Conf. Series: Journal of Physics: Conf. Series **824** (2017) 011001 doi:10.1088/1742-6596/824/1/011001

Prof. Roberta Hunter

IOP Conf. Series: Journal of Physics: Conf. Series **824** (2017) 011001 doi:10.1088/1742-6596/824/1/011001

Prof. Takeshi Sakurai

IOP Conf. Series: Journal of Physics: Conf. Series **824** (2017) 011001 doi:10.1088/1742-6596/824/1/011001

Dr. Anuradha Mathrani

IOP Conf. Series: Journal of Physics: Conf. Series **824** (2017) 011001 doi:10.1088/1742-6596/824/1/011001

Dr. Arramel

Prof. Dr. St. Budi Waluya

IOP Conf. Series: Journal of Physics: Conf. Series **824** (2017) 011001 doi:10.1088/1742-6596/824/1/011001

Parallel session with invited speaker Runbir Singh

IOP Conf. Series: Journal of Physics: Conf. Series **824** (2017) 011001 doi:10.1088/1742-6596/824/1/011001

Parallel session

Poster Presentation

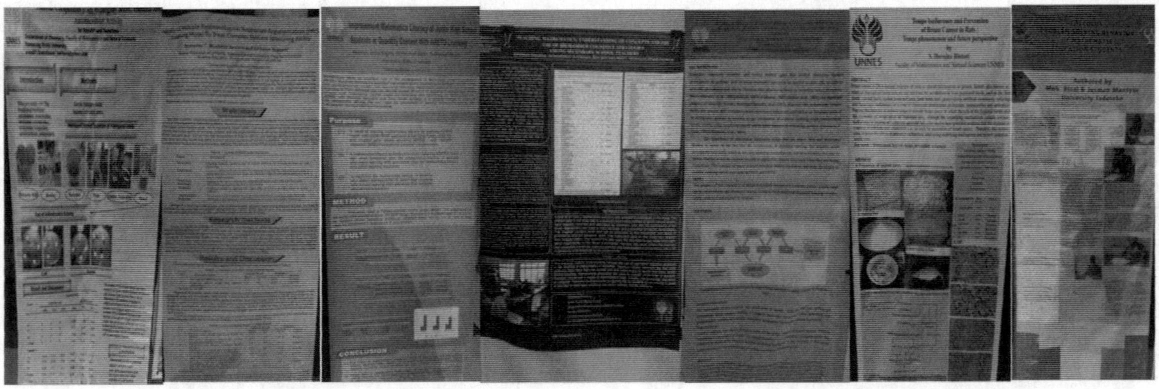

Peer review statement

All papers published in this volume of *Journal of Physics: Conference Series* have been peer reviewed through processes administered by the proceedings Editors. Reviews were conducted by expert referees to the professional and scientific standards expected of a proceedings journal published by IOP Publishing.

Stress Reduction of Pickup Truck Chassis Using Finite Element Method

O Kurdi[1, 2, *], M S Yob[1], S R Dasson[1], S Barrathi[1], A A Altayeb[1], and I Yulianti[3,]

[1] Mechanical Engineering Department, Faculty of Engineering, Universiti Selangor, Jalan Timur Tambahan, Bestari Jaya, Selangor Darul Ehsan, Malaysia
[2] Mechanical Engineering Department, Faculty of Engineering, Universitas Diponegoro, Jalan Prof Sudharto, Tembalang, Semarang, Central Java, Indonesia
[3] Physics Department, Faculty of Mathematics and Natural Sciences, Universitas Negeri Semarang, Kampus Sekaran, Central Java, Indonesia

*Corresponding author: ojokurdi@unisel.edu.my

Abstract. This paper performed stress reduction of the chassis of single cab pick truck using Finite Element Method (FEM). The first step was to evaluate and identify the critical area or maximum values of the static stress and its locations. The second stage of this project was dealing with designing a new chassis based on the Tata model, and then the final stage was a validation of numerical results with both experiments and analytical calculation to get the desired chassis that have strong and high-quality design.

1. Introduction

The pickup truck is one of the most important automobiles in the daily life all over the world. It is known for its use on tackling the difficulty of driving on damaged roads as well as in the rough desert land. One of the important parts of a pick up truck is chassis. The chassis of trucks is the backbone of vehicles and integrates the main truck component systems such as the axles, suspension, power train, cab, and trailer. Therefore it is important to have a strong truck chassis that can withstand the weight above it.

Stress is an important parameter of a chassis since high stress may lead to the failure of the chassis. The failure occurs since the magnitude of the highest stress is higher than the magnitude of ultimate stress or yield stress of the materials. There is avarious source of stress that is experienced by chassis which are a static, dynamic and cyclic load.

Numerous researches on thestatic characteristic performance of chassis have been done. Most of the works used FEM to perform stress analysis by employing commercial software such as NASTRAN, ABAQUS and ANSYS [1- 11]. In 2009, V. Veloso et al. [12] studied static characteristic to analyze the default of a longitudinal girder of a vehicle model and then optimized the prototype. The first step in the study was performing durability test on a prototype vehicle experimentally. The test result shows that failures occurred at the fenders fixation region of the suspension system on the vehicle. To overcome the problem, various reinforcement models were proposed, and the performance was analyzed numerically using finite element method (FEM). Modification of existing chassis was done by change the cross section of the chassis [2, 3, 8, 13, 14], reducing the thickness [1, 15] and modifying some part or region [7, 12, 15, 16, 17]. Validation of the simulation results was done by using analytical calculation [18-22]. However, this work proposed new design by change the certain region of the chassis, reduce the thickness and validate the simulation result by both analytical calculation and also experimental work.

This paper is focused on stress reduction of pickup chassis by modifying the design of the existing model which is Tata Telco line single cab (TATA TL SC). The first step has analyzed the stress due to static load using FEM to evaluate and identify the maximum stress values and its location. After the identification was performed on the existing model, then modification was done to reduce the magnitude of stress in the region. The last step was a validation of FEM simulation result of the modified model by using the experimental work and analytical calculation.

2. Method

The existing chassis type is a ladder frame and was depicted in Figure 1. It consists of two long sections or beams. One section is on the right side, and the other is symmetrical to that on the left side. These beams are called the longitudinal members where all load will be distributed to. There are various types of cross-sectional members that commonly used which are *I*, *C*, and rectangular beams. In TATA TL SC, the rectangular beams are used. There are other members called cross members which connect the longitudinal members together by welding or bolt. These cross members are used to help the longitudinal members to get good strength and stability. Rod members or normal plats are used for this type of members. The longitudinal members are upswept on the front and the rear sides of the frame to provide space for the axils movement caused by the springs.

In this study, the stress analysis is accomplished by the commercial finite element packaged ABAQUS. The simulation was started by assigning the material properties on the model, putting the load and choosing the boundary condition. The material of chassis is Steel with 552 MPa of yield strength and 620 MPa of tensile strength.

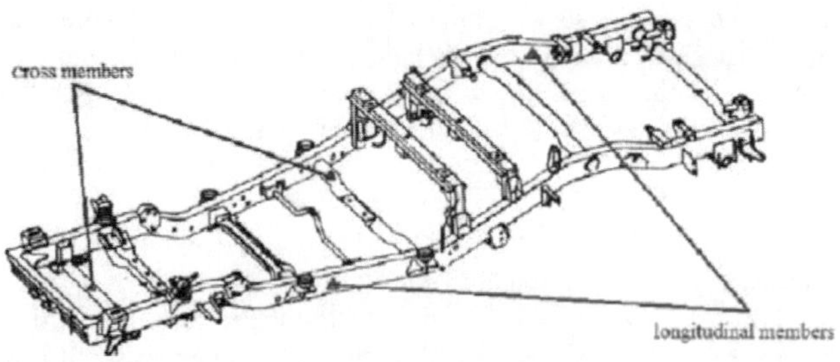

Figure1. TATA TL SC chassis frame (single cab, 4 x 2 and 4 x 4), (Ladder frame). (Tata Motors Limited)

3. Results and Discussion

Simulation result for the existing chassis is shown in Figure 2. The maximum Von Misses stress at the critical region is 227.2 MPa. For failures analysis, the magnitude of stress was compared to the ultimate tensile strength of the material. If the stress exceeds the ultimate stress of the material, the structure is considered as failed. The stress value obtained above are below yield stress. The safety factor for this condition is 2.42. Based on the data from the Department of Structure and Transport of Australian (Department of Structure and Transport), it was recommended by the manufacturer of trailer structure that their producthas the safety factor of at least 3 for highway usage and 5 for off road usage. The truck chassis also experience cyclic loading from road roughness. Structure with high stress which experiences cyclic loading will have alow life time. Therefore, it is important to reduce the maximum stress to improve the chassis life time. The stress can be reduced by optimizing the geometry in which the maximum stress occurred or by increasing the thickness of chassis.

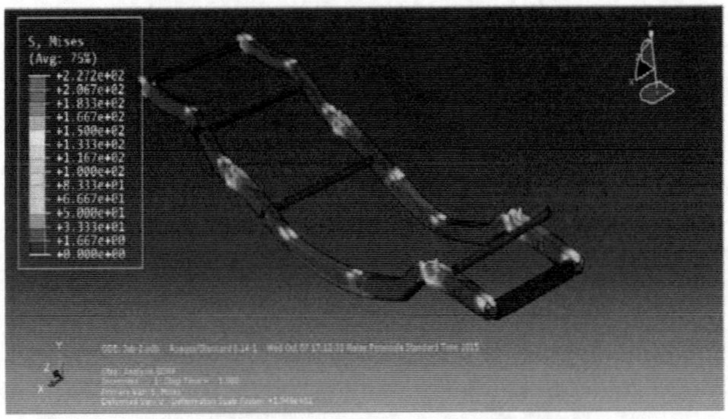

Figure 2. Von Misses stress of the existing chassis model

The modification was done by adding some cross members into the model to reduce the stress on the chassis (design 1) as shown in Figure 3. The cross section shape of the cross members also has been changed from the circular shape to the square shape. It will increase the stiffness and also inertia moment of structure that further results in stress reduction on the chassis. The simulation result of the modified model in term of Von Misses stress is shown in Figure 3. The Von Misses stress of the modified model was decreased by 23.42 % . The safety factor for this condition is 3.16 and the requirement for safety limit is satisfied.

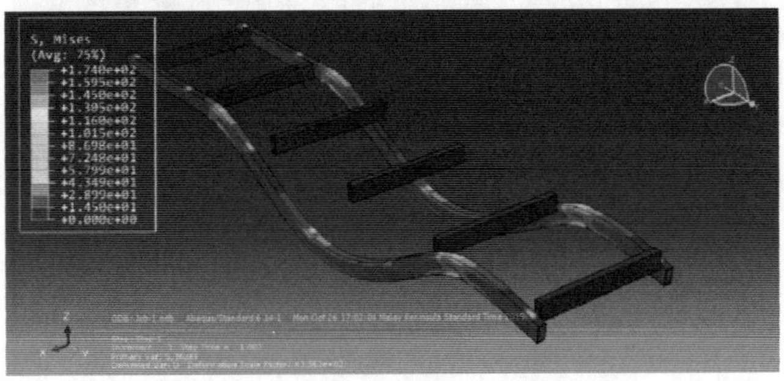

Figure 3. Von Misses Stress Of Design 1

Another modification was done to reduce the stress by proposed new design (design 2) as shown in Figure 4. The new proposed model contains two longitudinal members and six cross members. Both cross member and the longitudinal member has the square hollow shape. The model has the same type with the existing, but it is simpler since some complex shape in the existing design has been removed. The simulation result of design two at the front part of the chassis is shown in Figure 5. The magnitude of Von Misses stress at the front is 46.9 MPa that is much lower than that of the existing and the modified model. The design 2 shows the best performance compared to that of the existing and design 2.

Figure 4. Design 2 of the proposed model

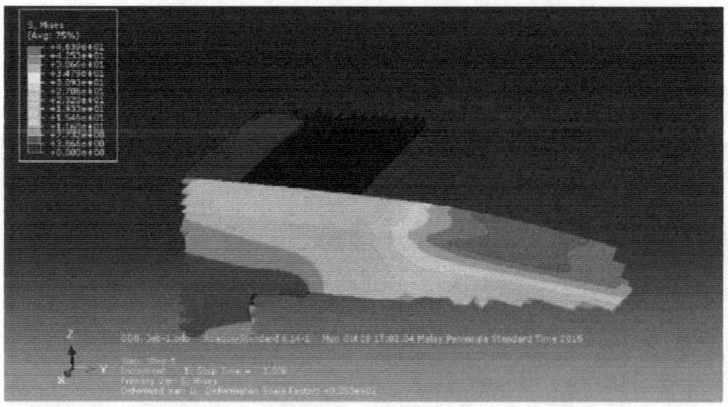

Figure 5. Von Misses stress for design two at the front part.

For validation purpose, the simulation results of the existing model were compared to the analytical calculation, while the simulation results of design two were compared to experimental results.

For analytical calculation, the structure can be assumed as simple beam subjected to several single loads. The bending moment distribution can be calculated after all of the reaction forces were obtained. Figure 6 shows the bending moment distribution of the existing model.

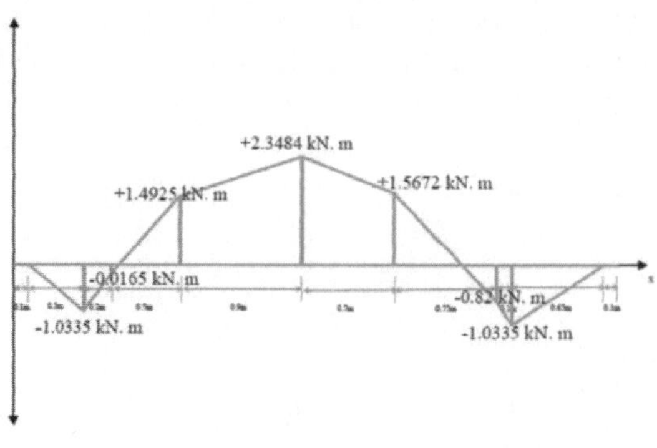

Figure 6. Bending moment diagram for the existing model

The value of Von Misses stress can be calculated using the formula for principle stress and Von Misses stress. The Von Misses stress obtained from the calculation is 240.5 MPa. The percentage different between simulation and analytical for existing model is 5.53%. It shows that the simulation result for design 1 is valid.

Comparison of the experimental results and the simulation results of design two is shown in Table 1. The comparison was made by comparing the Von Misses stress value at three locations which are point 1 at there are part, point 2 of the middle part and point 3 at the front part.

TABLE 1. Von Misses stress result comparison of design 2

	Von Misses stress (MPa) and point location		
	1	**2**	**3**
FEM result	5.9	17.6	29.2
Experiment result	12.43	21.5	46.5
Percentage different (%)	52.53	18.14	37.20
Average			**35.96**

Based on the percentage different on Table 1, it can be seen that the percentage different is quite high. The high different is due to assumption and simplification that was made in thesimulation. Meanwhile, the accuracy of the experimental results also depends on the setting of equipment, accuracy of the sensor and stability of the surrounding condition such as temperature and humidity. Some errors in measurement during the experiment most probably happened. However, the average percentage different of 35.96% can be accepted.

4. Conclusions

The Finite element method has been used to perform stress analysis of the chassis of single cab pick truck "Tata Telco line single cab" (TATA TL SC) successfully. The critical region of the chassis can be identified which was indicated by the highest Von Misses stress value. The modified models with better performance have been proposed. The modified model has asafety factor of 3.1 that higher than the required safety factor (3). The simulation results of the modified models are valid since they meet both analytical and experimental results.

Acknowledgement

We would like to thank Universiti Selangor (UNISEL) for funding the research. Our gratitude also goes to members of structure and materials laboratory, Civil Engineering Department UNISEL for providing the facilities for experimental work and the members of Mechanical Engineering Department UNISEL for their helpful discussion throughout the completion of this work.

References

[1] Agrawal M S & Razik M. 2013. *Int. J.Eng. Sci. Res. Technol.* 2(12), 3432-3438.
[2] Bhat KA, Untawale.SP, & Katore.HV. 2014. *Int. J. Pure. Appl. Res. Eng. Technol.***2(12)**71.
[3] ChandraM R, Sreenivasulu S& HussainS A. 2012.*Int. J. Model. Eng. Res*, **2(4)**2594.
[4] Rajappan R&Vivekanandhan M. 2013.*Int. J. Eng Sci*.**2(2)** 63.
[5] Dubey A & DwivediV.2003.The Vehicle Chassis Analysis by Load Cases & Boundary Conditions for Stress Analysis.
[6] R. Choudharya, T J. Bowsera, P. Wecklera, N. O. Manessb, and W. McGlynnb, 2009 Postharvest. Biol. Tec.**52**103.
[7] Si G M, Sun H, Zhang Q L, &Duan L. 2011b. *Appl. Mech. Materil.***55-57**. 110.

IOP Conf. Series: Journal of Physics: Conf. Series **824** (2017) 012001 doi:10.1088/1742-6596/824/1/012001

[8] Naveen Ala, K.Tejdeep Reddy , BVSS Bharadwaja. 2016. *International Journal of Innovative Reesearch in Science, Eng. and Tech.***5**326-337
[9] Rahul.V, Nirmal.R.U, Noraje Aniket Arunkumar, P.Baskar. 2016. *Int. Journal of ChemTech Research***9**384-391
[10] Gaurav Goel, Rajat Garg, Tarun Ranjan, Pratyanshu Soni and Baskar P. 2016. ARPN Journal of Eng.and Applied Sci. **11** 13599-603
[11] Bhuwanesh Sharma, Abdul Samad, Dr.Y.B.Mathur 2016 Int. Conf. on Emerging trends in Engineering & Management for Sustainable Development (Bikaner, Rajasthan, India)
[12] Veloso V, Magalhães H S, BicalhoG I & Palma E S.2009.Eng Failure Analysis.**16(5)** 1696.
[13] NayakD,Sharma K P& Parkhe A.2014.*Int. J. Adv. Technol.Eng.Sci*, **2(5)**292.
[14] Patil, Kamlesh Y & Deore, Eknath R.2015.*J.Mech.Civil.Eng.* **12(4 Ver. III)** 111.
[15] Patil H B, Kachave S D, Sharad, & Deorem E R.2013.*J. Mech. Civil. Eng.*6(1) 44.
[16] Bhoi A, Koushik L P, YadavN & Patnaik M.2013. Int. J. Eng. Res. Appl.**3(2)**1149.
[17] Patel V V, & Patel R I.2012. *Int. J. Eng. Res. Technol.* **1(3)** 1.
[18] Patel T M., Bhatt MG & Patel HK.2013. *Int. J. Emerg. Trend. Technol. Comput. Sci.***2(2)** 85.
[19] PatilN R, Kulkarni R R., Mane B R & Malve S H. 2014.**3(5)**661.
[20] Bajwa M S, Yatin R&Joshi A. 2013. *Int. J. Mech. Prod Eng.* 1(2).55.
[21] Sharma, P.K, Parekh, Nilesh J., & Nayak, Darshit.2014. *Int. J. Innov.Res.Sci.* 5(8). 182.
[22] Gaikwad A V & Ghawade P S. 2014.*Int. J. Technol Sci.***2(6)** 12.

Detection of Cadmium Ion by *Evanescent Wave* Based Chitosan Coated Optical Fiber Sensor

I Yulianti[1,*], S S Edy[1], B A Saputra[1], M P Aji[1], Susanto[1], and O Kurdi[2]

[1] Physics Department, Faculty of Natural Sciences and Mathematics, Universitas Negeri Semarang, Semarang, Indonesia
[2] Mechanical Engineering, Universitas Dipenogoro, Semarang, Indonesia

*Corresponding author: ianyulianti@gmail.com

Abstract. Evanescent wave based-optical fiber sensor to detect cadmium ion is proposed. Chitosan was used by using the *dip-coating* method. The sensor was fabricated in *U-bent* shape. *U-bent* optical sensor at aconcentration of 2ppm and 5ppm had asensitivity of 0.2067 dBm/ppm and -0.7995 dBm/ppm, respectively. At a level of 2ppm - 5ppm, the optical sensor has a linear response with asensitivity of -0.283 dBm/ppm. The sensor takes 9.5 minutes to reach steady stateat aconcentration of 1 ppm. Atalevel of 2ppm - 5ppm, the sensor takes 5 minutes to 10.45 minutes to reach steady state.

1. Introduction

The presence of heavy metal in water is a grave threat to human health and environment[1]. Heavy metal ions are very dangerous because it tends to accumulate[2], is not readily biodegradable (persistent) in the human body, and is toxic[3]. Among metal ions, cadmium is a contaminant that has a high toxicity, even at very low concentration[1]. The limit of cadmium ion concentration in drinking water is 0.05 ppm[1]. Medically, cadmium can affect human health, especially blood and renal system [4].

There are various methods to detect cadmium in the environment such as theconventional method by extracting the sample solution, solid phase extraction by adsorption or ion exchange, and electro-analytical techniques[1]. The methods have many disadvantages, such as: the measurement cannot be done real-time, in situ, and cannot be multiplexed. To overcome to disadvantages of the method described above, optical fiber sensorswere developed.

Optical fiber sensorisanother application of optical fiber technology which is initially applied in anoptical communication system to transmit light carrying anoptical signal. Optical fiber sensors have attracted significant attention due to its advantages such as high sensitivity, fast response, small size, electromagnetic interference independence, simple fabrication, and iseasy to implement[5].

Fiber optic sensor to detect heavy metal ions has been developed by various methods such as reflectance[6, 7], absorption[8], Fabry-Perot interferometer[9, 10], and fluorescence [11]. The reflectance method hasan intricate design that is not suitable to be used in multiplexed sensor systems. Meanwhile, thefluorescent method has the advantage of having high sensitivity. However, this technique has drawbacks such as complicated fabrication techniques and high costs. Whereas, absorption method has advantages regarding simple fabrication. However, it has alittle sensitivity. Another technique which exploits the evanescent wave phenomenon was also used for various application. Like absorption method, the evanescent wave technology also provides simple fabrication process.

An optical sensor which utilizes loss due to macro-bending has also been developed since it is easy to fabricate, has a linear response, and has a relatively good sensitivity[12]. One of the optical sensors which utilize macro-bending is a*U-bent* optical sensor. *U-bent* optical fiber sensor has been developed for many applications, such as optical bio sensor for detection of anti-IgG[13], optical fiber sensor for detection of explosive vapors [14]and optical bio sensor for detection of E.coli[15]. Therefore, in this work, an optical fiber sensor for cadmium ion detection was developed by using *U-bent optical fiber*using evanescent wave absorption technique.

2. Working Principle of the Sensor

To make an optical fiber serves as cadmium ion sensor, the optical fiber should be modified so that one of the light parameters, such as wavelength, intensity or phase, is modulated by the change of cadmium ion concentration. In evanescent wave absorption method, the variation of optical power due to absorption of the evanescent wave was used as anobserved parameter to detect the presence of the measurand.

In principle, theevanescent field is part of thelight that penetrate through the core-cladding interface. As the evanescent field penetrates into the cladding, the intensity decays exponentially as shown in Figure 1.

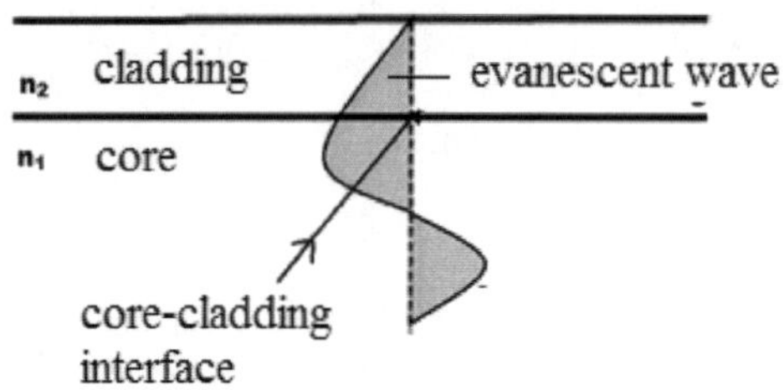

Figure 1. Illustration of evanescent wave

The optical power of the transmitted light through optical fiber $P(z)$ depends on the evanescent wave absorption coefficient as defined by

$$\boldsymbol{P(z) = P(0)e \quad (-\gamma\ )} \tag{1}$$

Where $P(0)$ is the input optical power, l is the optical fiber length, and γ is the evanescent wave absorption coefficient. Meanwhile, the evanescent wave absorption coefficient value depends on core refractive index (n_1) and cladding refractive index (n_2) and other parameters as given by

$$\gamma = \left(\frac{\lambda n_2 c \quad \theta c \quad \theta}{2\pi \ n_1^2 c \quad {}^2\theta_c \sqrt{s\ \ {}^2\theta - s\ \ {}^2\theta_c}}\right)\alpha \tag{2}$$

Where r is the optical fiber core radius, θ is the angle between the ray and the normal to the interface, $\theta_c$is the critical angle, and α is the absorption coefficient of thebulk material. Thus, the optical power of the transmitted light is profoundly affected by the refractive index of the core and cladding material. Therefore, by replacing optical fiber coating with other documents which its refractive index change with the presence of measurand, the optical fiber can serve as measurand transducer. To be used as aheavy metal sensor, various types of materials can be coated on the fiber, such as citizen for detection of lead and mercury [16]and Pyrrole and chitosan for the detection of cadmium, lead and mercury[17]. In this work, the material used to replace the cladding layer is chitosan.

3. Experiment

Silica multimode optical fiber with acladding diameter of 125 μm and acore diameter of 50 μm was used in the sensor fabrication. The first step of themanufacturing process was removing the jacket and the buffer of the fiber using Fiber opticstripper threeholes (H-119CC)at the region where the cladding layer would be replaced by chitosan. After the buffer layer had removed, the next step was etched the coating layer by immersing the fiber into 50% HF solution for 56 minutes. Before the etching process, the fiber was cleaned using alcohol. The chitosan was then coated on the cladding-removed area by using thedip-coating technique. Before coated on the fiber, the chitosan was first prepared by solving it using distilled water and 1% acetate acid. The process was done using magnetic stirrer at atemperature of 100°C for 3 hours. The chitosan coated-fiber was dried at room temperature for 24 hours. To improve the sensitivity, the fiber was bent at the chitosan coated region. The last step was attaching theoptical connector to both ends of the fibers.

4. Characterization of the Sensor

Characterization of the sensor was done by dipping an optical fiber that has been coated with chitosan into asolution of cadmium in various concentration values which are 0.9, 2, 3, 4, 5 and 6 ppm. Cadmium solution used in the characterization has a pH value of 7.One end of the optical fiber was connected to the optical light source, while the other end was attached to an optical power meter to measure the intensity of the light transmitted through the solution, as shown in Figure 2.

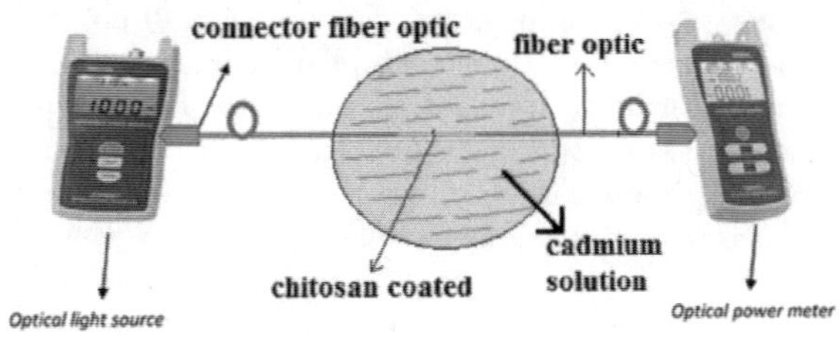

Figure 2. Characterization set-up

The response time of the sensor was measuredby recording the change of the output intensity for each value of the concentration. The response time is the time required for the sensor to reach 90% of the final value which is the valuein an equilibriumstate.

5. Results and Discussion

The developed optical fiber sensor is shown in Figure 3(a) and the image of chitosan-coated optical fiber taken by CCD microscope at amagnification of 400 times is shown in Figure 3(b).It is shown that the diameter of the coated fiber is 427 μm.

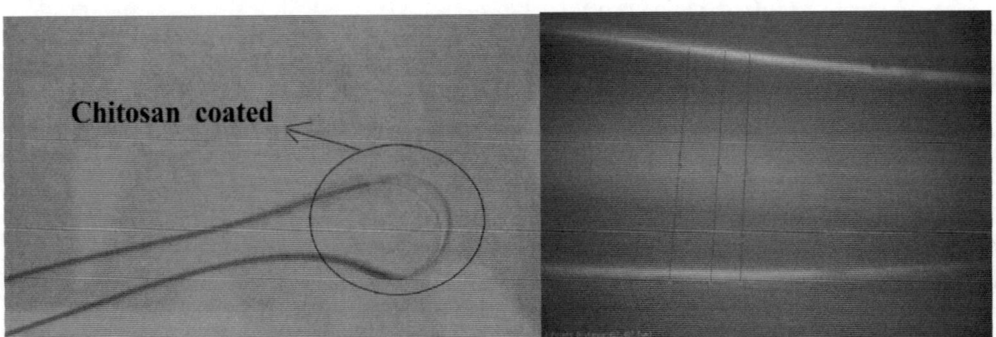

(a) (b)

Figure 3. The fabricated *U-bent* optical fiber (a) and the microscope image of the chitosan coating (b)

The output intensity was plotted against the solution concentration as shown in Figure 4(a). Polynomial regression of the curve showed that at aconcentration of 2 ppm and 5 ppm, the sensor has asensitivity of 0.2067 dBm / ppm and - 0.7995 dBm / ppm, respectively, with a correlation factor of 0.8072. It is also shown that the sensor provides liner response in the range of 2ppm to 5ppm with asensitivity of -0.283 dBm/ppm as shown in Figure 4(b). By using optical power meter with aresolution of 0.02dBm as adetector, the sensor can detect as low as 0.07ppm.

Contrary gradient value indicates that the light intensity decreased as theconcentration of cadmium ion increased. The decline of light intensity was caused by the increase of refractive index of the chitosan layer due to adsorption of cadmium ion. When the optical sensor was immersed in cadmium ion solution, an interaction between chitosan and cadmium ion occurred.The pores of chitosan were substituted by the cadmium ions and forming chelates structure and releasing H^+ ion [18].This process makes the chitosan surface smoother then that before it adsorbs cadmium ions and increases the chitosan density. The higher the cadmium ions concentration, the more chitosan pores substituted by cadmium ions, and thus the chitosan is denser. The increase of density results in the increase of refractive index. Therefore, as shown in equation (2), the optical power transmitted in the fiber decrease as the cladding refractive index increase and more power loss occurs during the light propagating through the fiber. However, when the dry sensor was first exposed to cadmium solution, the output intensity increased as shown in the graph. At dry state (0 ppm), the output power was -32.4dBm, while in 1ppm cadmium solution, the output intensity increased to -31.3dBm. The increase occurred since the chitosan at dry state was much denser than the chitosan in solution. Therefore, the refractive index of dry chitosan was much higher than that of wet chitosan. As a result, the output intensity of the sensor increased. It also can be observed from the results that great loss occurred as the light traveled along the sensor. The greatdamage occurred due to connection loss between fiber and laser source and optical power meter, connection loss at the fiber connector and loss due to macro-bending. However, the macro-bending loss gives advantage i.e. improving the sensor sensitivity by enhancing the absorption of the evanescent wave.

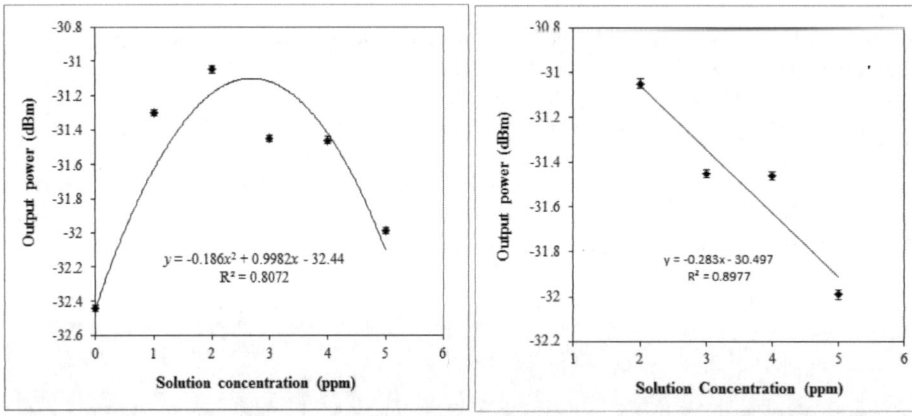

(a) (b)

Figure 4. Intensity vs. cadmium solution concentration: polynomial regression (a) and linear regression (b)

The response time of the sensor that indicates the time required for the sensors to reach steady state is shown in Figure 5. It is shown that at aconcentration of 1 ppm, the sensor requires 9.5 minutes to reach steady state. Meanwhile, at aconcentration of 2 ppm to 5 ppm, the sensor takes 5 minutes to 10.45 minutes to reach steady state.The pores of the membrane on the surface of the chitosan determine the time response of the sensor. The wider the pores, the faster the chitosan adsorb heavy metal ions[19]. The pore size of the chitosan depends on some of amine groups on the chitosan. It was also observed that the response time increased as the cadmium ions concentration increased. This increase of time response occurred since ahigher concentration of cadmium ions resulted in lower number of amine groups in the chitosan. Therefore, much longer time was required to reach steady state at ahigh concentration than that at low level. However, high response time was also observed also at 1 ppm which was caused by fluctuation of the optical light source that has not reach stable condition when the measurement was conducted.

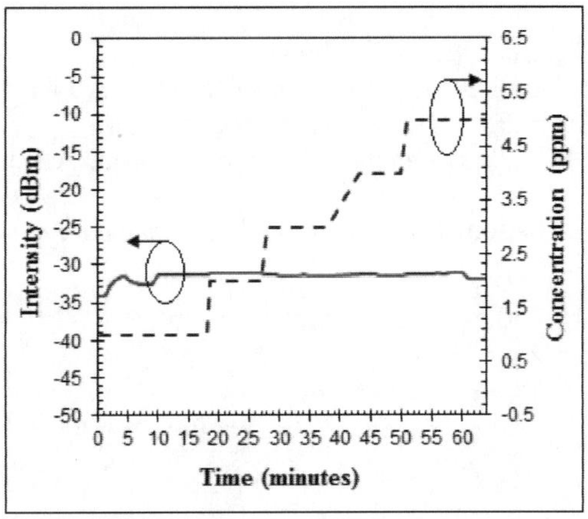

Figure 5. Response time of the optical sensor

6. Conclusions

U-bent optical fiber sensor to detect cadmium ions has been fabricated by modifying an optical fiber. Some part of optical fiber cladding was replaced with chitosan using thedip-coating technique. The sensor has alinear response in theconcentration range of 2ppm-5ppm. The sensor can measure cadmium concentration up to 0.07 ppm. This result still low compared to other published sensor [17]. However, the proposed sensor has anadvantage regarding thesimple fabrication process. The sensitivity can be improved by optimizing the coating thickness and the bending radius. About response time, the sensor needs to be improved which is the goal for our future work.

Acknowledgement

We would like to thank Universitas Negeri Semarang for funding the research. Our gratitude also goes to the members of Physics Department, Universitas Negeri Semarang for their helpful discussion throughout the completion of this work.

References

[1] Y. Tharakeswar, Y. Kalyan, B. Gangadhar, K. S. Kumar, and G. R. Naidu2012*Journal of Sensor Technology***2** 68-74.
[2] J. M. Unagolla and S. U. Adikary2015*Tropical Agricultural Research***26** 395-401.
[3] G. Zhang, R. Qu, C. Sun, C. Ji, and H. Chen 2008*Journal of Applied Polymer Science***110** 2321–2327.
[4] H. A. Yousef, A. Afify, H. M. Hasan, and A. A. Meguid 2010*Natural Science***2** 292-297.
[5] C. B. Mathews, T. M. Libish, P. Biswas, S. Bandyopadhyay, K. Dasgupta, and P. Radhakrishnan, "A Chitosan coated Fiber Optic Long Period Grating Biosensor for the Detection and Estimation of Cholesterol," in the 12th International Conference on Fiber Optics and Photonics 2014.
[6] R. Choudharya, T. J. Bowsera, P. Wecklera, N. O. Manessb, and W. McGlynnb 2009*Postharvest Biology and Technology***52** 103-109.
[7] H. Filik and Z. Yanaz 2009*Journal of Hazardous Materials***172**1297-1302.
[8] Y. Wu, X. Deng, F. Li, and X. Zhuang 2007*Sensors and Actuators* B **122** 127-133.
[9] L. H. Chen, T. Li, C. C. Chan, R. Menon, P. Balamurali, M. Shaillender 2012 *Sensors and Actuators B: Chemical***169** 167-172.
[10] J. Li, X. Qiao, R. Wang, Q. Rong, W. Bao, Z. Shao, et al., 2016 *Optics and Lasers in Engineering***79** 16-21.
[11] Avcı and İ. Kaya 2015*Tetrahedron Letters* **56** 1820-1824.
[12] P. Wang, Y. Semenova, Q. Wu, and G. Farrell 2011*Optics & Laser Technology***43** 922-925.
[13] V. V. R. Sai, T. Kundu, and S. Mukherji 2009*Biosensors and Bioelectronics***24** 2804-2809.
[14] R. Bharadwaj and S. Mukherji 2014*Sensors and Actuators B: Chemical***192** 804-811.

[15] R. Bharadwaj, V. V. R. Sai, K. Thakare, A. Dhawangale, T. Kundu, S. Titus, et al., 2011*Biosensors and Bioelectronics***26** 3367-3370.
[16] K. Zargoosh and F. F. Babadi 2014Spectrochimica Acta Part A **14** 1386-1425.
[17] R. Verma and B. D. Gupta 2015 *Food Chemistry***166** 568-575.
[18] Madala, S., Nadavala, S. K., Vudagandla, S., Boddu, V. M. and Abburi, K 2013*Arabian Journal of Chemistry*.
[19] W. Pitakpoolsil and M. Hunsom 2014 *Journal of Environmental Management***133** 284-292.

Interleaved Flyback DC-DC Converter Design with 350 W Power Output Using LT 3757 in LT Spice

S Rahayu[1,*], E Firmansyah[2], and M Isnaeni[2]

[1] Postgraduate Program of Electrical Engineering, School of Postgraduate Gadjah Mada University,Yogyakarta,Indonesia

[2] Department of Electrical Engineering and Information Technology Faculty of Engineering, Gadjah Mada University, Indonesia

*Corresponding author: sofitri.sie14@mail.ugm.ac.id

Abstract. DC-DC converter becomes one important part in micro-inverter used in solar panel application. Its function is to convert output voltage level of solar panel 42-48 V_{dc} to a voltage level of 350 V_{dc} before being converted into an AC voltage at the inverter. The proposed converter topology is a flyback because the number of components used is not too much which can suppress the production cost. In this paper, simulation of flyback converter on the interleaved operating mode with a maximum output power of 350 W using software Ltspicewas conducted. From the simulation results, obtained that by applying a switching frequency of 100 kHz, the obtained value of the components of the primary inductor (L_P) 3.3 μH, the secondary inductor (L_S) 27 μH, the output capacitor (C_{out}) 47μF and ripple voltage (V_r) 212.65 mV.

1. Introduction

Solar panel (photovoltaic) is used to convert solar radiation into anelectrical energy source. Micro-inverter (dc-ac power electronics converter) is an important component of thesolar panel system. Micro-inverter in photovoltaic conversion is a part of theinfrastructure that converts the power generated from photovoltaic panels into AC that can be transmitted to a network. There are three types of conventional inverter architecture: central inverter, string inverter, and micro-inverter [1][2]. Some researches on photovoltaic micro-inverter have been widely developed since micro-inverter can be directly applied in photovoltaic panel producing energy. Several other advantages of themicro-inverter system include: increasing reliability of the system, improve efficiency and reduce the cost[3].

Micro-inverter system can be implemented using two methods *i.e.* single power stage and dual power stage[4]. This research is focused on dual power stage method in the dc-dc converter topologies flyback. This dc-dc converter is highly important as it changes dc signal from thesolar panel into dc signal with a higher voltage. There are several ways to convert dc signal to a higher dc signal, using a flyback converter is one of them[3]. Flyback converter uses high-frequency transformer which is lighter and has asmaller size than lower frequency transformer, and it is also more efficient because it has asmaller size. Therefore, the overall shape of flyback converter is compact. This study provided the simulation of interleaved operating mode using Lt spice and LT 3757 as PWM controller, with a maximum power output of 350 W. The simulation results were then compared to the previous literature reviewed and it obtained 3.3 μH of primary inductor (Lp), secondary inductor (Ls) 27 μH, the output capacitor (Cout) 47 μF and ripple voltage (Vr) 212 mV.

2. Methods

2.1. Interleaved Flyback DC-DC Converter Circuit Design

Basic topology of the flyback dc-dc converter is shown in Figure 1. It consists of several parts: the input voltage from thesolar panel, high-frequency transformers, electronic switch, and capacitor filter that serves reduction of output voltage ripple.

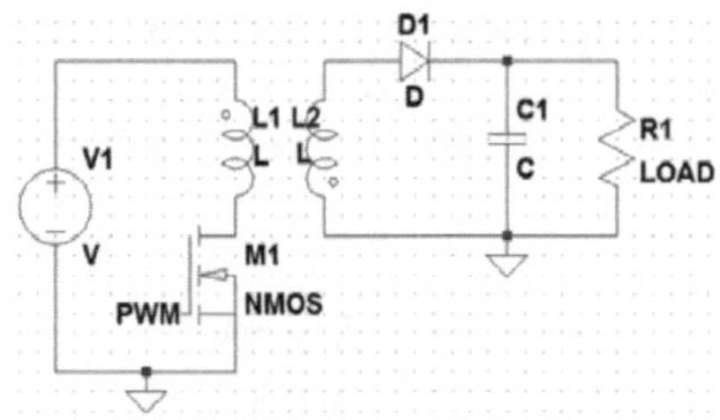

Figure 1. Topology Flyback Converter.

Input voltage from the solar panel to the flyback converter using a solar panel was adjusted to the sun module SW XL 340-350 MONO data sheet, which was 42-48 V with a power of 350 W. The interleaved flyback DC-DC converter circuit used LT 3757 as the PWM controller. PWM controller used as the electronic switch was MOSFET. MOSFET used in this research was Si7850DP with switching frequency of 100 kHz. This interleaved circuit changed the input DC voltage from the solar panel into a higher DC voltage through the switching process and thefeedback voltage from the output voltage. The activation of MOSFET was controlled by signals generated from the LT 3757.

This study used operating mode of interleaved flyback converter, as shown in Table 1 that specified the parameters used in the simulation of Lt spice. In this mode, flyback circuits were parallel to double the power of non-interleaved mode, and the output ripple voltage could be reduced due to the different phases when the electronic switch was on or off.

Tabel 1. Specifications of flyback dc-dc converter parameters in LT SPICE simulation

Parameter	Specification
Tegangansumber PV	42 – 48 V_{dc}
Frekuensi	100 kHz
N-Mosfet	Si 7850 DP
Kontrol PWM	LT 3757
DiodaSchottky	MBRS 360
Duty Cycle (D)	2,758 ms
Duty Cycle 2 (D_2)	1,223 ms

2.2. LT 3757

LT 3757 is a dc-dc controller capable of generating positive or negative output voltage with a wide input range. LT 3757 can be configured as flyback, SEPIC, BOOST or inverting converter. The LT 3757 drives an external power N-MOSFET from an internal supply of 7.2 V. In the current-mode architecture, the fixed frequency generates stable operation over a wide range of supply and output voltages. The operating frequency can be adjusted with an external resistor from 100kHz to 1 MHz, and synchronised to an external clock using the SYNC pin. The minimum operating supply voltage is 2.9 V, and the current is less than 1μA. LT 3757 features Soft Start and frequency foldback functions that are used to limit the inductor current during start-up and output short circuit[5].

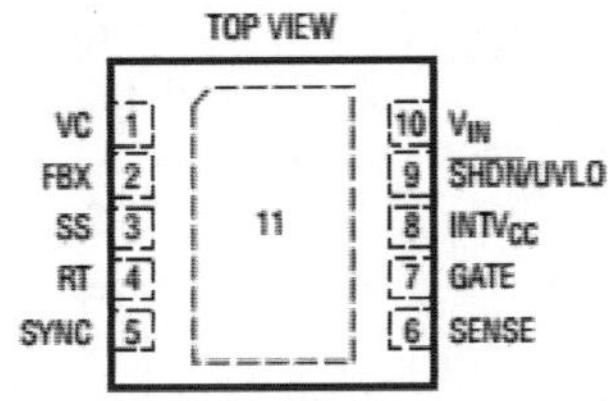

Figure 2. Configuration LT 3757.

2.3. Calculation of Winding Inductor

The process of switching on the circuit flyback dc-dc converter is determined by the signal pulse width modulation (PWM). The working principle is to set the duration PWM on and off electronic switch (t on and t off). The longer the duration of the switch on, the greater the output voltage. The switching process is conducted in a specific switching frequency. A comparison of the switch on the duration of the period known as the duty ratio switching output (Do) as showed by equation (1) [6].

$$D_o = \frac{t_o}{T} \quad (1)$$

The voltage produced by the switching process is then implemented at the high-frequency transformer primary. The transformer will then raise and generates voltage pulse electronic isolation between the input voltage source to the output voltage[7]. The transformer will raise the voltage level by the ratio of the transformer. The dc voltage would be enhanced by the transformer so as to generate dc output voltage is greater. Comparison of the number of winding primary (Np) and secondary (Ns) is equivalent to the ratio of the induced voltage in the primary and secondary transformer [8]. This comparison is called comparison transformer or trans ratio (n), which is also the value of a primary inductor and secondary inductor of the flyback converter transformer can be created by using $\frac{N_p}{N_s} = \sqrt{\frac{L_p}{L_s}} = n$.

The interleaved flyback converter is designed using a high-frequency transformer where can be applied to multiple winding inductors according to the operating frequency factors [9]. The flyback transformer design and calculation requires the duty cycle (D), in which to determine trans ratio (n), it can be calculated using equation (2) where the value of the output voltage (V_{ou}t) = 350 V, the value of input voltage (V_{in}) = 47 V, the value of duty cycle (D) = 2.758 and D2 = 1.223. Thus, the obtained value trans ratio (n) was three.

$$n = \frac{7{,}45}{2{,}255} = 3{,}3 \approx 3 \quad (2)$$

The dc-dc converter can be operated in continuous conduction mode (CCM) and discontinuous conduction mode (DCM), depending on the current of the inductor (iL). The discontinuous conduction mode (DCM) was used in this research, where iL was zero in several parts of the switching period [10]. Transformer design for the DCM operating mode is to improve the working operation; the main switching cycle can be calculated based on the output voltage (V_{out}) and the minimum input voltage (V_{in}). Maximum duty cycle (D_{max}) can be used as the starting point and calculated bythevalue of D_{max} = 0.88.

The used inductor design also depends on the mode of operation used. In the DCM operating mode, the values of primary inductor and secondary inductor on the flyback converter transformer can be determined by using equation (3) and (4), where the value of the maximum duty cycle (D_{max}) and lowest input voltage (V_{min}) are divided by the output power (P_{out}) and the applied frequency (f).

$$L_S = \frac{(0{,}88)^2.(47)^2.1}{2.350\ 100\ 10^3} = 24,44.10^{-6} \quad (3)$$

$$L_P = \frac{24{,}44\ .10^{-6}}{9} = 2{,}715\ .10^{-6} \quad (4)$$

From the equation (3) and (4), it was found that the value of L_S = 24.44 μH and the value of L_P = 2.715 μH. Therefore, to approach the actual model in the simulation, the components for L_S was set at 27 μH and L_P was set at 3.3 μH by the components available in the market.

2.4. Output Voltage Calculation

The calculation for theoutput voltage of interleaved flyback converter is conducted to see whether the voltage remains constant even if the load changes because of the increased power. The constant frequency value used to control LT 3757 was set using R (load) which has been developed from the data sheet. With the value of R = 140 KΩ, the frequency generated by time constant was 100 kHz. By Table 1, the output voltage could be determined by using equation (5). The value was customised with LT 3757 configuration in the FBX pin where the internal bias voltage of 1.6 V with voltage divider where the resistance values (R1 and R2) were connected from the value of output voltage to the ground (GND).

$$V_o \;= 1{,}6\,V\left(1+\frac{3{,}5\,m\quad\Omega}{16{,}074\;K\Omega}\right) = 350\;V \;(5)$$

2.5. Ripple Voltage Calculation

Ripple voltgae (Vr) of the load is 1% of the desired output voltage (V_{out}). So as to limit the ripple voltage maximum value, the value of the filter capacitance (C) must be higher than C_{out} indicated by equation (6).

$$C_o \;\geq \frac{1}{0{,}01\,.\,350\;100.\,10^3} = 2.857\;x\;10^{-6}\;(6)$$

Thus, Cout should be more critical than $2.857 x 10^{-6}$, as to adjust the capacitors available in themarket then 47 μF was selected. Therefore, the value of ripple voltage (Vr) could be calculated with equation (7), that was the period (T) and the value of used capacitor.

$$V_r = \frac{10\,.\,10^{-6}}{47\,.\,10^{-6}} = 212{,}65\;m \quad (7)$$

3. Results and Discussion

The simulation was conducted to determine how waveform ofprimary inductor, secondary inductor and voltage ripple (Vr). In this study, circuit simulation using LT 3757 was applied to interleaved flyback dc-dc converter as shownin Figure 3, which was done on the LT spice software. The simulations werecarried out as close as possible to the value of design that has been thoroughly calculated.

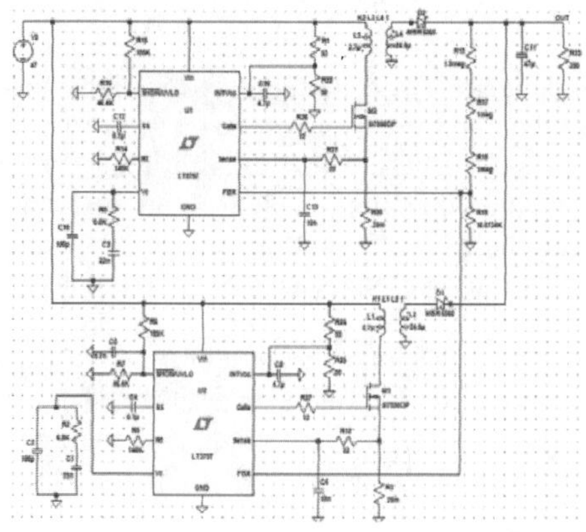

Figure 3. The circuit of interleaved flyback dc-dc converter.

3.1. Results of Transformer Wave

The comparison of thetrans ratio (n) of the transformer winding design was three. In Figure 3, with power load of 350 W, the average voltage primary is 94 Volt and the average current transformer is

36.75 A. Primary transformer voltage is a direct output voltage from the solar panel which is connected directly to the primary inductor (Lp), where if the switching process is ON, the primary winding will be connected to supply. As a result, the flux and the magnet on the primary side are increased and energy stored there in will be transferred. And when the switching is OFF, the magnet and the flux on the primary side is reduced, the secondary inductor generates positive voltage and therefore the diode moves forward. The simulation was conducted in the interleaved operating mode, so it resulted in two different stages of switching ratio as shown in Figure 4.

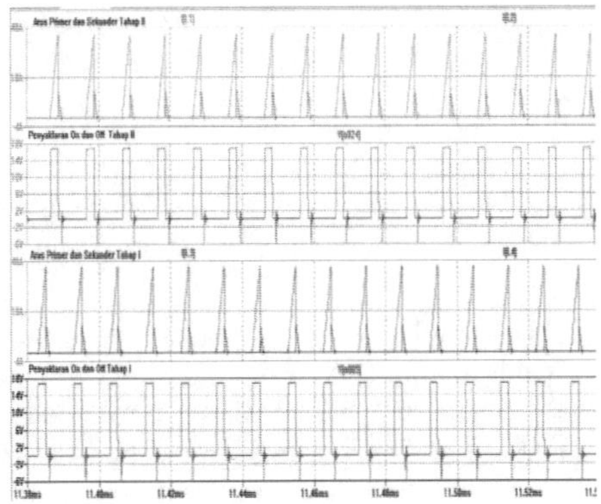

Figure 4. Theswitching process with transformer energy transferring process between primary and secondary sides.

3.2. Output VoltageResults

The desired output voltage in the circuit design of interleaved flyback dc-dc converter is 350 V with varying power load as shown in Table 2. The result of the circuit simulation with the output voltage is 350.73 V the power load of 350 W. The interleaved flyback circuit simulation with the lowest input voltage of 41.5 V and maximum power load of 350 W had a stable voltage at 350.70 V. By using equation (5), it was obtained the desired output voltage in the circuit design of dc-dc converter as 350 V.

Table 2. Results of fixed voltage simulation with varying power load

NO	Tegangan Input (V)	Beban Daya (W)	TeganganKeluaran (V)
1	47	10	350,79
2	47	75	350,76
3	47	100	350,75
4	47	150	350,71
5	47	200	350,70
6	47	250	350,78
7	47	300	350,79
8	47	350	350,70

3.3. Voltage RippleResults

The output ripple voltage of interleaved flyback dc-dc converter could determine the value of capacitor used, which was 47 μF. After mathematically modelling the corresponding data sheet and supporting theories, it was obtained a value of 212.65 mV. Also, the result of simulation performed using LT Spice software was 219.23 mV with anaverage voltage of 350.78 V as shown in Figure 5.

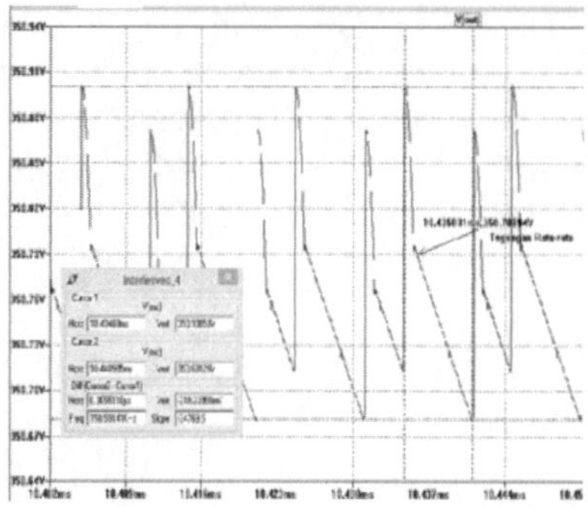

Figure 5. Results of voltage ripple.

4. CONCLUSION

Designing the interleaved flyback dc-dc converter circuit with thepower of 350 W have been exposed. By the specification in Table 1, interleaved flyback dc-dc converter circuit simulation could be used on a stable output voltage of 350.70 V with varying power load as shown in Table 2. When input voltage was at its lowest state (41.5 V) with a maximum power load of 350, the output voltage remained stable at 350.70 V. Therefore; this dc-dc converter design can be combined with an inverter in order to be a complete micro-inverter with dual power stage method.

Reference

[1] S. B. Kjaer, J. K. Pedersen, S. Member, and F. Blaabjerg," vol. 41, no. 5, pp. 1292–1306, 2005.
[2] F. Edwin and S. Member, pp. 821–827, 2012.
[3] S. Dumais, Alex., Kalyanaraman *Technol. Microchip*, pp. 1–54, 2012.
[4] L. Andres, G. Rodriguez, and J. C. Balda, pp. 2084–2091, 2013.
[5] L. T. Corporation pp. 1–36.
[6] U. and W. P. R. Mohan, N. T.M, *Power Electronics : Converters, Applications, and Design, 3rd ed.* Minnesota: John Wiley and Sons, 2003.
[7] S. Mode, P. Supply, and S. Topologies, "Switch Mode Power Supply (SMPS) Topologies," no. Part I, pp. 1–48, 2007.
[8] H. Berahim, Yogyakarta: Andi Offset, 1994.
[9] T. I. Incorporated, 2001.
[10] M. H. Rashid, "Power Electronics Handbook,2nd ed," California: Academic Press, 2007.

Developing Gamelan Model through Scientific Work

S Haryono[1,*], Wadiyo[1], I Sumpono[2] and L Handayani[2]

[1] Sendratasik Dept., Faculty of Language and Arts, Universitas Negeri Semarang, D7 Building, 2nd Floor, Sekaran-Gunungpati, Semarang, Indonesia
[2] Physics Dept., Faculty of Mathematics and Natural Sciences, Universitas Negeri Semarang D7 Building, 2nd Floor, Sekaran-Gunungpati, Semarang, Indonesia

*Corresponding author: haryonoslamet80@yahoo.com

Abstract. A gamelan trigger has been designed and developed to support *karawitan* learning. A research and development method involving a series of the process has been conducted to produce this scientific artwork. This paper examines the trigger production stageswhich mainly focus on the work of mechanical system setting of the gamelan model. The research finally comes to the conclusionthat the mechanical system of the model can meet the requirements needed for providing a real form of *gamelan* in a smaller size.

1. Introduction

A current digital era has brought a new shift of paradigm in all aspects including the development of an artwork. Gamelan is one of the artworks that has been shifted from the traditional form into the digital one for several purposes as reported by several studies [1,2,3, 4). This paper discusses part of the second year project result aiming to develop a mechanical system of gamelan trigger prototype which can be utilised to solve the problem of providing the real gamelan in a smaller size for supporting music learning in elementary school. A series of scientific work process has been carried out to continue the production of the trigger prototype which can assist as the learning media by reviewing and solving several questions raised during the first year of the project. One question raised and considered to be answered an unrealistic sound quality produced by the sound module.

2. Method

The main work of the study was done in physics laboratory Universitas Negeri Semarang (UNNES). The work in this period continued the previous oneby employing an adopted research and development method defined by Borg and Gallwith product revision stage as the main focus.

The product revision stage was started by reviewing the problems emerged from the previous year result. After carrying out the first year project, the product, in fact, showed unrealistic sound quality performed by the sound module. It seems that the sound produced was still in digital character and did not match with the sound produced by real gamelan. Because of that, in this stage the project was focused on improving this sound quality consisting of activities related to the hardware and software used, those were wiring and programming. Figure 1 shows the stage specifically.

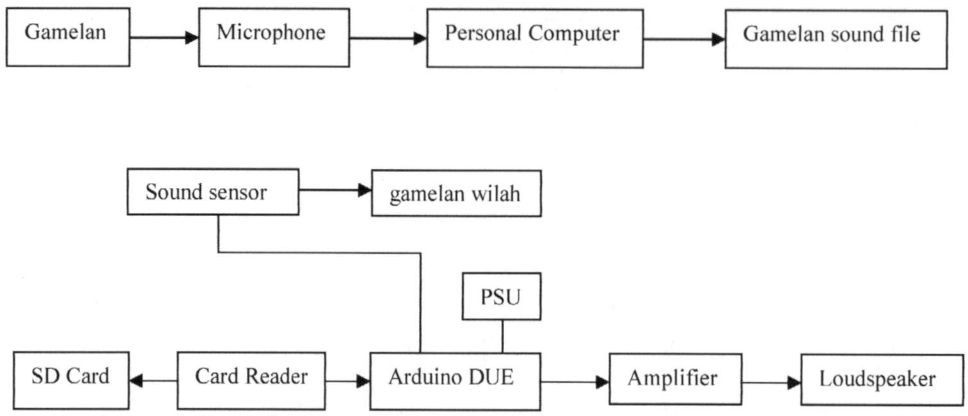

Figure 1. Stage of improving sound quality

3. Results And Discussion

The unrealistic sound quality was a major concern of the work in this second term. This sound quality has been tried to improve by carrying out several points of work. The first one was finding the cause of the uncomfortable sound. It was tried to find the cause from both hardware and software. Based on the track traced, it was detected that the cause of the unrealistic sound was the misuse of the development board causing a digital-formatted sound result. This problem cause finding was then followed by deciding the alternative solution that was the use of another kind of development board which enables it to accommodate, manage and show data of gamelan tone formatted in sound wave (filename.wav) in the larger size. This board also has different specifications including type, operating voltage, recommended input voltage, limits input voltage, digital I/O pins, analog input pins, analog output pins, total DC output current on all I/O lines, DC current for 3.3V Pin, DC Current for 5V Pin, Flash Memory available for the user applications, SRAM, Clock Speed, length, width, and weight. Figure 2 shows the development board used for the study.

Figure 2. The development board.

The work of improving sound quality was then continued by setting the new board on the computer. This setting process involving installation of a required software suitable for the next process, that was reading formated wave data of gamelan sound resulted from sampling operation. For this purpose, the gamelan sound should be saved first in a memory card having at least four Gb memory size. After software installation, the library program will also need to be installed. The steps were then followed by copying the library program folder into the memory card and restarting the board program. The process of sound quality improvement in this study period ended with a series of program inputting activity to gather the data reading by using a programming language referring to the number of gamelan tones used.

4. Conclusion

The sound quality improvement process gave the main result of analogue formated sound quality through the use of new kind of arduino development board. By using the new arduino, the new sound has a characteristicof resembling the real gamelan sound which enables the gamelan model to meet the mechanical system requirements needed for providing a real form of gamelan in a smaller size.

Acknowledgments

We thank Kemenristek-Dikti for supporting the study through the "Hibah Bersaing 2016" grant and Sendratasik and Physics Departments for permitting the use of the laboratory. Authors' special thanks also goes to laboratories officers, Sofi, Budi, Margi and Ismail for their assistance, work and discussion time.

References

[1] Schiemer G, Alves B, Taylor S J and Havryliv M, *ICMC 2003 Boundaryless Music: Proc.Int. Comput. Music. Conference.* Ed. B. Tan. California: USA: National University of Singapore and The International Computer Music Association. 329.
[2] Teng L S "Robust Gamelan Bonang Hammer Actuator," in eprints.utem.edu.my.
[3] Pramudi Y T C, Sunardi, Budiman F, Utomo S D.2003.*Jurnaldwi_ICSSEA_20111_3* , 231
[4] Haryono S, Wadiyo, Sumpono Iand Handayani L.2015.*Proc of ICMSE.*

Study of Lead as a Source X-ray Radiation Protection with an Analysis Grey Level Image

Susilo*, I N Rahma, Mosik, and Masturi

Physics Department Faculty of Mathematics and Natural Science, Semarang State University

*Corresponding author: susilosumarto@gmail.com

Abstract. X-ray utilization in the medical field still has a potential danger for the human. This occurs when exposure to x-ray radiation received exceeds the dose limit value. It required a radiation shielding to prevent the hazard, and lead is one of the metals usually used as x-ray radiation shield. This work aims to determine the metallic lead properties to find out of the step wedge lead radiograph image. The instruments used are the plane x-ray, digital radiography system and personal computer installed by MATLAB, while the material is step wedge lead. The image of radiograph was analysed using GUI applications on MATLAB software to determine the values of grey level from the image and the optical density of the radiograph image. The results showed the greater optical density, the higher the image contrast, and the value of optical density in the image is inversely proportional to the voltage x-ray since the value of grey level at high voltage is smaller than that of at low voltage.

1. Introduction

X-ray is one of the radiation source used in medical field for diagnosing the presence of a disease in the form of anatomy portrait featured in the film of radiography. For using x-ray, one should pay some attention to work safe from its radiation, since x-ray still has a potential danger for humans. This happens if the x-ray radiation exposure received by humans exceeds the dose limit value, so that an attempt for x-ray radiation protection is required. The government has issued government regulation No. 33 of 2007 about safety radiation and security of sources radioactive. Determination of the limiting dose is exercised by Nuclear Energy Regulatory Agency of Indonesia (BAPETEN). Determination of the dose limit value based on ICRP (International Commission on Radiation Protection) at 2008 that the dose of radiation exposure must be below 20 mSv in one year [1].

Shield of radiation mandatory has two specific characters as the core of characteristics, i.e. it can absorb x-ray radiation, and it must not be too thick since it is worn by workers working in the field radiation [2]. Thick retaining walls the primary radiation is a brick wall with thick 25 cm or concrete ($\rho = 2.2$ g/cm^3) with thick 15 cm equal to lead having thickness of 2 mm. Lead equivalent of concrete at each of these voltages as a function of the barrier thickness [3]. This shows that lead is very potential for shielding radiation. Lead is a stable nuclear in the last product of the radioactive series uranium,actinium, and thorium where if it interacts with α, β, γ or x – ray, it cannot decay into other elements, but the energy of the radiation will be absorbed since it has a high mass numbers and density. Lead has a mass density of 11,34 g/cm^3 so with the thickness of 2 mm lead can absorb the entire X-ray energy of 100 keV [4].

These days metallic lead has many benefits in life. One of its benefits is the cover of electricity installation on the rooftop. Along with the development of technology, electricity installation on the housetop is no longer using lead as the cover. Therefore, the trace lead plate which is no longer used will be the waste and contaminate the environment. Lead is one metal that endures corrosion. The other benefit of lead is as shielding material of x-ray radiation. The x-ray's room uses a lead plate as a radiation protection material. Thus it is needed a test toward trace lead plate as the material of x-ray radiation protection, so that the waste of trace lead plate can be decreased. The used method to find out the absorption of a material was by measuring radiation intensity before and after passing through the material [5]. One of the researches about metallic lead as the material of x-ray radiation protection was measuring dose rate of X-ray radiation, by measuring radiation dose rate, it was known the spare power of that material. This work did a test on radiograph image of step wedge lead, so that it can be known that trace lead can be used as the material of x-ray radiation protection.

2. Methods

The instruments are plane x-ray Mobile Media SF-100BY, digital radiography, and PC using matlab, while the material is a plate of the lead with the size of 11 cm x 12cm. This work used six lead plates tested with different thickness. After that, the sample was photographed using a system of digital radiography. The digital radiography system using X-ray gives exposure to the object, then the unit light-tight tube based of intensifying screen. The intensifying screen is sensitive to X-ray function that converts X-ray into visible light; next image is formed on the object can be captured by DSLR camera [6]. In the present work, we used exposure factor current 16 mA and exposure time of 0.10 s. while the x-ray tube voltage varied at 50 kV, 60 kV, 70 kV and 80 kV to determine the comparative value of grey level. Testing was done by arranging reciprocal resembles the step wedge, exposure using digital radiography system and the results were analysed using software GUI Matlab 2013. For make radiograph image analysis easier, it was GUI Matlab developed by entering some source codes so that the user only use it without knowing how the command work (Figure 1).

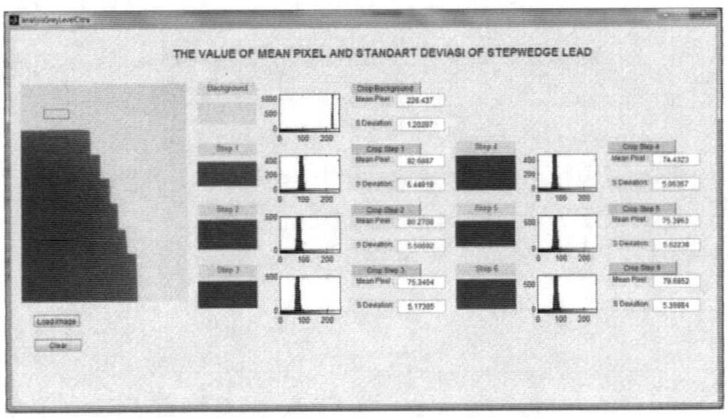

Figure 1. Display GUI Matlab Analysis

3. Result and Discussion

The factor affecting radiograph quality is exposure factor which consists of tube voltage (kV), tube current (mA) and exposure time (s). The right arrangement of exposure factor can produce optimum radiograph contrast which can show the grey level between objects having different density. The result of step wedgelead radiograph image of the different voltages can be seen on Figure 2.

(a) (b) (c) (d)

Figure 2. Radiograph image of step wedge lead on voltage (kV) variation value : (b) 50 kV,(c) 60 kV, (d) 70 kV, and (e) 80 kV

Figure 2 shows that the amount of x-ray voltage influences the result of radiograph image. On the voltage of 50 kV, the image tends to be dark compare to that of the voltage of 60 kV. The image on the voltage of 60 kV tends to be dark compare to the image on the voltage of 70 kV and 80 kV. There is no significant difference from the step wedge lead image. It needs an analysis using software MATLAB to find out the grey level from radiograph image. After doing analysis using software MATLAB, we can see the difference of grey level of step wedge. According to [7], the grey level image is comparable to the intensity of the x-ray beam, so the grey level value 0-255 can be used to determine the X-ray intensity I_o and I_x. Measurement of Optical Density (OD) value uses Grey Level (GL) can be done as in the determination of optical density with x-ray intensities Io and Ix, so the DO value using grey level can be written as [8]:

$$O = \log \left(\frac{I_O}{I_x}\right) \quad (1)$$

$$O = \log \left(\frac{G_O}{G_x}\right) \quad (2)$$

Digital Radiography (DR) system used in work, the background part was brighter than the object, whereas in conventional radiography and computed radiography (CR) background is darker than that of the object. Graph of the relationship between the thickness of lead towards optical density of radiograph image with the variety of the X-ray voltage is shown in Figure 3.

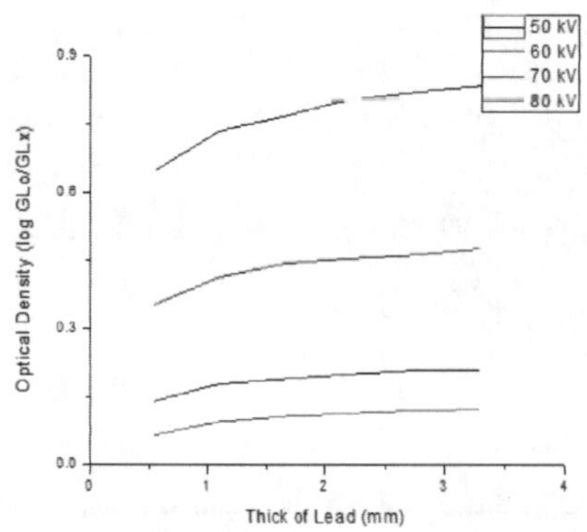

Figure 3. Graphic of relations the thickness lead against optical density radiograph image

The magnitude of the x-ray tube voltage affects the amount of x-ray energy. The higher the voltage used, the greater the produced x-ray energy. X-ray tube voltage can also affect the produced penetrating power of x-ray, and therefore it affects to the produced image quality of radiographs (contrast). Rated voltage will affect the quality and quantity of x-ray because the change affects the produced wavelength. X-ray wavelength is inversely proportional to the x-ray tube voltage. The higher the tube voltage value, the lower the wavelengths.

$$c = \lambda \quad (3)$$

$$E_k = hf - Q = h\frac{c}{\lambda} - Q \quad (4)$$

The kinetic of energy (E_k) equal to electron production (e) and voltage (V), so equation be (Meredith, 1997)

$$E_k = e \quad (5)$$

$$e \ = hf - Q = h\frac{c}{\lambda} - Q \quad (6)$$

$$\lambda = \frac{hc}{e\ +Q} \quad (7)$$

According to [9] the greater the optical density, the higher the contrast of the image. Therefore, it can be concluded high grey level image sample has a high optical density. The graph shows optical density is higher towards accretion step. The greater the x-ray voltage, the smaller the optical density values. Because the value of grey level on a high voltage is smaller than the value grey level at low voltage. This happens because the voltage of 50 kV and 60 kV has a higher contrast image than the image at a voltage of 70 kV and 80 kV.

Interaction of x-ray and the material was influenced by the production of X-ray and atomic material number [10]. This work is to determine the intensity of radiation by radiograph image, using GUI Matlab on a radiograph image analysis. As visually, qualitative analysis is indistinguishable from the dark to the light. Using Matlab, radiograph image analysis by taking some (crop). Background and all steps ROI's (Region of Interest). Results of ROI's crop were indicated by the mean pixel of grey level. ROI cropping process all steps were performed on the middle object. Since the central of ROI object is perpendicular to the x-ray tube, the x-ray percentage to the object is 100%. Then, the background ROI was cropped parallel to the step.

The results show that a reduction in the intensity of X-ray radiation significantly as an increase of the lead absorbent material thickness. The photons interact with material, some of the photon energy is taken by an electron as kinetic energy, and some are used to fight the atom binding energy or change into a new photon. Primary photons do not interact by crossing the material and out of the material [11]. Radiation absorption of a material is the ability of a material to absorb and hold the radiation to pass through the material. However, the absorption ability of the material depends also on the type of radiation that the absorber material [12]. The number of primary photons that escaped from the material decreases exponentially defined in Equation 8.

$$I = I_0 e^{-\mu} \quad (8)$$

By the ALARA (As Low As Reasonably Achievable) principle which states that the exposure dose of radiation received by humans should be as low as possible [13]. In the field of radiology, the used radiation protection materials are lead. Lead is a metal that has high linear absorption coefficient of 0.506 m-1 [14]. So that the lead can absorb x-ray radiation energy. As a result of X-ray radiation intensity is significantly reduced in lead thickness 0.55 mm and on the thickness of 1.1 mm and so the radiation intensity is reduced slightly which will form an exponential equation according to Equation 8. The results showed grey level image analysis could replace the method of radiation dose rate analysis that is commonly used as a method of testing the radiation absorbent material. Radiation intensity is proportional to the grey level image, so by looking at the grey level of the image of the radiograph, it can determine the intensity of the radiation from the image. This study used the trace lead with the mass density of 10.53 g / cm3. With a high mass density, trace lead can absorb the energy of the x-ray radiation. Compared to other radiation protection material such as barium which has mass density smaller than lead so that the energy absorption potential is also lower [15].

4. Conclusion

Grey level analysis proportional to the intensity of radiation can be used to find out the contrast of the image, and may know the linear coefficient absorbency of the material. Optical Density was inversely proportional to the voltage of the x-ray. Because of the value of grey level on high voltage smaller than the value of grey level at lower voltages.

IOP Conf. Series: Journal of Physics: Conf. Series **824** (2017) 012005 doi:10.1088/1742-6596/824/1/012005

References

[1] Djarwanti, Suhaedi, Nugroho A I and Barron B 2012. *Pros. Sem. Nas. VIISDM Teknologi Nuklir*. (Yogyakarta:Batan) 100
[2] Prayitno and Gunawan 2009. *Jurnal Perangkat Nuklir*. 05**(03)**
[3] Singer, George et al. 1947
[4] McCaffrey, J.P., H. Shen, B. Downtonand E. Mainegra Hing (2007.*Canada: Ionizing Radiation Standards, National Research Council of Canada*
[5] Kim Seon-Chil and Hong-Moon Jung 2013.*International Journal of Technology***2**: p 178
[6] Susilo, Supriyadi, Sutikno, Sunarno and Setiawan R 2014. *Jurnal Pendidikan Fisika Indonesia*
[7] Susilo, Sunarno, Setiowati and Lestari L 2012.*Jurnal Mipa Universitas Negeri Semarang*
[8] Susilo, Wahyu S B and Kusminarto 2013. *Jur. Pen. Fisika Indonesia(Indonesian Journal of Physics Education)* 9.1
[9] Bushong, S.C 2001.*A Harcourt Health Science Comp*. U. S of America
[10] Sutrisno D.A, Susilo and Masturi 2015.*International Conference on Mathematics, Science, and Education*
[11] Wiryosimin and Suwarno 1995. *Institut Teknologi Bandung*. Bandung
[12] Fitriawan and Margi, 2014. Volume 2. ISSN 1411-1349
[13] Daniel, Cupido and Elizabeth Furey 2008. *Journal of Medical Imaging and Radiation Sciences.***39** 189
[14] Kristiyanti and Atmojo S M 2005. *Prosiding PPI-PDIPTN Puslitbang Teknologi Maju-BATAN*.
[15] Kim, Seon-Chil 2012. *Journal of the Korean Physical Society*, 1**(60)**

Effect of Fibreglass Reinforcement on Compressive Strength of Teak (*Tectona grandis* L.f.) Leaves Waste Composite

Masturi[1,*], Sunarno[1], W N Jannah[1], R M Maulana[1], D Alighiri[2], A Purwinarko[3], Susilawati[4], U Amri[1], and S Rustad[5]
[1] Physics Department, Faculty of Mathematics and Natural Sciences, Universitas Negeri Semarang, Indonesia
[2] Chemistry Department, Faculty of Mathematics and Natural Sciences, Universitas Negeri Semarang, Indonesia
[3] Computer Science Department, Faculty of Mathematics and Natural Sciences, Universitas Negeri Semarang, Indonesia
[4] Physics Education Department, Universitas PGRI Semarang, Indonesia
[5] Dian Nuswantoro University, Semarang, Indonesia.

*Corresponding author: masturi@mail.unnes.ac.id

Abstract. We have successfully fabricated a leaves waste/PVAc composite reinforced by fibreglassusing simple mixing and hot-press methods. This work was done by varying the composition of PVAc/leaves composite/fibreglass to obtain the effect of fibreglass addition in improving the composite compressive strength. From the compression strength test, we received the increasing of fibreglass fraction tends to improve the compressive strength, where for the 4.8% (w/w) of fibreglass the compressive strength of composite produced reached up to 33.9 MPa, or increased up to 51% compared to that of without fibreglass. It showed that fibreglass can increase significantly mechanical strength of the composite.

1. Introduction

Indonesia is a tropical country that has a wide variety of plant species. In each area, the trees grow so rapidly it also impacts the high trash. The issue of garbage being debated by themselves in the midst of society, because the public has not had an innovative way to process waste. If these to be allowed to continue, it will become a serious problem. On the other hand, the problem of deforestation to be a device manufactured furniture from year to year continues to increase. It certainly would reduce Indonesia's forest area.

One of the innovations that can be developed to overcome this problem is an engineering material. The design leaves waste material into other wood boards with a composite technique that is light, robust and durable. Design material derived from leaves waste can reduce the accumulation of trash. It also can increase the value of leaves waste that has not been utilised properly while such engineering minimises the number of timber production is so high. Masturi et al [1] conducted research manufacture using polymer composite leaves waste Polyvinyl acetate (PVAc) as a binder and silica as a reinforcing (filler).

In this work we usedteak leaves waste as composite materials, Polyvinyl acetate (PVAc) as polymer and fibreglass as reinforcement (Filler). Availability of a considerable amount in nature makes teak leaves selected as composite materials. PVAc was chosen because it has a strong adhesive so used as a raw material in the production of glue, cloth, paper, and wood. PVAc used because it has properties smelled, non-flammable, and faster condense. Masturi et al [2] states that the PVAc is also used to improve the strength of a composite.Sriyanti and Merlina has also used the PVAc as binder of nanocomposite material [3].

Fibreglass is chosen as filler to make stronger composites. Fibreglass can be used to reinforce composite because it has mechanical properties and can store energy very well [4]. Additionally, fibreglass is also suitable for use as a filler because it has excellent thermal conductivity [5]. Fibreglass also adapted to the pressure at high temperature or low temperature, have good elastic energy for the various stress [6].

2. Methods

In this research, leaves waste used is teak leaves (*Tectona grandis*) from leaves waste in Semarang, PVAc and fibreglass. The leaves residue was dried then smoothed using a blender machine. Teak leaf powder sieved to produce a powder in the same size. An FT-IR characterization was used to investigate the content of cellulose in teak leaves powder.

In the other hand, the fibreglass which still forms sheet cut into small pieces with a size 3-5mm while PVAc was solved with water and stirred using magnetic stirrer until homogenously for 15 minutes so easily mixed with a powder of leaves waste. The water was chosen because PVAc is a hydrophilic polymer. The PVAc solution mixed with teak leaves powder with a certain ratio as in Table 1. The mixture is stirred until uniform and then continued to press process. The pressed process has done by putting the dough into a cylindrical mould and pressed at 3 tonnes for 15 minutes. The sample pressed removed from the mould and then dried for 24 hours at room temperature. The work was continued with mixing fibreglass into the mixture of teak leaves powder and PVAC solution to obtain the effect of fibreglass on compressive strength of the composite.

Tabel 1. Composition of teak leaves powder,PVAc

Teak leaves powder (g)	PVAc(g)	Fibreglass (g)
16	4	0
16	4	0.2
16	4	0.4
16	4	0.6
16	4	0.8
16	4	1

Characterization of compressive strength performed to measure the mechanical strength of composite before and after mixed with fibreglass by cutting the sample in small cubic with a size of 2x2x2 cm. The compressive strength is expressed as

$$\sigma = \frac{P}{A}$$

where P is maximum load required to press the composite until destroyed and A is the area was depressed.

3. Result and Discussion

FTIR characterization results are shown in Figure 1. Measurement of IR spectrum only in the range of 4000-400 cm^{-1}. IR spectrum area divided into two areas of functional groups and regions fingerprint. The area of the functional group lies in the wave number 4000-1450 cm^{-1} while the fingerprint region lies in wave numbers below 1450cm^{-1}.

FTIR results of teak leaf powder compared with the results of FTIR on pure cellulose micro-granular cellulose powder from SIGMA) of high purity grade reagents [7]. For pure cellulose appears spectrum peak in the range of 3400 cm^{-1} to 3500 cm^{-1}, 2800-2900, 1160 and 1035 cm^{-1} to 1060. In the area, there is a fingerprint for the cellulose peak in the range of 1300-1400 cm^{-1} indicating the C-H bond [8].The FTIR spectrum of teakleaf powder showed a peak in the area of 780 cm^{-1}, 1075 cm^{-1}, 1233 cm^{-1}, 1437 cm^{-1}, 1631.29 cm^{-1}, 2929.05 cm^{-1}, and 3401.04 cm^{-1}. Overall peak existing on teak leaf powder is almost similar to the peak in real cellulose shift only slightly there.

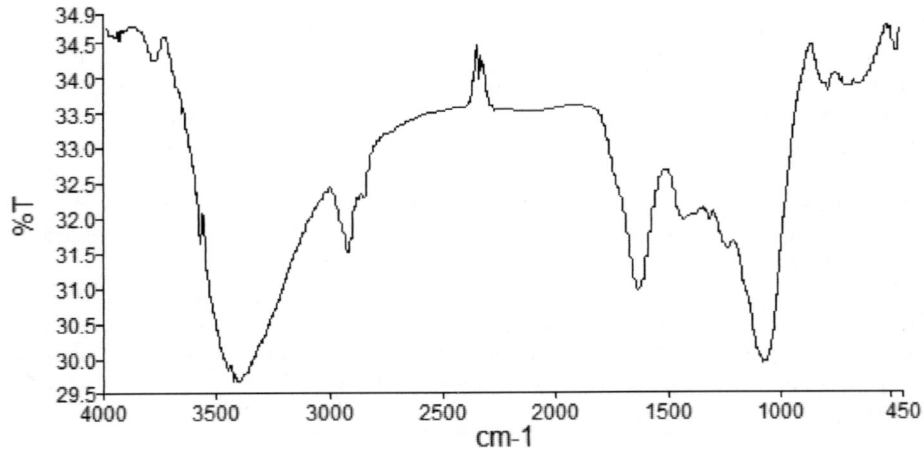

Figure 1.FTIR spectra of teak leaf powder

The peak of 2929.05 indicates the C-H bonds are strong and peak at 3401.04 indicates stretching group O-H bond. Meanwhile, the peak of 1631.29 cm^{-1} shows the absorption of water where hydrogen bonds formed between atoms of hydrogen and oxygen atoms of hydroxyl groups of the glucose monomers [8]. This bond indicates the presence of cellulose fibre formation. The peak of 780 cm^{-1} and 1437.7 cm^{-1} shows C-H bonding linking glucose units in the cellulose. Further, the peak of 1233.7 cm^{-1} indicates that there are groups C-O-C which is a structure of lignin. Absorption area happens to 1075.96 cm^{-1} shows the C-O bonds contained in the cellulose component [1,6,9].

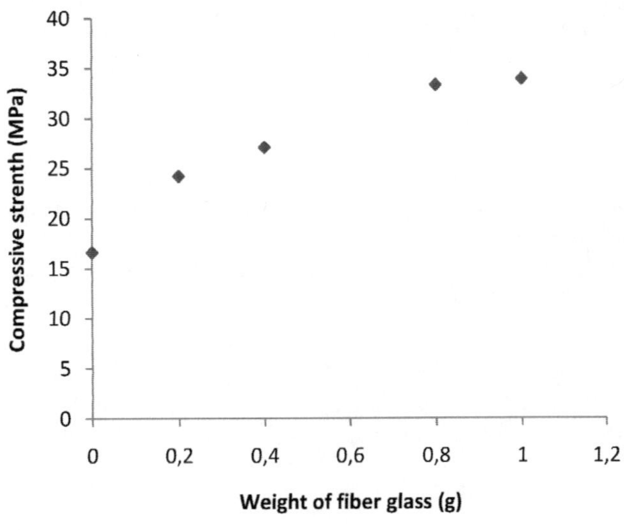

Figure 2. Compressive strength of composite

Figure 2 shows that the compressive strength of composite waste teak leaves before added the fibreglass has a compressive strength value of 16.6 MPa. The compressive strength is slight because only a sample composed of teak leaf powder and PVAc. PVAc is working to minimize the pores and form a strong bond with teak leaf powder [10]. Improving the compressive strength research, it continued with the addition of glass fibre. After added fibreglass, the compressive strength increased to 51% compared to the compressive strength without fibreglass.

The fibreglass can improve the compressive strength because these fibres play an important role as support the maximum load. The tension was applied to the composite initially accepted by the matrix then forwarded to the fibreglass so that the fibres will hold the amount up to maximum load. The increase in the compressive strength in composites with fibresglass do not occur continuously but stops at 33.9 MPa compressive strength. Composites in this condition are said to be in a state of saturation so that if the study continues, the compressive strength generated will be in the range of particular value or can go down. It is like the Masturi research's where a composite with silica powder was experiencing saturation. It occurs due to the addition of fibresglass is no longer effective in supporting the maximum load [10].

4. Conclusion

We have successfully made composite waste teak leaves with PVAc and the addition of fibresglass which had reached 33.9 MPa compressive strength or an increase of 51% compared to without the addition of fibreglass.

Acknowledgement

This work was financed by Excellent Research Grant of University for fiscal year 2016-2017.

Reference

[1] Masturi, M Abdullah, and Khairurrijal, 2011. *J. Mater. Cycles Waste Manag.*, vol. 13, no. 3, pp. 225–23.

[2] Masturi, Mikrajuddin, and Khairurrijal, 2010. *Berkala Fisika*, vol. 13, no. 2, pp. 61–66.

[3] I Sriyanti, and L Marlina. 2014. *Kajian Fisika dan Inovasi Pembelajaran Fisika*vol. 1, no. 1, pp. 69–73.

[4] EA Stefanescu, X Tan, Z Lin, N Bowler, and MR Kessler, 2011. *Polymer (Guildf).*, vol. 52, no. 9, pp. 2016–2024

[5] B Baudouy and J Polinski, 2009. *Cryogenics (Guildf).*, vol. 49, no. 3–4, pp. 138–143

[6] AS Khojin, R Bashirzadeh, M Mahinfalah, and RN Jazar, 2006. *Compos. Part B Eng.*, vol. 37, no. 7–8, pp. 593–602.

[7] JR Flores, E Rubio, VR Lugo, and VM Castaño. 2009. *Reviews on Advanced Materials Science* 21: 211-216.

[8] N Halib, MC Iqbal, and I Ahmad, 2012. *Sains Malaysiana*, vol. 41, no. 2, pp. 205–211.

[9] Masturi, AP Swardhani, ESustini, MBukit, Mora, Khairurrijal, and M Abdullah. 2010. *AIP Conf Proc*, vol 1284, pp.59-63

[10] Masturi, H Aliah, M.P Aji, AA Sagita, M Bukit, E Sustini, M. Abdullah.2011. *AIP ConfProc*, vol 1415, pp.90–93.

Applying an MVC Framework for The System Development Life Cycle with Waterfall Model Extended

W Hardyanto[1,*], A Purwinarko [2], F Sujito [2], Masturi[1], and D Alighiri[3]

[1] Pysics Department, Faculty of Mathematics and Natural Sciences, Universitas Negeri Semarang, Indonesia
[2] Computer Science Department, Faculty of Mathematics and Natural Sciences, Universitas Negeri Semarang, Indonesia
[3] Pysics Department, Faculty of Mathematics and Natural Sciences, Universitas Negeri Semarang, Indonesia

*Corresponding author: hardy@mail.unnes.ac.id

Abstract. This paper describes the extension of the waterfall model using MVC architectural pattern for software development. The waterfall model is the based model of the most widely used in software development, yet there are still many problems in it. The general issue usually happens on data changes that cause the delays on the process itself. On the other hand, the security factor on the software as well as one of the major problems. This study uses PHP programming language for implementation. Although this model can be implemented in several programming languages with the same concept. This study is based on MVC architecture so that it can improve the performance of both software development and maintenance, especially concerning security, validation, database access, and routing.

1. Introduction

Web application development is always growing rapidly through the years. It requires the system to work more quickly and accurately. Many developers are trying to create new and different technologies, especially for building a faster and more powerful web applications. Until now, web application like front-end or client-side still using HTML, CSS, and JavaScript to present data. While the back-end or server-side refers to the storage and processing of data that normally handled by PHP.

Today, information technology is one of the fastest growing businesses areas. The software contains three major modules, namely the interface, business logic, and data. All three are closely related so that software developers have to write all the code for production applications. Currently, with the MVC design, developers do not have to do that because it serves to separate the input, business logic, and output and it also makes model classes reusable without modification.

SDLC model consists of a sequence of steps that must be followed by system developers. Therefore, many companies engaged in software development chose this model to cause it is easy to make applications with high quality. One of the SDLC models, namely the Waterfall, which consists of five subsequent stages, there are business analysis, design, implementation, testing, and maintenance. The success of the Waterfall model makes many software development companies adopting the main framework for the design, build, and maintain an application.

This paper proposes the concept of MVC for software development using SDLC techniques, especially the expansion of the Waterfall model. The goal is to optimize the Waterfall models with

MVC architecture, so it will get higher quality applications regarding safety and minimizes processing time.

2. Methods

2.1. Waterfall Method.

In 1970, one of the articles Walter Royce [1] is the first publication that discusses the waterfall model. Older models widely criticized and modified so as to produce new models and different. The waterfall model of designing things related to software development before starting coding and to successfully build the system. The main function of the waterfall model is to ensure the success of the project. Waterfall method consists of several phases of the software development process, and there are requirement analysis, design, implementation, testing, and maintenance.

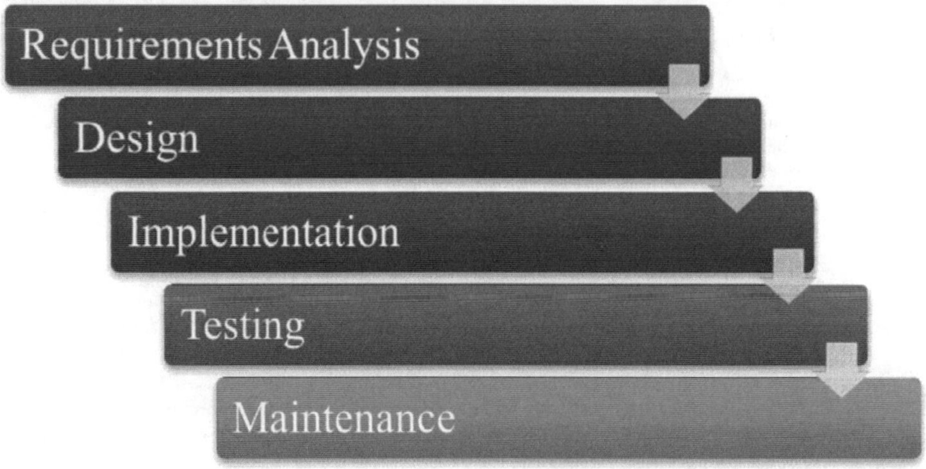

Figure 1. The stages of Waterfall Model

2.1.1.Requirements Analysis

This stage called Software Required Specification (SRS).The main focus at this stage is a complete description of the software to be developed. The project manager should discuss with users of the system to analyze the functional and nonfunctional requirements for the fulfillment of this stage. Functional requirements typically include the purpose, scope, perspective, function, user characteristics, interface, and database. Whereas the nonfunctional requirement is referring to criteria, constraint, limitation, and software performance[2].

2.1.2.Design

At this stage not only about interface design, but also algorithm design, software architectural design, database, concept design, and structural design. System and software design are made of the requirement specification that is learned in the first phase. The system design help in determining hardware and defining the overall system of architecture.

2.1.3.Implementation

This stage is the realization of business requirements and system design into the application program. Database starts with coding and deploys it. At this stage, the real code was written and compiled into an operational application, where the database and other needs have been learned in the previous stage. The work is divided in modular form to facilitate the development as teamwork. The focus at this stage is the development of software so that this stage is the longest stage of the SDLC[11].

IOP Conf. Series: Journal of Physics: Conf. Series **824** (2017) 012007 doi:10.1088/1742-6596/824/1/012007

2.1.4.Testing

It is also known as the verification and validation that is a process to check that the software meets the original requirements and complete the objectives that have been set. In fact, verification is a software evaluation process to determine whether a product of a given development phase satisfy the conditions imposed at the beginning of that stage. Meanwhile, validation is the process of evaluating software during or at the end of the development process to determine whether it fulfill the requirements specified or not[3]. Also, the testing phase is an outlet for debugging where bugs and system disorders discovered, corrected and improved.

2.1.5.Maintenance

It is the process of modifying the software after it was created and deployed, correcting errors, and improve performance and quality. Additional maintenance activities can be done in this phase includes software adaptation to its environment, to accommodate the needs of new users, and improve software reliability.

2.2. MVC Framework.

MVC design pattern was first proposed by Trygve Reenskaug in 1970 at Xerox Parc. According to him, "an important goal of MVC is to bridge the gap between the user model and the digital model on the computer[4]. Then, in 1988, the MVC paradigm described in detail by Krasner and Paul in their articles "A cookbook for using the model-view controller user interface paradigm in Smalltalk-80", which was published in the Journal of Object Oriented Programming.

They emphasized there be great benefits that can be obtained when people build applications with modularity. "Isolating functional units from one another as much as possible makes it easier for application designers to understand and modify any unit without having to know everything about the other units"[5]. MVC pattern is divided into three main roles allowing for more efficient collaboration. The main role here is the development, design, and integration.

The role of development taken by experienced programmers who are responsible for application logic. They take care of the data query, validation, processing, and others. The role of designers to developers who are responsible for the look and user-friendly applications. They display data supplied from the developers who worked on the first role. MVC design pattern is very suitable for the development of web applications because they combine some of the technology that's usually divided into a set of layers[6]. Figure 2 shows the relations between the components of the MVC architecture.

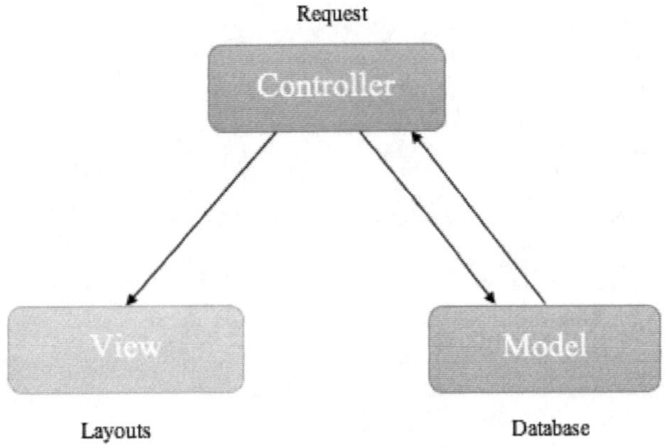

Figure 2.MVC Pattern [4]

2.2.1. Model

Model is part of a system that manages all things that associated with the data, validation, control, and structure of the data source (database). The model greatly reduces the complexity of code developers in writing[7]. Model responsible with the application business logic. This will serve to access the data (database, file, etc.) and will create a class that can be reused. The model usually builds with the data abstraction in mind, validation, and authentication[8].

2.2.2. View

The view is responsible for the management graphical user interface. This means all forms, buttons, graphic elements and all other HTML elements that are within the application. Views can also be used to generate the RSS content to the Flash presentation. By separating the application design from application logic, it greatly reduces the risk of human error when the designer decides to change the interface of the application. At the same time, job application developers greatly reduced because they no longer need to see the HTML code elements, design elements and graphic elements[7]. A view is usually called a web design or template. It controls the way data is displayed and how the users interact with it. It also provides a way to collect data from users. The technology that mainly used is HTML, CSS, and JavaScript.As a general rule, a view should not contain elements that include application logic, to facilitate designers to work with it. The logic blocks should be kept to a minimum.

2.2.3. Controller

The controller is responsible for handling the event. This can be done by one of the users who interact with the application. A controller receives the request and prepares the data to respond. It is also responsible for establishing the response format. The controller interacts with the model to retrieve the necessary data and produces View. This process is also known as action or verb[7]. When a request arrives at the server, the MVC framework sends to a method on a controller based on the URL[9]. The controller has all of the application logic and incorporates display in View with functions in the model. It is responsible for the retrieval of data from the View and to build a path for application execution. The controller will access the function model and interprets the received data that can be displayed by the View. It is also responsible for error handling[8]. A Controller manages the relation between the View and the Model. It responds to user requests, interact with the Model and View to decide what should be created and displayed[10]. As what mentioned above, we need an approach to Controller where all data processing applications should be dealt with at the level of the Controller. Our control systems can support JSON, text and XML data format for requests and responses.

2.3. Design and Implementation.

In this section, it describes the architecture of the design and operation of the framework. This is a three-level application framework that arranged into three major sections.

1. Users
2. MVC Components
3. Database

The first level consists only of users. Users can send a request and a response by components of the "View". The second level consists of three components called Model, View, and Controller. These components can communicate and process data with each other. Finally, the third level only database where data is stored permanently. Figure 3 illustrates how data flows throughout the system using MVC.

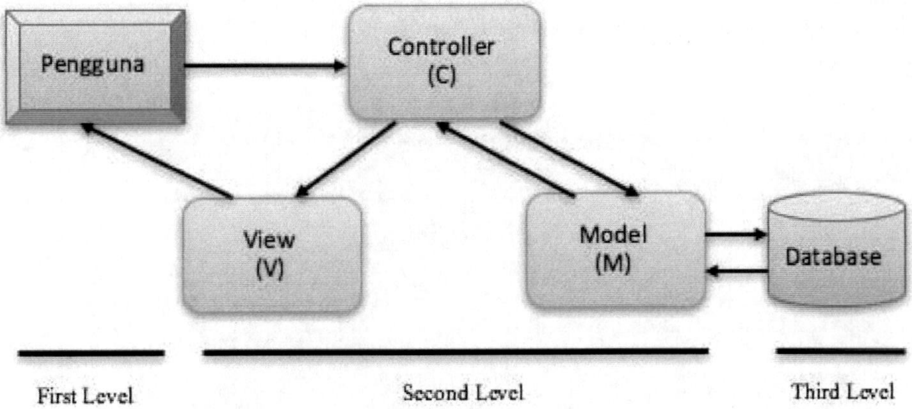

Figure 3.MVC Arsitektur [4]

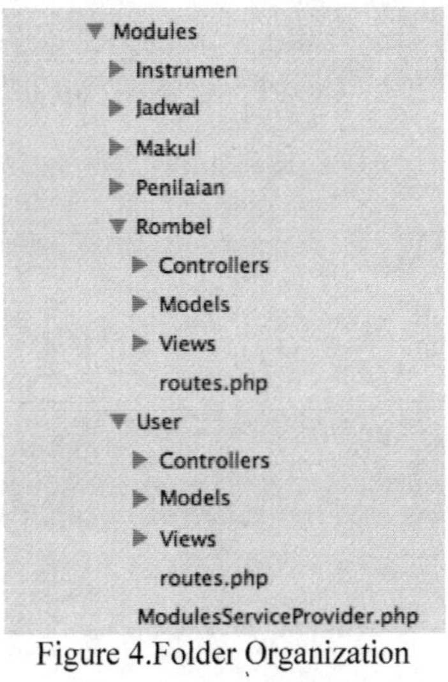

Figure 4.Folder Organization

Here is the explanation of the operational of the application.

1. The user provides a request via typed URL. Routing in the system works to continue to the Controller. Controller processes user requests. Based on user demand, the Controller will call the View and the Model. The Model is used if the user requests regarding data stored in a database. The Controller adjusts the demand for the model. The model represents the entire data and application logic.
2. Then the Controller continues it to the View. The view is responsible for the output of the model. The graphical display of pure contained in this process. View displays a request given by the user. So the users get the results they want.

2.4. *Waterfall Model Extended*

The main focus of this research is on the expansion of the waterfall model. The waterfall phase, in general, is divided into five stages. But the concept that we propose is the expansion of the waterfall model remain in 5 stages but has a different groove. Also, there is a stage in its development using MVC architecture which is implementation stage.Gambar stage 5 shows how the lane of the waterfall model with the expansion of the use of MVC architecture.

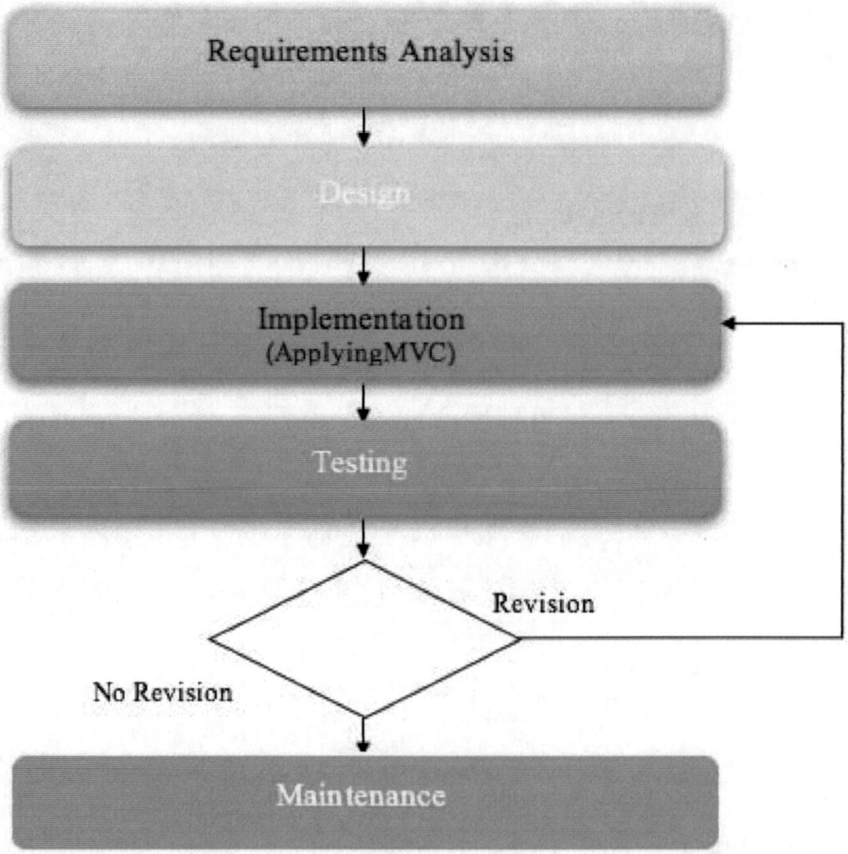

Figure 5.Expansion of the waterfall model and the application of MVC architecture

The expansion lane of the waterfall model starts from the analysis of needs. After analysis of needs is assessed, then the data will be designed from the user interface, application logic, database, and so on. Later in the implementation phase with the software is written through coding. In its development, we apply the MVC architecture at this stage. Because at this stage of development, there definitely be many obstacles around the business logic, interface, and database. So with the MVC architecture, the problem will be solved. After the application is completed, then we can do a test. If the applications were built without any flaws, then the system can be directly to the maintenance phase without revised. However, if the system is still uncompleted, then the system will return to the stage of implementation and do the revised. Once the system is back to be tested to maximize the application functionality.

3. Results and discussion

Implementation stage is the stage of making the information system according to the results at the design stage. We build a simple application and its called "Microteaching" to test this model.

Implementation of this system is using the frameworkLaravel. Framework laravel facilitate the implementation of the design stage into the system because it uses MVC architecture. The ease of using frameworkLaravel is the separation of program code based on the Model-View-Controller pattern, implementing the functional requirements into separate modules, and the availability of build-in library so as to accelerate the implementation process.

Here, we will show you some pictures of the application. We perform different types of operations, such as entering data into a database, retrieving data, change or update the data, searching data and so on. We get the appropriate results for each step.

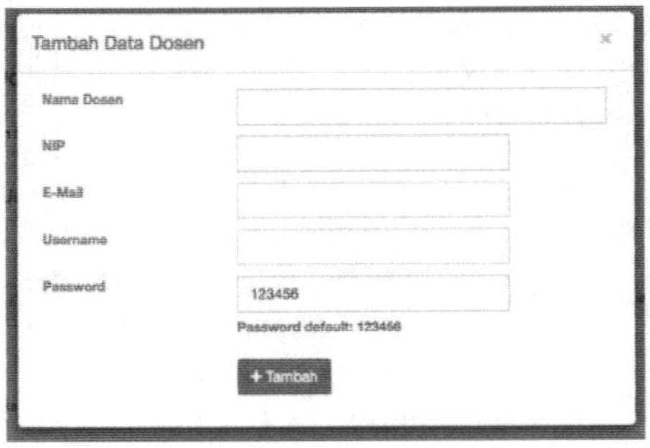

Figure 6. Insert Data

Figure 7. Update Data

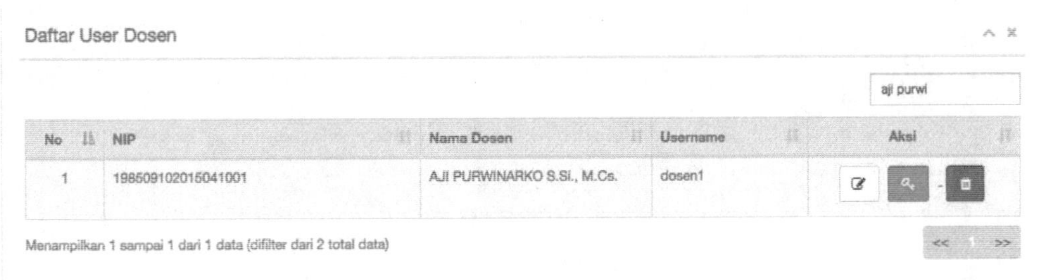

Figure 8.Result Of Data

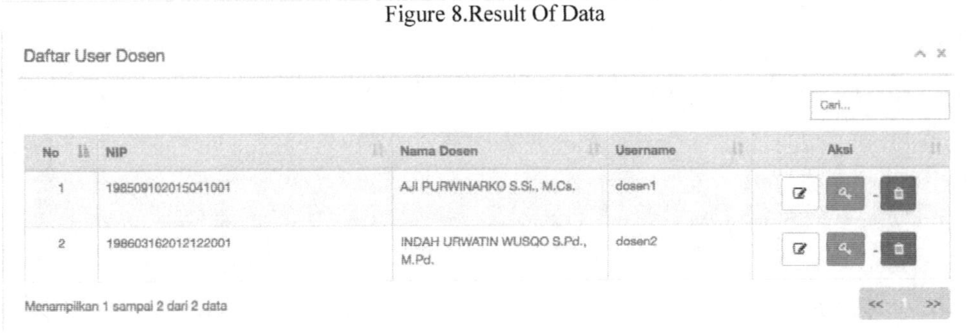

Figure 9.Data Search

4. Conclusions

Based on the proposals above, it says that the expansion of the waterfall model is used by many large companies for their internal projects. Due in large companies most clients can change their requirements at any specified time. Not only time, usually the client wants to change the graphical display totally at any time. So it does not waste time and would hold up the process of software development. However, with the application of MVC architecture in software development, application developers can easily change the parts that need to be changed suddenly. Because this framework separates View, Controllers, and Model. Therefore, this framework is very stable, efficient and able to create high-quality applications.

References

[1] Patel, U. A., 2013. New Idea In Waterfall Model For Real Time Software Development. *International Journal of Engineering Research & Technology (IJERT),* April, 2(4), pp. 114-119.
[2] Bassil, Y., 2012. A Simulation Model for the Waterfall Software Development Life Cycle. *International Journal of Engineering and Technology,* May, 2(5), pp. 742-749. Patel, U., 2013. New Idea In Waterfall Model For Real Time Software Development. *International Journal of Engineering Research & Technology (IJERT),* April, 2(4), pp. 114-119.
[3] S. D. Mone, "Comparative study of Waterfall model with RAD model," *International Journal of Modern Trends in Engineering,* pp. 481-484, 2015.
[4] Trygve Reenska, http://heim.ifi.uio.no/~trygver/themes/mvc/mvc-index.html
[5] Glenn E. Krasner, Stephen T. Pope, "A cookbook for using the model-view controller user interface paradigm in Smalltalk-80", Journal of Object-Oriented Programming, vol. 1, no. 3, 1988, pp. 26-49.
[6] Badurowicz,M."MVC Architectural PatternIn Mobile Web-Applications", Actual Problems Of Economics, 2011, pp.305-309.
[7] W. J. Gilmore, Easy PHP Websites, Columbus, Ohio: W.J. Gilmore, LLC, 2009.
[8] A Freeman, S Sanderson, Pro ASP.NET MVC 3 Framework, Apress, 2011.
[9] Stratmann, E., & Ousterhout, J.,"Integrating Long Polling with an MVC Web Framework", 2nd USENIX Conference on Web Application Development, 2011, pp113-124.
[10] J. Galloway, P. Haack, B. Wilson şi K. S. Allen, Professional ASP.NET MVC 3, John Wiley & Sons, Inc., 2011.
[11] Purwinarko A and Y. L. Sukestiyarno, "Model Expertise Management System di Universitas Negeri Semarang," *Sci. J. Informatics*, vol. 1, no. 2, pp. 177–184, 2014.

IOP Conf. Series: Journal of Physics: Conf. Series **824** (2017) 012008 doi:10.1088/1742-6596/824/1/012008

Development of Open-Ended Problems for Measuring The Higher-Order-Thinking-Skills of High School Students on Global Warming Phenomenon

Fianti[1,*] , F L Najwa[1], and S Linuwih[1]

Physics Departement, Mathematics and Natural Science Faculty,Universitas Negeri Semarang, Indonesia

*Corresponding author:fieanty@yahoo.com

Abstract. Higher-order-thinking-skills can not be developed directly, except by training which is employing open-ended problems for measuring and developing critics, creativeness, and problem-solving thinking-skills of students. This study is a research and development producing open-ended problems. The purpose of this study is to measure the properness and effectiveness of the developed product and to observe the profile of higher-order-thinking-skills of students on global warming phenomenon. The result of properness test of open-ended problems according to the experts is 92,59% on the first stage and 97,53% on the second stage, so we can assume that the product isvery proper. The result of effectiveness test shows the coefficient of correlation between student's midterm test scores and open-ended questions is 0,634 which is in the category of strong. Higher-order-thinking-skills of SMA Negeri 1 Salatiga students is in the category of good with the average achievement scores 61,28.

1. Introduction

An international study which measures the performance on mathematics and natural science for thestudent, Trends in Mathematics and Science Study (TIMSS), shows that in 2011, Indonesia's student was on the ranking 40 of 42 countries surveyed in the sciences. Indonesia score was 397 in Physics. This value is below the international average, 500. Based on the data of science content and cognitive domain, especially in physics, the percentage of Indonesian participant swith the correct answer in a matter of understanding are higher than that of in the matter of application and reasoning. Furthermore, anaspect of understanding and application is in the lower-order-thinking-skill,and aspect of reasoning is in the higher-order-thinking-skill [1]. Therefore, higher-order-thinking-skill of Indonesia's student is still low.

Higher-order-thinking-skill is the ability to solve problems where there is no algorithm that has been taught that requires justification or explanation. Also this problem has more than one solution to be accepted [2-3]. Higher-order-thinking-skill is the highest level in the hierarchy of cognitive processes [3]. Therefore according to Bloom's Taxonomy, higher-order-thinking-skill is one of the ability to analyze, evaluate, and create.

Higher-order-thinking-skill can not be developed directly and instantly. The test can be used as a tool to determine the profile of student's thinking skill and to train students for better thinking skills.

Teachers should use an open-ended problem to determine the student's thinking skill; this isa problem with more than one answer [4]. Open-ended problems consist ofmain concepts, processes, and skills beyond the specific instructions that can determine the students thinking skills [5]. Surf*et al.* [6] and Intrasite [7] revealed that the advantage of using open-ended problems is to encourage in developing the creativeness and analytic-thinking-skill. By using open-ended problems, the basic concepts can be explored, so the students do not only recall the certain facts but almost all possible facts to solve the problems. This gives a chance for the students to be asuccess in answering the open-ended problems, even they cannot remember a certain fact because there is always some alternative way of solving the problem in nature.

2. Research Methods

This studyisconducted byresearch and development method that produces and develops open-ended problems for measuring higher-order-thinking-skill of students. This study is conducted using formative research proposed by Tessmer [8] and modified as shown in the following flow chart.

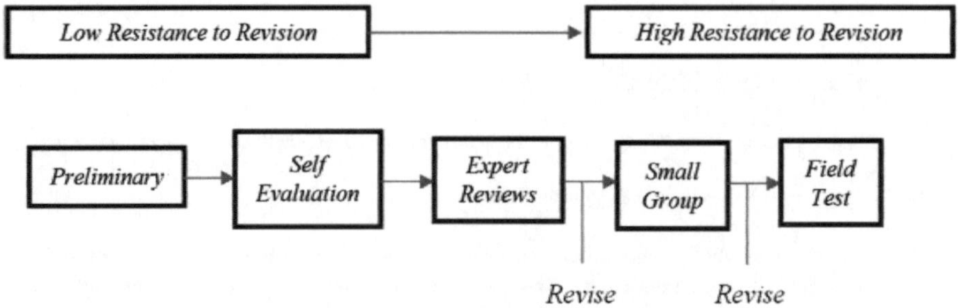

Figure 1. Flow chart of formative research design.

Figure 1 shows that the research method consisting of several stages: (1) preliminary or preparatory phase, (2) self-evaluation that includes analysis and design stage, (3) expert reviews or properness test of product, (4) small group or small scale trials, and (5) field test or a large-scale trial. The research was done in SMA N 1 Salatiga who had received the material on the global warming phenomenon of physics learning, namely MIA 5.4 as a research subject of asmall group, MIA 1.4 and MIA 2.4 as a research subject of thefield test.

The purpose of this study is to produce and develop open-ended problems, to determine the properness and effectiveness of the product, and also to determine the profile of higher-order-thinking-skills of students on a material of global warming phenomenon. Properness test is done by giving product properness questionnaire to the experts and students. Product effectiveness test is done by calculating the correlation between the student's midterm test scores and open-ended problems. The profile of higher-order-thinking-skillsis determined by analyzing the student's answer to the problems.

3. Results and Discussion

The results of this study include the design, properness, and effectiveness of open-ended problems, also profile of student's higher-order-thinking-skills. The developed open-ended-problems are consists of 10 problems on a global warming phenomenon. These problems represent the indicators of the higher-order-thinking-skill. In fact, higher-order-thinking-skill of students consists of several aspects, *i.e.* the ability to analyze, evaluate, and create. Moreover, the ability to analyze, evaluate, and create can be represented by three indicators as follows; first in analysis, students can recognize and distinguish between the causes and the result of a complicated scenario, students can analyze the incoming information and share it into smaller sections to find patterns and relationships, and students can identify and formulate questions. Second, in the evaluation, students can give an assessment of the idea and the methodology by using match and standard criteria to ensure the effectiveness, students can hypothesize, criticize, and do the test, and students can accept or reject a statement based on predetermined criteria. Third, in creation, students can make a generalization of ideas or ways, students can determine a way to solve the problem, and students can organize the parts of something into new structures that have never existed before [2].

Open-ended problems are given to experts and students to measure the properness. Properness questionnaire prepared with some aspects of the assessment criteria, *i.e.* the aspect of content, construct, and language. Properness test of open-ended problems to measure the higher-order-thinking-skills of students is done by experts in two stages. The first stage did before the open-ended problem sare tested in a small scale trial. The second stage did before the open-ended problems are tested in a large scale trial. The test result about the properness of open-ended problems at the first stage is presented in Table 1 and the second stage is presented in Table 2.

Table 1. The result of properness test open-ended problems at the first stage.

No	Evaluator	Content		Constructs		Language	
		Score	in %	Score	in %	Score	in %
1	Validator 1	9	100	9	100	9	100
2	Validator 2	8	88.89	8	88.89	8	88.89
3	Validator 3	8	88.89	8	88.89	8	88.89
Average		8.33	92.59	8.33	92.59	8.33	92.59
Criteria		very proper		very proper		very proper	

Table 2. The result of properness test open-ended problems at the second stage.

No	Evaluator	Content		Constructs		Language	
		Score	in %	Score	in %	Score	in %
1	Validator 1	9	100	9	100	9	100
2	Validator 2	9	100	9	100	9	100
3	Validator 3	8	88.89	9	100	8	88.89
Average		8.67	96.29	9	100	8.67	96.29
Criteria		very proper		very proper		very proper	

The distributions of the student responses questionnaire are also performed in two stages. The first stage is done by small-scale trial participants and the second stage is done on a large scale trial participants. Student responses questionnaire to open-ended problems is given to the students after the test. The results of student responses are presented in Table 3.

Table 3. The result of student responses to open-ended problems.

	Acquisition of Score per Aspects (%)				
Respondents	Language	Display	Systematic	Function	Testimony
Small scale trials	83.04	79.46	77.68	89.28	77.68
Large-scale trial	83.33	82.41	79.40	85.95	78.24
Average	83.19	80.94	78.54	87.62	77.96
Criteria	Good	Good	Good	very good	Good

The results of properness test by expert and student responses show an average percentage > 80%, its means that the open-ended problems are very proper to be used to measure higher-order-thinking-skills on global warming phenomenon.

The effectiveness of open-ended problems can be seen from the score of students mid-semester test that is compared to the score obtained from open-ended problems. The correlation coefficient between the score of mid-semester test and open-ended problems with a level of error of 1% is 0.634, a strong category. The result shows that the open-ended problems can be used to describe the higher-order-thinking-skills effectively.

Profile of student's higher-order-thinking-skills can be seen from the answers and the obtained scores on open-ended problems where the ability to analyze, evaluate, and create can be observed from it. Ability profile in analyzing, evaluation, and creation are presented in Table 4, 5, and 6.

Table 4. Ability profile in analyzing.

Interval values	Frequency	Percentage (%)	Category
$75 < t \leq 100$	6	11.11	very good
$58.33 < t \leq 75$	18	33.33	good
$41.67 < t \leq 58.33$	16	29.63	enough
$25 < t \leq 41.67$	13	24.08	low
$0 < t \leq 25$	1	1.85	very low
The average value	56.17		enough

Table 5. Ability profile in theevaluation.

Interval values	Frequency	Percentage (%)	Category
$75 < t \leq 100$	25	46.30	very good
$58.33 < t \leq 75$	15	27.78	good
$41.67 < t \leq 58.33$	11	20.37	enough
$25 < t \leq 41.67$	3	5.55	low
$0 < t \leq 25$	0	0	very low
The average value	70,68		good

Table 6. Ability profile in creation.

Interval values	Frequency	Percentage (%)	Category
$75 < t \leq 100$	6	11.11	very good
$58.33 < t \leq 75$	20	37.04	good
$41.67 < t \leq 58.33$	21	38.89	enough
$25 < t \leq 41.67$	7	12.96	low
$0 < t \leq 25$	0	0	very low
The average value	57		good

The developed open-ended-problems raise very diverse student answers, so it can show the profile of student's thinking skill well. Table 4 shows that ability profile in analyzing is good enough. It can be seen from the student's answers that a large number of students can recognize and distinguish between the causes and consequence of global warming process, analyze information about the earth's climate which is unstable, begin to identify the patterns, and able to identify the problems that led Indonesia became contributor of greenhouse gases.

Table 5 shows that the ability to evaluate is good. It can be concluded from the student's responses. The students can give an assessment of the vegetarian community to ensure the effectiveness and benefits, make hypotheses the kind of greenhouse gasses which is most influential on global warming then do the testing, and accept or reject the opinion in stopping the global warming.

Table 6 shows that the ability in creation is good. It can be seen from the student's answers which show that a large number of students can generalize an idea in controlling the global warming, determine a way to overcome the problem of organic waste in Indonesia, and developing some ideas to create an alternative energy or environmentally friendly technology.

The results show that the best ability of students of SMA Negeri 1 Salatiga is the ability to evaluate, creative, andthen analyze. These results are not in line with research conducted by Istiyono *et al.* [9] which states that the best ability of high school students is the ability to analyze, evaluate, and then to create. Higher-order-thinking-skill can be measured by summing the scores obtained by students from aspects of the analysis, evaluation, and creation. Profile of higher-order-thinking-skill of students stated in Table 7 below.

Table 7. Profile of higher-order-thinking-skill.

Interval values	Frequency	Percentage (%)	Category
$75 < t \leq 100$	11	20.37	very good
$58.33 < t \leq 75$	17	31.48	good
$41.67 < t \leq 58.33$	23	42.59	enough
$25 < t \leq 41.67$	3	5.56	low
$0 < t \leq 25$	0	0	very low
The average value	61.28		good

The table shows that 20.37% of the students have a very good higher-order-thinking-skills, 31.48% of students have a good higher-order-thinking-skills, 42.59% students have a good enough higher-order-thinking-skills, and the other have a low good higher-order-thinking-skills. The average value obtained by all students is 61.28 in the category of good, so it can be concluded that in general, SMA Negeri 1 Salatiga has a good higher-order-thinking-skills.These results are in line with research that has been done by Rahayu and Sonata [10] which states that a higher-order-thinking-skills of students of SMAN 18 Surabaya on Acid Alkali Titration material included in good categories. These results are also in line with the data from Balitbang Kemdikbud on the results of a national examination SMA/MA 2014/2015 stating that SMA Negeri 1 Salatiga has a good academic record.

4. Conclusion

The open-ended problems in measuring the higher-order-thinking-skills of high school students on global warming are very proper and effective. Higher-order-thinking-skills of the students of SMA Negeri 1 Salatiga are good, covering the ability to analyze good enough, the ability to evaluate is good, and the ability to be creative is good enough.

References

[1] Rofiah E, Aminah N-S, and Ekawati E Y 2013, *in Jur. Pen. Fisika*, 1**(2)**: 17.
[2] Lewy, Zulkardi, andAisyah N 2009, *in Jur. Pen. Matematika*, 3**(2)**: 14.
[3] Yee M H, Yunos M D, Othman W, Hassan R B, Tee T K, dan Mohama M M 2015, *in Prosedia-social and behavioral sciences*, 204 : 143.
[4] Emilya D 2010, *in Jur. Pen. Matematika*, 4**(1)**: 8.
[5] Husain H, Bais B, Hussain A, and Samad S A 2002, *in Procedia-social and behavioral sciences*60: 4556.
[6] Surif J, Ibrahim N H and Dalim S F 2013, *in Prosedia-social and behavioral sciences*116 (2014): 4955.
[7] Inprashita M 2006, *in Tsukuba Journal of Educational Study in Mathematics,* 25: 169.
[8] Tessmer M. 1993. *Planning and Conducing: Formative Evaluations*. London: Kogan Page.
[9] Istiyono E, Mardapi D, and Suparno 2014, *in Jur.Penelitian dan Evaluasi Pendidikan*, 18(**1**):1.
[10] Rahayu T and Yonata B 2013, *in Jur. Unesa Journal of Chemical Education*, 2(2): 12.

Problem-Based Learning Model Used to Scientific Approach Based Worksheet for Physics to Develop Senior High School Students Characters

D Yulianti

Physics Education Study Program, Faculty of Mathematics and Natural Science, Universitas Negeri Semarang Indonesia

*Corresponding author: yulifis04@yahoo.com

Abstract. The purpose of this study is to explore the application of Problem Based Learning(PBL) model aided withscientific approach and character integrated physics worksheets (LKS). Another purpose is to investigate the increase in cognitive and psychomotor learning outcomes and to know the character development of students. The method used in this study was the quasi-experiment. The instruments were observation and cognitive test. Worksheets can improve students' cognitive, psychomotor learning outcomes. Improvements in cognitive learning results of students who have learned using worksheets are higher than students who received learning without worksheets. LKS can also develop the students' character.

1. Introduction

To improve Indonesian Education, Curriculum 2013 has to undergo constant evaluation. According to the Minister of Education Regulation (*Permendikbud*) No 69, 2013, Curriculum 2013 aims to create Indonesia citizen who possesses excellent life skills and characters. For example, religious, creative, productive, innovative and efficient both as an individual and citizen and are also able to contribute to the society, nation, state, and world civilisation. The success of Curriculum 2013 is everyone's responsibility. Character education implemented in Curriculum 2013 through the process of integrating learning objective, synergizing educational institution roles, strengthening teachers' competence and model [1]. Curriculum 2013 adopts the scientific approach which not only promotes cognitive development but also character development such as life skills. It is in line with a study by Machin [2] stating that scientific approach affects students' cognitive, affective and psychomotor positively. Students Worksheet (LKS) assist the process of teaching and learning. However scientific approach and characters integrated LKS has not been available in the community. Problem Based Learning (PBL) is a method applied in Curriculum 2013. Based on Rizkianingsih *et al.*[3], PBL can encourage students to actively solve problems so they can get the concept of the material independently. This involvement will increase students' achievement. It is in line with Astuti's and Junaedi's study [4] discovering that PBL can improve students' activities and achievement. It can also increase students' creativity. Sungur et.al's [5] research also shows that the students taught using PBL have more skill score rather than the one taught and in a conventional method. Besides, students' right character nurtured through PBL. Okinoglu and Tandogan [6] discover that the implementation of PBL in science program classroom contributes to the change of students' attitude. Teaching and learning model used in Curriculum 2013 is Problem Based Learning. This study aims to describe the implementation of PBL classroom using scientific method and characters integrated student worksheet (LKS) with the material temperature and thermal expansion and to investigate students' cognitive, affective and psychomotor improvement after the treatment.

2. Method

This study was a trial from LKS classroom created in 2015. The subjects were students grade X in eight different schools in Semarang. The method was the quasi-experiment. The PBL using scientific method and characters integrated student worksheet (LKS) with the material temperature, and thermal expansion was implemented in experiment classes while the control one gave conventional method. All classes were given pre-test and post-test to discover students' increase of cognitive learning outcome. The test underwent the test of validity, reliability, a level of difficulty, and concurrent validity. Questionnaires related to characters given to student before and after the teaching and learning process. Data gathering instruments were the test, questionnaires, and documentation.

3. Results And Discussion

3.1. Cognitive Learning Outcome

The effectiveness of the material discovered through the increase of students cognitive test outcome. Effectiveness test can be analysed by using gain test and t-test. The result of the gain test shown in Table 1.

Table 1. Result of Improvement of Cognitive Learning Outcomes

Character	Result of Gain Test	Criteria
Control Class	0,472	Middle
Experiment Class	0,795	High

The table shows that experiment class shows better achievement in cognitive learning outcome rather than the control class. It happens because who gave two different treatments which are the availability of LKS in experiment class and conventional method of teaching in control class. Septiani *et al.*[7] study support this idea with the finding of the fact that the use of LKS can increase the average score of gain related to students' understanding of the concept. Another research by Ristiyani & Yulianti [8] also shows that students gain a better understanding of material's concept after suing scientific approach integrated worksheet.

The result of post-test score t-test analysis is available in Table 2.

Table 2. Result of T-Test Analysis

T-test	Result
t-counted	32,922
t(0,95)(68)	1,9955

It seen that $t_{-counted} > t_{(0,95)(68)}$ thus H_0 is rejected (Sugiyono, 2009 : 197). From t-test, it concluded that experiment group is better than the control group. It is in with the study from Ni'mah *et al.*[9], average learning outcome from the student who given LKS a is better than the one not proven by $t_{-counted} > t_{(0,95)(68)}$.

The results from both gain test and t-test showed that is the worksheets developed are effectively improved students understanding of a concept. That is in line with research from Yusuf [10] claiming that LKS use can increase students' achievement. This achievement is a result of the growth of students' interest in learning physics. The output from Isnaini *et al.*[11] supports the idea that physics worksheet developed can increase students' activities and learning motivation. Students' interest comes from the simple language use and the format of the LKS.

Another reason behind the result of LKS effectiveness in helping students understanding the material is that it also assists optimising students' scientific process. The LKS will contribute nurturing students' scientific attitude through experiment and discussion. Subagyo *et al.*[12] say that the use of LKS inquiry can assist students understanding the concept and bring out scientific attitude such as preparing, experimenting, observing, data analysing and concluding. Research from Hidayati & Endryansyah [13] also claims that scientific learning can improve students' achievement because students become active and communicative in discussion and are challenged to solve the problem in experiments. The LKS contains the example of physics' implementation in daily life which can help the students to grasp the real concept of physics.

3.2. Character Development

The characters such as discipline, communicative and curious have been entrenched among the students while creativity is starting seen. The result of observation shown in Table 3.

Table 3. Result of Character Observation

Characters	Observation Result	Criteria
Discipline	93,14%	Entrenched
Creative	73,71%	Developing
Communicative	86,29%	Entrenched
Curious	93,14%	Entrenched
Total	86,57%	Entrenched

Those developments were analysed using gain test and gave a result that presented in Table 4.

Table 4. Analysis of Character Development

Characters	Gain Test Result	Criteria
Discipline	0,662	Middle
Creative	0,146	Low
Communicative	0,361	Middle
Curiosity	0,336	Middle

The table above presents the increase average result between students' character before and after treatment. It means that LKS can effectively improve students' character development. By answering the questionnaires, students show their attitude when facing things related to discipline, creativity, communicativeness and curiosity. As explained by Azwar [14] that the immediate stimulus will bring out the consistency of response related to affective, cognitive, and attitude when facing their surroundings.

Although there is indeed increase in students' character, the degree is not that much. Some characters such as discipline, communicativeness, curiousness belong to middle category while creativity is only low. Discipline gains the most enhance result because of some reasons. The first one is the fact that the school has applied the regulation well with the punishment system which makes students accustom to follow the rules. One example is by coming to school punctually. The low level of creativity is the result of students' lack of developed characters, characters which originate from the first education before school and community, family. Samani and Hariyanto [15] state that characters education occurs in three pillars, school, family and community. Another reason is that the LKS only gave in limited occurrences, only to the discussion of temperature and thermal expansion. According to Ministry of Education (2010: 11), the main principal of character education is a continuous process that starts from the initial until the end of education level.

Character education integrated into students' worksheet in the form of specific suggestions and guidelines in doing discussion and experiment. The guideline encourages students to actively carry out the activities that can develop students' character. This guideline is given continuously to nurture students' habit. It was supported by the declaration from Ministry of Education (2010: 86) that character education should be sustainable, routine, within all subjects. Thus the values will be learnt in an active learning process. Character education integrated into every aspect of learning in every subject. According to Benninga et al. [16], if character education incorporated in, every aspect of school's life both inside and outside a classroom, it can improve students' character.

Discipline scores entrenched based on the percentage and scores middle from an again test. The indicator for such coming on time has been done by the students well (*Kemendiknas*, 2010: 26). LKS provides suggestion to the students sink their teeth in discipline by returning their lab appliance properly, and finishing their assignment on time. Students are accustomed to performing self-discipline because the regulation is applied well in the school.

Students' creativity is low. The indicator is creating learning situations that foster creative thinking and acting and giving challenging tasks to encourage new creation either authentic or modified (*Kemendiknas*, 2010: 26). This indicator achievement showed from students' performance in developing steps in the experiment without any guideline, students' variation and their creativity in creating simple thermometer. Next, character communicative is entrenched and get a middle category in the gain test. It happens because the learning process is done in a group where students work together discussing and experimenting with their friends. Curiosity belongs to the category of entrenched with a mean increase. Curiosity arises between students because the worksheets (LKS) provides a problem that hooks students' curiosity and the students gave the opportunity to find further information from various sources such book and the internet. If the students gave problem, they would be motivated to solve it actively investigate the subject in question to fulfil their curiosity (Pluck & Johnson [17].

This increase proves that LKS developed can be effectively used as a medium to teach integrated character in the process of teaching and learning. It is supported by Amelia et al. [18] claiming that students' character like honesty, discipline, confidence, curiosity, desire for reading, cooperation and communicativeness increase along with the present of character education in physics using LKS.

3.3. Psychomotor Learning Outcome

Psychomotor aspect in this study is about students' performance in preparing discussion/experiment, doing discussion/experiment, writing the report and presenting. The gain test was done to analyse the increase of observation result of students' performance before and after the

Problem Based Learning (PBL) process using a scientific approach and character education integrated worksheet. The result of a gain test presented in Table 5.

Table 5. Analysis Result of Psychomotor Learning Outcome

Group	Average Before	Criteria	Average After	Criteria	*Gain*	Category
Experiment	70,97	Good	83,87	Good	0,34	Middle
Control	69,34	Good	76,44	Good	0,22	Low

The increase presented shows that PBL using a scientific approach and character education integrated worksheet can enhance students learning skills. It is in line with research by Sungur et al. (2006) describing that student who learns using PBL have higher performance scores rather than the one who get to sue traditional model.

4. Conclusion

Application of PBL Model-aided with an integrated scientific approach and character education integrated LKS can improve learning outcomes. Cognitive learning outcomes (knowledge), psychomotor (skills) and the characters in the experimental group increased in medium criteria. Further research is required in sequence to determine other the character development and improvement of learning outcomes.

Acknowledgement

Thank you to *DP2M Kemenristekdikti* that has funded this research through *Hibah Bersaing* program in 2015-2016.

References

[1] Kaimuddin 2014. *Dinamika Ilmu*, 14 **(1)** : 47
[2] Machin A 2014. *Jurnal Pendidikan IPA Indonesia*, 3 **(1)** : 28.
[3] Rizkianingsih, Sukisno M and Susilo. 2013. *Unnes Physics Education Journal*, 2 **(3)** : 48
[4] Astuti Y and Setiawan B 2013. *Jurnal Pendidikan Indonesia*, 2 **(1)** : 88
[5] Sungur S, Tekkaya Cand Geban O 2006. *Journal of Biological Education*, 40 **(4)** : 155.
[6] Okinoglu, O and R. O. Tandogan 2007. *Eurasia Journal of Mathematics, Science & Technology Education*, 3**(1)**: 71.
[7] Septiani, D., S. Ridlo and N. Setiati 2013. *Unnes Journal of Biology Education*, 2 **(3)**: 359
[8] Ristiyani D. and Yulianti D 2014. *Unnes Physics Education Journal*, 3 **(3)**: 54
[9] Ni'mah, L. H., Saptorini and S. D. Pamelasari 2013. *Unnes Science Education Journal*, 2 **(1)**: 149
[10] Yusuf, M. 2010. *Jurnal Pendidikan Matematika*, 4**(2)**: 34
[11] Isnaini, M., Marwoto P and A. Yulianto 2012. *Journal of Innovative Science Education*, 1 **(2)**: 98
[12] Subagyo, Wiyanto Y and Marwoto P 2009. *Jurnal Pendidikan Fisika Indonesia*, 5: 42
[13] Hidayati, N and Endryansyah 2014. *Jurnal Pendidikan Teknik Elektro*, 3 **(2)**: 25
[14] Azwar S 2013. *Yogyakarta: Pustaka Pelajar*.
[15] Samani, M. and Hariyanto 2012. *Remaja Roesdakarya Offset*.Bandung
[16] Benninga, J. S., M. W. Berkowitz, P. Kuehn and K. Smith 2003. *Journal of Research in Character Education*, 1**(1)**: 19
[17] Pluck, G. and H. Johnson 2011. *Education Science and Psychology*, 2 **(19)**: 24
[18] Amelia, O. T., Yurnetti and Asrizal 2013. *Jurnal Pillar of Physics Education*,.**(2)**

Project Based Learning Multi Life Skill for Collaborative Skills and Technological Skills of Senior High School Students

Susilawati[1], S Ardhyani[2], Masturi[3], Wijayanto[4], and N Khoiri[5]

[1,2,4]Department of Physics Education, Semarang University of PGRI, Semarang, Indonesia
[3] Physics Department, Faculty of Mathematics and Natural Sciences, Universitas Negeri Semarang, Indonesia
[5] Department of Information Technology of Education, Semarang University of PGRI, Semarang, Indonesia

*Corresponding author: susilawati.physics@gmail.com

Abstract. This work aims to determine the effect of Project Based Learning containing Multi Life-Skills on collaborative and technology skills of senior high school (SMA) students, especially on thestatic fluid subject. The research design was aquasi-experiment using Posttest-Only Control Design. This work was conducted in SMA Negeri 1 Bae Kudus, with the population is all students of class X, while the sample is students of class X MIA 2 as an experimental class and X MIA 3 as a control class. The data were obtained by observation, test, and documentation. The results showed this model significantly affects the collaborative and technology skills of students of SMA 1 Bae Kudus, where the average result of collaborative and technology skills for the experimental class is higher than that of the control class. This is also supported by the remark of the post-test experimental class is higher than that of the control class.

1. Introduction

The development of information technology such as the Internet very rapidly and expanded to parts of the world has been used by many countries, institutions, and experts for various purposes including for education. The learning approach uses to explore knowledge, experience and synthesize[1]. The Internet is used as learning media in the process of teaching and learning activities in the classroom. Use of the internet can broaden the students to be more sensitive to the use of the internet as a medium to increase creativity and inspiration.

The times will change human needs as well as education although the essence of education itself has not changed, but the stock of knowledge in a person must be more complex. In the current era of cognitive abilities assumption, everyone is equal. Therefore, one must have a greater ability to become a superior human. There are three core capabilities that students need to have associated with 21st-century skills such as knowledge, meta-knowledge, and humanistic knowledge. The duration of intervention was more cost-effective to reduce the number of sessions in the intervention trajectory [2].

21st-century skills want the current knowledge oriented learning paradigm projects, problems, investigation, discovery, and creation. These activities provide opportunities for students to improve learning capabilities in the areas of cognitive, affective, and psycho motor. This affects the development of emotional intelligence, spiritual, and social students. Skills obtained in the process of learning physics in the classroom is still low. Skills that can be directly used in the world of work has not been able to be integrated into the learning process in the classroom. The demands of organizing the learning process consist of theory and practice, capable of the only limited theory held while the practice of only one to two times in one semester. One effort to complement and augment the skills can be given an alternative that is to equip skills through project activities. The project can be done by the students in the learning process and the outside class hours.

Related skills needed to cope with change in the century two-one as well as the experts noted the need for mastery of skills. Skills of learning and innovation consist of critical thinking and problem solving, communication and collaboration, creativity and innovation. Skill digital literacy includes information literacy, media literacy, literacy information and communication technology. Skills career and life skill consist of flexibility and adaptability, initiative and self-regulation, social interaction and cross-cultural, productivity and accountability, leadership and accountability. Empirical evidence on the impact of instruction on the development and enhancement of critical thinking skills[3]. One model of learning that emphasizes collaboration skills and technological skills of students is a project-based learning model. This learning model allows students to develop their creativity in designing and creating a project that can be used to overcome the problems. All studentswill need opportunities to learn from high-quality curriculum materials where they engage in science practices[4].The projectbased learning model is a model of learning centered on student activity, student teachers learn together. During to do the project, students can train and develop their communication skills skills of observing, using tools and materials, interpret, plan projects, applying the concepts, ask questions and communicate well through technology.

The project-based learning model students will work collaboratively. Students express a problem as a challenge or a question that must be answered and manage their own time to complete the project. Based on this background, this study will use project based learning model that will be studied in senior high school. The study was conducted to determine the effect of project-based learning model multi-life skill learning to collaborative skills and technology skills of students on the topic of static fluid.

2. Methods

This research conducted in SMAN 1 Bae Kudus, the population of all grade X MIA in SMA 1 Bae Kudus in the academic year 2015/2016. The sample in this research is the students of grade X MIA 2 as an experimental group and X MIA 3 as a control group. The sampling technique is cluster random sampling. In this study using Quasi-Experimental Design with the form Posttest-Only Control Design [5].Methods of data collection in this study is the test, observation, and documentation. Analysis of data from student test consists of two phases initial and final stages. The homogeneity test using Barlett test [6]. The data analyze using the Kolmogorov-Smirnov normality test, t-test and correlation test.

3. Results And Discussion

The initial data used in the analysis of form replay remark elasticity of grade XMIA2 and XMIA3. The sample is homogeneous used Barlett test. Table 1 a calculation for homogeneity test.

Table 1. Normality test, homogeneity test for technology skill and collaborative skill

Multi Lifeskill	**Group**	***N***	***Mean***	***P-value***	**Description**	***P-value***	**Description**	**Uji t**
Collaborative skill	Experiment	32	89.53	0.805	Homogen	0.589	Normal	0.000
	Control	32	68.86			0.148	Normal	
Technology Skill	Experiment	32	89.45	0.801	Homogen	0.512	Normal	0.000
	Control	32	57.96			0.167	Normal	

Table 1 shows that the probability value homogeneity test of grade XMIA2 and grade XMIA3 is 0.805. If probability value of homogeneity test larger than 0.05. This means that the ability of the two grade is homogeny as the same for initial conditions grade. Normality test from final data used Kolmogorov-Smirnov test. The probability value grade XMIA2 is 0.589. The probability value grade XMIA3 is 0.148. The probability value of normality test more than 0.05. This means that the data is normal distributed collaborative skills. The probability remark grade XMIA2 is 0.512 and grade XMIA3 is 0.167. The probability value of normality test larger than 0.05. This means that the data is normally distributed technology skills. Analyze data using at-test to determine which grade higher collaborative skill and technological skill.

The probability value t-test both groupswere 0,000. The probability value collaborative skills t test was 0.000 larger than 0.05. This means that Ho rejected, collaborative skills experimental group XMIA2 were treated using project based multilineskill higher than the control group XMIA3 in the implementation using project-based learning.Final data analysis using t-tests were used to determine which grade have higher technology skills. The calculation of the value t-test technology skills in Table 1.

The probability value t-test of technology skill was 0.000 higher than 0.05. This means that Ho rejected, the value of technology skills experimental group XMIA2 were treated using project based multilineskill is higher than the control group X MIA 3 that the implementation using project-based learning model.In the analysis of data using correlation test is used to determine the significance of the relationship project based learning multi-life skill with collaborative skills and technology skills of students. The following calculations correlation project based learning multi-life skill with collaborative skills and technology skills. Table 2 show R-square between collaborative skill and technology skill.

Table 2. Correlation test for project-based learning multi life skill, technology skill, and collaborative skill

Group	Aspect	Mean	R-Square	P-Value
Experiment	PjBL Multilife skill	86.31	0.431	0.000
	Collaborative skill	89.53		
	PjBL Multilife skill	86.31	0.173	0.018
	Technology skill	89.45		
	Collaborative skill	89.53	0.202	0.010
	Technology skill	89.45		
Control	PjBL	69.78	0.173	0.027
	Collaborative skill	68.86		
	PjBL	69.78	0.139	0.036
	Technology skill	59.96		
	Collaborative skill	68.86	0.057	0.187
	Technology skill	57.96		

Table 2 R-square indicates how large a percentage of the influence of project based multi life skill against collaborative skill 43.1%. Probability value informs the significance level of project-based learning multi life skill against collaborative skill, hence 0,000 / 2 = 0.000 with 0.000smaller than 0.05. There is a significant influence of project based multi life skill against collaborative skill. R-square indicates how large a percentage of the influence of project-based learning multi-life skill model against technology skill that is 17.3%. Probability value informs the significance level of project-based learning multi-life skill model for technology skill, hence 0,000 / 2 = 0.000 with 0.000 smaller than 0.05. There is a significant influence of project-based learning multi-life skill model against technologyskill.

R-square indicates thepercentage of influence collaborative skill and technology skill against that is 20.2%. Probability value collaborative skill inform significance level of skill against technology skill, hence 0,010 / 2 = 0.005 with 0.005 smaller than 0.05. There is a significant relationship between collaborative skill and technology skill. Analysis of data using correlation test is used to determine the significance of the relationship project based learning with collaborative skills and technology skills of students. The following calculations correlation of project-based learning with collaborative skill and technology skill. R-square indicates how large a percentage of the influence of project based multi-life skill against skill collaborative is 17.3%. Probability value informs the significance level of project-based learning on collaborative skills, hence 0,027 / 2 = 0.0135 with 0.0135 smaller than 0.05. A significant difference between the project based learning and the collaborative skills.

R-square indicates how large a percentage of the influence of project-based learning to technology skills is 13.9%. Probability value informs the significance level of project-based learning on collaborative skills, hence 0.036/ 2 = 0.018 with 0.018 <0.05. A significant difference between the project based learning and technology skills. R-square indicates thepercentage of influence collaborative skill and technology skill against is 5.7%. Probability value collaborative skill inform significance level of technology skills, then 0.187 / 2 = 0.005 where 0.0935 smaller than 0.05. There is a significant difference between collaborative skill and technology skill.

At the time of the learning takes place is treated with an experimental group using project-based learning model of multi-life skill. In the project-based learning multi-life skill model students are requiredthe skills of rational thinking that are finding its own information about the projects to be undertaken, then from the information that has been obtained before students have to take the decision to determine the tools to be created, the students are required academic skills that student. Rational thinking skill must develop for performance and lifeskill of thestudent. The shift in science education toward deeper, more integrated learning of domain content and scientific practices requires that teachers steer clear of strategies that promote the steady accumulation of more superficial knowledge and capabilities [7]. If a person has the ability and knowledge of the complex, the ability to think is relevant to high-level thinking skills. The need for early intervention, the bulk of current empirical is focused on single component, individual and small groups interventions[8].

Create a tool that has been agreed upon by each group. In groups of students must have social skills which should be able to cooperate with other group members and after the appliance has been so students must present the work steps in front of the group.Argumentation in scientific activity supporting students to contest both what the know and their means of knowing, building more carefully from students resources and attending to the development [9]. Also, students have the technological prowess to explore their work. Based on the discussion of these results, collaborative skills and technology skills group students experiment by project based learning multi-life skill model better than project-based learning model. The correlation between the project based learning multi-life skill with collaborative skills and technology skills significantly more than project-based learning model.

The results show the collaborative skills significant influence students technology skills when learning to use the project-based learning amulti-life skill.There is a link between the collaborative skills and technology skills of students and student learning outcomes using project-based learning multi-life skill model. The results of the study consist of three aspects such as affective aspect, cognitive aspect and psychomotor aspect [10]. Collaborative skills of students including the technology skill aspects of the psychomotor skills while including the aspect of affective aspect and student learning outcomes included on cognitive aspects [11]. Students are taught using project-based learning multi-life skill model turned out to have good learning outcomes as well. In the project-based learning multi life skills students are required to be able to find information themselves so that students' memories will be longer given the formula that has been acquired [12]. In addition to learning outcomes, the project-based learning multi-lifeskill is also more effective aspect if seen from the assessment process and product assessment has been done.

Studies conducted develop 21st-century skills regarding cooperation and technology capabilities using model project based learning in the subjects of English with web-based learning. Besides learning outcomes and attitudes of the students who have been educated with multiple bits of intelligence project based learning amultiline skill is more successful than theproject-based learning model.

4. Conclusion

Based on the results of the discussion is concluded that project-based learning multi life skill affect the independence of students of SMA 1 Bae Kudus. It is the results of the analysis differences the average remark of the experimental group and control group using t-test, there is a difference collaborative skills and technology skills of students on the remark of the observation sheet the experimental group and control group applied project-based learning multi life skill with a group that implemented the project-based learning model. Correlation test to see the effects ties dependent variable and independent variables saw that the influence of the more significant collaborative skills and technology skills by using project-based learning amulti-life skill. The students' response to the project based learning model charged multi-life skill that students were more active and enthusiastic during the learning process compared to the project-based learningmodel.

Acknowledgments

We express our gratitude to RISTEK DIKTI for financial support of this research. LPPM of Universitas PGRI Semarang on administrative support and motivation in the conduct of this study. SMAN 1 Bae Kudus provide opportunities limited test of this research activity.

References

[1] Morris B J, Crocker S, Masnick A M, Zimmerman J 2012. Intech. doi. org/10.5772/53885
[2] Boer H D, Donker A S, Werf M P C 2014. *Review of Educational Research.* **84 (4)** 509
[3] Abrami PC *et al* 2015.*Review of Educational Research.* **85 (2)** 275
[4] Harris C J, Penuel W R 2015. *J Res Sci Teach.* **52(1)** DOI: 10.1002/tea.21263
[5] Gall M D, Gall J P, Borg W R 2003. *Education Research: an Introduction (7th ed.).* Boston: Allynn and Bacon
[6] Fraenkel J R, Wallen NE 2007. *Res in Educ* (Sixth ed). New York: McGraw-Hill,
[7] Shemwell J T, Avargil S, Capps D K 2015.*J Res Sci Teach.* **52(1)** 1163
[8] Freeman J, Simonsen B 2015.*Review of Educational Research.* **85 (2)** 205
[9] Manz E 2015. *Review of Educational Research.* **85 (4)** 553
[10] Chetty N 2015. *Procedia Social and Behavioral Sciences* **174** 1886
[11] Zadeh M Y & Satir S 2015. *Procedia Social and Behavioral Sciences.* **182** 20
[12] Hande S, Mohammed C A,& Komattil R. 2014.*J. of Taibah University Medical Sciences,* **10(1)** 21

Physics Structure Analysis of Parallel Waves Concept of Physics Teacher Candidate

S Sarwi[1,*], K I Supardi[2], and S Linuwih[1]

[1]Physics Department, Faculty of Mathematics and Natural Sciences, Universitas Negeri Semarang
[2]Chemistry Department, Faculty of Mathematics and Natural Sciences, Universitas Negeri Semarang

*Corresponding author: sarwi_dosen@mail.unnes.ac.id

Abstract. The aim of this research was to find a parallel structure concept of wave physics and the factors that influence on the formation of parallel conceptions of physics teacher candidates. The method used qualitative research which types of cross-sectional design. These subjects were five of the third semester of basic physics and six of the fifth semester of wave course students. Data collection techniques used think aloud and written tests. Quantitative data were analysed with descriptive technique-percentage. The data analysis technique for belief and be aware of answers uses an explanatory analysis. Results of the research include: 1) the structure of the concept can be displayed through the illustration of a map containing the theoretical core, supplements the theory and phenomena that occur daily; 2) the trend of parallel conception of wave physics have been identified on the stationary waves, resonance of the sound and the propagation of transverse electromagnetic waves; 3) the influence on the parallel conception that reading textbooks less comprehensive and knowledge is partial understanding as forming the structure of the theory.

1. Introduction

Teachers have a strategic role in constructing the national character and develop the potential of learners. Education of physics teachers prospective, which will produce a physics teacher has become an important topic related to human resources in education. In connection with the civilizing principles of conservation and development of character values, during the learning process teachers understand what is happening to the students and what students can do, to achieve competence designed. Through understanding the thinking of learners can help the teacher directs learning, forming a more productive epistemology [1]. Learning should consider how mechanisms of thinking of students in constructing knowledge through experiential learning.

Referring to the general issue, will be carried out qualitative research focused on the analysis structure and conception parallel in the mastery of knowledge wave to equip prospective teachers of physics. Studies structure in the physics concept of science teaching according to Andersson *et al.*[2] may help explain the concept of a state structure that was built learners as part of the outcome, and provides input for improvement of curricula to be more effective. Observing the strategic role of teacher candidates who have the responsibility to educate the children of the nation, then it is worth the attention specifically on matters related to the competence of the candidate briefing physics teacher.

Learning Wave is presented in the form of verbal language for the definition of a concept, a mathematical language to explore the association between variables, and illustrations for explanatory variables, which are intended to be easier to understand the concept of waves. Research on sound waves to improve the curriculum of physical education colleges has been conducted [3]. Research on wave learning, especially wave phenomena for the students is continued. Results of research Hartmann [4]states that high school students answered duty mechanics with suggested topics more than one explanation of the so-called parallel conception. Conception parallel in this physics needs to get a serious note of teachers and lecturers because students have a conception of more than one, but there is only one correct concept. This qualitative study focused on the analysis and conception concept of parallel structures in the prospective teachers of physics is still a little wrestle or researching. This qualitative study explored based on three aspects which include how the pattern structure owned concept, description of the conception of the parallel, and the factors that influence conception in parallel to obtain donations epistemology theory of wave physics. This information will be an important input in the improvement of learning and curriculum development in Physics Education Study Program.

Formulation of the problem in this research is howdoes the physics teacher candidates to form the structure of concepts and conceptions parallel wave physics. The purpose of this research is to find a pattern structure formed the wave concept of physics teacher candidates, obtain information conception parallel wave physics, as an input to improve the learning of waves, and discover the factors that influence the formation of a parallel conception formed physics teacher candidates. The expected benefits of the research are: 1) the contribution of the theory of the formation of the wave concept of physics teacher candidates; and 2) gave a briefing wave physics knowledge as fundamental knowledge for prospective physics teachers to master the advanced physics course.

The results of the investigation gradually student views and analysis of the content structure contributes importantly to build student knowledge, i.e., the exact meaning of the illustrations that facilitate students in understanding the concept of wave [5]. Research on the use of mechanical waves pedagogy components has been developed, including Mamolo *et al.* [6].Results of research on the properties of mechanical waves through the analysis of the subject matter for understanding the structure of physics (The Constructing Physics Understanding) has been carried out [5]. The conclusion of the research is the use of dynamic transfer model analysis, and constructivist-based curriculum can help to map the elements of knowledge of the properties of waves.

Niedderer [7] stated that the concept is a brainchild of the knowledge structure interaction with the situation at hand. Someone is possible to have more than one conception of a context, where the conception that competes with one another in one's mind [7, 8]. Answer individually in the minds of students on the one task that has not completely controlled, sometimes showing more than one possible explanation; this indicates that more than one conception. Conception more than one happens to someone about a context be regarded as a parallel conception [9].Another research stated that in a physics lesson, students are thinking about the absence of the relationship between facts and formulas, or relationships between concepts which are represented in the form of the equation [10].

In a simple existence of parallel conception can be explained as follows. When a student asked a question about the concepts of physics, and they are given enough time, it is possible there will be more than one answer in the minds of students. Among the answers were there by the conception of the physicist, and said as a scientific conception. If the answer can not be by the conception of the physicists, then say the answer as alternative conceptions [9].Parallel conception can occur in the form of thinking and the scientific conception of alternative conceptions that appear simultaneously and be competitive. Parallel conception can be formed of two kinds of alternative conceptions competing. At the thought process occurs fusion between input variations of the concept and context of the structure of knowledge and ideas so as to form a conception. Concepts are formed can be a scientific conception, or alternative conceptions. Sometimes conception that formed in the form of more than one kind of conception is still competing in one's mind, this conception is said to parallel conception. Linuwih [8] found that students from the Department of Physics still understand the concept of mechanics through parallel thinking with conception, so the results of this research contribute to improving the way of thinking to form a concept science. How the wave pattern and conception concept structure parallel understood physics student teachers, will be the provision of important information later taught school.

2. Methods

This study was designed using qualitative research approach; the study aimed to obtain a capital-oriented scientific explanation of the theory, rules, models, patterns or new postulates that support a process of applied research and technology [11]. Subjects were students of the department of physical education, the five students who take the basic physics of the second half and six college students participating in a wave of the fourth semester. The research subject is determined based on the theoretical and purposive sampling. This study used a qualitative research approach to the cross-sectional design that aims to discover the essence of phenomena that occur at the same time. Steps in the study are shown in Figure 1. In qualitative research, the role of the researcher as well as a research instrument, can collect data in more depth. Data collection tools include thinking aloud, written essay tests, interviews, and discussions.

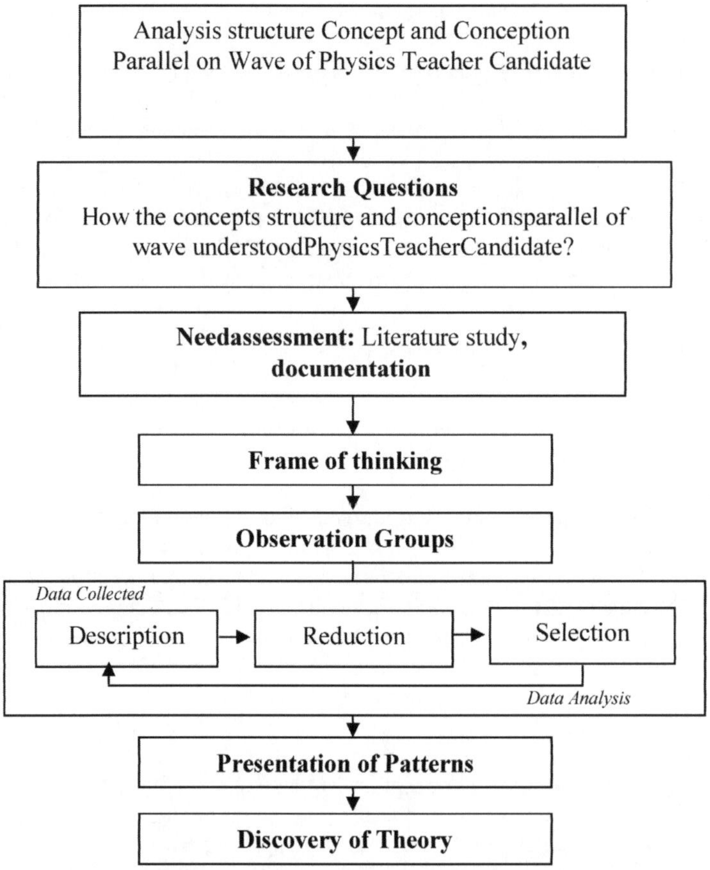

Figure 1. Procedure of research development

Table 1.Criteria CRI by Hasan [12]

Answers Criteria	Low CRI (<2.5)	High CRI (> 2.5)
True Answer	Answer correctly, and a lower CRI value means not understand the concept	Answer correctly, and high CRP value means to understand the concept well
False Answer	Answer incorrectly and the average low CRI means not understand the concept	Answer wrong, but high CRI means no understanding of the concept is wrong

The test structure of concepts analyzed using analysis of the concept. The variable part of the conception of parallel through the tracking of misconceptions developed using techniques Certainty of Response Index (CRI). CRI is usually based on a fixed scale, and the scale used in this study is the CRI scale of six (0-5) developed by Hasan [12]. The criteria proposed is presented in Table 1.Data analysis techniques used, namely a) descriptive percentages for quantitative data, b) analysis unit concept, cognitive structure analysis, and analysis of conception parallel to the qualitative data.

3. Results And Discussion

The results of the study were classified into two parts, namely structural analysis and parallel conception of the wave concept in which students in the science of wave formation. Reports are grouped based on data sourced from students of the second semester of Physical Education (group I) and fourth semester (group II) in the second semester 2015. The results are grouped into two parts: 1) analysis of the structure of the wave concept, and 2) the conception of parallel wave physics teacher candidates.

The structure of the wave concept regarding the core aspects of theoretical, proposition complements, and examples of the phenomena and objects in everyday life. Initial data processing is done by scoring each number on the material rope waves, sound waves and electromagnetic waves are then calculated the average and the percentage of the total score. The next stage is done grouping percentage of respondents in each group I and group II were presented in Table 2 and Table 3.

Table 2. The results of the essay written test Group I of the 3rd Semester

Subject	Respondent					Average	%
	R1	R2	R3	R4	R5		
GT	9	10	9	6	5	7,8	78
GB	7	14	11	11	6	9,8	49
GEM	14	10	8	11	9	10,4	69

Table 3. The results of the essay written test Group II of the 5th Semester

Subject	Respondent						Average	%
	R6	R7	R8	R9	R10	R11		
GT	9	6	10	8	10	6	8,17	82
GB	8	18	14	11	14	13	13	65
GEM	6	12	14	13	12	14	11,83	79

Table 2 and Table 3 show the highest percentage yield of Group I on waves of string that is equal to 78 and the group II 82%. On the wave of the rope, almost all respondents in group I answered very well while writing equations and drawing harmonic wave propagates, especially when $t = 0$ deviation waves reaching the maximum deviation is positive, i.e., where the sin of the angle is 1 so that large deviations will be equal to the amplitude, but respondents still difficulty in explaining the deviation of the medium particles which have the initial phase of $\pi / 3$. Then in suggesting waves of thestring as declarative knowledge or procedural, two out of five respondents thought waves of thestring as declarative knowledge, for their discoveries of physicists declared. However, regarding equations and theory of functions used to experiment simply with certain stages, waves of thestring is knowledge of a procedural nature, although there is some declarative knowledge, such as the quantities of amplitude, frequency and period.

Table 4. Comparison of Percentage of written test Group I and II

Group	Semester	Subject		
		GT	GB	GEM
I	3	78	49	69,3
II	5	81,7	65	78,9

Table 4 presents a comparison of the percentage of the results of written test group I (solid) and group II (hatching). Overall, the percentage of group II was higher than group I, both in concept rope waves, sound waves and electromagnetic waves. The difference in the percentage of Group I and II on the concept of sound waves and electromagnetic waves differently than the waves of string. Figure 2 presents a comparison of the percentage of the results of a written test group I and group II in histograms.

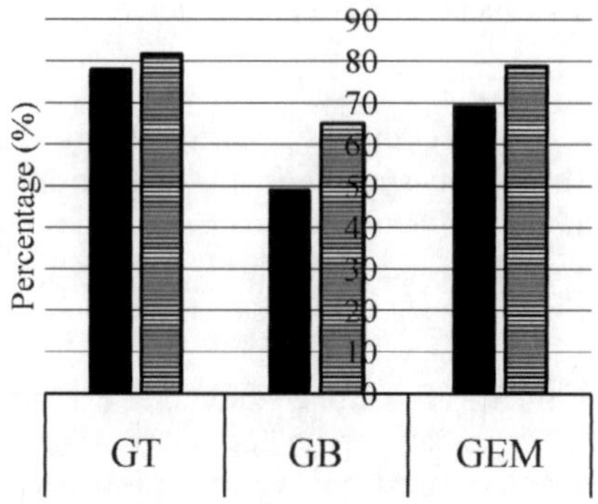

Figure 2 Comparison of Percentage of Written Test Group I (solid) and II (hatching) (GT: string wave, GB: sound wave, GEM: electromagnetics wave)

The results of the written test in the form of multiple choice complexes with more than one correct answer on a parallel conception is presented in Figure 3. The results that the maximum score achieved the fifth semester is 83, while the third semester of 67 (scale 100). The lowest score of each semester of five and three was 29 and 21; while the average score of each is 47 and 40 (scale 100).

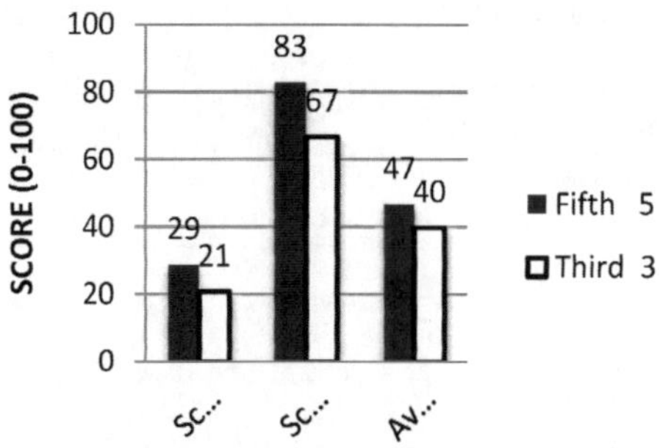

Figure 3. Histogram scores achieved candidate of Physics Teacher Semester 5th and 3rd

The student has completed a complex multiple-choice test, which requires critical thinking. The lower scores may be caused by the second-semester students' knowledge is not comprehensive. The concept of a wave given during the second semester has not yet reached the concept in depth and sequential. Parallel conception appeared on the issues that demanded answers more than one correct choice. Such issues are more open. In the third semester of research objectives, based on the discussion of the results conducted by researchers obtained information that more students understand partially on the waves concept.

Results of research on the conception parallel to express beliefs answers are used. Data belief answer in the form of histograms presented in Figure 4.

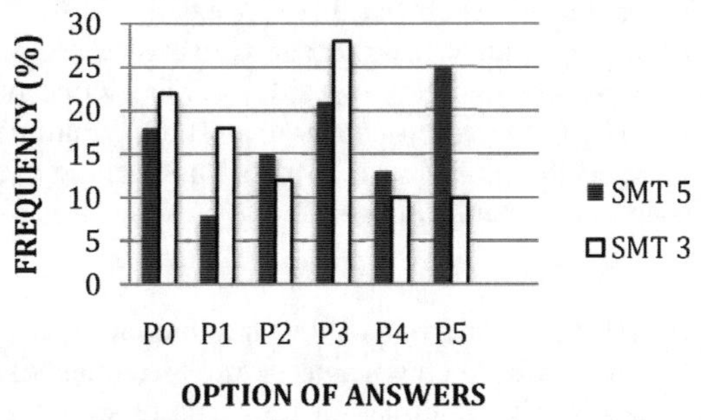

Figure 4. Histogram belief the answers givenby physics teacher candidates
Description: P0 (guesses), P1 (not sure), P2 (doubt), P3 (pretty sure),
P4 (almost sure), P5 (sure right)

The selected answer choices students have also impacted on the steps when solving physics problems that require higher-level thinking (critical, creative, problem-solving). Problem-solving becomes natural that the higher and the number of students taking courses then the confidence will be the correct answer is also greater. The psychological development also supports the development of college students when answering the question of physics. The problem of harmonic waves, mechanical waves and wave superposition can be answered with almost sure. Why can happen so? It is based on the assessment [13] stated that effective cooperation in small groups was able to resolve the issue with the best solution.

Illustration true is a sound wave formed in the resonance phenomenon is a longitudinal wave. The behavior of air molecules during the wave propagates shows sealing and stretching airborne particles. Conversion sealing or particles less moving are the node region; while the molecule is stretching the abdominal area (particle moving freely). Research line, Sarwi and Nugroho [14] claimed 2nd and 4th-semester students have difficulty explaining the wave propagation in the medium of air and characteristics of stationary waves. Both the student opinion is an alternative concept that can be accepted by logical thinking that includes a parallel conception. We found many introductory physics textbook illustrations with supporting text for sound standing waves of air columns in pipes inhibit student understanding of sound standing wave [15;16].

Hammer [9] demonstrated conception parallel to the students to basic physics. On an inclined plane slick, laid out a beam that is pulled by a rope. A rope connected to a pulley on the end of the inclined plane and connected with a weight hanging vertically in the form of a load with a mass greater than the mass of the beam. Parallel in student self-conception as stated happen the same acceleration and the same style on both bodies. Studied from the cognitive aspects of some of the causes of conception parallel students in understanding the concept of waves comes from the everyday environment that raises intuition, reading textbooks less comprehensive, understanding that knowledge is partial as the structure theory, the understanding of knowledge as a separate flake. Intuition owned student can be a source and contribute to science formed states. The impact of anincorrect understanding of the concept or less comprehensive and can contribute to the formation of alternative conceptions, which has the potential to produce a parallel conception. Likewise, the understanding of knowledge by students as separate fragments have an impact on the formation of a parallel with the building's conception of science is not true. Interviews revealed that increased content difficulty in college was major factors that negatively influenced incoming freshmen's views about physics, whereas a role change from student and easy content in high school positively impacted end-of-year seniors' views about physics and learning [17].

4. Conclusion

Based on the discussion presented in the next section, concluded several aspects, namely: 1) The structure of the concept of wave can be shown by way of illustration a map containing the core concepts, prepositions, sub-concepts with horizontal and vertical networks; besides the structure of the concept can also be shown by way of illustration map semicircle containing the theoretical core, supplements theory and examples phenomena that occur daily; 2) the tendency of the conception of the parallel wave physics has been identified on the stationary waves, resonance of the sound and the propagation of transverse electromagnetic waves; 3) parallel conception of students emerged as influenced factors: the everyday environment that raises intuition, reading textbooks less comprehensive, understanding that knowledge is partial as forming the structure of the theory, and understanding of knowledge as a separated parts.

Acknowledgments

The fundamental research titled Analysis of the structure and conception of the parallel wave of aprospective physics teacher, has been completed. Acknowledgments submitted to the Rector of the University in Semarang through the Chairman of the Institute for Research and Community Service awarded grants to fund research DIPA.

References

[1] Lising, L, &Elby, A 2005 *Am. J. Phys.* 73 (4)
[2] Andersson, B. Bach, F. Hagman, M. Olander, C. & Wallin, A 2005 p. 221-230
[3] Wittmann, M.C., Steinberg, R.N., & Redish, E.F 2003 25 (8) : 991-1013
[4] Hartmann, S. & Niedderer, H 2005 *p.471 – 481;* Springer AA, The Netherlands
[5] Hrepic, Z. 2005 *Proceedings of the National Association for Research in Science Teaching (NARST),2005*
[6] Mamolo, C.B., Fletcher, P.R., & Rebello, N.S 2006 United States: San Francisco, CA.
[7] Niedderer, H. 2001. "Physics Learning as Cognitive Development". In Evans R.H., Andersen A.M., SBrensen H.: Bridging Research *Methodology and Research Aims. Student and Faculty Contributions from the 5th ESERA Summerschool in Gilleleje*, Danmark. The Danish University of Education. p. 397-414 (ISBN :87-7701-875-3)
[8] Linuwih, S. 2011 UPI Bandung, Indonesia
[9] Hammer, D. 2004 *Proceedings of the Enrico Fermi Summer School*, Course CLVI: Italian Physical Society.
[10] Osewold, D. 2005 *Proceedings of the Seventh International History, Philosophy and Science Teaching (IHPST).* Winnipeg: Conference Winnipeg,2005
[11] Creswell, J.W. 2014 Singapore: SAGE Publications Asia-Pasific, Pte, Ltd.
[12] Hasan, S., Bagayoko, D. & Kelley, E.L. (September, 1999 *Phys. Educ.* 34(5)
[13] Laughlin, P.R., Hatch, E.C., Silver, J.S., & Boh, L. 2006. *Journal of Personality and Social Psychology,* 90 (4): 644-651
[14] Sarwi & Nugroho, S.E. 2013. Epistemological analysis of the concept of waves on the Physics Teacher Candidate. *Fundamental Research Reports,* LP2M Universitas Negeri Semarang
[15] Ozcan, O. 2015 *Eur. J.Phys.* 36, doi:10.1088/0143-0807/36/6/065042
[16] Zeng, L., Smith, C., Poelzer, G.H., Rodriguez, J.,Corpuz, E. & Yanev, G. 2014 *Physical Review Special Topics - Physics Education Research* 10, 020110
[17] Ding, L. & Zhang, P. 2016 *Physical Review Physics Education Research* 12, 020137

IOP Conf. Series: Journal of Physics: Conf. Series **824** (2017) 012012 doi:10.1088/1742-6596/824/1/012012

Study on Latent Misunderstanding on Electrical Current Concept and Its Impact

N Hindarto[1*] and Wiyanto[1]

[1]Science Study Program Post Graduate, Universitas Negeri Semarang

*Corresponding author: nathanhindarto@mail.unnes.ac.id

Abstract. Misunderstanding of the electrical current concept has been identified and is known since many years ago, but a similar error occurs still in the students today. This research will try to uncover why the misconception that ever happened was still going on today. What important factors that contribute to the occurrence of such misconceptions and its impact on understanding and mastery on the subject of electricity.

1. Introduction

The concept of electric current is one of the important and fundamental concept in understanding the problems of electricity, the misunderstanding of this concept will have an impact on difficulty and even failure in the next understanding of the concepts associated with electric current, in terms of learning about the electricity always related to the concept of electric current.

An electric current is defined as a flow of electric charge. A current may be a flow of positively charged particles, negatively charged particles, or both positive and negative charges simultaneously. By international agreement, the direction of *conventional electric current* is taken as the direction of positive charge flow; if negative charge particles, i.e. electrons move one way in a material, the conventional current is in the opposite direction [1]. The voltage of a battery causes a movement of charge, or electric current, around a conducting circuit. In most cases, the charges are carried by electrons. The greater the voltage produced by the battery, the greater will be the electric current in the circuit.

Usually, teachers use an analogy to get down the concept of electric current and potential to the student in their class. A battery is analogized as a water-pump that gives energy to the water to flow in the pipe. This analogy should be used carefully, because the situation is not quite the same. The water can flow in the open end pipe, but the electric current only flows in the complete circuit. The analogy should be chosen properly, otherwise will be contra productive [2,3]

IOP Conf. Series: Journal of Physics: Conf. Series **824** (2017) 012012 doi:10.1088/1742-6596/824/1/012012

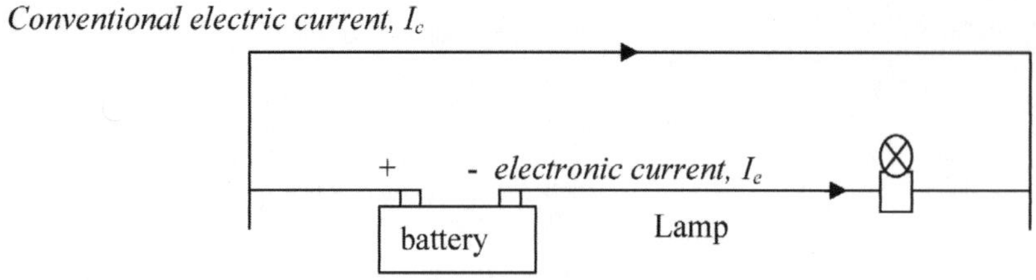

Fig. 1. A simple closed circuit current.

Misunderstanding of electric current has been recognized by physics teachers since decades of years ago [4]. When SEQIP (Science Education Quality Improvement Project) program has been implemented through an upgrading and development a subject matter tutor in Primary School Science in some provinces in Indonesia during the 2000's this problem were widely considered and discussed [5], however problem of misunderstanding is still happening, even with some prospective Post-graduate students of Science S2 Unnes and is also found in some of the final semester students.

Latent misunderstandings that continue to emerge are encouraging an attempt to question why this fault persists. It becomes a challenging study for those who engaged in physics education - science.

2. Methods

Problems revealed by a test, given to 25 students who choose study in science, Post-graduate program at Unnes 2016, and also to 3rd-semester students of Science program mayor in biology. Item tests related to electric current were based upon some test items developed by DIRECT program [6] as follows:

1. The electric charges flowing in a closed circuit are consumed by the lamp to produce light.

   a. Yes, the charge is consumed by the lamp filament to produce light

   b. Yes, charges are emitted

   c. No, thechargeis conserved. It is simply converted to another form such as heat and light

   d. No, thecharge is conserved. A charge moving through the filament produce "friction" which heats up the filament and produces light.

2. Compare the electrical current strength at point 1 and point 2. Which point has the larger electric current is....

   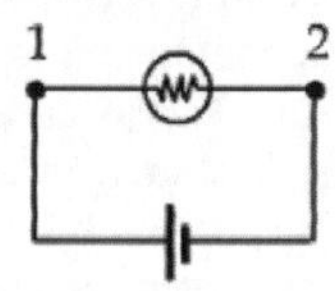

   a. Point 1
   b. Point 2
   c. Neither, they are the same
   d. Both points have a volatile, strong electric current

3. What happens to the brightness of lamps A and B when the switch is closed?

   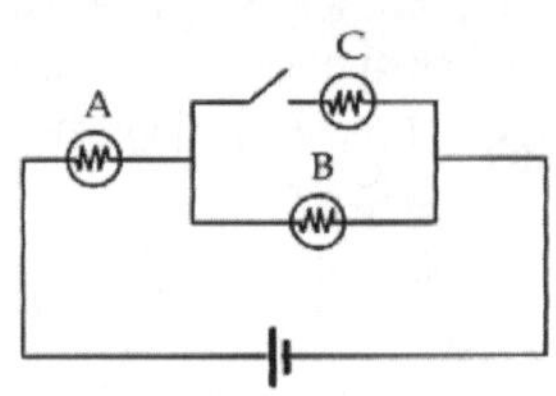

   a. A stays the same, B dims
   b. A and B increase
   c. A and B decrease
   d. A brighter, B dims

   (Note: A, B, C are identical lamps)

The answer to these questions is obtained as follows

No. item test	options				Group respondents
	a	b	c	d	
1	1	9	3	12*	Candidates Science Postgraduate students (25 students) (Group I)
2	3	6	15*	1	
3	12	2	6	5*	
1	7	0	1	1*	Third semester Science Postgraduate students, mayor in Biology. (9 students) (Group II)
2	2	0	5*	2	
3	5	2	0	2*	

* Choice of correct answers.

3. Results and Discussion

Presumably understanding of the concept of electric current in a closed circuit is still an issue in the discussion on electricity. More than half (13 out of 25) of respondents group I did not answer correctly. Nine students from this group consider electric charge is consumed by thelamp and converted into light then emitted by the lamp. They may assume that the light has energy, and this energy is produced from electric current. The Even more astonishing answer was given by the respondent of group II. Seven out of nine respondents answered theelectric charge that flows consumed by the lamp and then converted into light. This choice is interesting to study further, whether the answer choices are influenced by the biology subject studied? A creature must consume something before producing anything. It could be misunderstanding caused by intuition [4] which is affected by the disciplines studied, that create a background knowledge [7]

In general, it can be said that students had misconceptions in understanding the energy carried by electrical currents. This is clear from the answers that flow is reduced when passing through the light because some current turned into energy. In explaining this case, a teacher can use an analogy [2] of water flowing from an elevation. Water was pumped into a water tower, and then the water flowing down turning a waterwheel and collected in a tank to be pumped back. Moving of circulated water can be analogous to the electric current. The water pump which pumps water up to the tower it can be considered as a battery, it will give energy (namely mechanical energy) to the water. When the water is flowing down from the water tower, it has a kinetic energy and therefore able to rotate the waterwheel. The waterthen ends up in the tank; the mechanical energy was reduced because it has been used to slide down and drive the waterwheel, however, the amount of water remains unchanged. The same thing happens when the current flows through the lamp resistant. The energy carried by electric charges obtained from the battery as a voltage source encounter a friction with resistivity on the lamp, causing heat and light, but the charge in itself is not reduced.

Referring to the answers given by the respondents, both in group I and II, it can be concluded that in general the students have understood that a strong electric current in a closed chain is equally same. Respondents who answered point 1, currents larger than 2 points, we need to track closer their understanding of current and potential. Moreover, who answered point 2 is larger, need to be tracked if they are not able to distinguish between electric current and flow of electrons. Both those who chose point 1 or point 2 had larger current; it seems that students to solve the problem by way of sequential reasoning. Students reasoning by way of sequential assume that the current flowing in the circuit is affected whenever encountered element in the circuit [6,8]

Observing the respondents' answers to question no 3, it was only a small part of the group that can answer correctly, 5 among the 25 in group I and 2 out of 9 in Group II. Most of the respondents chose answer lamp A stays the same and lamp B dims. They think that after the current go through the lamp A, there is no longer effect against the lamp A, but because the current meet a branching, the flow will be divided. Consequently the current through the lamp B is smaller, and the lamp dims. In solving this problem, the students are likely to think using the pattern of sequential reasoning and then the local reasoning [6]. These patterns of reasoning may cause misunderstandings in resolving the question no 3. They are supposed to analyze in a comprehensive manner by looking at the circuit as a whole.

When the switch is closed, then there is a parallel relationship between B and C lamps, replacement value (B *) will be smaller than the lamp B, so that the total value of resistivity in closed circuit A + B > A + B *. Thus, in the condition of the switch is closed, the current in the circuit through the lamp A larger and brighter.

4. Conclusion

Misunderstanding the concept of electric current turned out to be a problem that always occurs on the topic of electricity in physics instructions. Various factors may play a role in the cause of this misunderstanding. This misunderstanding needs to be considered and addressed in order not to be a misconception that more difficult to overcome and lead to learning difficulties and might also in the development of scientific literacy [9]. There are some efforts to overcome this misunderstanding such as: using an analogy that is well structured in carrying out the teaching learning activity [2], problem-solving strategy based on conceptual and procedural knowledge [10], inquiry-based learning through inquiry lab instruction [11], but it is important to recognize that there is no singular method of science [12], This study is an initial study to detect and find solutions to the various misconceptions that repeatedly occur in a variety of discussion topics of physics

References

[1] Serway, RA,1990 Philadelphia: Saunders College Publishing
[2] Harrison A G & Coll R K. 2013 Jakarta: PT Indeks
[3] Kuo E & Wieman C E. 2015 *Physical Review Special Topics—Physics Education Research 11,* 020133
[4] Suparno, P, 2013 Jakarta: PT Grasindo.
[5] Tim SEQIP, 2000 Jakarta: Science Education Quality Improvement Project
[6] Engelhardt PV&Beichner RJ. 2004. *Am. J. Phys*. **72(1)**98
[7] Odegaard M, Haug B, Mork S, & Sorvik G V, 2015. *Procedia : Social and Behavioral Sciences,* 167. 274
[8] Dermott M C, Lillian C. 2013. *Euro. J. Sci.Math. Edu,* **1(1)** 1.
[9] Holbrook J & Rannikmae M. 2009.*Int. J. Environ. Sci. Edu.***4(3).** 275
[10] Surif J, Ibrahim N H & Mokhtar M. 2012.*Procedia- Social and Behavioral Sciences,* 56,pp 416-425
[11] Gormally C, Brickman P, Hallar B & Armstrong N. 2009.*Int. J. Scholarship. Teach. Lear.***3(2).**16
[12] Osborne J and Dillon. 2010.*Good Practice in Science Teaching*, England: Mc Graw Hill Open Univesity Press.

IOP Conf. Series: Journal of Physics: Conf. Series **824** (2017) 012013 doi:10.1088/1742-6596/824/1/012013

Subsurface Interpretation for Investigating The Landslidein HR Hadiyanto Street, Sekaran using Resistivity Method

I H Santoso[1,*], Supriyadi[1], Khumaedi[1], H Sulistiawan[1], and P Nugraha[1]

Physics Departement, Faculty of Mathematics and Science, Universitas Negeri Semarang, Indonesia

*Corresponding author: ivan.studentjap27@gmail.com

Abstract. Semarang city has a landscape of hills, valleys and the sea. That makes the city of Semarang potential to experience some disasters such as floods and landslides. Landslide-prone areas in the city of Semarang one of them is in the HR Hadiyanto street, Gunungpati districts. Based on the research that has been conducted to describe the lithological subsurface structures to map landslide prone zones and determine the positions of landslides based on resistivity measurements. The method used in this research is the 3D resistivity method in pole-pole configuration. The results of resistivity measurements indicate that there was a sliding plane along the road HR. Hadiyanto particularly is shown in cross-section image of the track 11 (ramp to the front of the north gate Trangkil Residential) and track 16 (the ramp ahead AKBID) at a depth of about 10 m allegedly due to differences in altitude layer of rock.

1. Introduction

Semarang is a beautiful city with its natural resources, but some areas prone to natural disasters such as floods and droughts to landslides [1]. One of the causes of landslides is the presence of the vulnerability of the soil. According to the Java Land Movement vulnerability map sheet Magelang and Semarang showed Figure 1, Gunungpati located in areas with medium vulnerability to ground movement occurs. Even though the old ground motion but can still be active again mainly due to high rainfall and strong erosion. Movement of land is also influenced by the slope. The range of natural slope ranging from ramps (5-15%) and up a very steep or steeper (greater than 70%) that depends on the physical and engineering properties of rocks and soils as a slope forming material [2].

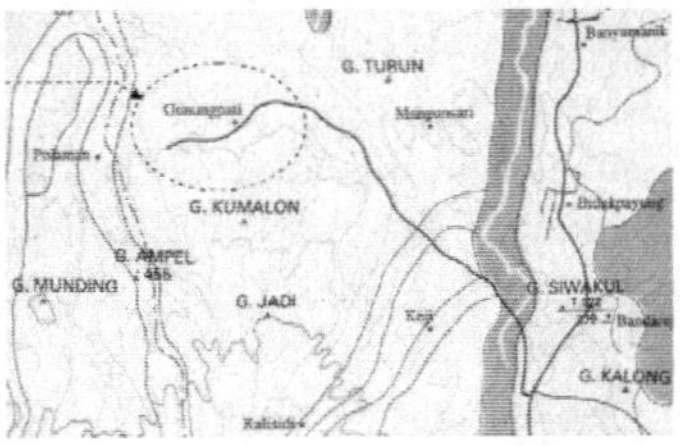

Figure 1. The vulnerability of ground movement map area Magelang Semarang

Conditions of road HR Hadiyanto in some stretches of the past few years as shown in Figure 2. Some roads have cracks and landslides on the west side. Based on the test area sondir Trangkil, hard soil depth reaches 15-20 meters with a curved planar sliding plane and land in the form of translational movement at a depth of 10-13 meters [3].

Figure 2. Landslide on roads HR Hadiyanto sekaran Gunungpati

Avalanche slope on residual soil, especially on steep slopes has a sliding plane different from the type of landslide that circular. On residual soil slope sliding of a relatively shallow depth of field, often by forming a nearly planar curve shown in Figure 3 with the volume of materials is still relatively large landslide [4].

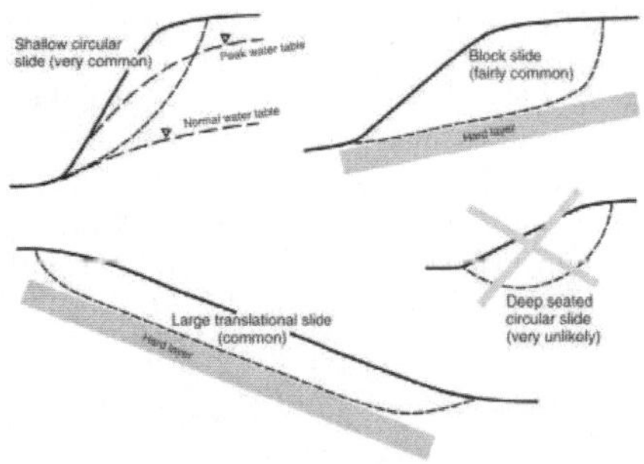

Figure 3. Model forms the sliding plane of landslide events on residual soil [4]

The Factors that cause avalanches can be divided into two groups, namely internal factors and external factors. Internal factors are characteristics of the topography, or the mass of material, among other things: the type and nature of the rocks, structural geology, stratigraphy, type and texture of the soil, the vegetation cover, height and angle of the slope. External factors avalanches among others: (1) Rain or addition of water, resulting in a decrease in shear strength, (2) Human activity is causing the addition of height and angle of the slope, (3) removal of trees or other activities that change the mechanical properties of the material become lower.

Apart from the above factors, there are other factors that sliding plane. Sliding plane can be obtained from the resistivity contrast between the two rocks that are close together. When the top layer resistivity is much lower than the resistivity of the bottom layer, it is possible avalanches occur this is because the layer will easily erode and flowing, especially when supported by a fairly steep fields and precipitation in the region is quite high [5].

To determine the potential for avalanches due to the sliding plane can investigate by resistivity measurement. The basic principle of resistivity method is to inject direct current in the earth through electrodes and measuring the potential response generated through an electrode potential [6].

Earth actually means 3D, so it needs to be a method that is capable of providing an overview resistivity below the surface in the form of 3 D [7]. However, 3D resistivity method is rarely used in geophysical surveys cause is the high cost and the need much power especially over a large area. At this time there is a solution developed to overcome this problem is the development of multi-channel resistivity meter with that expected to be able to read the results at the time of injection of current, thus the time, cost and effort required are not many [8].

In the resistivity method, there are several configurations. However Configuration pole-pole electrode configuration is most commonly used for 3D resistivity surveys. Basically, pole-pole configuration utilising only two electrode, namely electrode current (C1) and the other electrode in the form of the potential electrode (P1) as shown in Figure 4.

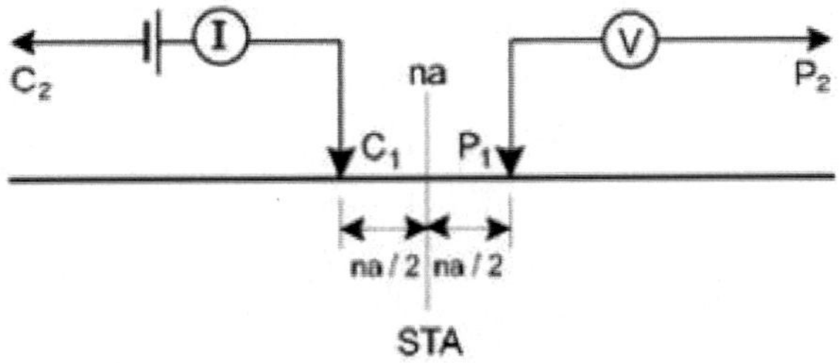

Figure 4. Configuration electrode Pole-Pole

Apparent resistivity obtained with pole-pole configuration is by the following equation.

$$\rho_a = 2\pi \quad (1)$$

With is apparent resistivity, is the electrode spacing (distance between electrodes C1 and P1), and the resistivity is measured directly in the field. Based on the equation 2, it is a factor of the geometry of the pole-pole configuration [9].

2. Research Methods

Research carried out by using resistivity method to determine the subsurface along HR Hadiyanto street, Gunungpati districts, Semarang as on land suitability map Gunungpati districts shown in Figure 6. In this study takes 19 tracks or stretch. Data collection was performed two different locations namely at the first location, there are 9 tracks or expanse that covers an area at UTM X 432 492 and UTM Y 9,223,846 to UTM X 432 508 and UTM Y 9,222,854 and at a second location covers an area in UTM X 432 560 and UTM Y 9,222,754 to UTM X and UTM Y 432 821 9221720. the reason for choosing a place to study at both locations is because the area has a topography that is fairly flat and straight so appropriate when used resistivity method.

Data acquisition carried out along the road Sampangan-Banaran which serves to determine the soil. Data collection techniques using resistivity method of pole-pole configuration in the field is carried out as follows:

- Placing electrodes on the track measuring 16 units along the 150 meters, with two potential electrode P1 and P2 and two current electrodes C1 and C2, to electrodes P1 and C1 (the first electrode and the sixteen) placed outside the track with no distance and C2 and P2 electrodes are used to measure the value of current and potential on the track
- Adjust the distance between the electrodes along the 10 meters and connects the cable liaison with electrodes
- The Cables connecting to the first electrode until the eighth electrode to the holes on resistivity tool that reads multichannel electrode 01-08
- The Cables connecting to the ninth electrode until the sixteenth electrodeto the holes on resistivity tool that reads multichannel electrode 09-16
- The rest of the holes used for connecting cables with a current source or battery and connecting cable with USB multichannel resistivity tool to be connected to a laptop

IOP Conf. Series: Journal of Physics: Conf. Series **824** (2017) 012013 doi:10.1088/1742-6596/824/1/012013

- Measure and record the coordinates of the location of each electrode by using GPS (Figure 5)

Figure 5. Location of the study along space. HR Hadiyanto street sekaran

3. Results and Discussion

Location data acquisition has varied height 48 m (north) - 208 m (south) from the surface of the sea. Here is the basic geological map of the study area is shown in Figure 6. Figure 6 is a geological map of the study area, it appears that track 1 located on the Formation Damar (tuffaceous sandstones, breccias, conglomerates), track 2 to 8 with Cipluk Formation (marl), track 9 and 12 is a resin formation, track 13 to 16 have tabled Kerek formations (clay stone, marl, sandstones and tuffaceous), and tracks 17 to 19 constitute Kaligetas Formations (volcanic breccia, clay stone, tuffaceous sandstones). The results of resistivity data processing on the track 11 to 16 using the software RES2DINV as following.

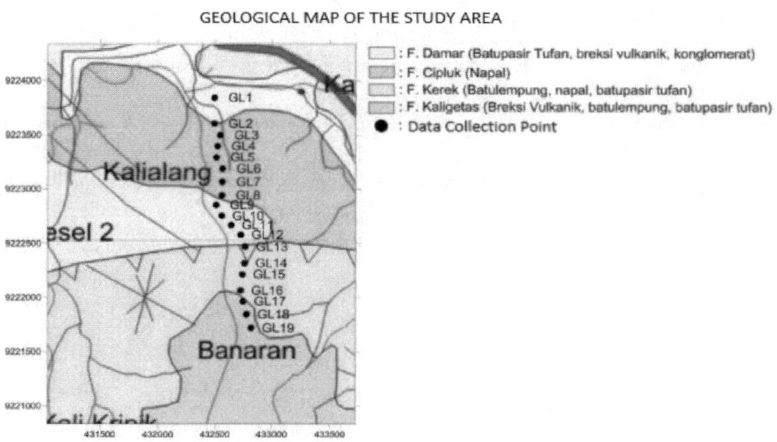

Figure 6. Map Basic Geological Research Areas

3.1. Tracks 11

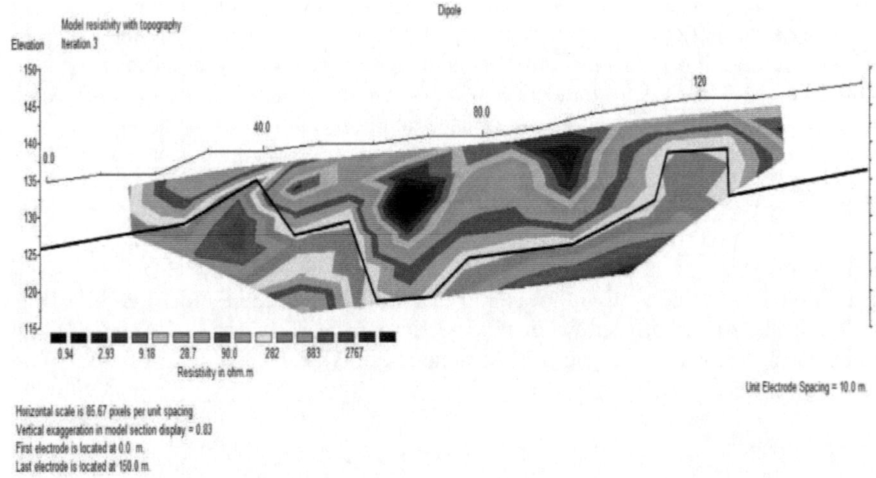

Figure 7. 2D images of subsurface resistivity values at UTM X and UTM Y 432 643 9.22267 million

Tracks 11 are on the edge HR. Hadiyanto street, track 11 generally has an altitude of 135-145 meters above sea level. Resistivity values were found to have a variation of 0.94 Ωm up to 2767 Ωm. Resistivity value of 0.94 - 28.7 Ωm believed to be the soil at a depth of 0-10 meters, the value 28.7 - 282 Ωm believed to be the rock layers marl at a depth of 10-15 meters, while the value of resistivity 282-2767 Ωm at a depth of 15 onwards allegedly is a rock conglomerate, Based on Figure 7, at a distance of 50-100 meters from the track with a depth of 10 meters there is a horizontal resistivity change significantly. This is presumably due to differences in altitude layer of rock.

3.2. Tracks 16

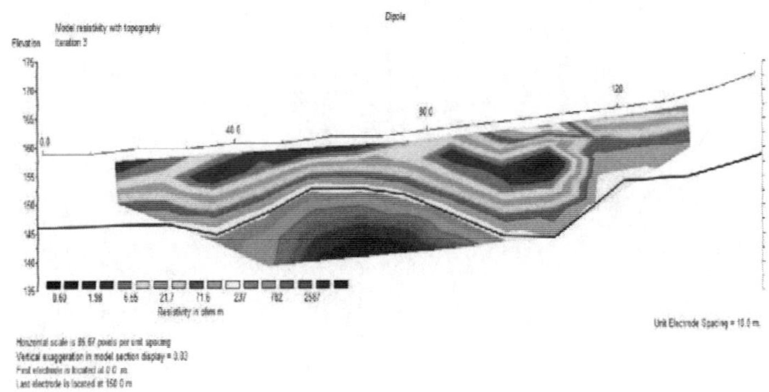

Figure 8. 2D images of subsurface resistivity valuesat *UTM X and UTM Y 432 724 922206*

Tracks 16 are on the edge HR. Hadiyanto street, track 16 generally has altitude of 160-172 meters above sea level. Resistivity values were found to have a variation of 0.60 Ωm up to 2587 Ωm. Resistivity value of 0.60 - 21.7 Ωm believed to be the soil at a depth of 0-7 meters, the value 21.6 - 237 Ωm believed to be the rock layers marl at a depth of 7-15 meters, while the value of resistivity 237-2587 Ωm at a depth of 15 meters onwards is thought to be rock breccias. Based on the Figure 8, at a distance of 80-120 meters from the track with a depth of 10 meters, there is a horizontal resistivity change significantly, it is suspected because of the height difference of the rock layers. Based on the geological map of the city of Semarang, the location of the track 16 there is contact between F. Kerek with F. Kaligetas.

4. Conclusion

Based on the results of resistivity measurements indicate that there are areas of sliding along the road HR Hadiyanto particularly shown in cross-section image of the track 11 (ramp to the front of the north gate housing Trangkil) and track 16 (the ramp ahead AKBID) at a depth of about 10 m allegedly due to differences in altitude layer of rock.

Acknowledgement

Thanks to Geophysics community who help me to the data acquisitions and especially Mr. Andya Satya who have helped me at last.

References

[1] Windraswara Rand Widowati E2010, *Rekayasa*. **8(2)**: 1.
[2] Mochtar I B 2012, *Proc. HATTI 16th Annual Scientific Meeting*, (Jakarta)
[3] CahyoH T, Nugroho U. and Purnomo M, *Proc. of the Civil Engineering National Conference* 7, UNS, 24-26 October 2013).
[4] Wesley L D 2010, *Geotcehnical engineering in residual solids*
[5] Prastiawan A 2013, *Universitas Negeri Semarang Press*:Semarang
[6] Iryanti M, Taufik R R, and NanangD A 2014, *Institut Teknologi Bandung* : Bandung
[7] Singgih P 2003, *Undergraduate thesis,* Institut Teknologi Sepuluh Nopember
[8] Loke M H 2000, *Electrical imaging survey for environmental and engineering studies*, (Penang)
[9] Jochymczyk K, PierwolaJ and Staporek G,. *Publications of the Institute of Geophysics Polish Academy of Sciences,*(Poland), 2006), M-29. 395.

Developing Physics Textbook Based on Cognitive Conflict for Deeper Conceptual Understanding and Better Characters

S Linuwih[1*], N W Lurinda[1], and Fianti[1]

[1]Physics Education, Mathematic and Science Department, Universitas Negeri Semarang, Indonesia

*Corresponding author: suhartolinuwih@gmail.com

Abstract. These study aims are to develop a textbook based on cognitive conflict approachment, to know theproperness of textbook, the legibility of textbook, and also the effect of using the textbook for increasing the conceptual understanding and improving the character of the students. This study was conducted by research and development method employing non-equivalent control group design to test the product. The subjects wereten-grade students of SMA N 1 Gubug in thesecond semester of 2015/2016. The properness test used properness-questionnaire, while the legibility test used themost closet. The data of conceptual understanding was taken from thepretest-postest result and the data of characters was taken from direct observation. By analysing the data, we concluded that the textbook based on cognitive conflict approachment was very proper to use with high legibility. By applied this textbook, students would be helped to get a deeper conceptual understanding and better characters.

1. Introduction

Physics is a part of science which is about of knowledge, facts, concepts, theories, and principles, which then referred as the product of science. *Permendiknas No 22, year 2006*explains that physics not only converting to develop the technology, but also educating the students to learn and to act based on critical, analytical, logical, rational, and systematical thinking. Therefore, physics shouldteach the conceptual understandingand implement the character education.

The observation in SMA N 1 Gubug showed that the physics learning process tended to be conventional. The teacher gave anexplanation and then the student did an exercise. The discussion tended to focus on the formula and the calculation. Addition, the sources of study for the student were limited.

Learning process requires strategy andmedia for achieving the learning's goals thatwere expectedbefore. One of strategy which can beapplied is cognitive conflict approachment. Cognitive conflict is a perceptual state in which one notice the discrepancy between one's cognitive structure and environment (external information) or between the components of one's cognitive structure (*i.e.* one's conceptions, beliefs, sub-structures, etc., which are part of the cognitive structure) [7]. Cognitive conflict approachment in learning process helps the student to build the knowledge by giving cognitive stimulation to make an internal conflict and then creates a prosses of assimilation and the accommodation in their mind. By this process, most students can build a conceptual understanding which is better than before.

Cognitive conflict must challenge student's existing concept, while it happened, thestudent will try to make sure the true one and build the new concept by themselves [8]. Makinguncertaintycan be stimulated through the providing of problems that rise conflict in the student's cognition.

IOP Conf. Series: Journal of Physics: Conf. Series **824** (2017) 012014 doi:10.1088/1742-6596/824/1/012014

The other hand, cognitive conflict approachment provides an easiness to understand the physics concept, train to think critically and creatively, also improve the students learning activity. It's a chance to improve student character by this approachment. Based on the observation and theliteraturestudy, this study develops a physic textbook based on thecognitive conflict which discusses dynamic electricity for ten grade in thesecond semester. The textbook based on cognitive conflict is expected to increase the conceptual understanding. Besides that, the writing of this textbook guides the student to study in demonstration and discussion-methods. It is an attempt to improve the student's characters, such as discipline, communicative, curiosity, and independently.

The purpose of this study is to develop a textbook based on cognitive conflict approachment, to know theproperness of textbook, the legibility of textbook, and also the effect of using the textbook for increasing the conceptual understanding and improving the character of the students.

2. Methods

This study used to research and development method. The procedure covers; (1) finding the problems and potentials, (2) collecting information, (3) designing the product, (4) validation the product, (5) revision the product, (6) testing the product in small-scale, (7) revision the product, (8) testing the product in a large-scale. Validation of product aims to assess the properness of the textbook. The validation-instrument was a questionnaire which was filled out by three teachers as the practitioners. Then, the small-scale test aims to assess the legibility of the textbook. The instrument for thelegibility test was a cloze-test. The data of properness and legibility was analysed by the following percentage-score formula,

$$P = \frac{f}{n} x\, 100, \tag{1}$$

where P : The percentage of properness,f : obtain total scores, and n : the maximum total scores.

The large-scales test of the textbook aims to know the effect of using atextbook in increasing the conceptual understanding and in improving the character of the student. This test used non-equivalent control group design. The subject is ten-grade students of SMA N 1 Gubug in thesecond semester of 2016/2017. The sample was chosen by purposive sampling method, X-A as the experimental class and X-B as the control class. The data of conceptual understanding was taken from pretest-posttest by using essay question that has passed the test of validity, reliability, thelevelof difficultness, and differences-question ability. The data of student character was taken from direct observation. Both of data were analysed by using gain-test as the following formula,

$$\langle g \rangle = \frac{\langle s_p \rangle - \langle s_p \rangle}{1 - \langle s_p \rangle}. \tag{2}$$

Where,

$\langle g \rangle$: gain factor,

$\langle s_p \rangle$: mean scores of postest,

$\langle s_p \rangle$: mean scores of the pretest.

3. Result and Discussion

3.1. The Description of Textbook

The developed product was physics textbook based on cognitive conflict. This textbook was discussed dynamic electricity for grade ten that was adjusted with thecurriculum of 2016. The draft of textbookconsists of: cover page, competency standard and basic competency, textbook using instruction, chart of concept, and learning activities. Each learning activity servesseveral sections to support the activity. Among them are "Problem", contains the goals of learning and the indicators. To stimulate the students for thinking, this textbook gives a section as "hypothesis"; it contains a problem which must be answered or be discussed by the students. They must write the answer on section named "Prediction" which is already available. This section aims to train the students to be discipline in finding solutions for the problem.When the student used this textbook, they will be guided to find the conclusion through a section named "Go Explore". The student was asked to do an experiment also to discuss the observation-result. On each ending of the lesson, the student was expected to have a better explanation about the problem at the beginning of the activity. This textbook also provides any example-questions and exercises.

3.2. The Properness of Textbook

There are four components that need to be validated, that are the contents, language, presentations, and graphics properness.Each component consists of several aspects, and each aspect consists of several indicators. The complete assessment-result is presented in Table 1.

TABLE 1. The result of the properness-analysis of the textbook (scored-range is 0-100).

Aspect	Average-score of each aspect	The criteria of properness
Contents	93.75	Very proper
Presentations	91.96	Very proper
Languages	87.81	Very proper
Graphics	89.72	Very proper
Total averages	90.81	Very proper

The calculation above shows that the average score was closer to ideal (100). It is corresponding to the research by Zemenu[12], that the textbook with theaverage index value for all categories was closer to the ideal standard was good in promoting the students' engagement in active learning.The developing of this textbook was referring on competency standard and basic competency that was predetermined. The textbook provides any problem in section "Problem" that was followed by aninstruction to do experiments and discussions, also any exercise to help students in increasing their ability for mastering the topic. The similar opinion was revealed by Daluba[2] that student who learns by demonstration would think about more activity oriented to the problem. The activity made the student to provide arelevant answer by their understanding. Besides that, their research finds that academic achievement score of astudent by demonstration method was better then that conventional method.

The textbook was developed to help the learning process, so it prioritised the student active-involvement, the improving of characters, and the best interaction to make the student enjoy. The Izzati *et al.* find that the innovation of learning-presentation can improve the student activity such as to ask a question, to answer, also to discuss, so that students will enjoy the learning process. This textbook also contains illustrations either pictures or photos to attract the readers as well as supporting the information. It aims to stimulate the interest and the curiosity of the student to study with this textbook. According to Yazdanmehr and Shoghi [11], the using of contextualised visuals can turn the student's interest and motivation on, activate their creative mental powers in discussion and even predict what the related contentdoes focus on.

3.3. The Legibility of Textbook

The result of the small-scale test showed that the average score of the legibility is 78.2. So we can say that the textbook has high legibility, so that it is easy to understand. The sentence which was written in the textbook was simple and adapted for the student as a reader. Besides that, the writing concern about the structure (*i.e.* subject, verb, object, and adverb or adjective) so that the reader can understand the topic easily. The legality is depended on font-type and font-size. Dimtry *et al.*[3] revealed that the mean value of legible line lenght is about 100-120 mm, while the mean value of the most legible font size is close to 12 points but without specification of thetypeface. The textbook which was developed in this study had 128 mm of line length and 11 points of font size. It was close to the ideal criteria.

3.4. The Using Textbook in Increasing Conceptual Understanding and in Improving Character

The level of conceptual understanding was measured by writing test with essay questions. The questions consist of seven items which are covering all of the competencies. The average score of the pretest-posttest is illustrated in Figure 1 and Figure 2.

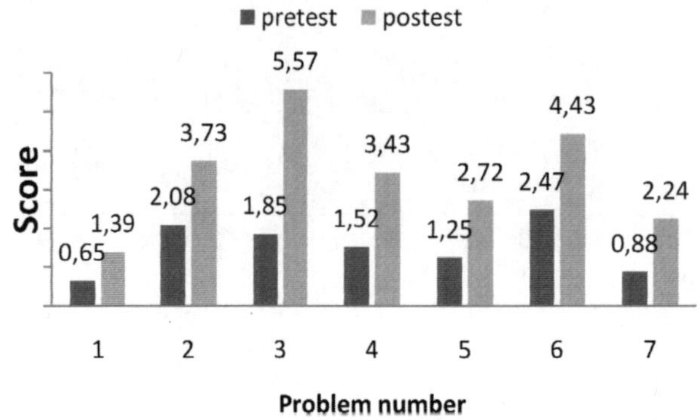

FIGURE1. The average-score of conceptual understanding in experimental class.

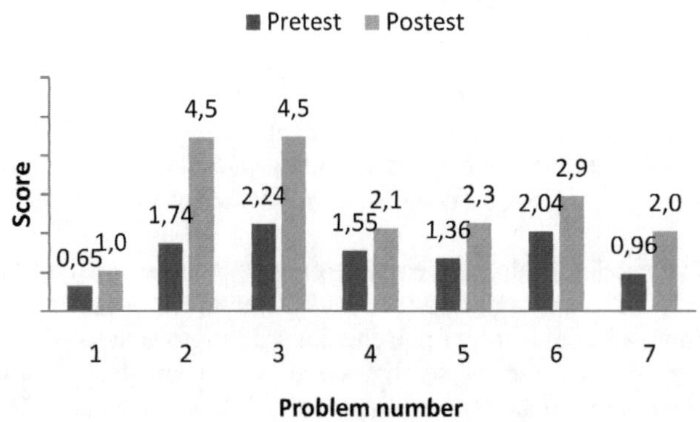

FIGURE 2. The average-score of conceptual understanding in controlled class.

Both Figure1 and Figure 2 show the increasing of conceptual understanding. Actually, the increased score in experimental class is higher than controlled class, so we conclude that the using of textbook based on the cognitive conflict in experimental class can increase the conceptual understanding better than other in controlled class. Gain-test was done to make sure about that conclusion. The result of the gain-test is shown in Table 2.

TABLE2. Results of gain-test the class-sample.

Class	Average of pretest	Average of posttest	Gain factors	Criteria
Controlled	30	55	0.36	Medium
Experimental	31	65	0.50	Medium

Both of these gain factors are not different far because the process of conceptual changes requires a long time, while this research only takes four meetings in a month. According to Visniadou as cited by Taber [9], that the conceptual changes are a slow process, where any observed apparent sudden changes are hard-won and simply offer the surface evidence of extended, preconscious processes influenced by many months of classroom experience.

To prove if there is a significant difference between conceptual understanding scores in experimental class and controlled class, we have donea hypothesis test by using t-test. The result showed that there is a significant difference of conceptual understanding between the two samples. It is appropriate to research by Amin [1], that the cognitive conflict will make an interaction between the phenomenon and the different existing concept.It will give a chancefor the students to build the better conceptual changes.

The assestment about student character was done by direct observation. There are four characters which were observed *i.e.* discipline, curriousity, communicative, and independent. The average-scores of those characters were analysed by gain-test which is the result as the following Table 3.

TABLE 3. The result of gain-test of the characters.

Character	Controlled-class				Experimental-class			
	Initial	Final	Gain	Category of gain	Initial	Final	Gain	Category of gain
Discipline	69.44	95.14	0.84	High	72.22	98.61	0.95	High
Curiosity	45.83	54.86	0.17	Low	25.00	59.72	0.46	Medium
Communicative	47.22	60.42	0.25	Low	46.53	81.94	0.66	Medium
Independent	23.61	25.00	0.02	Low	25.00	76.39	0.69	Medium
Average	46.53	58.85	0.32	Medium	42.19	79.17	0.69	Medium

This result shows that the character of experimental-class has a higher increased than the controlled-class. Although overall the increasing both in experimental-class and controlled-class was in the same category, but actually it was different enough from each character. So we can conclude that the using of textbook based on cognitive conflict can improve the student character. Although there are some students who did not show the increasing of characters significantly, it is a natural thing because some students were familiar with existing learning method, learning by using the cognitive conflict textbook may have been uncomfortable [5].

By using this textbook, the student will find a cognitive conflict situation so that they can explore their self to study about that situation then improve their character. Similar to a study by Iancu[5],that a situation which rises a conflict in students cognition field will make them be more active also to build their knowledge enthusiastically. The experimental class was given any stimulation through problems which are close to their daily life. That raises student interest and enthusiastic to ask, to discuss, and to find the solution to the given problem. This is supported by another report that the curiosity related to the student abillity to ask and to respond in academic, practical, and discussion activities [6].

4. Conclusion

This research developed a physics textbook based on cognitive conflict which was very proper to use. This textbookwas easy to understand with high legibility. The using of atextbook for physics learning could increase the conceptual understanding of the student also improve their characters *i.e.*,.discipline, curiosity, communicative, and independent.

References

[1] Amin T G, Smith C, and Wiser M 2012. *Handbook of Research in Science Education.* **2** .**(4)**.New York: Routledge.
[2] Daluba NE 2013. *World Journal of Education.* **3 (6)** 1.
[3] Dmitry A T, Alexander P S, and Victor VF 2015. *Procedia: Social and Behavioural Science.* **174**1300.
[4] Hikmat, Tayubi Y R, Purwana U, and Suhandi A 2014. *Pros. Pertemuan Ilmiah XXVIII HFI Jateng & DIY.*Yogyakarta. 324
[5] Iancu and Mariana 2014.*Procedia: Social and Behavioural Science.* **127** 68.
[6] Izzati N, Hindarto N, and Pamelasari S D 2013.*J. Pend IPA Ind.* **2 (2)**183.
[7] Lee GandYi J 2013. *Int. J. of Sci. and Math. Edu.***11**601.
[8] Radhiah AR, Norah M N, and Norasyikin M Z 2015. *Int. J. Edu. Stud.* **8 (13)** 73.
[9] Taber and Keith S 2011. *Education Review.***14 (1)** 1.
[10] Watson M K, Pelkey J, Noyes C, and Rodgers M2016. *J. of Cleaner Prod.***63 (4)** 1.
[11] Yazdanmehr 2014. *Theory and Practice in Language.* **4 (3)** 473.
[12] Zemenu MZ 2014. *American Journal of Education Research.* **2 (1)** 44.

The Scientific Approach Learning: How prospective science teachers understand about questioning

Wiyanto[1,*], **S.E. Nugroho**[1], **and Hartono**[1]

[1] Pysics Department, Faculty of Mathematics and Natural Sciences, Universitas Negeri Semarang, Indonesia

*Corresponding author: wiyanto_fis@yahoo.com

Abstract. In the new curriculum, questioning is one of theaspects of scientific approach learning. It means teachers should facilitate students to ask their questions during science learning. The purpose of this research was to reveal the prospective science teachers' understanding about questioning and how the science teachers implement of that in the scientific approach learning. Data of the prospective science teachers' understanding was explored from their teaching plan that produced during microteaching. The microteaching is an activity that should be followed by students before they conduct partnership program in school. Data about theimplementation of questioning that conducted by theteacher was be collected by video-assisted observation in junior school science class. The results showed that majority of the prospective science teachers had difficulty to write down in their teaching plan about how to facilitate students to ask their questions, even majority of them understood that questioning is not students' activity, but it is an activity that should be done by teachers. Based on the observation showed that majority of teachers did not yet implement a learning that facilitates students to ask their questions.

1. Introduction

The Curriculum 2013 has been implemented in schools. The curriculum is developed by perfecting some thought patterns, including the strengthening of inquiry active learning. The active inquiry learning is reinforced by scientific approach learning, also called the scientific process-based approach.

On the scientific approach learning, the active-student is not just students do an activity, such as students solve a set of problems, but more than that the students do the scientific process. For example, students actively researching the effect of the length of rope to the period of a simple pendulum experiment. In these activities, students can be guided by a student worksheet.

The Scientific approach is not entirely new in the context of science teaching in Indonesia. In the previous curriculum, the government has introduced inquiry learning, discovery, and process skills in science lessons. In inquiry learning, students are facilitated to find their concepts. Even in America inquiry learning has been intensively discussed and implemented since the 1960s through education reform, which is triggered by the successful launch of Sputnik by the Soviet Union [1].

The experts stress the importance of facilitating student to learn science as science is done by experts [2][3]. In this case, the science learning is oriented to facilitate students actively search. This active learning facilitates students to do thescientific process as scientists discover or invent scientific products. Such learning is called inquiry-based learning [1].

Scientific inquiry learning includes: exploring the phenomenon and defining problems, proposing explanations (hypotheses), designing and implementing a way of testing the hypothesis, organising and analysing the data, drawing conclusions [1]. Based on the steps of inquiry learning, the authors have developed an inquiry-based student worksheet (IBSW) [3]. The IBSW guides students to observe phenomena, to propose hypotheses, to design experiment to test the hypothesis, to conduct anexperiment, to collect and to process the data, to draw conclusions, and to communicate the results.

In the IBSW, students are guided to ask questions after observing phenomena or information that contain problematic. The student's question is then expressed in the form of problem formulation. So, the problem arises from observation of phenomena that disturbed the students' curiosity. This kind of learning can improve students' motivation and engagement [4][5].

According to the Curriculum 2013, the scientific approach learning puts questioning at the beginning of learning, i.e. after observing. This scientific approach includes activities: observing, questioning, associating, concluding, and communicating. However, in the syllabus that issued by the government is not explicitly stated that questioning should be associated with theprevious observation. For example, in the junior high school science syllabus as listed in theattachment of Regulation of the Ministry of Education and Culture number 58/2014, an activity of questioning is written: "Students ask a question about the difference between solids, liquids, and gases." This activity is not related to the activity of observing which is written in the previous step.

Questioning activities listed on the syllabus can be understood differently by teachers or by prospective teachers. Therefore, the question is how the prospective science teachers understand about questioning and how teachers implement the questioning in the lesson?

2. Method

The purpose of this study is to reveal anunderstanding of the prospective science teachers about questioning in the context of the scientific approach appropriate to the Curriculum 2013. As research subjects were 12 prospective science teachers (i.e. students of Physics Education Study Program), UniversitasNegeri Semarang (Unnes), who participated in microteaching and four science teachers from two junior high schools at Purbalingga Regency.

Microteaching activity was held before the prospective teacher students carry out the school-internship in odd semester 2016/2017. The prospective teacher students were individually assigned to write a teaching plan and then demonstrate it into practice on microteaching activities.

One of the characteristics of the teaching plan by the Curriculum 2013 is containing the scientific approach learning which consisted activities: observing, questioning, associating, concluding, communicating. The teaching plan produced by the prospective teachers were analyzed, especially on the questioning activities to see their understanding about the questioning in the context of the scientific approach learning.

Also, to address the second issue, this study has observed the science learning in junior high school. The observations were carried out to record the science learning, which practised by four junior high school teachers in their classroom. The observations were conducted with the help of video recording. A total of 21 teaching-learning processes was successfully observed and recorded from the four teachers.

The results of the video recordings were written in the form of a transcript containing the sequence of activities of teachers and students as well as the duration for each type of activities. All activities are then grouped into two categories, namely active-students and passive-students. One of the sub-activities about the active-students is the questioning, which is a part of the scientific approach. The results of these observations are then discussed and linked to the analysis results of the prospective teachers' teaching plan.

3. Results and Discussion

Each prospective teacher-student who will do aninternship should follow microteaching program that conducted in the campus ofUniversitasNegeri Semarang. Microteaching program organized in small groups with participants an average of 12 students per group. Each participant is required to make a teaching plan; then he should demonstrate it in the group. When a prospective teacher acts as a teacher, the role of the other as students.

Based on theevaluation of the teaching plan, the prospective teachers' understanding about questioning in the context of the scientific approach can be revealed. The results are shown in Table 1.

Table 1. Understanding of the prospective teachers about "questioning."

No.	Understanding about questioning	Number of the prospective teacher
1	Questioning is activity that students do, so teachers should facilitate students to ask question	5
2	Questioning is teachers' activity, i.e. teacher ask question to students	6
3	Teaching plan has not applied the scientific approach	1

Table 1 shows that 5 of the 12 prospective teacher students wrote in their teaching plan that questioning is an activity of students, i.e. the students ask questions. From 5 prospective teacher students, only two students who wrote that questioning is the teacher facilitate students to ask questions related to what has been observed in the previous step. While three other prospective teachers wrote that questioning is an activity to facilitate students ask questions, but the facilitations are not associated with the activity observed in the previous step.

Table 1 also shows that six prospective teacher students (or more than half of all students) write that questioning is a teacher's activities, i.e. the teacher ask aquestion to his students. And, the rest (one person) of the prospective teacher students, his teaching plan has not written in the scientific approach.

Based on the analysis results of the teaching plan, the majority of prospective teacher students do not understand the concept of questioning in the context of the scientific approach as stated in the curriculum. The questioning is regarded as a task of the teacher to ask question or problem that should be answered by his students. The design of this kind of learning will tend to lead to teacher-centered learning.

From the five prospective teacher students who already understand that questioning is a student activity, only two prospective teachers who understand that questioning is related to observation in theprevious step. In the context of inquiry learning, questioning step is then followed by proposing hypothesis [1]. The other prospective teachers regard that questioning the student activities facilitated by teachers to ask things that are not yet understood. So far, such as activity is usually practised in school science learning, even before the Curriculum 2013 implemented, and such as student question is usually not as an implication of a curious observation.

Two prospective teachers have already designed activities to facilitate students to ask the things that disturb their curiosity after they observed phenomena in the previous step. However, there has not been one of them demonstrated the questioning steps in front of the class (when they demonstrated their teaching plan). This indicates that questioning is a difficult step to be implemented on the scientific approach learning.

So, the majority of prospective teacher students still have difficulty in understanding and practising the scientific approach learning. How about the teachers at school?

The results of the video-assisted observation show that teachers have not explicitly facilitated their students to ask questions. The teaching-learning processes still tend to the passive-students or still centred on the teacher. The results of these observations are shown in Table 2.

Table 2. Student activities in implementation of the Curriculum 2013

Teacher	Education Background	Number of Observations	Student Activity	
			Active (%)	Passive (%)
Teacher #1	Undergraduate of physics education	3	67,92	32,09

Teacher #2	Undergraduate of biology education	6	43,64	56,36
Teacher #3	Undergraduate of biology education	6	12,82	87,18
Teacher #4	Undergraduate of biology	6	32,47	67,53
Average			39,21	60,79

The data in Table 2 was obtained from a video-assisted observation of the learning process carried out by four science teachers at different Junior High School. A total of observations was 21 times teaching-learning processes. On average, the results show that only 39.21% of learning time that is used by the teachers to facilitate active students. Or in other words, teaching-learning processes organised by teachers (60.79%) tend to teacher-centred learning.

From 39.21% of the learning time, there is no time for teachers explicitly facilitate students to ask questions, especially student questions as theeffect of presenting the phenomenon or information which disturb the students' curiosity. This case is similar to the prospective teacher students that have difficulty in understanding the questioning, the teachers are also still experiencing problems in implementing the questioning strategy in the context of scientific approach learning.

Table 2 also shows interesting data, i.e.Teacher #1 has already held a student-centered learning. It was due to the teaching materials at the observation time was the topic of physics, while his educational background was physics education. Because of his educational background according to the topic being presented, so he could perform better teaching-learning process, including he was also able to integrate laboratory activities that allow students to be more active. However, Teacher #1 has not explicitly facilitated the students to ask questions by the Curriculum 2013.

According to the Curriculum 2013, materials of science in ajunior high school packed in one subject, namely Science. Every science teacher, whatever his educational background, present the material biology, physics, and chemistry in an integrated learning. Therefore, the educationalbackground of the teachers become a determining factor in the implementation of the Curriculum 2013 in junior high school [6].

4. Conclusion

Most prospective teacher students have not understood the sense of questioning, and the majority of them do not know how to facilitate students to ask questions. These findings are also strengthened by the results of the learning observation at schools. The observations show that the teachers have not explicitly facilitated students to ask questions by scientific approach learning.

Therefore, efforts to improve the ability of the prospective teachers and the practising teachers in facilitating students to scientific work, especially that related to ask thequestion, are still needed. It can be pursued through an advanced research which followed up by training for teachers. On the other hand, continuous improvement is needed in preparation program of the prospective teachers, including therevitalization of some courses.

References

[1] Lawson, A.E. 1995. *Science teaching and the development of thinking*. California: Wadsworth Publishing Company.
[2] McDermott, L.C. 2013. *Eurp. J. Sci. Math. Edu.***1(1)**1.
[3] Wiyanto. 2008. *Menyiapkan guru sains mengembangkan competence laboratories*. Semarang: Units Press.
[4] Davis, T. A. 2013. *Bioscene: Journal of College Biology Teaching*, **39(2)** 32
[5] Hynes-BerryM and BerryG. 2014.*Eurp. J. Sci.Mat. Edu.*, **2(2)**87
[6] Wiyanto and Widiyatmoko A. 2016. *J. Edu. Hum. Dev.***5(2)** 169

Encapsulation of Citronellal from Citronella Oil using *β*-Cyclodextrin and Its Application as Mosquito (*Aedes aegypti*) Repellent

A Pujiastuti[1], E Cahyono[1,*], and W Sumarni[1]

[1] Department of Chemistry, Faculty of Mathematics and Natural Science, Universitas Negeri Semarang, Semarang 50229, Indonesia

*Corresponding author: edkim@mail.unnes.ac.id

Abstract: Mosquito (*Aedes aegypti*) is a threat to human health due to its capability to spread dengue fever. Citronellal in citronella oil is one ofnatural active compound that has repellent activity. Essential oil is a sensitive material whichiseasy to degrade. Encapsulation is coating technology use to avoid essential oil from degradation problems. β-Cyclodextrin is frequently used as acoating material in encapsulation. The aims of this study wereto prepare the citronellal encapsulation and to evaluate its control-released and repellency. In this study, encapsulated citronellal was prepared using 83.65% citronellal and encapsulation were prepared with the theemulsion-based method and dried using freeze-dryer. The best-controlled release was performed in citronellal encapsulate with a weight ratio of 1:1 (citronellal : β-Cyclodextrin). The morphology of encapsulated citronellal was analyzed using SEM. SEM result showed it has three dimensions random shape and agglomerate in some part with thebrighter spot. Citronellal encapsulate showed the highest repellent effect at 84,67% for 5 minutes in mosquito repellency test although it has lower result compared with citronellal inliquid form.

1. Introduction

Mosquito (*Aedes aegypti*) is a serious threat to human health around the world. It is the major vector for dengue fever disease. Controlling mosquito population and using insect repellent to prevent mosquito bites are usually done to control this disease. DEET (N,N-diethyl-meta-toluamide) is a synthetic chemical substance that widely adds to repellent products because of its repellent activity. However, if DEET is used in long period, it may cause irritation problems and occur resistance tothemosquito [1, 2, 3]. Researchersutilized essential oil from plants that have repellent activity already conducted by several researchers. Essential oil of citronella and basil with additional of vanillin have repellent activity and their protection time could up to 6.5 hours. The other research reported that essential oil from citronella had repellent activity and its protection time was 98 minutes [5]. Citronellal could be one of theactivecompounds in citronella oil that had repellent activity [6].

Essential oil is a sensitive material that degradation easily occurs because of the existence of oxygen, light exposure and high temperature. Encapsulation is a coating technology that frequently used to avoid essential oil from degradation problems and able to release its content under certain condition [7]. The evaporation of citronella oilencapsulatedshowed can be controlled in gelatin [8]. A substance that is frequently used as acoating material in encapsulation is β-Cyclodextrin, thyme and geraniol trapped in β-Cyclodextrin cavity could be protected from oxidation [9]. One of the most-used technologies to support encapsulation process is freeze-drying. A blend of essential oil with coating material was frozen and dried under certain pressure until itssolvent sublime [10], this technology is best applied to dry compounds that sensitive to heat. Another advantage of this technology is the product has high quality and stability [11].

The aim of this study was to preparedencapsulated citronellal and investigated the influence of weight ratio citronellal and β-Cyclodextrinto the product's controlled release. Repellency of product with best-controlled releasewould be testedto *Aedes aegypti*.

2. Material and Methods

2.1 Materials

Citronellal obtained by distillation in the low pressure of citronella oil purchased from Indesso (Indonesia) and the standard citronellal and n-hexane were provided by Merck and β-cyclodextrinwas provided by Sigma-Aldrich.

2.2 Methods

GC analysis was performed on Agilent 6820, equipped with HP-5 (30 m x 320μm i.d.; film thickness 0.25 μm) capillary column. The inlet temperature was 250°C. The carrier gas was helium. The column oven temperature was programmed as follows: initial temperature 65 °C, 1 minute; ramp to 170 °C at 5 °C/minutes. Citronellal was identified through spiking method using citronellal (Merck) as standard. GC peak areas were used to determines the percentage of citronellal. Qualitative analysiswas conducted by spike method to confirm that product from distillation process was citronellal.

Encapsulation citronellal with β-Cyclodextrin at aweight ratio of 1:1, 1:2 and 1:3 were prepared by theemulsion-based method and dried using freeze-dryer (FD-1-EYELA). The emulsionwas prepared by mixing citronellal, β-Cyclodextrin and ethanol. β-Cyclodextrin was weighed and placed in beaker glass with acapacity of 100 mL. 12 mL of ethanol 70% was addedto the beaker glass filled with β-Cyclodextrin and stir it at 500 rpm. The mix heated at 55 °C for 15 minutes then 2 grams of citronellal was added into the beaker glass and continued to stir for 4 hours. The emulsion was frozen at -80°C for 4 hours then vacuumed freeze dryer for 29 hours to sublime the solvent. The product obtained from this process was awhitepowder with citronella scent. Store the citronellal encapsulate in therefrigerator.

The morphology of β-Cyclodextrin and encapsulated citronellal were assessed by JSM 6510 and examinates at 15kV. Controlled release of encapsulated citronellal was investigated to determines the concentration of citronellal at day 0, 2, 4, 6 and 8. five closed glass bottle with acapacity of 25 mL was prepared and labeled by the day when the test has taken. Each bottle filled with 0.1 gram of encapsulated citronellal. One mL of n-hexane was added to the bottle associated with each label of theday the sample was extracted.The bottle was vortexed for 5 minutes to extract citronellal. Extracted citronellal samples were analyzed by GC with the same method as citronellal to determines the percentage of citronellal in encapsulated citronellal each day of extraction.

2.3 Mosquito repellent test

Citronellal encapsulated with thehighest controlled release was tested its mosquito repellency against *Aedes aegypti*. The repellency test was conducted in a cage that divided into three compartments [12]. Cage test illustration image shows in Figure 1.

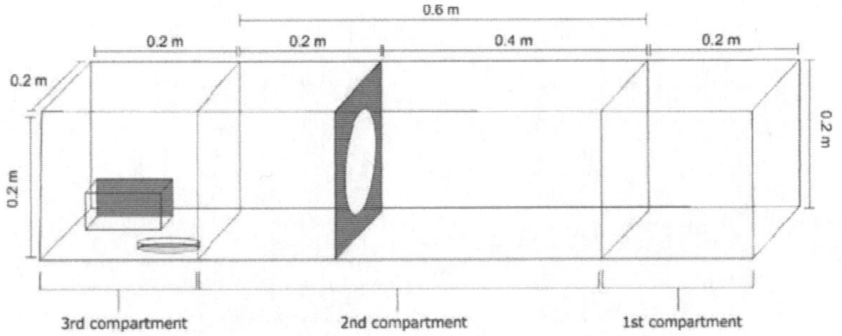

Fig. 1 Cage illustration for mosquito repellent test.

First compartment (0.2 m x 0.2 m x 0.2 m) was made from thepolyester net and woven with wood frames, use to release 15 non-blood-fedmosquitoes. Second compartment (0.6 m x 0.2 m x 0.2 m) was an acrylic cage, and disposable piece of cardboard with acenter hole (0.09 m) was inserted. The third compartment was identically designed with the 1st compartment. A rat was restricted to wire net and 1 gram of encapsulated citronellal was placed in the first compartment. The number of mosquitoes that migrated into the thrid compartment was counted at 5, 10, 15, 45 and 60 minutes. This test was also conducted to β-Cyclodextrin and citronellal in liquid form. All the test was carried out three times.

3. Result and Discussion

3.1 Identification of isolated citronellal

The concentration of citronellal from distillation process was determined by GC. Figure 2 shows the GC chromatogram of citronellal from distillation. The concentration of citronellal from distillation process was 83.45%. Qualitative analysis with spike method used to confirm product from distillation process was citronellal.Figure 3 shows the GC chromatogram of citronellal before and after citronellal standard was added. The highest peak appears at retention 7 minutes in both chromatogram. It indicates the product from distillation process was citronellal.

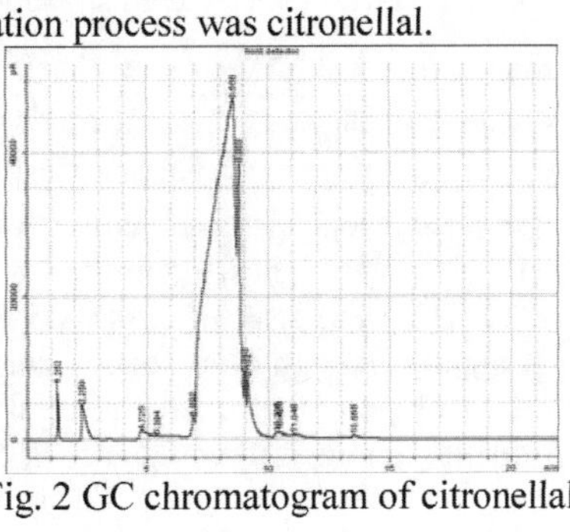

Fig. 2 GC chromatogram of citronellal

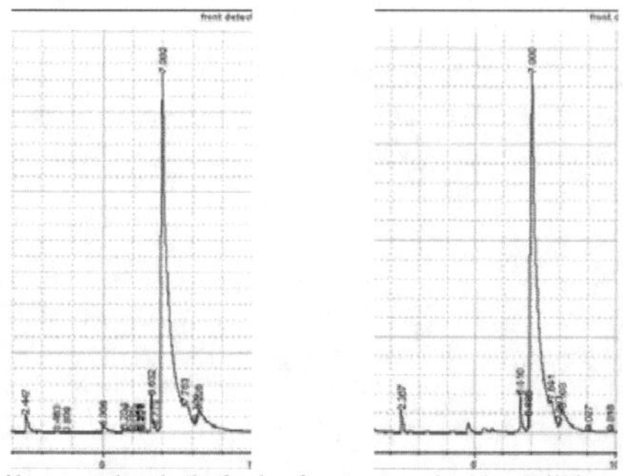

Fig. 3 GC chromatogram of spike method (left: before standard addition, right: after standard addition)

3.2 SEM analysis

β-Cyclodextrin and encapsulated citronellal's morphology were investigated by SEM. Figure 4 shows the SEM image of β-Cyclodextrin and encapsulated citronellal. Based on Fig. 4 β-Cyclodextrin was composed of 3 dimensions random shape with different size. Meanwhile, the encapsulated citronellal was showed agglomeration in some part of the image with thebrighter spot. Based on thedifferencein both SEM images, the formation of encapsulated citronellal was demonstrated.

Fig. 4 SEM image of β-Cyclodextrin (left) and encapsulated citronellal (right)

3.3 Controlled release of encapsulated citronellal

Controlled release of all sample encapsulated citronellal were determined at day 0, 2, 4, 6 and eightis illustrated in Figure 5. It shows that encapsulated citronellal made from theweight ratio of citronellal:β-Cyclodextrin is 1:1 (B1), has the highest controlled release compared to encapsulated citronellal made from weight ratio 1:2 (B2) and 1:3 (B3).As seen from Table 1, sample B1 had the lowest evaporation of citronellal. It demonstrates by the concentration of citronellal until day 4 was 9.95%. These results indicate that encapsulation of citronellal using β-Cyclodextrin as coating material could reduce the evaporation of citronellal although it isn't significant yet. This condition happens due to the unsuitable characteristic of materials used in this study. Encapsulation of apolar compound in β-Cyclodextrin would have abetter result by modified the β-Cyclodextrin to elevate its hydrophobic site. β-Cyclodextrin basically is a semi-polarcompound that has both hydrophobic and hydrophilic site. Elevate β-Cyclodextrin hydrophilic site would enhance its interaction with apolar compound. Acetylated-β-Cyclodextrin showed abetter result in thebindingapolar compound. Also, some researcher had been added emulsifier while preparing anemulsion. The addition of emulsifier by 2% would enhance load capacity and encapsulation efficiency [13].

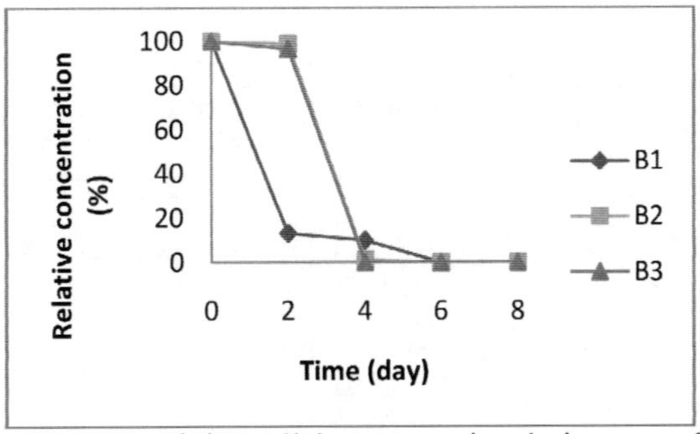

Fig. 5 Relative percentage of citronellal concentration during controlled release test

Table 1. Controlled release test result

Sample	Consentration percentage at day				
	0	2	4	6	8
B1	100 %	13,06 %	9,95 %	0 %	0 %
B2	100 %	98,73 %	0,98 %	0 %	0 %
B3	100 %	96,45 %	0 %	0 %	0 %

3.4 Mosquito repellency of encapsulated citronellal

Mosquito repellency percentage was calculated by this following equation: repellency percentage = [(C – T)/C] x 100% where C was the number of released mosquito and T was the number of released mosquito found in the third compartment.Repellency percentage is illustrated in Figure 6. Based on data in Table 2, the highest repellency percentage of encapsulated citronella was 84,67% obtained after 5minutes and the repellency effect remained until 30 minutes with thepercentage of 48,87%. Meanwhile, the highest result fromcitronellal in liquid formwas 91,33% obtained after 5 minutes and the repellency effect remained until 60 minutes with thepercentage of 71,33%. These results from both samples were different. It indicates there was citronellal remainedin the solvent after emulsion making process. This condition caused the concentration of citronellal in 1 gram of encapsulated citronellal was different with the concentration of 1 gram citronellal in liquid form and cause encapsulated citronellal had lowerrepellency.

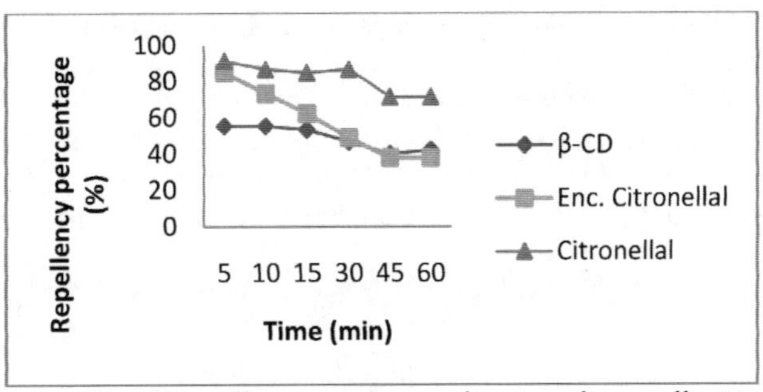

Fig. 6 Repellency percentage in mosquito repellent test

Table 2.Mosquito repellent test result

No	Sample	Repellency percentage at minutes					
		5	10	15	30	45	60
1	β-Cyclodextrin	55,33	55,33	53,33	46,67	40,00	42,20
2	Encapsulated Citronellal	84,67	73,33	62,20	48,87	37,80	37,80
3	Citronellal	91,33	86,67	84,67	86,67	71,33	71,33

4. Conclusions

Encapsulated citronellal showed best-controlled release in mix composition ofcitronellal and β-Cyclodextrin in aweight ratio of 1:1 and the highest repellency percentage was 84,67% obtained after 5 minutes. Based on this study, encapsulation of citronellal could be an alternative to increasing the stability of thevolatile compound, reduce evaporation rate and extend the repellent activity.

Acknowledgements

The authors would like to thanks to Balai Besar Penelitian dan Pengembangan Vektor Reservoir Penyakit (B2P2VRP), Salatiga, Indonesia for use its facility on mosquito repellent test.

References

[1] Kumar, S., N. Wahab, and Radhika W 2011*Asian Pac J Trop Biomed* **1**(2) 85-88.
[2] Hsu, W., J. Yen, and Y. Wang 2013*J Env Sci HealthB* **48** (11) 1014-1019.
[3] Stanczyk, N.M., J.F.Y. Brookfield, L.M. Field, and J.G. Logan 2013*PlosONE* **8** (2) 1-6.
[4] Tawatsin, A., S.D. Wratten, R.R. Scott, U. Thavara, and Y. Techadamrongsin 2001*J Vector Ecol* **26** (1) 76-82.
[5] Sritabutra, D., M. Soonwera, S. Waltanachanobon, and S. Poungjai 2011*Asian Pac J Trop Biomed* **1** (1) 124-128.
[6] Kim, J.K., C.S. Kang., J.K. Lee., Y.R. Kim., H.Y. Han., and H.K. Yun 2005*Entomol Res* **35** (2) 117-120.
[7] Vaidya, S., Rajesh Boshale, and Rekha S. S 2007 *J Dry Technol* **24** (8) 983-992.
[8] Solomon, B., F.F. Sahle, T. Gebre-Mariam . K. Asres, R.H.H. Neubert 2012*Eur J Pharm Biopharm* **80** (1), 61-66.
[9] Mourtzinos, I., N. Kalogeropoulos, S. E. Papadakis, K. Konstantinou, and V. T. Karathanos 2008*J Food Sci.***73** (1) 89-94.
[10] Zuidam, N.J. and V.A. Nedoic 2009*Encapsulation Technologies for Active Food Ingredients and Food Processing*. Springer, New York, U.S.A.
[11] Garti, N., and D. J. McClements 2012*Encapsulation Technologies and Delivery System for Food Ingredients and Nutraceuticals*. Woodhead Publishing Ltd., Cambridge, UK, pp. 93.
[12] Kim, S.I., J.S. Yoon, S.J. Baek, S.H. Lee, Y.J. Ahn, and H.W. Kwon 2012*J Med Entomol* **49** (4) 876-885.
[13] Hwang, JS., Kim, JN., and Wee, YJ. 2006Biotechnol. Bioprocess Eng. **1**(4) 293-300.

The Potency of White Rice (*Oryza sativa*), Black Rice (*Oryza sativa L. indica*), and Red Rice (*Oryza nivara*) as Antioxidant and Tyrosinase Inhibitor

I Batubara[1,2,*], M Maharni[1], and S Sadiah[2, 3]

[1] Department of Chemistry, Faculty of Mathematics and Natural Sciences, Bogor Agricultural University, Campus IPB Darmaga, Bogor, 16680 Indonesia.
[2] Tropical Biopharmaca Research Center, Bogor Agricultural University, Jalan Taman Kencana No 3, Bogor, 16128 Indonesia
[3] Department of Anatomy, Physiology, and Pharmacology, Faculty of Veterinary Medicine, Bogor Agricultural University, Campus IPB Darmaga, 16680, Indonesia.

*Corresponding author: ime@apps.ipb.ac.id

Abstract. Rice is known to have many beneficial biological activities and is often used as "bedak dingin", a face powder. The content of vitamins, minerals, fiber, and several types of antioxidants, such as ferulic acid, phytic acid, tocopherol, and oryzanols [1-2] are predicted to be potential as a tyrosinase inhibitor. The purpose of this study is to determine the potency of extracts from there types of rice, namely white, red, and black rice as an antioxidant and tyrosinase inhibitor. The rice was extracted with three different solvents, n-hexane, ethyl acetate, and methanol. The results showed that the highest antioxidant activity using 1,1-diphenyl-2-picrylhydrazyl method was found in the methanol extract of black rice (IC_{50} 290 µg/mL). Meanwhile, ethyl acetate extract of white rice has the highest antioxidant activity withphosphomolybdic acid method (41 mmol α-tocopherol equivalents/g sample). Thus, methanol extract of black rice and ethyl acetate extract of white rice are potential as an antioxidant. For tyrosinase inhibitor, *n*-hexane extract of red rice (IC_{50} 3156 µg/mL) was the most active extract. The active component for radical scavenging is polar compound and for antioxidant by phosphomolybdate method is less polar compounds in black rice methanol extract based on TLC bioautogram. In conclusion, the black rice is the most potent in antioxidant while red rice is for tyrosinase inhibition.

1. Introduction

Rice is the staplefood for the majority of Indonesia's population and it is an important component in the system of national food security. In addition to being a staple food, our ancestors also use rice for skin care and it is believed that rice could bleach the skin. Rice is not only a source of energy and protein, but also a source of vitamins, minerals, fiber, and several types of antioxidants, such as ferulic acid, phytic acid, tocopherol, andoryzanol [1].

Today the use of rice as a bleaching agent is widely used in the market, especially the Japanese rice. Beside that, in Indonesia the rice has been traditionally used as a whitener for generations in various areas such as Sumatra, Banjar, Madiun, Medan, and other areas in Indonesia. People in the region believe that the nutritious rice powder whitenand soften the skin especially on the face. Rice powder or known as "bedakdingin" traditionally made from white rice with a mixture of different types of extracts such as lemon, olive oil, extracts of jasmine, roses, red betel, ylang-ylang, tuberose, frangipani, and others [2].

There are three types of rice, commonly known in Indonesia, namely white rice, brown rice, and black rice. The color of the rice is different which related to the compounds in the rice. The color in rice is believed could give antioxidant activity. An antioxidant is a compound that able to delay or inhibit the oxidation reaction [3]. Antioxidants may protect cells from the harmful effects caused by free radicals. These free radicals are derived from the metabolism of the body or outside the body [4].

Besides containing antioxidants, rice is thought to have anactivity to inhibit tyrosinase, which related as awhitening agent. Tyrosinase inhibitors are compounds that can inhibit the formation of melanin. Tyrosinase inhibitors currently used in cosmetics as awhitening agent. This study aims to determine the potency of the three types of rice, namely white rice, brown rice, and black rice as an antioxidant and tyrosinase inhibitors.

2. Methods

Research methodology includes several stages, sample preparation, extraction, phytochemical test, bioassay test, and TLC bioautography. Samples of white rice, brown rice, and black rice were dried and milled. The moisture and ash content of the rice powder is determined by AOAC [5] method. The rice powder was extracted by maceration method with three different solvents, *n*-hexane used first continue by ethyl acetate, and by methanol for the last extraction process. The phytochemical content [6], antioxidants [7], and tyrosinase inhibitors [8] activities of all extracts were determined. The active spot of the most active extract then was determined by thin layer chromatography bioautography.

2.1. Extraction of White Rice, Red and Black Rice [9].

A total of 15 g rice powder extracted by the maceration method with three different solvents. The first solvent was a nonpolar solvent (*n*-hexane). The residue of thefirst extraction was then extracted with semipolar solvents (ethyl acetate), then finally the residue was extracted with a polar solvent (methanol). Each extraction was performed for 24 hours, and the ratio between rice powder and thesolvent was 1: 5 (w/v). Each filtrate obtained concentrated using
rotary evaporator at 30 ° C and the yield is calculated.

2.2. Antioxidant Activity Test [10].

Extracts were dissolved in 96% ethanol, then phosphomolybdate reagent (0.6 M sulfuric acid, 28mm sodium phosphate, and fourmM ammonium molybdate) was added and incubated in a water bath at 95 ° C for 90 minutes. The mixture was cooled to room temperature prior tomeasured the absorbance at a wavelength of 695 nm. BHT is used as a positive control and the antioxidant activity expressed as equivalent α-tocopherolis based on the following equation:

$$A = 0.011\,C + 0.0049$$

Where: C is α-tocopherol equivalence (μmol/mL)

2.3. Anti-Oxidant Activity Test [7].

The antioxidant assay used in this study adopted a free radical-scavenging activity using the 2,2-diphenyl-1-picrylhydrazyl (DPPH) test from Batubara et al[7] by slightly modifying. Briefly, samples were diluted in ethanol. A 100 μL aliquot of sample and 100 μL DPPH 125 μM in ethanol was added to each well of a 96-well plate. After 30 min, the absorbance of the mixture was measured at 517 nm. The positive control was ascorbic acid while ethanol was used as the blank. The inhibitory activity was calculated according to the following equation:

$$\text{Inhibition (\%)} = 1 - \frac{Ab \quad -Asc}{Ab} \times 100\%$$

Where *A* sample is the absorbance of the sample, *A*blank is the absorbance of ethanol as the blank. Each sample concentration of the samples and thepositive control was tested in triplicate.

2.4. Tyrosinase Inhibitory Activity

Inhibition of tyrosinase activity (monophenols) and DOPA auto-oxidation (diphenols). This assay was performed using methods as described earlier [7].

3. Results and discussion

The rice powder quality that was used in this study was determined by moisture and ash content. Determination of moisture content is useful to determine the resistance of samples against fungi and microbes during storage [11]. The results of moisture content of samples are 12:23%, 13:15% and 11:55% for black rice, red rice and white rice, respectively. This indicates that the sample can notbe stored for long periods, because of the moisture content of the sample is higher than 10%. The optimal stability of the material will be achieved when the moisture content of a sample is less than 10%, so that microbial growth can be reduced [12]. Based on SNI [13], the rice powder has good quality because the maximum moisture content of rice that has a good quality is 14%.

Determination of ash content is the way to determine the presence of amineral or inorganic compounds in a material. The ash content of black rice, brown rice, and white rice was 2.63%, 8:59%, and 7:46%, respectively. The smallest ash content was found in the black rice. This indicates that the mineral content that remains in the black rice powder is less than brown rice and white rice.

The extraction method used in this study is maceration. This method requires a lot of solvents and a long time in the process, but can decrease the heat damaged and can be extracted in large quantities. The yield of the extraction is shown in Table 1. The methanol extract on each type of rice has the greatest yield compared with the ethyl acetate extract and *n*-hexane extract. This indicates that the compounds contained in rice mostly the polar compounds that can be dissolved in methanol. Harborne[6] states that methanol can dissolve the polar and non-polar compounds. The lowest yield was found in the ethyl acetate extract, it's because rice contained only a small amount of semi-polar compounds. Based on the rice type, white rice has the lowest extraction yield either in ethyl acetate extract, *n*- hexane extract, and methanol extracts.

Table 1. Extraction yield of three types of rice with three different solvents

Rice name	Yield of extraction in solvent (%)		
	n-hexane	Ethyl acetate	methanol
Black rice	1.21^{b}	0.53^{b}	1.74^{b}
Red rice	1.35^{c}	0.52^{b}	1.92^{c}
White rice	0.46^{a}	0.20^{a}	0.77^{a}

Table 2. Phytochemical content of rice extracts

Rice name	Solvents	Phytochemical groups					
		alkaloid	saponin	Flavonoid	tannin	triterpenoid	steroid
Black rice	*n*-hexane	-	-	-	-	+	+
	Ethyl acetate	-	-	-	-	+	+
	methanol	-	+	+	-	+	-
Red rice	*n*-hexane	-	-	-	-	+	+
	Ethyl acetate	-	-	-	-	+	+
	methanol	-	+	+	-	+	.
White rice	*n*-hexane	-	-	-	-	+	+
	Ethyl acetate	-	-	-	-	+	-
	methanol	-	+	+	-	+	-

(-) not detected; (+) detected

The phytochemical analysis is one way to determine the content of secondary metabolites in a plant. In this study, phytochemical analysis was performed on the *n*-hexane, ethyl acetate, and methanol extracts of black rice, brown rice and white rice. The methanol extract of black rice, brown rice and white rice contains saponins, flavonoids and triterpenoids (Table 2). This result is consistent with the report of Al-Farsi and Lee [14], that the rice has phytochemical components of flavonoids and phenolic acids. Flavonoid compounds in the sample indicates that the sample suspected of potential as an antioxidant and inhibitor of tyrosinase. Flavonoids are one of the active compounds as inhibitors of tyrosinase[15]. Extracts of ethyl acetate and n-hexane generally contain triterpenoids and steroids (Table 2).

Tyrosinase or phenol oxidase is the main enzyme involved in the synthesis of melanin [16]. Tyrosinase inhibitor activity of black rice, brown rice and white rice extracts are expressed in IC_{50}, the concentration of the extract that can inhibit 50% of tyrosinase activity. Tyrosinase catalyzes two reactions: *o*-hydroxylation L-tyrosine became 3.4-dihydroxyphenylalanine or DOPA (*o*-diphenol) known as monophenolase and oxidation of DOPA into dopaquinon (*o*-quinones) called diphenolase, before becoming eumelanin or pheomelanin[16]. The reaction between LTirosin and L-DOPA as substrate with tyrosinase enzyme produces dopachrome and continued with the formation of melanin. Dopachrome formation can be inhibited by a tyrosinase inhibitor. Extract which has tyrosinaseinhibitory activity can inhibit melanin production. Dopachrome formation in this study characterized by the formation of brown color. The tyrosinase inhibitory activity of all rice extracts is shown in Table 3.

Table 3. Tyrosinase inhibition activity of rice extracts

Rice name	IC_{50} (mg/mL)					
	Monophenolase for extract in solvent			Diphenolase for extract in solvent		
	n-hexane	Ethyl acetate	methanol	*n*-hexane	Ethyl acetate	methanol
Black rice	3.40^{b}	3.99^{b}	3.23^{b}	3.56^{b}	4.23^{b}	3.71^{b}
Red rice	3.16^{b}	3.96^{b}	4.23^{c}	3.66^{b}	4.83^{b}	4.21^{c}
White rice	3.66^{b}	4.61^{c}	3.22^{b}	3.71^{b}	4.24^{b}	3.77^{b}
Positive control kojic acid		0.05^{a}			0.04^{a}	

The tyrosinase inhibitory activity of all samples via monophenols and diphenols have not many different IC_{50} values, but the activity is not as good as kojic acid as apositive control. The high value of IC_{50} showed that the tyrosinase inhibitory activity of rice is very small. The *n*-hexane red rice extract is most potent tyrosine inhibitor for monophenols with anIC_{50} value of 3.16 mg/mL compared with other rice sample extract but not significantly different with most of the extracts.

Table 4. Antioxidant activities of rice extracts

Rice name	IC_{50} (mg/mL) of DPPH for extract in solvent			Antioxidant activity (mmol α-tocopherol/g extract) for extract in solvent		
	n-hexane	Ethyl acetate	methanol	*n*-hexane	Ethyl acetate	methanol
Black rice	2.93^{b}	1.89^{b}	0.29^{b}	31.37^{b}	27.13^{b}	27.71^{b}
Red rice	2.72^{b}	1.54^{b}	0.39^{c}	32.95^{b}	21.71^{b}	19.89^{c}
White rice	nd^{c}	nd^{c}	1.29^{b}	29.19^{b}	41.13^{b}	29.43^{b}
Positive control		0.006^{a}(ascorbic acid)			28.58^{a}(BHT)	

nd: higher than 5.00mg/mL

The antioxidant activity of black rice, brown rice and white rice extracts are determined by two methods, namely DPPH and phosphomolybdate. DPPH method is the most commonly used to determine the antioxidant activity because DPPH is stable in the radical form and make the measurement accuracy. The antioxidant activity by DPPH method is summarized in Table 4. It shows that the methanol extract of all extracts has good activity compared to ethyl acetate and *n*-hexane extracts, but the activity is not as well as ascorbic acid as apositive control. The methanol extract contains flavonoids, which are groups that reported acts as anantioxidant. The most active DPPH scavenger is found in black rice methanol extracts. Kaneda*et al*[17] reported that the black rice has antioxidant activity which is better than white rice because it contains phenolic acids, flavonoids, and anthocyanins are more than white rice.

The phosphomolybdateMethod is used to determine the antioxidant activity. During the reaction, the hydroxyl group of the phenolic compounds will react directly with aphosphomolybdate reagent which forms a molybdenum complex [18]. The antioxidant activity of black rice, brown rice and white rice are expressed in the equivalent of α-tocopherol with BHT (butyl hydroxy toluene) as a positive control. The antioxidant activity of all extracts is summarized in Table 4. The ethyl acetate extract of white rice has the greatest antioxidant activity, followed by the *n*-hexane extract of red rice. This activity is better compared with BHT as a positive control.

The antioxidant activity of the rice extracts with phosphomolybdate methods is different with DPPH. In DPPH method, extracts that have the highest activity is methanol extract of black rice, while in phosphomolybdate method the highest activity is ethyl acetate extract of white rice. This is caused by the difference in the active compound each extract and the differences of the principle methods used. For the next step, only a methanol extract of black rice is selected because this extract also has tyrosinase inhibitory activity.

Thin layer chromatography (TLC) bioautography is a qualitative method to determine the activity of the sample that combines with thin layer chromatography. This method requires a minimum sample and a shorter time. Qualitative test by TLC bioautography for methanol extract of black rice was performed using methanol: chloroform (1: 9) as the eluent, whereas DPPH and phosphomolybdate used as a spray reagent.

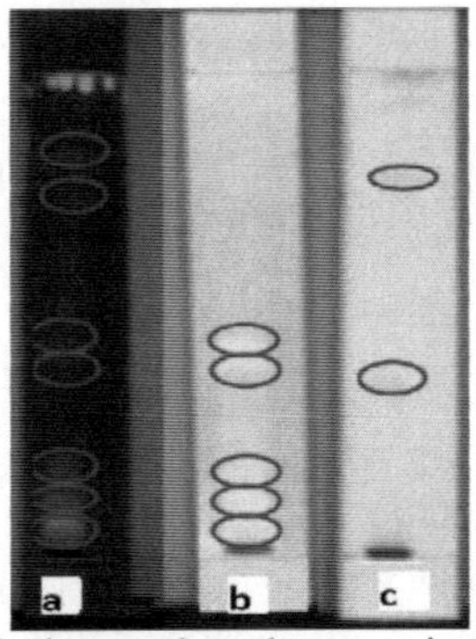

Figure 1. Bioautography of black rice methanol extract in 366 nm detection (a), DPPH (b), andphosphomolybdate (c) staining

Figure 1 shows the TLC bioautogram of methanol extract of black rice with DPPH reagent and phosphomolybdate to determine the spot that has antioxidant activity. The chromatograms before spraying for antioxidant activity test produces seven spots on TLC (Figure 1a). After spraying using DPPH, only fivewhite, yellow spots at the bottom of the TLC plate was found (Figure 3b). It showed that the compound's potential as an antioxidant with DPPH reagent in the methanol extract of black rice is polar compounds because these compounds can not be eluted by the methanol: chloroform (1: 9) and retained on the silica plate.

TLC bioautogram methanol extract of black rice with phosphomolybdate reagent shown in Figure 1c. After spraying using phosphomolybdate, only twoblue-green spots are found that are at the top of the TLC. It shows that the compound's potential as an antioxidant with phosphomolybdate reagent in the methanol extract of black rice is less polar because it can be eluted by the eluent and not retained on the silica plate.

In the chromatogram TLC bioautography using a reagent spray DPPH and phosphomolybdate (Figure 3), there is one spot/stain extract that is active as an antioxidant either using DPPH and usesphosphomolybdate, it indicates that the spot/stain is a fraction of the best as an antioxidant than the spot/stain other. According to Chang [15] class of compounds that act as antioxidants are flavonoids and triterpenoids. Flavonoids in nature are compounds that are polar because it is in the form of glycosides (bound with sugar). Thus might compounds that act as antioxidants using DPPH reagent is flavonoids. As with the triterpenoids nonpolar, these compounds are thought to act as antioxidants using reagents phosphomolybdate. The statement was supported by the results of phytochemical test (Table 2) which showed that the methanol extract of black rice contains flavonoids and triterpenoids positive thus might compounds that act as antioxidants in the methanol extract of black rice are flavonoids and terpenoids

4. Conclusions

The methanol extract of black rice has the highest potential as radical scavenging, with IC_{50} of 291 mg / mL using DPPH, while the ethyl acetate extract of white rice has the highest potential as anantioxidant by phosphomolybdate method (41.13 mmol α-tocopherol equivalents / g sample). Rice extract that has the highest activity of tyrosinase inhibitor is *n*-hexane extract of red rice on monophenols (IC_{50} 3.16 mg/ml). The TLC bioautogram inform that the active radical scavenging is the polar compound and active antioxidant by phosphomolybdate method is the less polar compounds of black rice methanol extract. Further separation is necessary to determine the active compounds in samples of rice to obtain the antioxidant activity and tyrosinase inhibitors.

References

[1] Soi-apronful, Rungtipet al, 2012 *Food and Nutrition Sci* 3 135-140.
[2] Femina2013 Merawatwajahdenganbedakdingin.http://www.femina.co.id/article/merawat-wajah-dengan-bedak-dingin
[3] J. Pokorny, N. Yanishlieva, M. Gordon, *Antioxidant in Food: Practical Application.* New York(US): CRC Pr. (2001)
[4] B. Halliwell, R. Aeschbach, J. Loliger, O. I. Auroma, *Food ChemToxicol.* 33: 601-617. (1995)
[5] [AOAC] Association of Official Analytical Chemist. 2nd Edition. Vol 1. Maryland (US): Association of Official Analytical Chemist (2006)
[6] J. B. Harborne, *Metode Fitokimia Penuntun Cara Modern Menganalisis Tumbuhan.* Padmawinata K, Soediro I, penerjemah. Bandung (ID): ITB Pr. Terjemahan dari: *Phytochemical Methods*(1987)
[7] I. Batubara, T. Mitsunaga, and H. Ohashi, *J. Wood Sci***55**, 230-235 (2009)
[8] I. Batubara, L. K. Darusman, T. Mitsunaga, M. Rahminiwati, E. Djauhari, *J Biol Sci.* **10**:138-144 (2010).
[9] [BPOM] Badan Pengawas Obat dan Makanan. 2004. *Monografi Ekstrak Tumbuhan Obat Indonesia.* Jakarta (ID): BPOM RI
[10] Z. Zarai, E. Boujelbene, N. B. Salem, Y. Gargouri, A. Sayari, LWT - Food Sci Technol. 50(2):634-641 (2012)
[11] W. Harjadi, *Ilmu Kimia Analitik Dasar.* Jakarta (ID): Gramedia.(1993)
[12] F. G. Winarno, *Kimia Pangan dan Gizi.* Jakarta (ID): Gramedia Pustaka Utama (2002)
[13] [SNI] Standar Nasional Indonesia. *Beras.* Jakarta (ID): Badan Standardisasi Nasional. SNI 6128:2008
[14] M.A. Al-Farsi MA and C. Y. Lee, *Food Chem.* 108:977-985. (2008)
[15] T. S. Chang, *J Mol Sci.* 10:2440-2473 (2009)
[16] K. Likhitwitayawuid, *J Curr Sci.* 94:44-52 (2008)
[17] I. Kaneda, K. Iwaki, H. Kubo, H. Sakurai, *J Health Sci.* 52:495-511 (2006)
[18] G. Zengin, A. Aktumsek, G. O. Guler, Y. S. Cakmak, E. Yildiztugay, *Rec Nat Prod.* 5:2 (2011) 123-132 (2010)

Purification of Methyl Ricinoleate on Producing of Cetane Improver

Abdullah[1,*], Triyono[2], W Trisunaryanti[2], and W Haryadi[2]

[1] Department of Chemistry, Faculty of Mathematics and Natural Sciences, Universitas Lambung Mangkurat, Banjarbaru, Indonesia
[2] Department of Chemistry, Faculty of Mathematics and Natural Sciences, Universitas Gadjah Mada, Yogyakarta, Indonesia

*Corresponding author: abdullahunlam@gmail.com

Abstract. Purification of methyl ricinoleate on a mixture of methyl ester that was derived from castor oil has been conducted. Methyl ricinoleate with high purity is required in the manufacture of additives for improving the cetane number of diesel oil. Purification of methyl ricinoleate was performed by fractional distillation under reduced pressure at 4.5 mmHg. In order to produce of cetane improver, methyl ricinoleate was nitrated with a mixture of HNO_3 and acetic anhydride at 11-13°C for 4 hours. Based on the analysis using gas chromatography (GC) was known that purity of methyl ricinoleate can be increased from 87.2% to 99.5%, with a yield of 30.6% (v/v). FTIR analysis showedthe absorption spectrum at 3440 (1 /cm), which is derived from the vibrations of the -OH (hydroxyl) on the ricinoleate methyl molecules. The results of characterization showed an increase in density of 0.9161 (mg/L) to 0.9171 (mg /L) and the refractive index of 1.4619 to 1.4627. FTIR spectra of nitrated methyl ricinoleateshowedan absorption spectrum at 1627 (1/cm), in which indicates a new compound with nitrate group (-NO_3).

1. Introduction

One important parameter for determining the fuel quality is a cetane number. Cetane number is a measure of the combustion quality of diesel oil in a machine [1, 2]. Low-quality of diesel oil is indicated by low the cetane number. Malaysian's diesel hasa cetane number of 50, Thailand of 50, and Singapore of 51[3-4]. Diesel oil with a high cetane number will have a positive impact on engine performance and exhaust emissions[5-7]. Therefore, Indonesian's diesel oil still require the addition of additives in order to get better quality.

Type of additives (cetane improver) that is widely used today is a class of organic nitrate compounds, namely 2-ethyl-hexyl-nitrate (EHN). EHN compound is a derivative of petroleum (nonrenewable). Therefore, some researchers [8-11] tried to find another additive as an alternative. The researchers used triglycerides or fatty acid as a raw material. Firstly, triglyceride was converted into esters, then nitrated into nitrate esters. In acquiring this nitrate ester compound, Porier et al. [8] and Kalam[4]wereinitiallyhydration of the double bond ofcarbon atoms (C=C),then followed by areductionin order to obtain the –OH group (hydroxyl). The –OH group was then nitrated using HNO_3 in order to get mononitrate compound. If -OH group is not ready, nitration will be occurred on C=C bonds and produced thepolynitrate compound.The polynitrate compoundhas high polarity, so solubility in diesel oil is low and emissions of NO_x from the exhaust gas become high[10, 11].

In order to get mononitrate compound, Rabello et al.[11] choose a shorter way by using castor (*Ricinus communis*) oil as raw material. Castor oil was chosen because of the fact that the oil hasricinoleate acid that contains ahydroxyl group (-OH).Ricinoleateacid is the largest component(87-88%)incastor oil. According to Munack[12], there are any otherfatty acids in castor oil, such as palmitate (1-2%), oleate (3-4%), linoleate(5-6%), and stearate (1-2%).In order to produce mononitrate compound as afuel additive, Rabello et al. [11]conducted transesterification of castor oil, then be followed by nitration of methyl ester.

The high diversity of acid in a mixture of castor oil leadhigh types of nitrationproduct.There are many types of ester that derived from the acids tend to increase thepolarity of additive molecules.Beside of hydroxyl group (-OH), C=C is aninteresting site for electrophiles. Therefore, purifying of methyl ricinoleate before being used on nitration processis important.Purification of methyl ricinoleate can be carried out by fractional distillation. Distillation can be conducted because of the differences in boiling point of thecomponent.

2. Methods

2.1. Materials

The materials used werecastor oil (Brataco Chemical Ltd.), methanol (Merck), Na (Merck), HNO_3 65% (Merck), HCl (Merck), acetic acid anhydride(Merck),diethyl ether (Merck), Na_2SO_4 (Merck), universal pH paper (Merck), andaquadest (Chemistry PhysicsLaboratory ofUGM).

2.2. Equipment

Some of the main equipment used are three-necksdistillation flask (Pyrex), reflux condenser (Pyrex), magnetic stirrer (Stuart SB 163),thermometer, magnetic stirring bar, separation funnel (Pyrex), viscosimeter, pycnometer, gas chromatography (GC HewlettPackard 5890 series II, FID, HP5 30 m, column temperature of 300°C, injection temperature of 150°C, carrier gas: He, burner gas: $H_2 + O_2$), GC-MS (Shimadzu GC 17A-QP 5000, Impact Electron, Rtx-5MS (5% diphenyl, 95% dimethyl polysiloxane) 30 m, d: 0,25 mm, maximum temperature: 350 ^{o}C, Injector mode: split 1:60, detector temperature: 300 ^{o}C) andFTIR spectrophotometer (Shimadzu Prestige 21).

2.3. Procedure

2.3.1. Transesterification of castor oil.

Two hundred millilitres of methanol and 0.2 g of sodium metal (Na) were added successively into three-necks distillation flask (500 mL). The solution was stirred slowly (100 rpm) for 5 minutes, then 100 mL of castor oil was added into the flask. After all of thecastor oil was added, the stirring speed was increased to 600 rpm and stirred for 24 hours.After the reaction runs for 24 hours, stirring was stopped and the mixture was added into separation funnel (1L). Then, into a separating funnel was added 600 mL of HCl that contains 2 mL of HCl 6 M. The mixture was allowed to stand for ±20 hoursso that form two layers. The upper layer was washed with aquadest until neutral, dried over Na_2SO_4, and filtered. The filtrate (diethyl ether) was then evaporated. The product of transesterification was characterised by determination of density, refractive index and viscosity. Analysisof product was also conducted with GC-MS and FTIR.

2.3.2. Purification by fractional distillation.

Purification of methyl ricinoleate was conducted by fractional distillation using reduced pressure distillation at 4.5 cmHg and the distillate was collected in the three-necks flask (3 fractions). Distillation in this work used 1300 mL of methyl ester. Distillate of each fraction was analysed by gas chromatography (GC), FTIR and measured by volume, density and refractive index.

IOP Conf. Series: Journal of Physics: Conf. Series **824** (2017) 012018 doi:10.1088/1742-6596/824/1/012018

2.3.3. Nitration of methyl ricinoleate.

Nitration was conducted at low temperature (11-13 °C), using a mixture of ice and water in awater bath. Firstly, 0,3 mL HNO_3 was added in athree-necks flask, followed by 1.5 mL acetic acid anhydride. Furthermore, amagnetic stirrer was operated slowly (200 rpm), while waiting for the temperature in the flask constant. After a constant temperature was reached, then into the flask was added 2.3 mL methyl ricinoleate quickly. Nitration was conducted for 4 hours, then the reaction was stopped by pouring the mixture into a separation funnel, in which already contains 20 mL diethyl ether and 200 mL of cold water. The upper layer (diethyl ether) was washed with aquadest until neutral, then was dried by theaddition of Na_2SO_4anhydrous. Diethyl ether was evaporated by heating at 80 °C for 30 minutes. The product then analysed by GC and FTIR.

3. Results and Discussion

Transesterification of castor oil produces a mixture of methyl ester compounds. Characterization of methylesterscovers density, refractive index and viscosity. Table 1 showsthe data of physical properties of castor oil before andafter transesterification. The data of density, refractive index, and viscosity of castor oil become lowerafter transesterification. This circumstance was theimpact of reducing on interaction force in the molecules.Originally, fatty acids bonded in glycerol of triglyceride molecule. However, after triglyceride molecules change to esters, one molecule of triglyceride turns into three ester molecules. The interaction between methyl ester molecules will not as strong as on fatty acid.The lower interaction on methyl ester molecule makesthe density of the ester become lower too. The lower of density will impact on reducing of refractive index properties. Transesterification process makesthe molecular size of triglycerides become smaller. The molecular size of triglyceride is about three times of methyl ester molecules. Decreasing on viscosity is as theimpact of decreasing in molecule size.

Table 1. Data of physical properties of castor oil before andafter transesterification

	Density (g/mL)	Refractive Index (n)	Viscosity (cPs)
Before	0.9537	1.4785	297.4
After	0.9161	1.4619	19.8

Chromatogram of transesterification product was presented in Figure 1. Transesterification of castor oil produced a mixture that consists five kinds of methyl esters compound. Based on the analysis using mass spectroscopy (mass spectra are not presented in this article), it was known that these compounds are methyl palmitate (0.8%), methyl linoleate (3.2%), methyl oleate (6.8%), methyl stearate (1.9%) and methyl ricinoleate (87.2%).

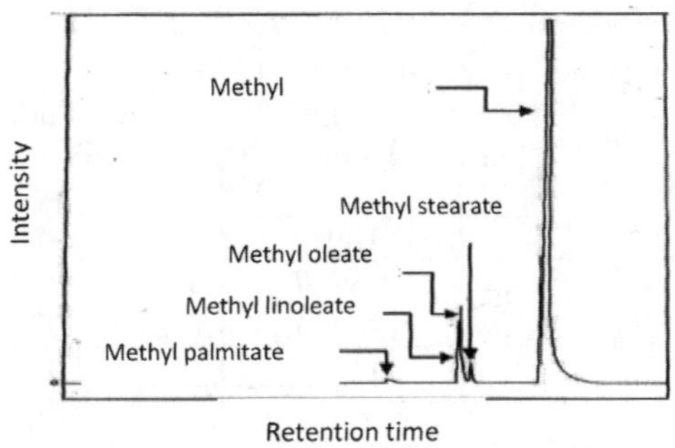

Figure 1.Chromatogramof methyl esters from castor oil.

Based on GC data showed that methyl ricinoleate still mixed with another methyl ester, where the content of methyl ricinoleate still relatively low in order to produce the additive. This circumstance is necessary to beincreasedby fractional distillation. On the process of fractional distillation was observed temperature data for every 2 minutes. This data can be used in predictionwhen distillate of methyl ricinoleate with high purity already commenced.Figure 2shows(area Fraction I and II) a rapid of temperature change for each observation time, so that the graph like resembles a staircase. Such circumstances indicate that the purity of methyl ricinoleate in the area is still low. It has been proved by the data of chromatogram as shown in Figure 3 (FI and FII).

Figure2 shows that in the 46thto 58th minutes, the distillation temperature finished at 186 °C. Starting with next minute (60th), distillation takes place at a temperature of 188°C for a long time up to the 345thminutes (for 275 minutes). Figure 2shows clearly that initial distillate in fraction III is not yet optimal. Therefore, the third fraction of distillatewas separated into IIIa and IIIb. Fraction IIIa was stopped at 110th minutes, and collection of distillate was continued for fraction IIIb. After all distillate of fraction III was run out, distillation temperature decreased. This suggests that the distillation finished.

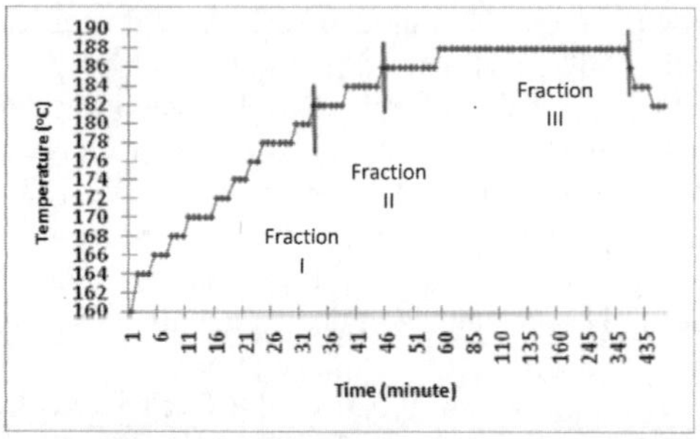

Figure 2. Graph of distillation temperature versus time

Table 2 shows thevolume of distillation, where total for Fraction III (FIIIa + FIIIb) was 761 mL or by 58 % of its initial volume (1300 mL). The density and refractive index for FIIIa and FIIIb seem larger than thevalue in FI and FII. Increasing of density will be accompanied by increasing on refractive index, in which consistent with the data of density and refractive index. The purity of methyl ricinoleate increase from FI to FIIIb. Methyl ricinoleate has a boiling point of 412 °C at 760 mmHg (1 atm), while the other has a much lower boiling point. Methyl stearate = 295 °C, methyl palmitate = 338 °C, methyl oleate = 352 °C, and methyl linoleate = 349 °C [13]. Methyl linoleate has a boiling point equal to the boiling point of methyl linoleate [14]. Thus, by fractional distillation under reduced pressure method, purification of methyl ricinoleate can be conducted easily.

Table 2. Physical data properties of methyl ricinoleate fractionation results inpressure reduction

Fraction	Volume (mL)	Density (g/mL)	Refractive Index (n)	Purity (%; GC)
FI	149	0.8952	1.4578	38.82
FII	211	0.9107	1.4611	70.06
FIIIa	363	0.9168	1.4622	97.86
FIIIb	398	0.9171	1.4627	99.52

According to Yao et al. [14], purity of methyl ricinoleate can be increased by extraction using hexane, then washed with a solvent. The solvent is a mixture of methanol, water and acetic acid (90:10:1). The result of purification showed that purity of methyl ricinoleate was originally 88% can be improved to more than 98.5% with a yield of 53 % (w/w). Although this method produced high purity of methyl ricinoleate, it needs many steps in the process.

The castor oil methyl ester from purification with the highest purity (FIIIb) was then analysed by FTIR spectrophotometer. Figure 4a shows there issome absorption spectrum of the methyl ester of castor oil. The main absorption spectrum of functional groups present at 3440, 1743 and 1658 (1/cm).The spectrum of absorption at 3440 (1/cm) is derived from the vibration of O-H (hydroxyl group). The hydroxyl group in the ester compound resulted from methyl ricinoleatemolecule.Absorption spectrum at 1743 (1/cm) is derived from the range of C=O contained in all the methyl ester, whereas the spectrum in 1658 (1/cm) is derived from the range of C=C of the methyl ester of castor oil (except methyl palmitate and methyl stearate). Other absorption spectra provide additional information aboutester compound of the result of esterification.

Castor oil methyl ester from FIIIb then was analysed by FTIR spectrophotometer. Figure 3a shows some absorption spectrum of methyl ester castor oil. The main absorption spectrum of functional groups present at 3440, 1743 and 1658 (1/cm). The spectrum absorption at 3440 (1/cm) is derived from the vibration of O-H (hydroxyl group). The hydroxyl group in the ester compound resulted from methyl ricinoleate molecule. Absorption spectrum at 1743 (1/cm) is derived from the range of C=O contained in all methyl ester, whereas spectrum at 1658 (1/cm) is derived from C=C of the methyl ester of castor oil (except methyl palmitate and methyl stearate).

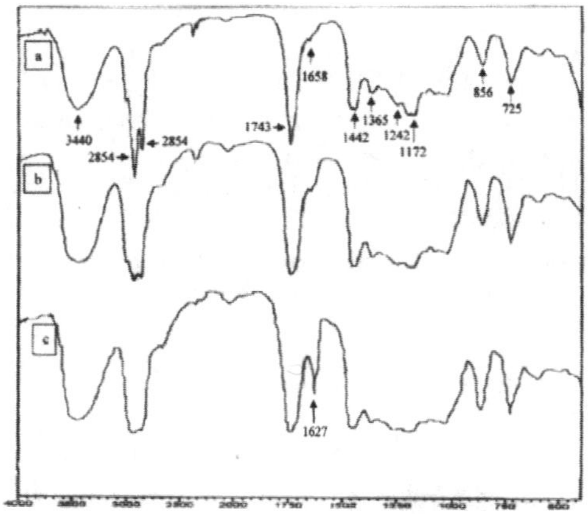

Figure 3. a) FTIR spectra of methyl ricinoleate before purification), b) FTIR spectra of methyl ricinoleate after, c) FTIR spectra of nitrated methyl ricinoleate at low temperature (11-13 °C) for 4 hours

Figure 3.b is FTIR spectra of purified methyl ricinoleate.Spectrain Fig 3b shows an increment of theareaat 3440 (1/cm). This increment is estimated to have acorrelation with thelevel of methyl ricinoleate purity. Figure 3c is FTIR spectra of nitrated methyl ricinoleate. Figure 3c shows a new spectrum at 1627 (1/cm), as a result from N-O stretching vibration of nitrate group(-NO_3). Nitration on methyl ricinoleate with high purity used a mixture of HNO_3 and acetic anhydride. This mixture will produce acetyl nitrate that acts as a carrier of nitronium ion. Mechanism of nitronium ionforming had been shown in the formerly literature [10]. Nitration on methyl ricinoleate will take place at the –OH group [15], then producemethyl ricinoleate nitrate.However, nitration can also occur at the C=C of methyl ricinoleate molecule. Nitration at C=C of methyl ricinoleate molecules can be minimised by conducting the nitration at low temperatures and a limited amount of HNO_3 [10].

4. Conclusion

Purification of methyl ricinoleatefrom oil methyl ester mixture can be conducted by fractional distillation under reduced pressure at 4.5 mmHg. Methyl ricinoleate purity can be increased from 87.2% to 99.5%. The distillatefrom fractional distillation of methyl ricinoleate was collected at 186-188°C.Nitration of methyl ricinoleate at temperatures of 11-13°C for 4 hours using a mixture of HNO_3 and acetic anhydride has produced nitrate group compounds of methyl ricinoleate

Acknowledgements

The authors would like to thank the Committee of the International Seminar ICSME 2016 has giventhe opportunityto present the result of this work.

References

1. Bamgboye, A.I. and A.C. Hansen 2008 *Int. Agrophysics***22** 21-29.
2. Knothe, G. 2005 *Fuel Processing Tech.***86** (10) 1059-1070.
3. NasikinM. and A. Makhdiyanti 2003 *Journal Teknologi***1** (XVII) 45-50.
4. Kalam, M.A. and H.H. Masjuki 2005 *J Sci Ind Res***64** 920-927.
5. Raghavan, V., et al. 2009 *Combust Sci Technol***181** (6) 877-891.
6. Moiseev, I.I. 2013 **82** (7) 616-623.
7. Pulkrabek, W.W. 2014. *Engineering Fundamental of The Internal Combustion Engine.* Second edition. Pearson Education Limited. England.
8. Poirier, M.A., D.E. Steere, and J.A. Krogh 1995 Google Patents
9. Suppes, G., J. Heppert, and M. Mason 2001 Google Patents
10. Canoira, L., et al. 2007 **86** (7-8) 965-971.
11. Rabello, C.R., B.G. Siqueira, and R.B. DeMenezes 2009 Google Patents
12. Munack, A. 2006 *Books: Biodiesel - A comprehensive handbook. Martin Mittelbach, Claudia Remschmidt (Ed.)* in *Biotechnology Journal*. 2006. 102-102.
13. Yuan, W., A. Hansen, and Q. Zhang 2005 **84** (7-8) 943-950.
14. Yao, L., et al. 2010 **87** (8) 937-945.
15. Abdullah, Triyono, W. Trisunaryanti, and Winarto H. 2013 *Indonesian Journal of Chemistry***13** (1) 36-40.

Mass Transfer Coefficientin Stirred Tank for *p*-Cresol Extraction Process from Coal Tar

D S Fardhyanti[1,*], D S Tyaningsih[1], and S N Afifah[1]

Chemical Engineering Department, Faculty of Engineering, Universitas Negeri Semarang, Semarang, Indonesia

*Corresponding author: dewiselvia@mail.unnes.ac.id

Abstract. Indonesia is a country that has a lot of coal resources.The Indonesian coal has a low caloric value. Pyrolysis is one of the process to increase the caloric value. One of the by-productof the pyrolysis process is coal tar. It contains a lot of aliphatic or aromatic compounds such as*p*-cresol (11% v/v). It is widely used as a disinfectant. Extractionof *p*-Cresol increases the economic value of waste of coal. The aim of this research isto study about mass tranfer coefficient in the baffled stirred tank for *p*-Cresolextraction from coal tar. Mass transfer coefficient is useful for design and scale up of industrial equipment.Extraction is conducted inthe baffled stirred tank equipped with a four-bladed axial impeller placed vertically in the vessel. Sample for each time processing (5, 10, 15, 20, 25 and 30minutes) was poured into a separating funnel, settled for an hour and separated into two phases. Then the two phases were weighed. The extract phases and raffinate phases were analyzed by Spectronic UV-Vis. The result showed that mixing speed of *p*-Cresol extraction increasesthe yield of *p*-Cresol and the mass transfer coefficient.The highest yield of *p*-Cresol is 49.32% and the highest mass transfer coefficient is4.757×10^{-6} kg/m^{2}s.

1. Intoduction

Indonesia is a country that has a lot of coal resources.However, the Indonesian coalhas a low caloric value. Pyrolysis is one of the processes to increase caloric value. This process produces aliquid by-product which is called coal tar. It's smell is sharp and unpleasant, it is often considered as waste.It needs to be able to produce compounds that have high economic value. Previous research suggests that coal tar contains many components of organic compounds either aliphatic or aromatic(1).

Coal tar contains about 348 kinds of chemical compounds. Some of them are very valuable chemical compounds. Aromatic compounds benzoid (benzene, toluene, xylene, naphthalene and anthracene) and phenolic compounds (phenol, cresol, xylenol, catechol, and resorcinol) that has been used as raw materials or intermediates in various chemical industries (anti-oxidant, antiseptic, resins, plasticizers in the plastic industry, paint, perfumes, drugs, etc.) are widely found in coal tar.In addition, coal tar also contain heterocyclic nitrogen compounds (pyridine, quinoline, iso-quinoline, and indole), homocyclic hydrocarbon compounds (benzene, phenol, toluene, ethyl benzene, xylene, and naphthalene) and heterocyclic oxygen compounds (dibenzofuran)(2)

Table 1. The largest compounds of the Indonesian coal tar(3)

Compounds	Percentage v/v (%)
Phenol	8,06
2-methyl phenol (o-cresol)	3,45
4-methyl phenol (*p*-cresol)	11,06
3-ethyl phenol	5,55
1,8-dimethyl naphtalene	5,50
1,6-dimethyl-4-isoprophyl naphtalene	8,65

Table 1 shows that the largest compounds of the Indonesian coal tar are*p*-Cresol (11.06% v/v). Liquid-liquid extraction method is used to increase the economic value of the coal tar especially for the extraction of *p*-Cresol.It is widely used as a disinfectant. A researcher who studied about coal tar especially phenolic compounds extraction from coal tar is rarely. Fardhyanti (2016) studied the extraction of phenolic compounds from coal tar and the liquid-liquid equilibrium for this extraction process. The extractor equipment for theindustry are not only designed by equilibrium process, but it also designed by the mass transfer process.

The extraction of *p*-cresol from coal tar israrely studied. Hismathet al (2011) studied the extraction of phenolic compounds from neem leaves and obtained the highest yield of phenol using acetone as a solvent.Putri (2014) studied the degradation of *p*-Cresol using bacteria Actinobacillus sp. and *p*-Cresol could be degraded almost 56.76% v/v.

This aims of this research are to study the mass transfer coefficient of *p*-cresol from coal tar in a baffled stirred tank.Mass transfer coefficient is useful for design and scale up of industrial equipment.

2. Method

2.1. Materials
The materials used in this study is coal tar (sub-bituminuous Kalimantan Coal) as the feed and acetone (e-Merck), and distilled water as the solvent.

2.2. Extraction Experiment
Instruments were used in this study:Electrically heated with stirrer (Yellow MAG HS 7, Yellow Line), Impeller diameter 3 cm and 5 cm, Motor stirrer (IKA RW 20 Digital), pycnometer 5 ml, stative, clamps, boss head, funnels separator 250 mL, pipette measuring 25 mL, measuring cup (10 mL, 25 mL, 50 mL), ball filler, Spectronic UV-Vis (Thermo, Scientific, Genesys 10UV), cuvettes, flasks measure (25 mL, 50 mL, 100 mL), beaker glass (100 mL, 250 mL). The experiment which used in this research is presented in Figure 1.

Each solution is extracted with the solvents of aqueous acetone and aqueous methanol in the baffled stirred vessels equipped with a four-bladed axial impeller placed vertically in the vessel.

Table 2. Material systems and conditions for the extraction process.

	Quantity
Feed	coal tar
mass of feed	200 mL
mass of solvent	200 mL
solvent	aqueous acetone
impeller diameter	3 and 5 cm
vessel diameter	9 cm
mixing time	5, 10, 15, 20, 25 and 30 minutes
mixing speed	210, 230 and 250 rpm
temperature	318 K.

200 mL of feed and 200 mL solvent were brought into contact in stirred tank. The solution was mixing in stirred tank with the operating conditions as in Table 2. Every sample for each time processing was poured into a separating funnel, settled for an hour and separated into two phases. Then the two phases were weighed. The extract phases and raffinate phases were analyzed by Spectronic UV-Vis.

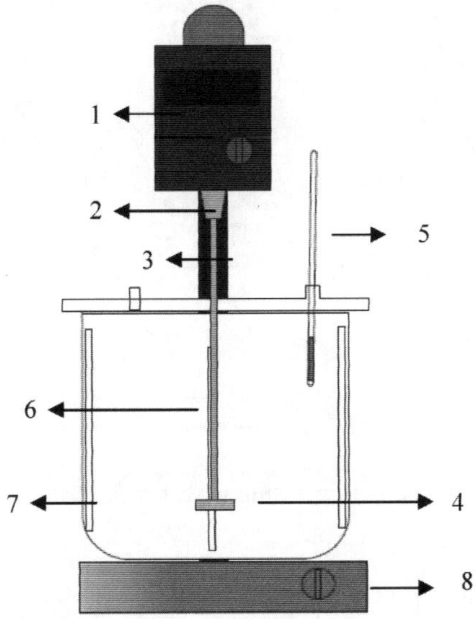

Explanation:
1. Motor stirrer
2. Buffer
3. Mechanical stirrer
4. Impeller
5. Termometer
6. Tank
7. Baffle

Figure 1. The Baffled Stirred Tank

3. Calcualtion

3.1. Mass transfer coefficient (overall) p-Cresol is determined by this equation :

$$k = \frac{r_t}{\alpha . V .(x_t - y_t)} \tag{1}$$

3.2. Specific surface area between phasesare calculated from the average drop diameter (Sauter-mean diameter) as follows:

$$A = \alpha.V = \frac{6.\phi}{d_{32}}.V \quad (2)$$

3.3. Sauter-mean diameter (d_{32}) empirically predicted by the correlation proposed by Mlynek and Resnick's(7). This correlation is represented as follows:

$$\frac{d_{32}}{d_p} = 0.058(1+5.4\phi)\,N_{we}^{-0.6} \quad (3)$$

4. Results

Figure 2 shows the effect ofmixing speedand the operation time for the concentration of *p*-Cresolin the extract phase. It shows that incresing of mixing speed and operation willincrease the concentration of *p*-Cresol in the extract phase. Because the effect of turbulencebymixing speedwilldecrease the thickness of the layer liquid film,so that the contact time between solute and solvent are getting longer(8).

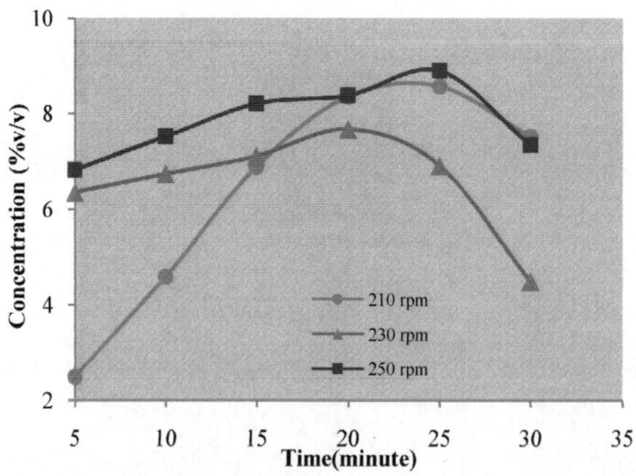

Figure 2.The effect of mixing speed and operating time for the concentration of*p*-Cresol in the extract phase.

Table 3. Yield of p-Cresol for each mixing speed.

Mixing speed (rpm)	Yield(%)
210	24,36
230	35,89
250	49,32

Table 3 shows thatthe increasing of mixing speed increases the yield of *p*-Cresol.It was caused by the increasing of turbulence in the extraction process (9).

Table 4.k value for each agitation speed

Mixing speed (rpm)	k (kg/m^{-2}s)
210	1,865. 10^{-10}
230	2,665. 10^{-9}
250	4,757. 10^{-9}

Table 4 and Figure 2 show the effect of mixing speed for the mass transfer coefficient of *p*-Cresol extraction from coal tar. The increasing of mixing speed increases the mass transfer coefficient of *p*-Cresol extraction. It was caused by the increasing of turbulence in the extraction process (9).

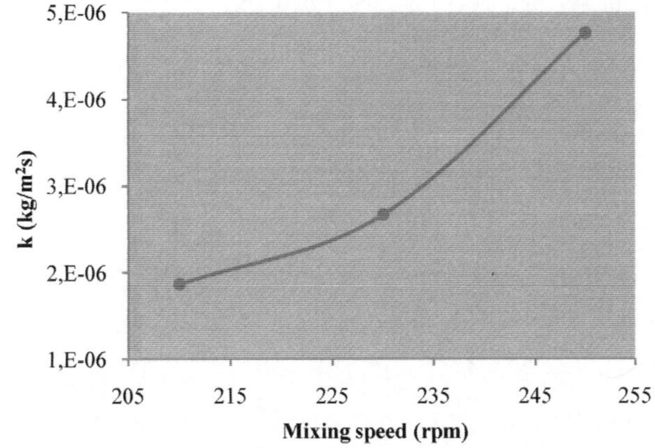

Figure 3. The effect of mixing speed for mass transfer coefficientof*p*-Cresol extraction.

5. Conclusion

5.1. The result showed that mixing speed of p-Cresol extraction increases the yield of p-Cresol and the mass transfer coefficient

5.2. The highest yield of p-Cresol is 49.32%.

5.3. The highest mass transfer coefficient is 4.757 x 10-6 kg/m^{2}s.

Nomenclature

α	= specific surface area between phases per volume unit (1/m)
A	= specific surface area between phases(m^2)
d_p	= diameter stirrer (m)
d_t	= diameter tank (m)
k	= mass transfer coefficient (kg/m^2.s)
N_{we}	$= \frac{N^2 . d_p^{\ 3} . \rho}{\sigma}$, stirrer weber number
r_t	= the rate of mass transfer at time t (kg/s)
V	= total volume (m^3)
x_t	=weight fraction of *p*-Cresol in equilibrium
y_t	= weight fraction of *p*-Cresol in the extractphase at time t
ϕ	= holdup of dispersed phase

ρ = density (kg/m^3)
μ = viscosity (kg/m.s)
N_p = mixing speed (rpm)

Acknowledgments

The authors would like to acknowledge to Ministery of Research Technology and Higher Education of this through the research grant.

References

[1] A.A. Lappas, D.Papavasiliou, K.Batos, and I.A. Vasalos, 1990. Jurnal Fuel, Chem. **69**, pp. 1304-1308
[2] S. Jun 2004, Bachelor Thesis, Department of International Development Engineering, Tokyo Institute of Technology, Japan
[3] D.S. Fardhyanti,W.B. Sediawan, P. Mulyono and M. Hidayat 2015. Journal of Basic and Applied Research International **7(1)**, pp. 23-27.
[4] D.S. Fardhyanti, W.B. Sediawan, and P. Mulyono 2016, International Journal of Thermodynamics **19(2)**, pp. 110-119.
[5] I.Hismath, W. M. Wan Aida, C.W. Ho 2011, International Food Research Journal **18(3)**
[6] P.O. Putri, Kemampuan Isolat Bakteri Actinobacillus Sp. dalam Mendegradasi p-Kresol dan Sunset Yellow,(Jurusan Kimia, Universitas Negeri Malang, 2014) pp.1-9
[7] A.H.P.Skellandand J.S. Kanel 1992, Industrial and Engineering Chemistry Research, **31**. pp.908-920 (1992).
[8] P. Mardina, "Pengaruh Kecepatan Putar Pengaduk dan Waktu Operasi pada Ekstraksi Tannin
[9] A.Zainal,Jurnal Kimia **5(2)**, pp. 125-132 (2011). PendekatanTeoriLapisan Film(TeknikKimia : UNDIP, 2005), pp. 17-20.

Application of ZnO Nanoparticle as Sulphide Gas Sensor Using UV/VIS/NIR-Spectrophotometer

N Juliasih* , Buchari , and I Noviandri

Department of Chemistry, Bandung Institute of Technology, Indonesia

*Corresponding author : asihjuli@gmail.com

Abstract. The nanoparticle of metal oxides has great unique characteristics that applicable to the wide industrial as sensors and catalysts for reducing environmental pollution. Sulphide gas monitors and detectors are required for assessing safety aspects, due to its toxicity level. A thin film of ZnO as the sulphide gas sensor was synthesised by the simple method of chemical liquid deposition with variation of annealing temperature from 200 °C to 500 °C, and characterised by Scanning Electron Microscope (SEM), X-Ray Diffraction (XRD), and UV/VIS/NIR-Spectrophotometer. Characterization studies showed nanoparticle size from the range 62 – 92 nm of diameters. The application this ZnO thin film to sulfide gas, detected by UV/VIS/NIR Spectrophotometer with diffuse reflectance, showed specific chemical reaction by the shifting of maximum % Reflectance peak. The gas sensing using this method is applicable at room.

1. Introduction

There are many application of metal oxide as gas sensor used electrically method by conductance changing of the sample upon exposure to the gas. ZnO as a part of it, depending on the form and the shape of deposited thin films, has wide range technological applications including a gas sensor, photocatalysis, piezoelectricity and optoelectronic (1,2). Sulphide gas, like CS2 and H2S often produced in many industrial activities for example in carbon black, sulphur recovery, and viscose rayon production process. It also produced in large quantity from industrial as petroleum or natural gas drilling and refining, wastewater treatment, coke oven, tanneries, kraft paper mills, and landfills (3). At low concentration, it produces effects on the nervous system. The monitoring of these toxic gas emissions is extremely important for environmental protection. Sulphide gas monitors and detectors are required for assessing safety aspects, due to its toxicity level. Reliable detector or sensors with low cost and simple production process, low energy consumption, high sensitivity and selectivity, operable in a low concentration of sulphide gas, become high demand recently. ZnO nanoparticle was synthesised by simple chemical liquid deposition have great unique characteristics that applicable to the wide industrial as sulphide gas sensors and catalysts for reducing environmental pollution (4).

2. Experimental Details

A thin film of ZnO can be produced by deposition of this metal oxide on the glass substrate. Soda lime float glass with 1 – 1.2 mm of thickness (Yancheng Huida Medical Instruments Co.) was used in this experiment. The substrate was cleaned by rinsing into 40°C of acetone for 15 minutes, then rinsed with ethanol at room temperature, final washed with distilled water and dried. $ZnCl_2.2H_2O$ and NaOH were used as the precursors, and deposition was done by spraying. Annealing temperature was varied from 200 °C - 500°C. Thin film of ZnO as the sulfide gas sensor was synthesized by simple method of chemical liquid deposition, and characterized by Fourier Transform Infrared Spectrophotometer,

Scanning Electron Microscope (SEM), X-Ray Diffraction (XRD) with Cu[(Kα)] with $\lambda\alpha_1$ = 1.54060 nm and $\lambda\alpha_2$ = 1.54439 nm , and UV/VIS/NIR-Spectrophotometer (Shimadzu UV3101PC) at λ = 200 – 2400 nm. Application of thin film to sulphide gas was tested using 20 ppm of CS_2 cylinder gas standard with the flow of 0.2 L/min.

3. Experimental Results

The infrared spectrum of ZnO nano thin film showed specific peaks for annealing temperature 500 °C. It can be seen from the spectrum that the highest annealing temperature showed results as the clearest spectrum. Functional group –OH seems wider at the lower annealing temperature.

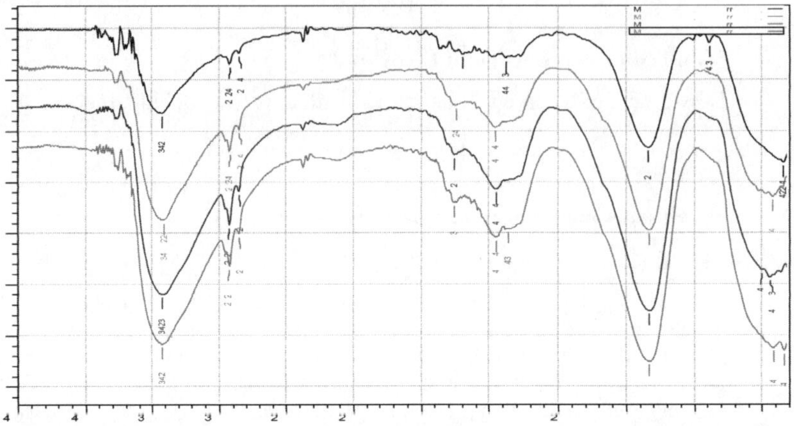

FIGURE 1. Infrared spectrum of ZnO nano thin film variance annealing temperature

200 °C (green), 300 °C (blue), 400 °C (red), and 500 °C (black)

UV/VIS-spectrophotometer analysis result showed increasing of annealing temperature can increase the transmittance.

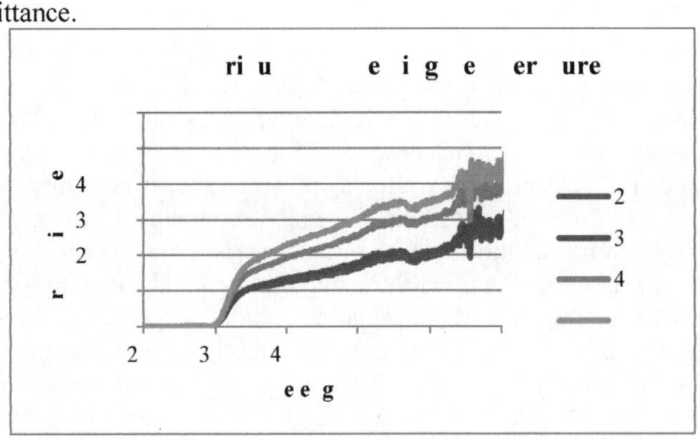

FIGURE 2. UV/VIS spectrum of ZnO nano thin film

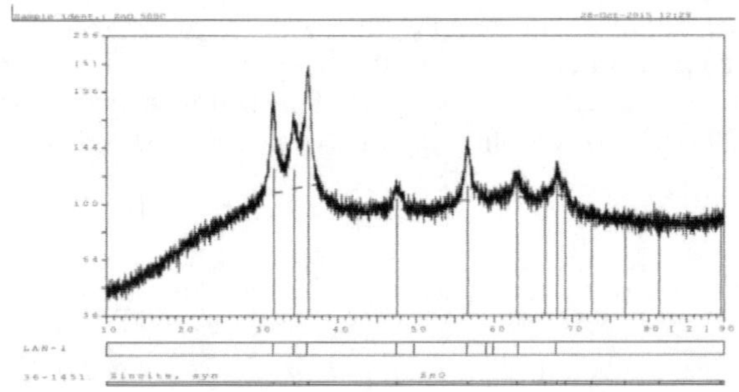

FIGURE 3. XRD spectrum of ZnO nano thin film

XRD pattern of ZnO variance 500 °C showed peaks corresponding mainly to the (100), (002), and (101) planes. Characterization studies showed nanoparticle size from the range 62 – 92 nm of diameters.

FIGURE 4. SEM morphology of ZnO nano thin film

Application of ZnO thin film for sulphide gas sensing was tested using UV/VIS/NIR-spectrophotometer showed the shifting of % Reflectance peak from wavelength 1566 cm^{-1} with 35.95 % of reflectance to wavelength 1768 cm^{-1} with 59.98%. This good response as a starting point for detecting gas quantitatively at wavelength 1768 cm^{-1}.

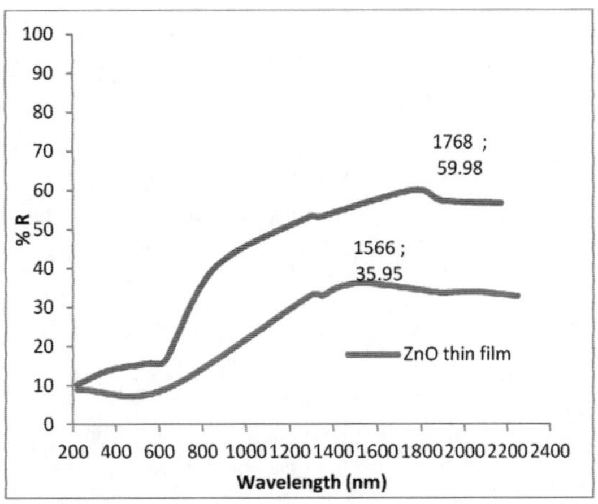

FIGURE 5. Application of ZnO to sense 20 ppm of CS2 gas standard

4. Conclusion

ZnO nanoparticle can be produced by simple chemical liquid deposition, and at an annealing temperature of 500 °C can get particle size from the range 62 - 92 nm of diameters. Application of ZnO thin film to sulphide gas, detected by UV/VIS/NIR Spectrophotometer with diffuse reflectance, showed specific chemical reaction by the shifting of maximum % R peak at a wavelength to 1768 nm.

5. Acknowledgments
This work was supported by Department of Chemistry, Bandung Institute of Technology, Indonesia.

References

[1] Farag H K, Hanafi Z M, Dawy M, and Abd El Aziz E M, 2010 Characterization of ZnO nanopowders synthesized by the direct precipitation method *Canadian Journal of Pure and Applied Sciences* Vol 4(3) p 1303-1309
[2] Alvarado J A, Maldonado A, Juarez H, Pacio M and Perez R 2015 Characterization of nanostructured ZnO thin films deposited through vacuum evaporation, Beilstein *J. Nanotechnol.* Vol 6 p 971-975
[3] Sudhir K P, Ki-Hyun K and Kea-Tiong T 2012 A review of sensor-based methods for monitoring hydrogen sulphide *Trends in Analytical Chem.* Vol. 32 p 87
[4] Zhanfeng Z 2009 *Synthesis and Modification of Metal Oxide Nanostructure and Their Applications* (Queensland University of Technology School of Physical and Chemical Sciences Inorganic Material Research Group)

Potential Development Essential Oil Production of Central Java, Indonesia

D Alighiri[1,2,*], W T Eden[1,2], K I Supardi[1,2], Masturi[3], and A. Purwinarko[4]

[1] Essential Oil Study Center, Faculty of Mathematics and Natural Sciences, Universitas Negeri Semarang, Indonesia
[2] Chemistry Department, Faculty of Mathematics and Natural Sciences, Universitas Negeri Semarang, Indonesia
[3] Physics Department, Faculty of Mathematics and Natural Sciences, Universitas Negeri Semarang, Indonesia
[4] Computer Science Department, Faculty of Mathematics and Natural Sciences, Universitas Negeri Semarang, Indonesia

*Corresponding author: alighiri.dante@gmail.com

Abstract. Indonesia is the source of raw essential oil in the world. Essential oils are used in various types of industries such as food and beverage, flavour, fragrance, perfumery, pharmaceuticals, and cosmetics. However, the development of Indonesian essential oil industry has not been encouraging for the production of essential oils, further it is unable to meet global demand. Besides that, the quality of volatile oil resulted cannot meet the international market standards. Based on the facts, the potential of Indonesian essential oils needs to be developed to provide added value, through increased production, improved quality and product diversification. One part of Indonesia having abundant of raw essential oil source is Central Java. Central Java has the quite large potential production of essential oils. Some essential oils produced from refining industry owned by the government, private and community sectors include cananga oils (Boyolali district), clove oils (Semarang district), patchouli oils (Brebes district, Pemalang district, and Klaten district). The main problem in the development of plants industries that producing essential oil in Central Java is low crops production, farming properties, quality of essential oils are diverse, providing poor-quality products and volatile oil price fluctuations. Marketing constraints of Central Java essential oils are quite complex supply chain. In general, marketing constraints of essential oils due to three factors, namely the low quality due to type of essential oil business that generally shaped small businesses with different capital and technology, domestic marketing is still a buyer-market (price determined by the buyer) because of weak bargaining position processors businessman, and prices fluctuate (domestic and foreign) due to uncontrolled domestic production and inter-country competition among manufacturers.

1. Introduction

Indonesia is a tropical country, so many types of plants that grow. According to [1,3] that not less than 17% of plant species are in Indonesia. Some parts of plant essential oil can be taken, both from leaves, roots, flowers, fruits, and even the trunk [7]. The development of Indonesian essential oil industry is very slow. The types of essential oils exported since colonial times are lemongrass, cananga [8], patchouli and fragrant roots. Until 1970, the type of essential oil that is exported is still limited that commodity. Since 1975, exported kinds of essential oils began to grow with clove oil, sandalwood oil, ginger, pepper and nutmeg, then massoia oil of Irian Jaya in the 1990s [5]. In 1993, the types of essential oils that exported amounted to 15 species [2,3].

The development of Indonesian essential oil industry has not been encouraging. According to the Chairman of Dewan Atsiri Indonesia [1], the primary production of Indonesian essential oils is clove oil (clove leaf oil and clove stem oil) [4], citronella oil, patchouli oil, cajuput oil, and turpentine oil. The world trade value essential oils is estimated USD 4 billion, total exports of Indonesian essential oil is USD 120 million [6]. The review of its kind there is 300 kinds of essential oils in world trade, while the kinds of essential oils that are likely to be developed in Indonesia are 40 types.

Table 1. The needs of world essential oils and Indonesian essential oil production [1]

Essential Oil	Global demand (Tonnes)	Indonesian Production (Tonnes)
Clove oil	5.000-6.000	3.500-4.000
Patchouli oil	1.200-1.500	800-1.000
Citronella oil	>2.000	500-600
Turpentine oil	>10.000	10.000
Cajuput oil	>1.000	350-400
Nutmeg oil	>400	350-400

Director General of IKM Data from 2003-2008 shows the export value of Indonesian essential oil is lower than the value essential oil derivatives products import. As an illustration of essential oil export and import imbalance at the end 2007 that value of essential oil exports rise to $ 101 million, while the value of essential oil derivatives imports is USD 389 million. That trend continues to this day. Large population and the ability of better economic to the market destination of various commodities. The concern if the commodities are essential oils [1].

2. Methods

Samples of essential oil industries are five districts based on the largest production activity of essential oils data that is used as a source of information. Data to be collected includes: (1) Type of essential oil produced, (2) The capacity and quality of production is already underway, (3) Technology processing/refining use, (4) The availability of raw materials and production support (5) Barriers and opportunities, and (6) Management and Marketing

3. Result and Discussion

Identify the Strengths, Weaknesses, Opportunities and Threats

In Boyolali distric, Central Java is a small industry producing cananga oil. The industry has a lower strength development than the weakness. The major constraint weakness on the Boyolali distric, Central Java is limited raw materials in the company resulting flowers supply from farmers who are also limited. Besides, cananga flower is seasonal nature also farmers prefer to sell cananga flowers as sow flowers because the price is higher than in the company.

Five factors of both the opportunities and threats can be described based on observations in Boyolali, distric, Boyolali. The most influential of the possibilities factor in that company is a high demand in the international market. This indicates that in the absence of the company's competitors which can develop production to meet worldwide demand. The company can increase profits from the optimal production. The threat is a lack of relationships with raw materials suppliers that would have an impact on the continuity of business because the raw materials are limited. Its continuity of production is hampered.

IOP Conf. Series: Journal of Physics: Conf. Series **824** (2017) 012021 doi:10.1088/1742-6596/824/1/012021

Table 2. Identification of Strengths and Weaknesses at Cananga Oil Development in Small Industries in Boyolali Distric

Internal Factors	**Strengths**	**Weaknesses**
Financial condition	-	Capitalization is still lacking
Human Resources	Long time experience	Low human resources quality
Production	Low business risk SNI standard (Indonesian standard) quality Accommodate cananga flower farmers	Limited raw materials owned companies Management of technology tools are still lacking Lesa waste management
Marketing	-	Marketing is through exporters

(Primary Data, 2016)

Table 3. Identification of opportunities and threats at canangga oil development in small industries, Boyolali distric, Central Java

Internal Factors	Opportunities	Threats
Competitor	Lack of competition	-
Customer	High demand in the international market	Demands demand of high-quality essential oils
Suppliers	-	the lack relationships with raw materials suppliers
Natural conditions	-	Cananga flower land diminishing
Government policy	Government's attention to the distiller	Caterpillar pests in the rainy season
Economic conditions	-	Volatile oil prices fluctuating
Socio-cultural	Environmental conditions are safe	-
Technology aspect	Advanced technology	-

(Primary Data, 2016)

General Analysis of Essential Oil Potential in Central Java

The main problem in the development of plants industry producing essential oil in Central Java, among others: (1) lower production plants, (2) The type of farming, (3) different essential oils quality, (4) Providing lesser products quality, (5) volatile oil prices fluctuation. Problems in Central Java on marketing did not become a significant constraint as a result of any number essential oils are always some accept. Enterprises essential oil industry in Central Java mostly cultivated by farmers or craftsman in the countryside as small industries. A Large industry exists only in Purwokerto, namely PT. Indesso Aroma and PT. Takasago. Then, the secondary industry such as PT. Java Plant, PT. Natura Aromatik Nusanputera in Surakarta. Farm management is a sideline with a small capital and technology potluck. Not all technology package (kind of superior varieties, cultivation and processing / post-harvest) are available for some aromatic plants commodities because many types and research priorities are still small compared with other plantation crops. Small-scale farming, production and low essential oils quality. This is due to limit technological capability that does not meet the technical requirements, the use of not superior varieties, equipment and less qualified processing methods.

General Analysis of Essential Oil Production in Central Java

Most of the oil refining industry is still continuing production in Central Java because of factors that support the business, such as location which adjacent to raw materials source, close to water sources, and easy access to obtain fuel.

The main production facilities are iron plate kettle, furnace, condenser, and receiver or essential oil container. Other necessary equipment such as 200 litres of a plastic drum to contain essential oils, rakes, spoons, chamber, oil funnel and filter cloth. Most of the essential oil refiners in Central Java has been used fairly standard distillation with stainless steel 304 equipment. However, they also found smaller-refiners still use goods used for the refining process.

The primary raw material used in essential oil Central Java, most of the leaves such as clove oil, patchouli oil, piper betle oil. Then from the leaves and stalks of local lemongrass oil and aromatic lemongrass oil. A small portion of the interest such as cananga flower oil and oil from rhizomes like ginger and turmeric. In general, the production process can be done 5-6 months in a year.

The human resources required in the manufacturing process is relatively not too much. Power to manufacturing process requires only 3 per distillation process. If the first day of the entrepreneur did two times distillation process then required six temporary workers per day per kettle (assuming the entrepreneur has two kettles).

The technology used in the production of essential oil processing in Central Java, including simple or traditional technology. The process commonly used is distillation with water vapour.

Competition Analysis and Market Opportunities

The level of competition Indonesian essential oil in the international market is mainly determined by essential oil quality produced by Indonesia and competitor countries, such as Madagascar, Tanzania and Sri Lanka. The state is producing essential oil not only from developing countries, such as China, Brazil, Indonesia, India, Argentina and Mexico, but also developed countries like the United States, France, Germany, Italy, and England. The difference is developing countries produce more semi-finished essential oils then export them to developed countries. Another case in developed countries, although they import semi-finished materials from developing countries to be processed into finished goods, they export it back to other nations, including developing countries to be finished products with higher added value. Nevertheless, clove oil market opportunity is still wide open, especially on the world market which increasing demand volume.

General Marketing Path Analysis

Essential oil marketing channels are no different from other agricultural commodities. In domestic marketing, producers sell their products to intermediaries or exporters agents. It was only then that product into the hands of an exporter. Most trading of essential oil for export.

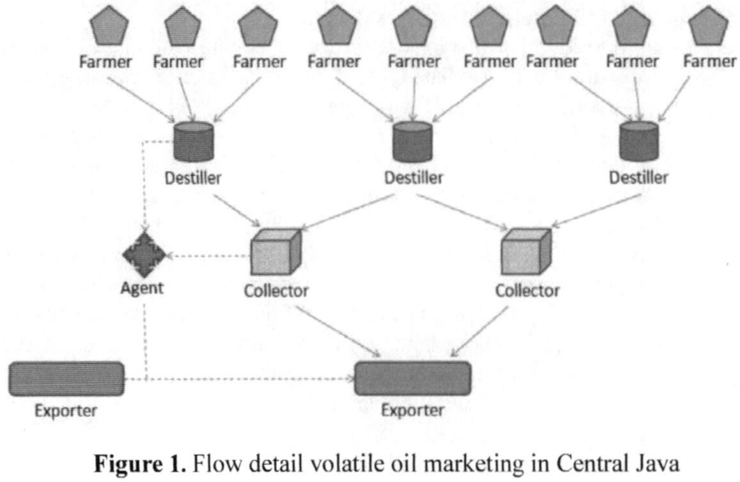

Figure 1. Flow detail volatile oil marketing in Central Java

The primary constraint marketing of this essential oil is trading supply chain quite complicated.The volatile oil processing business is still experiencing difficulties to supply directly to the exporter or the end-user. As a result of this complex trade, the chain is specified quality uniformity. Factors to be considered in essential oils marketing efforts, mainly for export purposes is attention quality, competitive pricing and production sustainability. In general, marketing constraints essential oils due to three factors, namely: (1) a low quality because of the type of essential oil producers that is generally small businesses with different capital and technology, (2) Domestic marketing is still a buyers market (price determined by the buyer) because of weak bargaining position processing entrepreneurs, and (3) Prices are fluctuating (domestic and foreign) due to uncontrolled production of domestic and inter-country competition among manufacturers.

4. Conclusion

This study has shown one part of Indonesia which are abundant of raw essential oil source is Central Java. Central Java has the potential production of essential oils are quite large. Some essential oils produced from refining industry owned by the government, private and community sectors include cananga oils (Boyolali district), clove oils (Semarang district), patchouli oils (Brebes district, Pemalang district, and Klaten district). The main problem in the development of plants industries that producing essential oil in Central Java is low crops production, farming properties, quality of essential oils are diverse, providing poor-quality products and volatile oil price fluctuations. Besides, marketing constraints of essential oils due to three factors, namely the low quality due to type of essential oil business that generally shaped small businesses with different capital and technology, domestic marketing is still a buyers market (price determined by the buyer) because of weak bargaining position processors businessman, and prices fluctuate (domestic and foreign) due to uncontrolled domestic production and inter-country competition among manufacturers.

Reference

[1] Dewan Atsiri Indonesia. 2015. Minyak Atsiri Indonesia. Dewan Atsiri Indonesia. Jakarta: Indonesia. Unpublished.
[2] Kusuma, Heri Septya and Mahfud Mahfud. 2016. Microwave hydrodistillation for extraction of essential oil from *Pogostemon cablin* Benth: Analysis and modelling of extraction kinetics. *Journal of Applied Research on Medicinal and Aromatic Plants* (JARMAP) 86 (2016) 9.
[3] Paul, Anamika, Ganesh Thapa, Adreeja Basu, Purabi Mazumdar, Mohan Chandra Kalita, and Lingaraj Sahoo. 2010. *J. Industrial Crops and Products* 32 (2010) 366-374.
[4] Jirovetz, Leopold, Gerhard Buchbauer, Ivanka Stoilova, Albena Stoyanova, Albert Krastanov, and Eric Schmidt. 2006. *J. Agric. Food Chem.* 2006, 54, 6503-6307.
[5] Marwa, J., Purnomo, H., Nurrochmat, D.R., 2010. Managing the Last Frontier of Indonesian Forest in Papua. AKECOP-IPB, Bogor.
[6] Ministry of Finance, 2014. Dana Bagi Hasil Sumberdaya Alam. The Ministry of Finance, Jakarta.
[7] De Guzman, C.C., Villanueva, M.A., Torres, R.C., Regros, R.A., Cosico, W.C., Aragones,E.G., 1997. Production and Processing of Citronella, Patchouli, and Ilang-ilang. Department of Science and Technology, Philippines, pp. 135–185.
[8] Kuspradini, Harlinda, Agmi Sinta Putri, Edi Sukaton, and Tohru Mitsunaga, 2016, *Bioactivity of* Essential Oils from Leaves of Dryobalanops lanceolata, Cinnamomum burmannii, Cananga odorata, and Scorodocarpus borneensis. *J. Agriculture and Agricultural Science Procedia* 9 (2016) 411-418.

Measuring Visual Literacy Skills on Students' Concept Understanding of Genetic Transfer Material

F Fibriana*, S D Pamelasari, and L S Aulia**

Department of Integrated Science, Faculty of Mathematics and Natural Sciences, Universitas Negeri Semarang, Semarang, Indonesia 50229

Corresponding author: *fibriana.f@mail.unnes.ac.id; **stephanipamelasari@mail.unnes.ac.id

Abstract. Visualization is an important skill for all students majoring in natural sciences. Also, the visual literacy skills (VLS) are essential for Microbiology learning. The lecturer can use the external representations (ERs) to visualize the microorganisms and its microenvironment. One of learning materials which are rather difficult to interpret in microbiology is genetic transfer. In this study, we measure the VLS on students' concept understanding of genetic transfer material using a simple test. The tests were held before and after the lecture on this topic employing a combination of talking drawing with picture and picture model. The results show that in the beginning, students showed their poor visual literacy. After the lecture, students were able to draw their understanding on the genetic transfer in bacteria. Most students' visual literacy ability improves in the level of acceptable. In conclusion, the students' ability was improved in the average amount of conceptual knowledge. This result reveals that some students comprehend in the correct level of ability, meaning that they have a high degree of conceptual (propositional) and visual knowledge.

1. Introduction

Microbiology is a discipline which studies all microscopic organisms in the form of unicellular, multicellular or acellular such as bacteria, microfungi, mold, microalgae, protozoa, and archaea [1]. In Indonesia, microbiology began to be introduced in the middle school and is a compulsory subject for higher level students of biology, chemistry, integrated science, pharmaceutical, food, health, medicine and the environment [2]. Nowadays, mastering microbiology by natural sciences students is important since it is directly related to everyday life and can be associated with life skills aspects [3]. In microbiology learning, students are required to translate the microscopic form of microorganisms and its environment in their mind, something that is hard and confusing for them. In fact, the biologists use the visualization tools for understanding and researching the molecular and cellular biosciences. The visualization tools by some psychologists are called by external representatives (ERs); for example: molecular models, photographs, micrographs, pictures, diagrams, illustrations, drawings, images, analogical representations, maps, symbolic pathways, genomic representations, graphs, icons, static visuals, dynamic visuals, animated visuals, multimedia, and virtual reality environments [4]. The use of external representations (ERs) such as diagrams and animations in science education, particularly in microbiology has rapidly increased over the past decades.

In addition, the best way to learn microbiology is by conducting the practicum and a direct observation under the microscope. However, not all schools and faculties have the facilities to support this activity. Therefore, teachers and lecturers can simply construct the picture and picture model in an attempt to explain the abstract phenomena. This model is effective for understanding theories and concepts of biology, biotechnology, and microbiology by visualization. Picture and picture learning model uses the pictures which are paired or sorted into a logical sequence. One of microbiology materials is genetic transfer mechanism of bacteria. Even though the undergraduate students have directly observed the form of bacterial cells in microbiology practicum; however, most of them have not observed the genetic materials inside the cells due to the limitation of microscope resolution capacity. Therefore, this material of study is rather difficult for students to interpret if only taught in a common lecture without any learning model application. In regards to this issue, the implementation of picture and picture model is promising to facilitate students to imagine the process of genetic transfer process in the simplest way. This particular ER is suitable for achieving desired learning outcomes. Moreover, it is also important for lecturers to teach students the visualization skills to interpret ER to ensure that the students are capable of visual literacy skills. Thus, we conducted a research on the undergraduate students in microbiology subject to measure their ability in visual literacy skills. The aims of this study are to discuss the use of picture and picture model in developing students' visual literacy skills, to measure the level of students' visual literacy skills and minimize visualization difficulties as well as enhancing the general literacy of future students.

2. Methods

The study was conducted in the Department of Integrated Science, Faculty of Mathematics and Natural Sciences, Universitas Negeri Semarang. The population of this study was 46 undergraduate students who registered in Microbiology subject (Course Code: D0014037). The students were taught the genetic transfer learning material in the 12th meeting using the combination of talking drawing with picture and picture (P&P) model [5]. The course was designed to help students comprehending the material by training their visual literacy because learning biology is considered difficult, especially in some areas like cell division, genetics, and hormones [6]. The course design is presented in Table 1.

Table 1. Course Design

Step	Action
1	Select the concept, topic, or object
2	Measure initial ability
3	Introduce the materials through picture
4	Invite students to share and discuss their work with their classmates in group
5	Instruct in a concept, topic or object selected
6	Encourage students to create another drawing based on the instruction
7	Facilitate discussion among students comparing and contrasting their drawing
8	Assess their concept mastery through drawing

The course design focuses on developing students' visual literacy through drawing since biology is the most visual of all sciences and has a long history of the use of imagery to define and link concepts in living systems [7]. The P&P strategy was used to train students to help them formulating the concept of material. The example of learning material is presented in Figure 1.

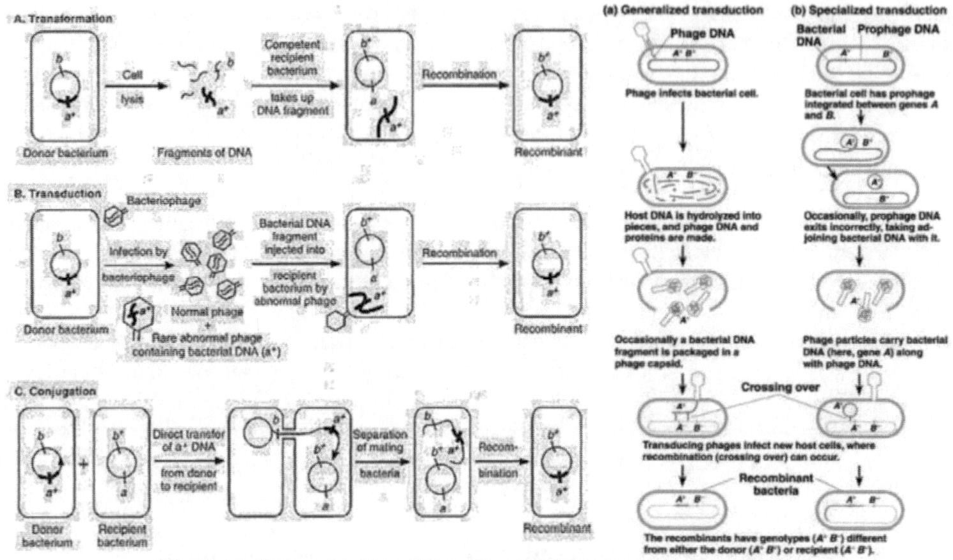

Figure 1. The example of genetic transfer external representation

After the implementation of P&P model, a test was conducted to measure the students' visual literacy skills. In this regard, their reasoning skills (R), conceptual knowledge (C), and mode of representation (M) were determined. These three areas were measured based on their response to the questions related with ER. Students were asked to re-draw the transformation, transduction, and conjugation method of genetic transfer in bacteria and explain it in words. The score on their answer was ranged from 0-3 on a 4-point Likert scale as shown in Table 2.

Table 2. Scoring guideline using 4-point Likert scale to measure the students' visual literacy skills

Grading	Score	Definition
Correct	3	High degree of conceptual (propositional) and visual knowledge used to provide a relevant response with a high amount of detail i.e. C – R – M.
Acceptable	2	An average amount of conceptual (propositional) knowledge, where one uses mainly conceptual knowledge and ER based knowledge to respond to the probe i.e. C – M, with little evidence of in-depth reasoning e.g. student regurgitate answers from conceptual knowledge.
Partially correct	1	Response based on reasoning with ER, little or no evidence of conceptual understanding i.e. R – M, or response based on reasoning only with regard to conceptual knowledge and no evidence of ER-based reasoning i.e. R – C. (i.e. R-M or R-C, not both)
Incorrect	0	No response or incorrect response based on lack of, or incorrect conceptual knowledge and/or reasoning ability in relation to the ER

3. Results and Discussion

Results

In the preliminary investigation, students showed their poor visual literacy where they exhibited the evidence of visualization difficulties that affect their ability to interpret and learn from external representation (ER). Figure 2 presents the comprehension level before they get material training, most of the students are at the level of little or no evidence of conceptual understanding and No response or incorrect response based on lack of, or incorrect conceptual knowledge and/or reasoning ability in relation to the ER. After an explanation by using picture and picture (P&P) model, students were able to draw their understanding on the genetic transfer in bacteria.

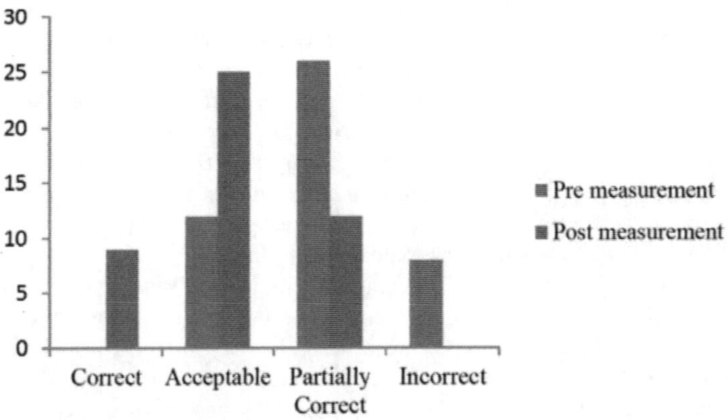

Figure 2. Students' Ability Level of visual literacy

From the figure 2, we can see that most students' visual literacy ability improves in the level of acceptable. It means that their ability is in the range of the average amount of conceptual (propositional) knowledge. In fact, some of the students comprehend in the correct level of ability, meaning that they have a high degree of conceptual (propositional) and visual knowledge used to provide a relevant response related to materials.

Discussion

In this study, we did indirect training on visual information interpretation and drawing. In fact, drawing is a powerful tool for thinking and communicating, regardless of the discipline. Also, drawing is a skilled process integrated into the practice of science, used in the generation of hypotheses, the design of experiments, the visualization and interpretation of data, and the communication of results [8]. Drawing has always helped artists to observe and reflect on their ideas; however, using drawing as a tool to help students to develop and document more complex understandings is not often used in science instruction. Images that have been made by a student was different from the other students, this issue has happened because the differences in the imagination, creativity, and ability to draw. Therefore, in one class, there was a variation of drawing results. However, their drawing explanation could reveal their understanding on the ER in this study.

The learning model by drawing is declared can improve long-term memory due to several reasons. One of them is citing the theory of "dual coding" initiated by [9]. In his theory, the memory consists of two separate theory systems. The first system is the Verbal System used to represent thinking with the language. The second system is a Non-Verbal System to represent and think of non-verbal information, such as images. When some information is received, one or both of these systems can be activated. An information form as sentences or words that entered will be stored in the memory system of verbal or non-verbal, or both. If memory can be encoded properly in both of the systems of verbal and non-verbal, the possibility that information stored in long-term memory will increase. In this study, the genetic transfer ER as visual literacy can provide information that cannot be given thoroughly only with words. When the students see the pictures, the students are more interested in learning, whereas the ability to draw an object can enhance students' understanding of the object and improve the memory of students as they "make" the object themselves.

Learning by a common lecture method that usually applied by teachers will indirectly impact the declining ability of scientific-creative thinking. Surveys of teachers and students indicate when students drew to explore understandings in science, they were more motivated to learn than conventional teaching and drawing activity seem like play can stimulate students' interests of to explore their scientific interests. To stimulate an interest in science is important if students are to be motivated to engage in scientific research in the long term [8] [10]. The use of drawing to individual learner differences is shaped by the learner's current or emerging ideas and knowledge of visual conventions [8].

Drawing can function as a communication tool even in non-interactive social contexts. For example, generating an explanation of a concept can have the effect of inducing deeper conceptual understanding in the explainer as well as the explainee [11]. By drawing, students can expand expression and imagination without limits. In this process, the child can develop ideas, channeling his emotions, foster interest in art and creativity. Students can show their skills, express their ideas, experiences, observations into a scratch line, shape, color, according to tools they use images that make students more creative [12]. Moreover, drawing can improve motoric ability, the premise is that drawing focuses on effective coordination of mind and muscle to culminate in the production of swift and meaningful patterns of movement that has to be learned [13].

In studying visual literacy, we need to understand what it meant. There are numerous definitions of visual literacy, each emphasizing various unique characteristics [14]. Research and studies on visual literacy began to be developed after Debes introduce visual literacy in 1969. Visual Literacy refers to a group of vision-competencies a human being that can develop by seeing and at the same time having and integrating other sensory experiences. The development of these competencies is fundamental to normal human learning. When it is developed, they enable a visually literate person to discriminate and interpret the visible actions, objects, symbols, natural or man-made, that he encounters in his environment. Through the creative use of these competencies, he can communicate with others. Through the appreciative use of these competencies, he can comprehend and enjoy the masterworks of visual communication [8].

In other words, visual literacy is the ability to communicate knowledge through imagery [7] [14]. Different types of mental image and visualization in science can serve different purposes. For example, realistic diagrams (such as an anatomical diagram) highlight the salient features of an object. Schematic diagrams (such as an electrical circuit diagram) illustrate relationships, assist in calculations, or provide descriptions of a phenomenon or process. Other visualizations in science education include photographs, simulations, astrophotography and scale drawings of equipment. Scientific visualizations can offer a means for imagining the unseen (such as the molecular, atomic and subatomic worlds) [15].

To develop visual literacy, students not only see the pictures provided in textbooks as it has been considered by the teacher but also they are guided to visualize concepts into images. Certainly making visualizations is integral to scientific thinking. Scientists do not use words only but rely on diagrams, graphs, videos, photographs, and other images to make discoveries, explain findings, and excite public interest [8]. Visual literacy skills can be learned in ways analogous to textual literacy. Through effective teaching, people can learn to develop the ability to recognize, interpret, and employ the distinct syntax and semantics of different visual forms. Researchers assert that the process of becoming visually literate continues through a lifetime of learning new and more sophisticated ways to produce, analyze, and use images [16].

Because science is so often visual and spatial in nature, drawing is one of the key activities in teaching and learning science, alongside practicum, modeling, role-play, and digital simulation [17]. The effectiveness of drawing and modeling to support rich learning we explain using the notion of affordance as a productive constraint. Drawings and models, because of particular visual spatial requirements, constrain and guide the learner into seeing phenomena in new ways [18].

Effective learning strategies will help learners overcome the limitations of the material presented, organize their knowledge more effectively, and integrate new and existing understanding; ultimately, they can be transformative by generating new inferences [8]. In this study, picture and picture (P&P) model is a strategy in the form of cooperative learning to deliver genetic transfer material through picture logical sequence. In P&P model, the lecturer asks the students to have pre-learning drawing phase. After that, the lecturer presents the learning material by showing the picture (ER) related to the material. Some students are asked to explain by their understanding and interpretation after the lecturers' explanation. After that, the lecturers ask the students to make the group where they have to re-draw the ER in a logical sequence based on their understanding and analyze the drawing results together.

Making visual images within the mind as one reads provide an effective framework for organizing, remembering, and constructing meaning from text [5] [8]. In the pre-learning drawing phase, students are introduced to the topic to be studied, and then the students were directed to share the knowledge they "think" they have about the learning topics. Before the lecturer explains the topic in expository text, lecturer leads the students to describe their imagination about the topic on a piece of paper through drawing mental images. During the pre and post-learning, students were asked to share their pictures with each other [5]. Their drawing in pre-learning can be different with their drawing in post-learning because students receive new knowledge during the learning process. Thus, the students' understanding can be seen through their drawing in post-learning.

The other way to introduce learning through drawing method is by directing the students to find the topic to be learned from their natural surroundings, books, or another image that can be given by the teacher, like the picture shown on the LCD, preserved animals or organs, artificial sectional, etc. Then the students were asked to draw the topic in detail because the topic drawn is a topic of their choosing, then students will pay more attention to detail of the material [19]. The difference from the strategy of "talking drawing" is students receive apperception and basic materials first and then draw according to their individual creativity, then the teacher guide students to present the results in front of the class to initiate discussions between students. During the discussion, students were asked to correct the image that they have made according to results of the discussion. The practice of this lesson requires a shorter time, and the teacher can control the discussion better. But overall, the two methods of the image has the same main steps, namely: 1) providing information about the material using a picture, (2) guiding students to visualize information by the understanding, imagination and skill of students, and (3) analyzing their drawing image. However, it does not mean drawing for learning is the only method that should be applied, but rather drawing is used to complement the writing and speaking activity [10].

Noteworthy by the teacher is when students share their knowledge through images, is important if the owner of the image is explaining the image he had made by himself rather than asking other students to interpret images owned by somebody else. When the teacher explains the students are drawing or asked the other students to interpret the work of other students, it could be the purpose and the meaning of the images can be extremely different, because every student has the distinction of imagination and their motor abilities [5].

4. Conclusion

Based on the analysis it can be concluded that the students' ability in comprehending material improves significantly in the level of the average amount of conceptual (propositional) knowledge. In fact, some of the students comprehend in the correct level of ability, meaning that they have a high degree of conceptual (propositional) and visual knowledge.

References

[1] Madigan M and Martinko J (editor) 2006 *Brock Biology of Microorganisms* (13th ed.). Pearson Education. p. 1096. ISBN 0-321-73551-X

[2] Fibriana F and Amalia A V 2016 Potensi *Kitchen Microbiology* Untuk Meningkatkan Keterampilan Teknik *Hands-On* Dalam Pembelajaran Mikrobiologi. *Unnes Science Education Journal 5(2)* p 1194-1201

[3] Kusnadi, Rustaman N Y, Redjeki S, Aryantha I N P 2012 Analisis Kemunculan Keterampilan Spesifik Lab Mikrobiologi Melalui Pembelajaran Mikrobiologi Berbasis Proyek Inkuiri "Mini-Riset" Mahasiswa Biologi. *Jurnal Pengajaran MIPA, 17 (1)* p 53-59

[4] Lohse G, Walker N, Biolsi K, and Rueter K H 1991 Classifying graphical information, Behav. Inform. Technol. 10, 419–436

[5] Paquette K R, Fello S E, and Jalongo M R 2007 The Talking Drawings Strategy: Using Primary Children's Illustrations and Oral Language to Improve Comprehension of Expository Text in *Early Childhood Education Journal* p 65-73

[6] Cimer A 2012 What makes biology learning difficult and effective: Students' views *Educational Research and Reviews* Vol. 7 no. 3. DOI: 10.5897/ERR11.205

[7] Bell J C 2014 Visual Literacy Skills of Students in College-Level Biology: Learning Outcomes following Digital or Hand-Drawing Activities *The Canadian Journal for the Scholarship of Teaching and Learning* vol 5(1)

[8] Ainsworth S, Prain V, and Tytler R 2011 Drawing to Learn in Science *Science 333*

[9] Paivio A 2001 Dual coding theory: Retrospect and current status *Canadian Journal of Psychology* vol 45
[10] Edwards L 2011 Drawing and doodling can help you learn science: study *Medical Xpress Nottingham*
[11] Fan J E 2015 Drawing to Learn: How Producing Graphical Representations Enhances Scientific Thinking *Translational Issues in Psychological Science* p 170-181
[12] Liu L 2007 The Relationships between Creativity, Drawing Ability, and Visual/Spatial Intelligence: A Study of Taiwan's Third-Grade Children *Asia Pacific Education Review* p 343-352
[13] Tachie-Menson A, Opoku-Asare N A, and Essel H B 2015 Teaching And Learning Of Drawing For Book Design And Illustration: A Study of Higher Education in Publishing *Global Journal of Human-Social Science: Arts & Humanities – Psychology*
[14] Brumberger E 2011Visual literacy and the digital native: An examination of the millennial learner *Journal of Visual Literacy* p 19-46
[15] Vavra K L, Janjic-Watrich V and Loerke K 2011 Visualization in Science Education *ASEJ* p 21-30.
[16] White C, Breslow L and Hastings D 2015 Exploring Visual Literacy as a Global Competency: An International Study of the Teaching and Learning of Communication *International Conference in Interactive Collaborative Learning (ICL) Proceeding*
[17] Tytler R 2016 Drawing to Learn in STEM *Science Conference* p.45-50.
[18] Tytler R, Prain V, Hubber P, and Waldrip B 2013 *Constructing Representations to Learn in Science*. Sense Publisher Roterdam, Australia
[19] Stein M, McNair S, and Butcher J 2011 Drawing on Student Understanding Using Illustrations to Invoke Deeper Thinking about Animals Science and Children

The Development of Teaching Aid in The Implementation of Natural Science in The Curriculum 2013 Junior School

Siti Patonah[1], Duwi Nuvitalia[1], Ernawati Saptaningrum[1], Khumaedi[2], and Ani Rusilowati[2]

[1]Universitas PGRI Semarang, Indonesia
[2]Universitas Negeri Semarang, Indonesia

Corresponding author: siti_blimbing9@yahoo.co.id

Abstract. The purpose of this study was to develop the teaching aid required in the curriculum implemented 2013 science subjects at the junior level.
The method used in this research is the Research and Development. In this article will be described in the description of the development of the teaching aids that have been generated. Teaching aids consist of three pieces, namely water filtration magical, electric convection box and Musschenbroek bimetal. Based on the expert validation, then all three devices have fulfilled the requirements as teaching aids. Where all three can be used to demonstrate the concept of separation of mixtures, air convection, and thermal expansion of the metal. All three are decent teaching aids used to learn science subjects at junior in implementing the curriculum in 2013.
Keywords: development, teaching aids, Natural Science, implementation, curriculum 2013

1. Introduction

Learning is a complex process that occurs in every person throughout his life **[1]**. One of the keys to success in the classroom is the availability of facilities and infrastructure to study one or several concepts. Moreover, to teach science subjects in junior high. According **[2]**, junior high school students still classified as children, where they are generally enjoying the activity, not the result of activities. Junior high school age children typically range from 12-15 years, during this period various problems must be overcome to develop commitment, so they have a good identity and mature **[3]**. Along with the increasing age of the child, it is a way of thinking of students will move from the concrete to the abstract **[4]** .

Natural science, is the science related to natural phenomena and systematic material systematically arranged, generally accepted that a collection of observations and experiments **[5]**. Natural Science also called the science, has three components that can not be separated, namely products, the scientific process, and scientific attitude. Science as a product has a meaning of science is an organisation of facts, concepts, procedures, and laws of nature. Science as science processes explain the findings obtained from the scientific or scholarly work. While science as a gesture has a meaning that the scientific attitude underlying scientific processes that are useful in producing science **[6]**. At the level of Junior School are expected to emphasize last learning Salingtemas (Science, environment, technology, and society) integrally geared towards learning experience to design and create a masterpiece through the application of the concept of science and competence of scientific work wisely **[7]**.

Natural Science does not only contain a collection of facts and concepts alone but Natural Science is also a mindset in dealing with natural phenomena around. Natural Science as a human attitude, is the application of process skills. Science process skills include: observing, measuring, draw conclusions, controlling variables, formulate hypotheses, make charts and data tables, create operational definition and conduct experiments. Natural Science as a process/method , Natural Science implementing stages in the implementation of the scientific method of learning Natural Science. Natural Science as a product, is merely a collection of the results of the empirical and analytical activities undertaken by scientists for centuries.

Meanwhile, the Natural Science should provide a meaningful learning experience for students, by studying the Natural Science, the students will get used to having the right attitude, skills properly apply the scientific method, which in turn will affect the student's personality, so it is very appropriate lessons to be taught taught in junior high. Natural Science is a field of science that studies the natural phenomena. So that the Natural Science is not just a collection of formulas but more than that is a way of knowing how nature works. So that is not enough just to teach science is taught in theory alone but also need to be taught how the theory was obtained. One unit of science as a process, product and attitude expected to be achieved through the implementation of the curriculum, 2013. For the existence teaching aids where teaching aids are equipped with supporting implementation. In this case, the research to develop Natural Science teaching aids that can be used to teach science in junior high.This research aims to develop the teaching aids required to implement the curriculum in 2013 in science subjects in secondary school.

2. Methods

The study design used in the form of Research and Development with Brogg patterns and Gall (2003) that has been adapted. To further the design shown in the following chart:

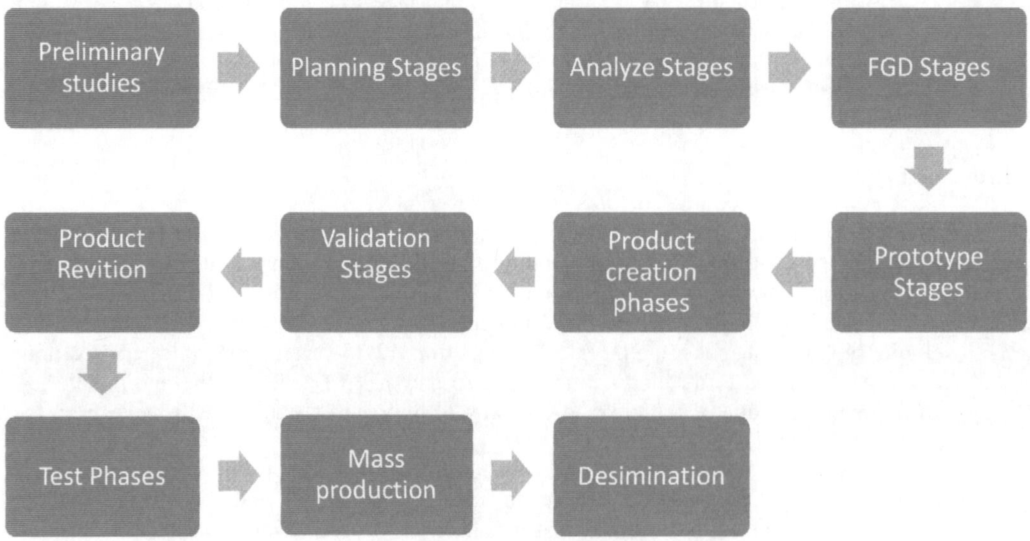

Figure 1. Design of Research

Analysis of the data used is the method mix analyze . That is using qualitative and quantitativeFormatting the title

3. Result and Discussion

Based on the analysis of the teaching aids needed to teach science had previously obtained material analysis and needs teaching aids in the table 1 [8]

Table 1. Requirement Viewer tool in the implementation of the natural science curriculum 2013 class VII junior school

No.	Subject Matter	CC	BC	Teaching aids
1	The object of science and observations	1	1.1	Caliper measurement tools, made of wood large size that can be affixed on the board / wall
		2	2.1	
		3	3.1	
		4	4.1	
2	Classification of objects,	1	1.1	No teaching aids
		2	2.1,2.2, 2.3,2.4	
		3	3.2,3.3	
		4	4.2,4.3	
3	Classification of Living Beings	1	1.1	No teaching aids
		2	2.1,2.2, 2.3,2.4	
		3	3.2,3.3	
		4	4.2,4.3	
4	Systems of Life Organization	1	1.1	Model of animal cells Model of plant cells
		2	2.1,2.2	
		3	3.4	
		4	4.5, 4.6	
5	Change objects around Us	1	1.1	Water purification, simple chromatography, simple microscope
		2	2.1, 2.2	
		3	3.5	
		4	4.6, 4.7	
6	Energy and Life System	1	1.1	No teaching aids
		2	2.1,2.2, 2.3	
		3	3.6	
		4	4.8, 4.9	
7	Temperature and Change	1	1.2	Thermistor heat
		2	2.2	
		3	3.7	
		4	4.10	
8	Heat and displacement,	1	1.1	Some teaching aids are developed are: heat transfer by conduction, heat transfer in a variety of materials, gas convection.
		2	2.1	
		3	3.7	
		4	4.10,4.11	
9	Interaction with the Environment Living Beings	1	1.1	A chart of interaction between living organisms in ecosystems and global warming scheme
		2	2.1, 2.2	
		3	3.8,3.9,3.10	
		4	4.12,4.13	

Based on the table then developed three teaching aids to teach the curriculum in 2013 at the junior level. All three teaching aids are water filtration magical, electric convection box, and Musschenbroek bimetal. All three of these tools have been validated experts and revision of the final product. Each device is shown in the following explanation:

a. Teaching aids filtering magic

Teaching aids are made of the main material in the form of transparent acrylic sheets and prison . This tool is the height of 1 meter. Box each level can be assembled, and students

IOP Conf. Series: Journal of Physics: Conf. Series **824** (2017) 012023 doi:10.1088/1742-6596/824/1/012023

can have a broad creative space to determine the best cleaning results by dismantling pair. In addition to a series of filtering tools are also provided water purification materials in the form of fibres , gravel, sand, zeolite. There are six boxes where the box above to enter the dirty water then the ultimate box is to accommodate the filtrate. The next four squares in the middle can be used to find the appropriate modifications to obtain the best filtrate. So in addition to cognitive of these tools can also be assessed in terms of affective and psychomotor. As a result of the validation of acquired expert advice to include indicators of clear water standards by using a pH meter (Figure 2).

Figure 2. Filtering Magic

b. Box convection electric

Known convection box because its shape is a box (Figure 3). The materials used to make this tool is made of acrylic and glass. The difference from previous convection box is usually used candles as a source of heat in the box. While on these tools candle bulb replaced with some three pieces. Electric convection box is the advice given by the validator expert that the new value is arising from pre-existing convection tubes. To use these teaching aids to learning, the variable heat source can be varied to obtain optimal results. The heat source (both lamp) connected to an electric current. As the materials used in these tools is smog bomb/mosquito repellent

Figure 3. Box Convention Electric

c. Musschenbroek bimetal

Musschenbroek there usually measure the expansion angle both horizontally and use three separate types of metals (Figure 4). While Musschenbroek bimetal teaching aids can be used to determine the expansion that can be viewed vertically. Validation of experts advises that the tools made as much as possible can have the durability and value of the other metal variants. So based on the results of the validation of the revisions of products as follows:

Injections are used should be made of metal, so it is not easy to melt when heated. The materials used to make these tools in the form of acrylic, metal and metal injection. Unlike the Musschenbroek tools in general, then the added metal teaching aids bimetal. Variables that can be used for experiments using this tool is the time variable, high elongation, and also the curvature of the heated metal.

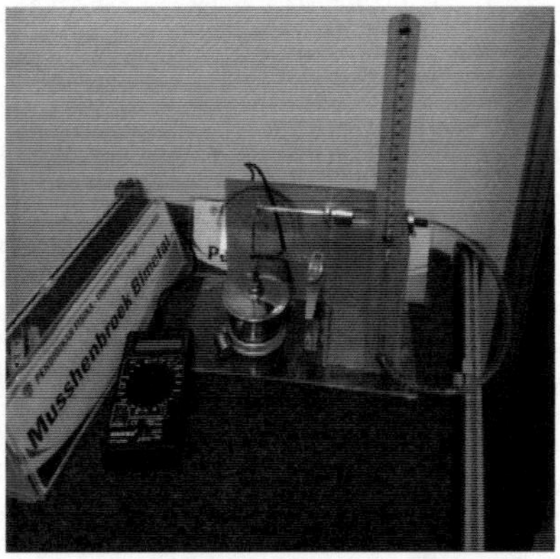

Figure 4. Musschenbroek bimetal

All three teaching aids are then tested in the class scale to determine the effectiveness of its use for subsequent mass produced and disseminated.

4. Conclusion

Based on the analysis needs of teaching aids, then developed the three teaching aids natural science in junior high. All three of these tools have been validated skilled and revised products. Each tool has a specification for use in science teaching.

5. Acknowledgments

The authors would like to Ministry of Higher Education that has funded this research well in skim Research PEKERTI. We would also like to thank from UNNES as research partners for their time and effort for their help with all aspects of this research study.

References

[1] A. Arsyad, Media Pembelajaran, Jakarta: Rajawali Press, 2013.
[2] O. Hamalik, Psikologi Belajar Mengajar, Bandung: Sinar Baru Algesindo, 2002.
[3] A. Dariyo, Dasar-dasar Pedagogi Modern, Jakarta: PT. Indeks, 2012.
[4] S. B. Djamarah, Psikologi Belajar, Jakarta: Rineka Cipta, 2002.
[5] U. Samotawa, Pembelajaran IPA di Sekolah Dasar, Jakarta: PT. Indeks, 2011.

[6] M. Education, Dokumen Kurikulum 2013, Jakarta: Ministry Education, 2013.

[7] M. Education, Buku Sumber untuk Dosen LPTK Pembelajaran IPA SMP di LPTK, Jakarta: Ministry Edukacion, 2014.

[8] D. N. E. S. K. A. R. Siti Patonah, "Concept Analysisof Content of Natural Sciences Subject to The Implementation Junior Class VII-Grade Curriculum 2013 (Pekerti Competition of DIKTI)," in *The 2nd International Conference of Science and Mathematics Education*, Semarang Indonesia, 2015.

IOP Conf. Series: Journal of Physics: Conf. Series **824** (2017) 012024 doi:10.1088/1742-6596/824/1/012024

Development of Ethnoscience Approach in The Module Theme Substance Additives to Improve the Cognitive Learning Outcome and Student's entrepreneurship

Sudarmin[1], R Febu[2], M Nuswowati[1], and W Sumarni[1]

[1]Department of Chemistry-Faculty of Mathematics and Natural Sciences, Universitas Negeri Semarang, Indonesia, Tel. (024) 70805795 Po Box 50229

*Corresponding author: sudarmin@mail.unnes.ac.id

Abstract. Ethnoscience approach is an interesting research today. The purpose of this research is to develop approaches ethnoscience and modules ethnoscience theme additives based ethnoscience; as well as assess the feasibility and effectiveness of module theme additives based ethnoscience to improve learning outcomes and the entrepreneurial character of students. This type of research is the Research and Development (R & D). In this research consist of four stages, namely define, design, development and implementation. The subjects of this study were students of the School of MTs Maarif NU Brebes. Data were analyzed by descriptive qualitative and quantitative. The results showed that ethnoscience approach and the module theme substance additives used declared worthy of National Education Standards Agency (BNSP) with an average percentage of validation on the feasibility aspect of the content, language feasibility, and feasibility of presenting respectively for 94.3%, 86 % and 92% and a very decent entry criteria. The effect of the application modules substance additive based ethnoscience can improve on the cognitive learning classical amounted to 90.63%, and increased learning outcomes category was based on the scores of N-gain. Influence ethnoscience approach application and module theme substances additives based ethnoscience able to improve the entrepreneurial character of students. Based on the results of this study concluded that the ethnoscience approach and module theme substance additives based ethnoscience effective to improve learning outcomes and students' entrepreneurship.

1. Introduction

Indonesia is currently implementing the curriculum in 2013 for primary and secondary school levels, thus for science subjects in secondary schools should refer to the curriculum [1]. On learning of science in junior high school or junior refers to the curriculum in 2013, it must be developed as an integrative science subjects and not as an educational discipline that has been happening. The content of the curriculum in 2013 on the subjects of science can support and engage the culture and local wisdom, it means that science teachers should be responsive to the development of culture and local wisdom, technology and art that are around it to build curiosity and the ability of learners to make the right in their science lessons. Learning science which utilizes culture and local wisdom called Ethnoscience Approach [2]. Learning with Ethnoscience approach is learning with the linking of local culture, indigenous science and scientific knowledge has been developed in Brazil [3] and Canada [4], and Tanzania [5]. In fact science learning that takes place in school when less attention to local culture that developed in the community, due to the limitations of teachers in linking concepts, processes and contexts of science and scientific knowledge [6].

Suastra [7] states that, most (90%) the purpose of learning science at school is directed on achievement of scientific knowledge in the form of concepts, facts, theories, principles, laws, and rules as the product of science, and the rest is aimed at developing the skills and scientific attitude . In the science curriculum in 2013 at this time, a scientific attitude developed referring to entrepreneurship student's.

At this time , Indonesia has a diverse culture and has not been widely used as a source of science learning [8]. Indigenous knowledge from society as the culture and local wisdom in order to be maintained, it was necessary to the conservation of the various noble values and local wisdom through science education based on local culture or ethnoscience, example culture farming [9], as well as African countries [10] and Brazil [11]. Thus the science teacher as a subsystem of education is responsible for the preservation and conservation of the noble values and incorporate them to the learners. A project-based learning model and Chemo-Entrepreneurship integrative Ethnoscience Approach been done by Rohayati, *et al* [12]. Pursuant to the study of research and theory, the approach can be applied to learning with Ethnoscience approach for theme substance additive in science curriculum 2013. In learning themes such additives, science teachers are expected to implement innovative approaches through Ethnoscience by focusing on the learning resources of local foodstuffs as life skills certain areas in Indonesia to improve entrepreneurship students. In this research, students were given the task of looking for a project to the public information indigenous science of the substance additive from natural ingredients that are often used as colourants, preservatives, and sweeteners in food and drink locally, then reconstruction of indigenous science become to the substance additive into a science knowledge. Thus this innovative learning will be more effective than learning by lecture, discussion, and frequently asked questions.

In the era of global and scientific literacy today, then students in Indonesia does not quite mastered the knowledge of science or theory, but also to scientific literacy that is capable of applying in the social life of society in solving a phenomenon of nature and everyday life [13]. Such education is directed to the formation of the entrepreneurial spirit that is education that has the courage and willingness to face life's problems in a reasonable and creative search for solutions, it should be developed to reduce unemployment [14]. According to Sakali [15] with their entrepreneurial spirit and creative thinking will make a solution to reduce the unemployment rate. So as to reduce the number of unemployed is need for a solution that is learning the concept of entrepreneurship among students. In this research, the development of entrepreneurial spirit in students through learning additives food ingredients with ethnoscience approach. The importance of this research is done, because the results of observations by Khaerunisa, R.F [16] in Madrasah Tsanawiyah (MTs) Brebes and Rohayati [12] in High School in Magelang demonstrate his learning has not developed the entrepreneurial character, linking scientific knowledge with the local culture, as well as learning science has not contextual. Though Brebes has a kind of local wisdom, such as salted egg, food rice yellow (*yellow rice*), drinks wine, ***cendhol [javaese]***, and well-known plant scallions, palm sugar, *turmeric*, and suji as a substance natural additives, while famous traditional in Magelang is foods ***gedhuk*** [*Javanese*] with using natural pigments so interesting food.

In the teaching and learning process in the classroom science teachers also rarely apply their science lessons integrated with culture and entrepreneurship character [12]. The results of interviews with teachers in the School of Science states that the limitations of the media or integrated science and learning resources and learning tools. Teaching books and learning source used are books and Student Worksheet is not all based science integrated. In some of the books held by teachers and students of both the book and the Student Worksheet for topics additives have not addressed the linkages theme additive dyes, preservatives, sweeteners with a variety of additives naturally found around the community Brebes, as well as indigenous knowledge society about substance natural additives and food and local beverages can be reconstructed or transformed into scientific knowledge and can develop the entrepreneurial spirit of students. The model of applied learning is project-based learning. Project-based learning approach is a scientific approach suggested in education in Indonesia today.

Ethnoscience is indigenous knowledge in the form of the language, customs and culture, morals; as well as the technology created by the community or a particular person that contain scientific knowledge [**17**]. Indigenous knowledge is one study in ethnoscience, which knowledge from indigenous in the community can test the truth through literature review, scientific explanation, work and scientific processes; so it can be used as a source of innovative and applied learning in science learning in the classroom [18]. On the theme of substance additive as part of the topic of science in junior secondary school. The teaching and learning can apply Ethnoscience approach, as the use of natural additives in the variety of traditional food and beverage, which traditional foods are one part of local wisdom and culture in Indonesia. Teaching and learning with ethnoscience approach, students are trained skilled to made and the use of additives have naturally evolved in Indonesian society and as a culture, meaning that the ingredients and use of the additive as a ways and habits that developed and shared by a group of people and passed down from generation to generation [19].

On learning in the classroom to provide students entrepreneurship, Susiana, N [14] to improve the entrepreneurial spirit of students using project-based learning approach to making local food products, and then sell them to friends and the community. Lee *et al*. (20) said that entrepreneurship education in each country is different because of the unique culture in each country. Teaching materials that integrate entrepreneurship education course contain more material to be taught, and thus require more instructional hours to be able to implement it. Therefore, the setup module can be an alternative teaching materials appropriate to integrate entrepreneurship education, because the modules students can learn independently so that more efficient allocation of instructional hours. Thus the modules developed in this research as a form of teaching materials in it have been equipped with instructions for self-learning so that students can perform the learning activities theme additives without the presence of a teacher directly. In this study, the approach is a strategy creation ethnoscience learning environment and planning learning experiences that integrate culture as part of the learning process of science [21]. The application of science teaching and learning with ethnoscience approach, the necessary ability of teachers to combine indigenous knowledge with scientific knowledge. However, not many science teachers who have this ability. Therefore, the main purpose of this study was to determine the feasibility of a theme additives module integrated ethnoscience, determine the effectiveness of modules developed to increase students' knowledge, and entrepreneurial students.

2. Methods

The method used is research and development (R&D). While that was developed in this work is a science module ethnoscience charged, along with learning materials, learning models and research instruments. The study population was all students in grade VIII (VIIIA and VIIB) Maarif NU 1 Brebes. The sample in this research is ten students to test readability and enforceability of learning modules with ethnoscience approach on the theme of food additives. While the experimental class is a class VIII A and Class B as the control class. The research design was pretest-posttest control group design. The independent variables in this study is a model of teaching and learning modules ethnoscience approach developed. Class experiments using project-based learning approach Ethnoscience, while the control class using learning lectures, discussions, and question and answer. The dependent variable is the result of learning and entrepreneurial spirit of students, while the control variables are learning materials, teacher, curriculum, 2013, and the number of lessons.

In this research, the stages of research and development in this study consisted of: (1) analysis of needs, (2) analysis of learning, (3) analysis of students and context, (4) design module, (5) design validation module, (6) design revisions (7) testing the readability module, (8) the revised module, (9) test field implementation. The technique of data collection through the test and non-test. The method collecting data in this study consists of (1) interview, (2) the method questionnaire (questionnaire), (3) the test method, and (4) documentation. Data analysis methods include the analysis of test instrument mastery of concepts through the test of validity, reliability, the level of difficulty and distinguishing features. The feasibility analysis modules include modules feasibility content, language, and display the indicator according to the National Education Standards Agency (BSNP); as well as an analysis of the responses of students and teachers. In this study to measure entrepreneurship students through observation, questionnaire, and an assessment of the product life skills of local knowledge with additional foodstuff additives.

In this study, as the data for entrepreneurial spirit obtained through observation during the learning process in the classroom and laboratory. Aspects of the entrepreneurial character assessment include teamwork, discipline, responsibility, communicative, confident, resilient, creative, and innovative. The data obtained were analyzed descriptive Qualitative and quantitative. The research data obtained from this study analyzed the results of the module development, the effectiveness of the module in its application teaching ethnoscience approach, learning outcome, analysis of entrepreneurial character, and classical completeness.

3. Result and Discussion

This research will develop science modules with theme-based additives is charged or ethnoscience. Module compiled based problem analysis, requirements analysis, then validated against the validation of content, language and display module substance additives based ethnoscience. Retrieving data using questionnaires legibility module, student and teacher responses. While the learning outcomes data using test instruments that measure learning outcomes aspects of knowledge, while the data retrieval code entrepreneurship students use observational data, questionnaires, and assessment of products. In the following description presented results of the research.

3.1. The Development of module theme substance additive based ethnoscience

3.1.1. Feasibility of Module theme substance additive based ethnoscience

Feasibility science module and ethnoscience theme additive substance is derived from an assessment instrument according to National Education Standards Agency (BSNP) were assessed by an educational expert and science teacher at Junior High School. Feasibility than science module theme Additive substances based ethnoscience and validated by expert content, language and presentation. Module theme substances additive based ethnoscience is feasible to use in learning, if when obtained percentage score of 62.50% <skor≤81,25%, while said to be very feasible if obtained percentage score of 81.25 <skor≤100% based on criteria BSNP [22]. The results of expert assessment and validation of the first and second are presented in Table 1.

Table 1. Results of validation by expert on the content, language and presentation of the module

No.	Validator [Assement]	Validation phase 1		Validation phase 2	
		Average (%)	Criteria	Average (%)	Citeria
1	Validator Contents I	100%	very feasible	97.1%	very feasible
2	Validator Contents I I	100%	very feasible	92.6%	very feasible
3	Validator Language I	100%	very feasible	82%	very feasible
4	Validator Language II	100%	very feasible	90%	very feasible
5	Validator Presentation I	100%	very feasible	95%	very feasible
6.	Validator Presentation II	100%	very feasible	89.3%	very feasible

The results show that the assessment by the National Education Standards module theme additives science and based ethnoscience very decent used in science learning in junior high schools because it has met the eligibility criteria BSNP. This can be seen in the results of the expert assessment of the eligibility criteria module that shows excellent results. As for how to integrate the content of substance additives into foodstuffs with based ethnoscience related indigenous science in substances additive naturally follow the patterns Michie [23] which is associated reconstruction sciences scientific of indigenous science of society Aburigin and through a participatory action research cycle, teachers contextualised and conceptualized the Cultures Cross-Curriculum Priority (CCP) in terms of social justice, pedagogy, and student engagement [24]. Related research has also been conducted ethnoscience approach Chowdhury [25] who implementation of The Module with approach-TechnologySociety Science-Environment (STSE) and

Socio-Scientific-Issues (SSI) in science education, and critically examines their contributions to science education and science teaching, and how they are underpinning values and ethics in society. The research Chowdhury [25] shows that the ethnoscience (socio-scientific) approach can improve values in science education; foster values and ethics in students' minds, and benefit the societies.

In the research module development through stages define, design, and implementation. Module theme additives based ethnoscience after design, then do validation with two stages of assessment or validation by experts. In the first phase of this assessment are suggestions of experts to revise the map section for more simple concept, the purpose of learning to be more detail and each sub-subject matter contained evaluation. After the expert advice given, the researchers could then proceed to revise the feasibility assessment module in the judging of the second stage. Assessment and validation modules in the second stage consists of three components, namely component content, language components, presentation components. Assessment module second stage, each component of the indicator evaluation by the two validators that of lecturers and school teachers. Each component consists of several aspects that are contained within the grains and the assessment indicators of the module. The results of the feasibility validation for component modules by each validator is described as follows

3.1.1.1. The components of the feasibility of contents

Assessment of the feasibility component content of the module on the second stage after the module passed the first stage assessment. Assessing the feasibility of the module contents are intended to assess the six aspects, which include the scope and breadth of study material, the accuracy of the study materials, novelty, stimulate curiosity, contextual and entrepreneurship. The sixth aspect contains seventeen grains assessment that has been validated by experts. The average score for overall feasibility components contents reached 3.79 with an average percentage of 94.85% and included into the criteria very worthy according to National Education Standards Agency or BNSP. These results were obtained from the validation by two experts with the material contents and feasibility assessment instruments both expert assessment results are averaged. In assessing the feasibility of the contents of the first validator giving an average score of 3.88 (97.1%) and the second validator giving an average score of 3.7 (92.6%). Thus concluded that the contents of the feasibility component module additive theme-based substances etnosains said to be very good and feasible for use in science learning in class, as determined in accordance with the provisions of the National Education Standards Agency (BNSP).

Assessment and validation of an experts show different scores, it is because each expert giving a different score on each nut. In the first expert and experts both assess that material in the modules thema substances aditive and based ethnoscience was sufficient and there was nothing to be fixed just first expert gives advice to fix the cover to make it more interesting, create a concept map that is simpler and gives examples natural additives based on local wisdom, such as leaves ***suji***, curcumin, or others. In the module, it appears that the goal of learning to be more detailed and include four aspects of Audience, Behavioral, Condition, and Degree, as demanded by the curriculum 20013, while for each sub subject that in the given evaluation questions and give feedback on each test question. At the material, researchers must add the charge ethnoscience adapted to the conditions that exist around the students. In assessing the feasibility of the contents of this module, then there is a difference in absorbing the contents of the module, it is in line with research conducted by Libman [26], implying that the lessons learned should be connected to real-life situations in which students are likely to use it. Advice from experts such material was then followed up for revision until the material experts claim that the module that was developed declared fit for use

Gupta [27] has developed science module and how the integration of a indeginous knowledge into a curriculum, modules, teaching materials, and school learning. On the development of the module, as teaching materials include public knowledge relating to agriculture, how to maintain health, ethno-toxicants, cutting, pruning, tissue culture and cloning, seed treatment in the nursery, symbiosis and Nitrification, fishing-cum-paddy cultivation, fuel collection, classification of soil types, classification of ground cultivation, cultivation on the fertile alluvial soil on the river islands, bush fallow cultivation, seasonal cultivation, cultivation yearly, annual and biennial cultivation, flowering in the first year of a biennial crop variety, mixed cultivation, identification of certain plant species on trees, or in soil causing severe harm to the crop production, weeds used in removal of pests and other harmful organisms, earth-warm in maintenance of soil fertility. In this research, only limited indigenous knowledge about natural additives, food and traditional drinks as the local wisdom that made the field of study.

3.1.1.2. The components of the language feasibility

The linguistic component consists of seven aspects of conformity with the development of students' cognitive, communicative, interactive, straightforward, logical, the use of terms and symbols and emblems in science, as well as compliance with the rules of Indonesian correct. The seventh aspect refers to the provisions of the National Education Standards (BSNP) and contains fifteen items assessment has received a positive response from the experts. The average score for the overall feasibility of the language component reached 3.43 and the average percentage of 86% that go into very decent criteria according to National Education Standards. These results were obtained from the validation by two expert content with instruments feasibility language component. The next two expert ratings averaged wherein the first validator giving an average score of 3.27 (82%) and the second validator giving an average score of 3.6 (90%). Thus the reference to the results of judging of the language in this module can be concluded that the components of the feasibility of the language in the theme substance additives and based ethnoscience to be very good and deserves to be used without revision, by the provisions set of BSNP.

The assessment of the experts showed different scores, it is because each validator experts give different scores on each nut. Based on the assessment of language experts, it is known that the value obtained in every aspect ranging from 3 to 4. Not all the aspects that received a score of 3 fixed. Improvements were made based on suggestions and feedback given by language experts. Neither validator expert I and expert II assesses that the language in the modules either only expert I provide suggestions for improvement in the use of punctuation, writing sentences on the matter, the writing module should pay attention to the spelling of the enhanced *Ejaan Bahasa Indonesia Yang Disempurnakan* or EYD, giving the quote image which can show the value of current that the source / reference and giving sentences that refer to each table to make it more easily understood. This is according to research Wagiran [28] which states that the accuracy of the word is a word to evoke the same idea in the reader's imagination, such as thinking and feeling author. Based on the advice of the researchers make improvements on the part suggested by experts to modules developed are feasible for use

3.1..3. The components of presentation feasibility

Presentation component consists of three aspects such as presentation techniques, supporting the presentation of the material, and presentation of learning. These three aspects are contained fourteen items of ratings that have received positive responses from the experts. The average score for the overall feasibility of the language component reached 3.70 and the average percentage at 92% which is very feasible criteria according to National Education Standards. These results were obtained from two expert validation on the material instruments, where the validator I gave an average score at 3.80 (95%) and validator II gave an average score at 3.60 (89.3%). It can be concluded that the components of the presentation feasibility in the science module is said to be very good and deserves to be used without revision.

Based on the advice the module improvements on the part suggested by experts were done to reach the feasible criteria for use. Eligibility is determined by a module other than expert validator is also determined from the responses of students and teachers to use module theme Substance Additive based ethnoscience. Student responses obtained to trials legibility and scale implementation of the field. Data were taken together while testing the readability of the module and also the application of questionnaire responses of students to the module. Results of student feedback on readability test modules and implementation of the field is presented in Table 2.

Table 2. Students recapitulation student response to validation modules and implementation in the field.

No	Trials	Percentage Result (%)	Information (criteria)
1	Legibility	98.25%	Very good
2	Field implementation	96.7%	Very good

The test trial was conducted on module legibility in VIIIB class with ten students. The module used was the module that has been corrected based on the feedback given by experts. Data taken while testing the readability is the questionnaire responses of students to the module. Legibility trial aims to determine the readiness of the module before use in the classroom application. Based on the results of trials on legibility, the response shows the average percentage at 98.25% and the students responded positively to the modules that have been developed. Based on the results of these responses, it is known that the students responded positively to the modules that have been developed. Students expressed that module the substance additive and based Ethnoscience is interesting, using understandable language; simple vocabulary election, drawing support materials, presentation of the material presented in a systematic and consistent, and the material in the module add the new insight.

The result of test implementation on students, their response was obtained at an average percentage (96.7%), it is at very good criteria. The percentage of student responses is also supported by the comments of students who were asked at the end of learning. According to the students, they mentioned that the teaching material in the form of this module is new and exciting and can add a reference to learn because the module is equipped with interesting pictures and content of the material that is easily understood. Moreover, the student and teacher's feedback on the questionnaire of the module was also assessed by teachers. The results of responses are shown in Table 3.

Table 3. The responses of teacher in MTs Brebes based on the field test implementation

No	Results of Response in Science Teacher	Percentage (%)	Criteria
1.	Science Teacher 1	91.6%	Very good
2.	Science Teacher 2	98.8%	
	Average Percentage	95.2%	

Based on the results of the questionnaire on modules theme substance additive based ethnoscience, the teacher responses were assessed. This is evident from the results of the poll by the acquisition of the teacher I who obtained an average percentage at 91.6% while the feedback questionnaire filled out by the teacher II obtained an average percentage at 98.8%. As a whole the average percentage was at 95.2% with the criteria very good. The result of teacher responses shows that the module can be broadly accepted to facilitate teachers in delivering learning materials. Teachers at MTs Maarif NU 1 Brebes felt happy and helped by the presence of the integrated science module. Teachers found it because the module enables teachers to deliver the particular theme. The ethnoscience-integrated module developed is new to both teachers and students where science learning is associated with local culture. Ethnoscience learning digs up the original views of students about the culture or habits of people in the use of natural additives; then, it is translated into scientific knowledge. This is consistent with research by Rist et al. [29] which helps to improve the assumption ethnoscience received by the community of local indigenous knowledge that proved to be true.

3.2. The development of learning approach based Ethnoscience

In this research, the development of learning approach based ethnoscience to improve entrepreneruship students using the module substance additives based ethnoscience. In this research, ethnoscience approach using project-based learning (PjBl). In The Project based learning based ethnoscience, the researchers gave students assignments MTs NU Maarif Brebes as research subjects for planning, implementation, and evaluation of projects relating to (a) interview and observation community about indigenous science related additives which are used in the manufacture of life skills traditional., (b) plan, implementation of manufacturing of the product life skills, as well as the end product evaluation life skills activities produced by students. In this research, product life skills, local knowledge generated is making *salted eggs* with traditional salt, food of yellow rice cone with natural dyes turmeric, and traditional beverages ***cendhol*** with natural dyes, sweeteners, and natural preservatives. In this reserach, students who produce local wisdom life skills assessment of creativity and the entrepreneurial character of students. Ethnoscience approach in this research is characterized by project-based learning integrated ethnoscience, as it has been developed Sumarni, W [30].

3.3. Analysis of the effectiveness from the module additives based ethnoscience

The effectiveness of module subtance additive based ethnoscience was obtained from the pretest, post-test values, results of classical completeness and also the observation sheet of student's entrepreneurship interest while learning. The module suntance additive bsed ethnoscience is declared to be effective as a supplement for teaching books to improve learning outcomes and improve entrepreneurial students. The Module subtance additive based ethnoscience developed is able to improve student learning outcomes, which proved on analysis of the calculation of student learning outcomes as shown in Table 4.

Table 4. The field Implementation test results in class VIIIA students

No	Information	*Pretest*	*Post test*	Score N-gain	Criteria
1	The number of students	32	32		
2	Average value	51.6	80.38	**0.6**	moderate
3	The lowest value	36	68		
4	The highest value	72	92		

The effectiveness of the module subtance additive based ethnoscience developed can be seen from the results of student learning and entrepreneurial students. The developed module subtance additive based ethnoscience is expected to be one of the teacher materials that serve to improve learning outcomes aspects of entrepreneur knowledge and improve interest in the learning process, especially for teaching in science.

3.3.1. Analysis of the application modules to the cognitive learning outcomes

Effectiveness module theme substance additive based ethnoscience validated based on the study of the experimental class is class VIII A through pretest and posttest. Before the modules that have been developed are implemented, students are given questions about additives, the results were analyzed as the value of pre test, and after the module is implemented, students return given the same questions but were randomly assigned a number and its double selection and the result is used as the value of posttest. Rate procurement concept with this method, called the pretest and posttest design assessment as conducted by Arifianawati, *et al* [31], which do analyzes results of studying chemistry with pre test posttest design as well. The results of pre-test and post test has been obtained, then further analyzed normality test, homogeneity test, t-test and N-Gain. This is done in order to really known experimental class learning outcome. Analysis of the study results of pre-test and post-test can be concluded that the average post-test results better than the average pre-test, it means learning model approach and modules developed ethnoscience effective. In this study, analysis of the N-Gain is used to determine the increase of student learning outcomes. From the analysis of N-gain seen that students who get high N-Gain as much as 8 students, N-Gain were as many as 24 students and no N-Gain obtain low.

While in Table 4 it can be seen that for an average increase in learning outcomes experimental class of the pre-test and post-test is obtained N-Gain of 0.6, namely the criteria of being. Student learning outcomes in addition to the value obtained from the pre-test and post-test were also obtained from the value of individual tasks and the value of the task group. The duty value averaged then combined with the value of posttest in order to obtain the final value. In the study, students can pass the study said if the final value of each student reaches a value of at least 75, or if converted to 3.00. Classical completeness of students can be seen in Table 5.

Table 5. Classical completeness of learning outcomes

No.	Source Data	Result test	Conversion	Classical Completeness
1.	Average value of assignment	90.56	3.62	
2.	The mean value of pretest	79.75	3.19	
3.	Average value of post est	83.85	3.33	90.63%
3.	Number of students completed	29		
4.	The number of students not completed	3		

In this research, thoroughness score set by the school is 75 or if converted to 3.00. In Table 5 it can be seen that the number of students who pass the study for the implementation of the class is 29 out of 32 students in order to obtain the percentage of classical completeness of the application class VIII A of 90.63%. The large percentage of more than 85% so that the module can achieve mastery of classical science. This is consistent with research Rosyidah [32], which states that by developing modules theme substance additives based ethnoscience can improve 93.75% of students achieving mastery learning.

3.4. The implementation Ethnoscience approach towards entrepreneurship students

The final assessment of students in this research not only on knowledge, but the assessment also emphasizes the aspects of students' entrepreneurial attitude. Competency assessment entrepreneurial attitudes of students performed at the students when making food manufacturing project life skills and sell food and traditional snacks made them. Rate the realm of attitudes interest in entrepreneurship students can be seen in Figure 1.

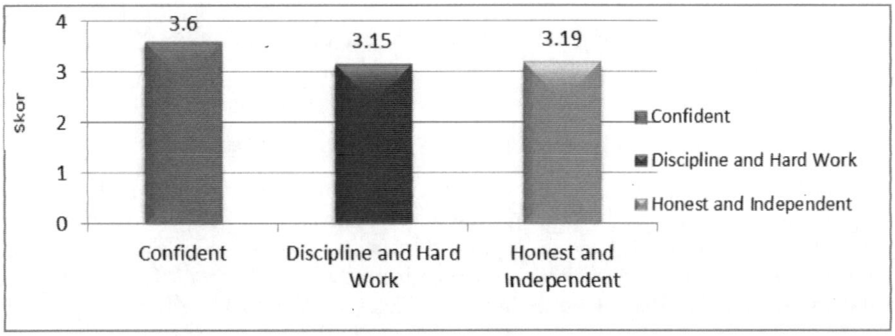

Figure 1. Assessment of entrepreneurship student's after learning approach ethnoscience

The effectiveness of science module ethnoscience theme-based additives in addition to the learning outcomes assessed mastery of concepts, but also the attitude and the entrepreneurial character of students. Rate realm attitude assessed is entrepreneurship students do when students sell food or traditional snacks made in those assessed by the researchers, science teachers MTs, and buyers. In this study, there are several aspects of the entrepreneurial character who rated namely self-confidence, discipline, hard work, honest and independent. After every aspect rated, then analyzed the average score of every aspect of entrepreneurship. The data obtained were analyzed and calculated the percentage to determine how the entrepreneurial character of students while selling traditional homemade food and snacks. The meals are made and sold a traditional food with the addition of additives, namely salted with salt preservatives, **drinks cendol** with natural dyes from ***Suji,*** and the manufacture of additives yellow rice with natural dyes from turmeric.

In this study, data showed that the confidence aspect average score obtained was 2.70 with a maximum score of three, and if converted to an average score of 2.70 was 3.60 with a value attitude. Aspects can confidently say that the confidence of students classified in the criteria very well. The students already have high confidence. The confidence of the students indicated boldly speaking in public, but in this case, there are still some students who offended in selling at the time on a question and given feedback by the buyer. When student practice of selling food and beverages, seen students feel that the buyer insulted; so impressed the students angry and less friendly when selling, nevertheless found some students are shy in selling; so that buyers buy food or snacks that students are selling because of pity. This is because the entrepreneurial character is learning and new experiences for students, so it still looks there are some students who have not yet formed an entrepreneurial attitude that is the lack of confidence in the students to sell angry and less friendly when selling, nevertheless found some students are shy in selling; so that buyers buy food or snacks that students are selling because of pity.

In the aspect of discipline and hard work the average score obtained was 2.36 with a maximum score of 3, which when converted an average score of 3.15 with a value of 2.36 is the attitude of B + so that it can be said discipline and hard work of students classified in the criteria good. Basically the students already have the discipline and hard work is good, it can be seen as they see them excited and tried to sell the food/snacks made in them by using speech polite and behaviour was polite and did not force the purchaser to buy food / snacks artificial they. However, there are still some students who still use the language when selling Brebes so impressed absence upload posting between seller and buyer. It makes the parents who will buy the feeling that students are less friendly and less polite in selling. Results of assessment of all aspects of entrepreneurial interest on average earn a score of 2.48 with the highest average score of 3, so when converted an average score of 2.48 was 3.31 with a value of B + attitude. 3.31 if the conversion value in percentage classical be 82.81, in which case it can be said that interest in entrepreneurship students included in the classification began to grow. This is consistent with the results of research Yenny [33] which states that the development of teaching materials or modules can increase entrepreneurship student's. At lately has been investigated relate to the relationship between people's knowledge of food conservation activities, which through learning integrated food ethnoscience able to foster environmental conservation and the conservation of traditional food or food local wisdom [34].

4. Conclusion

Based on the results of research and discussion, it was concluded The results showed that ethnoscience approach and the module theme substance additives used declared worthy of National Education Standards Agency (BNSP) with an average percentage of validation on the feasibility aspect of the content, language feasibility, and feasibility of presenting respectively for 94.3%, 86 % and 92% and a very decent entry criteria. The effect of the application modules substance additive based ethnoscience can improve on the cognitive learning classical amounted to 90.63%, and increased learning outcomes category was based on the scores of N-gain. Influence ethnoscience approach application and module theme substances additives based ethnoscience able to improve the entrepreneurial character of students. Based on the results of this study concluded that the ethnoscience approach and module theme substance additives based ethnoscience effective to improve learning outcomes and the entrepreneurial character of students.

Acknowledgements

On this occasion I would like to thank the team of researchers "ethnoscience" from the State University of Semarang. While this article is part of the main research with main theme "**Ethnoscience**", therefore on this occasion we thank you to (1) The Ministry of National Education that has funded this study, (2), Semarang State University who has helped fund research grant colleges (3) All science students undergraduate and graduate involved in this research,

References

[1] Kemendikbud. (2014). Materi Pelatihan Implementasi Kurikulum 2013 di tahun 2014 (*Material Implementation Training Curriculum 2013 in 2014 years*) **Accessed on Janury 2016**: https:// matematohir.files.wordpress.com/2013/07/materi-pelatihan-implementasi-kurikulum-2013-tahun-2014.pdf .

[2] Sudarmin.(2014). Model Pembelajaran Sains Berbasis Etnosains (MPSBE) untuk Menanamkan Nilai Karakter Konservasi dan Literasi Sains [*Model of Learning Science-Based Ethnoscience (MlSBE) for Embedding Character Values Conservation and Science Literacy*]. *Researh report*. Semarang State University.

[3] Battiste, M. (2002). Indegenous Knowledge and Pedagogyin First Nation Education A Literature Review With Recomenation. **Accessed on October 2016.** http://www.afn. ca/uploads/files/education/24._2002_oct_marie_battiste_indigenousknowledgeandpedagogy_lit_review_for_min_working_group.pdf.

[4] Ward, J. (2010). Grounding Curriculum and Pedagogies in Indigenous Knowledge and Indigenous Knowlegde Systems. **Accessed on October 2016:** *ejournal.narotama.ac.id*

[5] Ruheza, S, Mattee, Z.A; Chingonikaya, E.E, and, Zuena Kilugwe, (2013). Integration of The Indegenous Knowledge System (IKS) for Sustainnable Management and Use of Biodiversity in South Nguru Mountain Forest, Tanzania: The Influence of Socio-Economic and Politiccal Factors. *Journal of Sustainable Development in Africa, Volume 15, No.8: 94-114.* **Accessed on October 2016**: https://www.google.co.id/?gws_rd=cr,ssl&ei=bFKGWNGBM4SmvQSajI 4AQ#q= Ruheza%2C+S%3B+2 Mattee%2C+Z.A%3B+3Chingonikaya%2C+E.E+and+4Zuena+Kilugwe

[6] Fibonacci, A, and Sudarmin. (2015). Development Fun-Chem Learning Materials Integrated Socio-Science Issues to Increase Students Scientific Literacy. International Journal of Science and Research (IJSR), Volume 3 Issue 11, November 2014 , ISSN (Online): 2319-7064. **Accessed on August 2016**: http://www.ijsr.net/archive/ v3i11/ .pdf or www.ijsr.net

[7]. Suastra, IW, Tika, K. (2011). Efektivitas Model Pembelajaran Sains Berbasis Budaya Lokal untuk Mengembangkan Kompetensi Dasar Sains dan Nilai Kearifan Lokal di SMP [*Effectiveness Model-Based Science Learning Local Culture to Develop Basic Competence of Science and Local Wisdom Values in Junior High School*]. *Journal of Education Research and Development*, vol 5(3): 258-273. **Accessed on October 2016**. https://www.google.co.id/?g = Efektivitas+ Model+Pembelajaran+Sains+Berbasis+Budaya

[8] Sudarmin. (2014). Model Pembelajaran Kimia Berbasis Etnosains [MPKBE] untuk Mengembangkan Keterampilan Generik, Literasi Sains, dan Sikap Ilmiah [*Model of Learning Chemistry Based Etnosains [MLCB] for Developing Generic Skills, Literacy Science, and Scientific Attitude*]. ***Inaguration of Professor*, 23 Oktober 2014**. UNNES Semarang.

[9] Djulia, E. (2005). Peran Budaya Lokal Dalam Pembentukan Sains: Studi Naturalistik Pembentukan Sains Siswa Kelompok Budaya Sunda Tentang Fotosintesis dan Respirasi Tumbuhan Dalam Konteks Sekolah Dan Lingkungan Pertanian [Role of Local Culture In Science Formation *Naturalistic Study of Culture Formation Science Student Group Sunda On Photosynthesis and Respiration Plants in the*

Context of the School of Agriculture and Environment]. **Disertation**. UPI –Bandung Indonesia. **Accessed on December 2016**: http:// repository. upi.edu/7908/

[10]. Jegede,O.J, and Aikenhead,G.S *Transcending Cultural Border: Implications for Science Teaching*. Accessed on January 2017.http://www. Jegede@ouhk. edu.hk..

[11] UNNESCO. (2016). Ecological Sciences for Sustainable Developmen, **Accesed on January 2017**: http://www.unesco.org/new/en/natural-sciences/environment/ecological-sciences.

[12] Rohayati, Woro S, Nanik W. (2015). Kontribusi Pembelajaran Berbasis Proyek Terhadap Jiwa Kewirausahaan Siswa [*Contributions Project Based Learning Students Against the Spirit of Entrepreneurship*]. *Jurnal Inovasi Pendidikan Kimia*, 9, 2:1556 – 1565, **Accessed on December 2016**: https://doaj.org/article/97a0d11f69524508868bfcf59340a0e7

[13] OECD. (2015). *PISA 2015 Draft Science Framework March 2013*. Accessed on November 2016: http://www.oecd.org/pisa/ pisaproducts/Draft%20PISA%202015%20Science%20Framework%20.pdf

[14] Susiana, N. (2007). Program Pembelajaran Kimia untuk Menumbuhkan Sikap Wirausaha Siswa (*Chemistry Learning Program for Cultivating students' entrepreneurial attitude*). *Disertion-Graduate school UPI Bandung*. Accessed on November 2016: http://digilib.upi.edu/digitallist.php?start=281

[15] Sakalli, B. (2011). An Invetigation To Improve Entrepreunerhip Module Vocational and Technical High School in Northern Cyprus. *TOJNED : The Online Journal Of New Horizons In Education, vol 1, Issue 4*. **Accessed on November 2016**. http://www.tojned.net/

[16] Khaerunisa, R.F (2014). Pengembangan Modul Sains Terpadu Berbasis Etnosains Tema Bahan Kimia untuk Meningkatkan Hasil Belajar dan Minat Belajar Siswa [*Development of Integrated Science Module Etnosains Based on a Theme of Chemicals to Improve Learning Outcomes and Student Learning Interests*]. ***Thesis***. Semarang; FMIPA UNNES. Acceed on April 2016: http: //lib.unnes.ac.id/ view/creators/ index.R.html

[17] Sudarmin. (2015). Pendidikan Karakter, Etnosains dan Kearifan Lokal: Konsep dan Penerapannya dalam Penelitian dan Pembelajaran Sains [*Character Education, Ethnoscience and Local Wisdom: Concept and Application in Research and Science Education Character Education: Etnosains and Local Wisdom]*. FMIPA-Semarang: CV. Swadaya Manunggal, ISBN 978-602-1034-00-2.

[18] Okechukwu, S., Abonyi, Lawrence, A & Njoku. (2014). Innovations in Science and Technology Education: A Case for Ethnoscience Based Science Classrooms. *International Journal of Scientific and Engineering Research, Volume 5, Issue 1*. **accessed Dec 2017** : http://s3.amazonaws. com/academia.edu.documents/ 33498416/ Innovations-in-Science-and-Technology-Education.pdf?

[19] Abonyi, Okechukwu Sunday. Effect of an Ethnoscience Based Instructional Package on Student' Conception of Scientif Phenomena. **accessed on January 2017**. *https://www.researchgate.net/..*

[20] Lee, S.,Chang, D dan Bae Lim, S. [2005]. Impact of Entrepreneurship Education:A Comparative Study of the U.S. and Korea. *International Entrepreneurship and Management Journal 1, 27–43*, **accessed December 2016**: *https://cba.unl.edu/research/articles/1139/*

[21] Sardjiyo dan Pannen, P. [2005]. Pembelajaran Berbasis Budaya Model Inovasi Pembelajaran Dan Implementasi Kurikulum Berbasis Kompetensi [*Culture-Based Learning Model Learning Innovation and Implementation of Competency Based Curriculum*]. *Jurnal Pendidikan*. 6(2): 83-98. **Accessed on october 2016** : *uilis.unsyiah.ac.id › Paulina Pannen › Sardjiyo*

[22] Buletine BSNP. (2007). Penilaian Buku Teks Pelajaran [*Assessment of Module*] **Accessed on Dec** 2016: http://litbang.kemdikbud.go.id/index.php/15-ban-pt/116-penilaian-buku-teks-pelajaran

[23] Michie, M. (2002). Why indegenous science should be include in the school science curriuculum Australian Science Teachers . *Journal. Vol 48(2): 36-40*. **Accessed on December 2016**: *https://search.informit.org/documentSummary;dn=200206975;res=IELAPA*

[24] Baynes, R. (2016). Teachers' Attitudes to Including Indigenous Knowledges in the Australian Science Curriculum. The Australian Journal of Indigenous Education, Volume 45, Issue 1 , August 2016, pp. 80-90. **accessed on December 2016**: https:// doi.org/10. 1017/jie.2015.29.

[25]. Chowdhury, M.A. (2016). The Integration of Science-Technology-Society/Science-Technology-SocietyEnvironment and Socio-Scientific-Issues for Effective Science Education and Science

Teaching. Electronic Journal of Science Education, 20 (5) :19-38. **accessed on January 2017**: www.researchgate.net/publication/303759460_

[26]. Libam, Z. (2010). Integrating Real- Life Data Analysis in Teaching Descriptive Statistics: A Constructivist Approach, *J Stat Educ* 18 (1), **accessed on January 2017**: http:// ww2. amstat.org/publications/jse/v18n1/libman.pdf

[27]. Gupta, A.D. (2015). Way to Study Indigenous Knowledge and Indigenous Knowledge System. *Research Journal of Recent Sciences*. Vol. 4(ISC-2014),:6-29. accessed on January 2017: www.isca.in, www.isca.me

[28].Wagiran. (2006). Meningkatkan Aktivitas Mahasiswa Dan Mereduksi Miskonsepsi Melalui Pembelajaran Konstruktivistik Model Kooperatif Berbantuan Modul [*Improve Student Activities And Reduce misconceptions Constructivistic Learning Through Cooperative Model Assisted Module*]. *Journal of Science Education* , 13(1): 25-32. **accessed on December 2016**: *download.portalgaruda.org/article.php?*

[29] Rist, S. and F. Dahdouh-Guebas, (2006). Ethnosciences—A step towards the integration of scientific and indigenous forms of knowledge in the management of natural resources for the future. *Env Dev Sustain* 8 (4) 467-493. **accessed on January** 2017: http://link.springer. com/article/10.1007/ s10668-006-9050-7.

[30] Sumarni, W. [2015]. The Strength and Weakness of the Implementation of Project Based Learning: Review, International Journal of Science and Research, vol 4 issue3, 478-484. **Accessed on December 2016** : https://www.ijsr.net/archive/v4i3/v4i3.php.

[31] Arifianawati, S, Sudarmin, Woro Sumarni. (2016). Model Pembelajaran Kimia Berbasis Etnosains Untuk Meningkatkan Kemampuan Berpikir Kritis Siswa [*Model of Learning Chemistry Based Etnosains To Improve Critical Thinking Skills Students*] Jornal Pengajaran Matematika dan Ilmu Pengetahuan Alam, 21 (1): 46-51. **Accessed on Oktober 2016**: http://dx.doi.org/10.18269/jpmipa.v21i1.669.

[32] Rosyidah, A, N. (2013).Pengembangan Modul Sains Berbasis Etnosains Zat Aditif dalam Makanan Untuk Kelas VIII SMP Negeri 1 Pegandon Kendal [*Developing Modules Subtance Additive based Etnosains For Class VIII SMP Negeri 1 Kendal Pegandon*]. *Unnes Science Education Journal*, USEJ 2 (1) : **Accessed on December 2016:** http://journal.unnes.ac .id/sju/ index. php/usej/issue/view/223

[33] Yenny, K. (2010). Pengembangan Bahan Ajar Kelarutan dan Hasil Kali Kelarutan Berorientasi Chemoentrepreunership (CEP) untuk Menumbuhkan Minat Kewirausahaan Siswa *[Development of Instructional Materials solubility and solubility Kali Results Oriented Chemoentrepreneurship (CEP) to Growing Interest in Entrepreneurship student*]. **Thesis,** FMIPA Universitas Negeri Semarang.

[34] *Demi* , S. (2016). Indigenous Food Cultures: Pedagogical Implication for Environmental Education, *International Institute of Social Studies* (ISS), **acccessed on January 2017**: https://www.iss.nl/fileadmin/ASSETS/iss/Research_and/2-ICAS_CP_Demi.pdf

Integrating Character Education Model With Spiral System In Chemistry Subject

Hartutik[1], Rusdarti[2], Sumaryanto[3], Supartono[4]

[1, 2, 3, 4]Management Education Post Graduate, Semarang State University Indonesia.

[1] Corresponding author: irenehartutik@gmail.com

Abstract. Integrating character education is the responsibility of all subject teachers including chemistry teacher. The integration of character education is just administrative requirements so that the character changes are not measurable. The research objective 1) describing the actual conditions giving character education, 2) mapping the character integration of chemistry syllabus with a spiral system, and 3) producing syllabus and guide system integrating character education in chemistry lessons. Of the eighteen value character, each character is mapped to the material chemistry value concepts of class X and repeated the system in class XI and class XII. Spiral system integration means integrating the character values of chemistry subjects in steps from class X to XII repeatedly at different depth levels. Besides developing the syllabus, also made the integration of characters in a learning guide. This research was designed with research and development [3] with the scope of 20 chemistry teachers in Semarang. The focus of the activities is the existence of the current character study, mapping the character values in the syllabus, and assessment of the integration guides of character education. The validity test of Syllabus and Lesson Plans by experts in FGD. The data were taken with questionnaire and interviews, then processed by descriptive analysis.

The result shows 1) The factual condition, in general, the teachers designed learning one-time face-to-face with the integration of more than four characters so that behaviour changes and depth of character is poorly controlled, 2) Mapping each character values focused in the syllabus. Meaning, on one or two basic competence in four or five times, face to face, enough integrated with the value of one character. In this way, there are more noticeable changes in students behaviour. Guidance is needed to facilitate the integration of character education for teachers integrating systems. Product syllabus and guidelines validated by experts and the syllabus results averaging 4.37; guidebooks integrating character education in chemistry learning 4.36 with a maximum score of 5. Thus the device is declared valid. Through focus group discussions, each expert gave input for the improvement of learning modules of character education.

1. Introduction

Not yet deepen and well-cultured behaviour in daily life, caused by very few examples of exemplary and honesty from their parents, friends, teachers, authorities, or law enforcement. Improvements that can be made by education is by evaluating the educational curriculum that emphasises character education. Regulation of the Minister of Education and Culture Nomor 22 Tahun 2016 about Standard Elementary and Secondary Education which states that education units do lesson planning, implementation of a learning process and learning assessment to improve efficiency and effective achievement of competencies of graduates [5]. From the basic rules, researchers to develop a training plan development of teaching materials in the form of a syllabus and guide the integration of character education in the chemistry subject. The research objective 1) describing the actual conditions giving character education, 2) mapping the character integration of chemistry syllabus with the spiral system, and 3) producing syllabus and guidance integration system of education character in chemistry lessons.

Character education is a character related to the moral concept (moral knowing), moral attitudes (moral feeling) and moral behaviour. Based on the National Education curriculum centre there are 18 character values. These 18 character values of the Curriculum Center, namely: religious, honest, tolerance, discipline, hard work, creative, independent, democratic, curiosity, national spirit, love the homeland, rewarding achievements, friendly/communicative, love peace, fond of reading, care for the environment, social care and responsibility [1].

The integration with spiral system [2] explains that the curriculum development should go back to the basic idea repeatedly and profoundly. Spiral system as a model of character education integration is done by selecting the appropriate character values characteristic of chemistry material and student characteristics. The learning approach characterised through subjects combination,assuming that any learning process presupposes a broad range of knowledge [6].

Ministry of National Education [7] defines teaching character education that integrated into the learning process as the values introduction and the facility gained awareness of the importance of the value, and integration values into students behaviour daily through a learning process, both of which take place inside and outside the classroom in all subjects. The combination is very useful as part of the achievement of aspects of attitudes and skillsv [10].

System integration of character education designed as a learning strategy. A teacher requires effective teaching strategies to stimulate students to engage in material and severe thought [4,5]. The approach used to integrate character education in chemistry subjects are eighteen grades characters mapped to one school year, with step one or two basic competence, the 10th class materials assigned to one character value. This is repeated from 10th level to 12th.

Spiral system intended so that eighteenth-character value spread over a one-year lesson repeated according to the nature spiral. This spiral system is designed to prevent repetition and emphasis on only one character only.

2. Methods

This study uses a model of Research and Development (R & D) from [3]. Results of research and development in the product that has the characteristics blend of concepts, principles, assumptions, hypothesis, and procedure that found in basic research.

Of 10 steps of this research and development, in this study only apply the six steps, namely:

Stage 1: Research and data collection, analysis of chemistry learning problems, analytical needs of students to an education of character, levelling indicator character assessment, and spiral system integration guides.

Stage 2: Planning, mapping the character values in the syllabus, preparation of guidelines for the integration of character education in chemistry, determination of the format and guidance components.

Stage 3: Development of products draft, an arrangement of syllabus and guidelines (first draft) and instruments of research.

Stage 4: The initial field trials, conducted syllabus validation and guidelines (first draft) and research instruments. The device validation performed by a team of experts.

Stage 5: Reviewing the test results, the revised syllabus and guidelines (first draft) into draft two to improve based on a team of experts input. The study was made to obtain logical devices.

Stage 6: Field test, conducted limited testing to 20 chemistry teacher, to determine the practicality learning devices. At this stage, the learning model is done that integrates in-class character. Continued focus group discussions involving character education experts, chemistry education specialists and teachers as practitioners. At this stage, the refinement of the syllabus and guides are ready to be used for learning in the classroom.

Data analysis techniques: descriptive data analysis of quantitative and qualitative descriptive. The quantitative descriptive analysis method of grouping category score data based on the guideline-scale conversion of five (Likert scale). While descriptive qualitative data analysis techniques, use stage data display, data reduction, data verification and conclusion.

3. Results and Discussion

3.1 The factual conditions

In general, teachers integrate character education by incorporating 4 to 8 character values in one face-to-face meeting advance in the lesson plan (RPP). With such system, the teachers have difficulty by observing the change in the behavior of students during the learning process.

3.2 The character education integration system with spiral system

The charactereducationintegrationsystem with a spiral system is done by mapping one character value adjusting material characteristics and conditions of the students with one or two basic competence. With this system, the syllabus will be produced with the mapping of character values. The mapping system outlined in the character education integration guidance which contains indicator levels of character and a rubric character assessment as elemental assessment.

3.3 Syllabus validity test and practical guidebooks integrating character education

Syllabus validity test and practical guidebooks integrating character education are done through group discussion forum with experts. Based on interviews with the respondents, obtained input that integrating character education training focused on the character value responded very positively by 86.67% of respondents. A total of 90.67% respondents said that the spiral system used to integrate character education in chemistry is a solution with indication character levelling system.

3.4 The Analysis of Lecturer and Students' Activities

Fundamental changes in 10th class behaviour focused on the training process. This is characteristics by habituation, and accompanied by rules, obtained an average score of 67%, the 11th level leads to habituation with consciousness but not by coercion is gained 65%, while the 12th class leads to demands for habituation, with awareness so as to take the value gained 71%. Rating affective aspect of characters based on the guidelines in the indicators and real assessment rubrics approved by 88% of respondents. How to motivate the students to reflect the behaviour of the character value is by exemplary of the teachers, and all school component responded well by 83.33%, giving a direct example of doing good things approved 93.33%. For system control, as much as 90% of respondents agree at the beginning of the syllabus, lesson plans, and chemistry book.

The results of the validation syllabus, lesson plans, and conducted by Delphi obtained a mean value as Tabel 1.

Tabel 1. Learning Tools Validation

No.	Development Tools	Validation Result	Criteria
1	Syllabus	4,37	Very well
2	RPP	4,36	Very well
3	Indicator Levelling Character Values	4,56	Very well
4	Instructor Guidance	4,47	Very well
5	Character Education Integration	4,43	Very well

Learning tools validation applied in the learning process and the results show that teachers have been objective based on student behaviour changes assessment.

4. Conclusion

Analysis of actual conditions showed chemistry teachers integrate character education with four to eight characters in one face-to-face meeting. There are no levelling system indicator character values at class 10th, 11th, and 12th. Training with the system integration of character education in chemistry spiral generated learning device, a valid evaluation instrument based on experts and practitioners.

Selection of one character value focused in learning design needs to consider the characteristics of teaching materials and students. Observation of student behaviour changes based on the assessment sheet used by teachers as an effective evaluation base. Teachers have been able to manage learning integration with character education and the subject material.

References

[1] Balitbang Depdiknas. 2002. Kurikulum dan Hasil Belajar. Pusat Kurikulum Jakarta.
[2] Boehm, B. 1988, " A spiral Model of software Development and Enhacement' IEEE Computer 21,5,61-72.
[3] Borg, W.R, and Gall, M.D. 2007. Educational Research. Eighth Edition. Michigan: Pearson Education.
[4] Lickona.(1991).Educating for Character;How Our School Can Teach Respect and Responsibility.Bantan Books,New York
[5] Lickona Thomas. 2013. *Educating for Character: Mendidik untuk Membentuk Karakter*. Terjemahan Juma Abdu Wamaungo. Jakarta: Bumi Aksara.
[6] Peraturan Pemerintah Departemen Pendidikan dan Kebudayaan (Permendikbud) Nomor 22 Tahun 2016.
[7] Koesoema, D. 2012. Pendidikan Karakter Utuh dan Menyeluruh. Yogyakarta: Kanisius.
[8] Departemen Pendidikan Nasional. 2013. *Kurikulum 2013 untuk Tingkat Sekolah Menengah Atas/ Madrasah Aliyah.* Jakarta: Binatama Raya.
[9] Eliasa, Eva Imania. 2014. Increasing Value of Teamwork and Responsibility of The Students Through Games: Integrating Education Character in Lectures. Procedia-Social and Behavioral Sciences 123 (2014) 196-203.
[10] Kupperminc.2001.School Social Climate and Individual Differences in Vulnerability to Psychopathology among Middle School Students. Journal of Social Psychology.Vol 39.

IOP Conf. Series: Journal of Physics: Conf. Series **824** (2017) 012026 doi:10.1088/1742-6596/824/1/012026

Metacognitive Analysis of Pre-Service Teachers of Chemistry in Posting Questions

T Santoso[1] and L Yuanita[2]

[1] FKIP, Universitas Tadulako, Indonesia
[2] Study Program of Doctorate in Science Education, Postgraduate Universitas Surabaya, Indonesia

*Corresponding author: tri_palu@yahoo.co.id; yuanita@sby.dnet.net.id

Abstract. Questions addressed to something can induce metacognitive function to monitor a person's thinking process. This study aims to describe the structure of the level of student questions based on thinking level and chemistry understanding level and describe how students use their metacognitive knowledge in asking. This research is a case study in chemistry learning, followed by 87 students. Results of the analysis revealed that the structure of thinking level of student question consists of knowledge question, understanding and application question, and high thinking question; the structure of chemistry understanding levels of student questions are a symbol, macro, macro-micro, macro-process, micro-process, and the macro-micro-process. The level Questioning skill of students to scientific articles more qualified than the level questioning skills of students to the teaching materials. The analysis result of six student interviews, a student question demonstrate the metacognitive processes with categories: (1) low-level metacognitive process, which is compiled based on questions focusing on a particular phrase or change the words; (2) intermediate level metacognitive process, submission of questions requires knowledge and understanding, and (3) high-level metacognitive process, the student questions posed based on identifying the central topic or abstraction essence of scientific articles.

1. Introduction

Students who are learning then ask questions, they are thinking involving metacognitive knowledge. Activities students ask questions would trigger the use of deep thinking strategies, and if students do not ask questions, strategic thinking will not show up [1,2]. Also, questions students have substantial potential, which is to direct the investigation and guide the formation of students' knowledge. The question of students are thinking the tool to encourage meaningful learning, and as a useful feedback to teachers about student understanding [3]. Questions help us to understand the world and construct meaning from the data and information, as a psychological tool to think because questions help to explore and scaffolding ideas, directing specific thinking in a particular direction, and can enhance students' understanding of phenomena and scientific concepts [4]. Thus, asking is the key to active, meaningful learning, and as the heart to do scientific work in all disciplines of science and technology. Therefore, in teaching and learning, the skills of students to ask questions need to be developed and facilitated.

The findings of several studies illustrate that the questions are produced by teachers and posed to students or question that has been available in textbooks. The teacher assigns students to answer the questions so that students hardly have any questions during the learning takes place [5,6]. This finding is supported by Kaberman & Dori [7] in his research report, the phenomenon with increasing level of students education rarely ask the question had even stopped asking. The consequences of this condition will have an impact on the majority of learner's impaired ability to formulate questions and to conduct an investigation. The next consequence is certainly reducing the level of students' activity in constructing their knowledge based on what became an issue for him or her.

Several studies related to asking questions have been done. For an example, inquiry activities in the chemical laboratory [8], questions teachers to stimulate productive thinking [4], computer-based environmental laboratory activities [9], arguments in discussion groups in science classes [10], and the attitude and practice of teacher's questions in science classes [11]. The influence of external metacognitive activities to produce the questions in learning are less effective [7]. Meanwhile, effective questioning influenced by internal metacognitive activities such as the metacognitive process of student, this research is still not many to do. Based on these finding, the exploratory of the level and the metacognitive process of pre-service teachers of chemistry in posting questions will be done. The fundamental questions of research are: (1) what do questions profile of pre -service teachers' chemistry? (2) how do the characteristics of the metacognitive process of pre-service chemistry teachers in developing skills of posing questions?

2. Literature Review

2.1. Metacognition in learning

The search for meaning is the core of any educational activity. In the process of the search for meaning requires critical thinking and inquiry. Critical thinking and inquiry based on the awareness and ability of learners to take responsibility and control in constructing meaning and confirm knowledge. Awareness and ability is an attribute of metacognition [12].

Metacognition is defined as "thinking about thinking" involves the awareness and control of one's cognitive processes [13]. National Research Council (NRC) defines metacognition as the process of pondering and directing his thinking. Application of metacognitive skills in learning requires knowledge of learning strategies and awareness of when to apply the right strategy [14]. Also, the control or regulation is also necessary to examine what is currently known and determine what still needs to be studied [13]. Thus, it can be said that metacognition is someone's reflection and awareness on the cognitive processes to induce self-regulation and coordinating learning tasks.

Conceptualization of metacognition consists of two dimensions, i.e. knowledge of cognition and the regulation of cognition [12]. In general, knowledge of cognition refers to what people know about their cognition. Knowledge of cognition includes declarative knowledge, procedural and conditional. Declarative knowledge is knowledge of ourselves as learners and what factors affecting our performance. Conditional knowledge includes knowledge about why and when to use a particular strategy. Individuals who have a high level of knowledge of the conditional will be better able to assess the situation demands a certain learning, and in turn, choose the most appropriate strategy for the situation [14,15].

Regulation of cognition consists of three components, planning, monitoring, and evaluation [14,15]. Planning involves selecting the strategy and the right resources, goal setting, activate the relevant background knowledge and set the time and budgeting. Monitoring involves looking back and skills of self-examination, which is required to control the learning. Evaluation relates to assessing the regulatory process of learning and learning outcomes of a person. Example: re-evaluate one's goals, revise predictions, and consolidate intellectual acquisition.

2.2. Asking questions as Metacognitive Skills

In metacognitive perspective, an asking question serves to know what they know and what they do not know [16]. In asking the question requires the application of knowledge of cognitive and metacognitive strategies, but the extent to which this strategy can be applied vary widely depending on the area of content and knowledge students. When students practice asking questions will lead to self-awareness on the success of his understanding.

Asking questions directed to higher-level thinking plays a central role in understanding, monitoring comprehension, self-testing and self-control. To become active learners and independent thinkers, learners should generate questions that form, focus, and guide their thinking [17]. Several studies have found that metacognitive activities externally imposed, for example by teachers, generate questions that are less effective than those produced by students [7].

Students who have good metacognitive skills will become active learners that make learn more fun and effective [14]. Evidence suggests that metacognitive activities have a positive impact on students' thinking and awareness of student learning at both the secondary and college education [18]. Activities metacognition is intended to provide the knowledge, awareness and critical strategies in the learning process [12]. According to Schunk [19], the development of metacognitive skills correlated with achievement and success of student learning. These statement supported by Stewart et al. [20] who state that students with better metacognitive skills gained great academic success.

2.3. Asking Questions in Learning

Asking questions in learning activities has long been recognised as a useful strategy to improve the formation of knowledge and increased interaction of learners to share thoughts [11]. Almeida [21] suggest that the students involved in the process of asking questions can help the cognitive growth of students. Blonder [22] explains further that the most important lesson is to challenge students to take an active role to ask questions that can encourage independent learning. The results of questioning activity is to give students the opportunity to: (1) developing communication, teamwork and skills to self-evaluation among them myself and my fellow colleagues, (2) increase the higher order cognitive processes such as evaluation, reflection and critical thinking, (3) encourage active learning, (4) to encourage deeper learning, and (5) improving student feedback process [23].

Questioning skills have been recognised as an essential component to promote inquiry learning potential in developing scientific thinking and support the formation of critical thinking of students, therefore, need to be emphasised in learning, especially in science education [5]. If the teacher focuses on a student only to find the answers, it means teachers stop students to think, otherwise, if the teachers can teach students to ask questions and provide an opportunity for students to ask, it means the teachers create the stage of the student to think [6]. In another word, thinking is driven by the question, therefore, the creation of a learning strategy that supports an environment in which students questions can generate other students to ask question need to be done. The types of questions that generate more questions or productive are critical questions [2]. The critical question is induces to high-level cognitive processes, such as the analysis of ideas, comparison and contrast, inference, prediction, and evaluation [16]. When such as learning strategies are provided, so students and teachers will be actively involved in the process of critical thinking. This learning strategy in line with the standards of the learning process in Curriculum 2013 [24].

2.4. Categorization of Student Questions

The classifying types of questions students are difficult to establish. However, a certain criterion can be made to classify the types of student questions. Watts et al. [25] offer a way to analyse the kinds of questions that occur in science classes, into three categories: consolidation, exploration, and elaboration. Categorization of these questions is an effort to distinguish between questions that are seeking to understand, resolve conflicts, examine the situation, give a discourse (the issue), and keep track of ideas and their consequences.

3. Research Methodology

This research is a case study in learning basic chemistry. Study participants have involved as many as 87 students of Chemical Education FKIP UNTAD. The research instrument was questionnaire-based student worksheets. Questionnaire-based student worksheets (WS) was used to explore the skills students to pose questions. In WS, students were asked to read an article teaching materials and a scientific article, and then students asked to submit questions in writing. Questions students analysed their contents by using the classification of questions developed by Kaberman & Dori [7]. The classification of these questions consists of, (1) thinking level, a question that requires a response at the level of higher thought than knowledge; and (2) chemistry understanding levels, are questions that seek answers by understanding the chemistry symbolic, macroscopic, microscopic, and process. Each student questions were analysed separately for the level of thought and the level of understanding of chemistry using the rubric as presented in Table 1. The question was given a score. A score of the students is a combined score of the level of thinking and understanding the chemistry which is classified into three groups, i.e. high, medium, and low. Two students from the group were interviewed using the think-aloud method, to explain why they ask certain questions.

Table 1 Rubric to assess the skills of the students ask questions (adapted from [7]).

IOP Conf. Series: Journal of Physics: Conf. Series **824** (2017) 012026 doi:10.1088/1742-6596/824/1/012026

Score	Thinking levels	Chemistry understanding levels
0	The question requires an answer fully described in the article	The question does not require an answer with an understanding of one aspect of chemical thought (symbols, macro, micro, or process)
1	The question requires an answer with the level of knowledge and understanding.	The question requires an answer with only one aspect of understanding the chemical thought.
2	The question requires an answer to the understanding and application-level thinking.	The question requires an answer to the understanding of the two aspects of chemical thinking
3	The question requires an answer to the thinking level of analysis and synthesis, for example, analysis of information and applications, the ability to identify problems and make conclusions, prediction, assessment, took the position.	The question requires an answer to the understanding of the three aspects of chemical thought.

4. Results and Discussion

Exploration inquiry carried out by the students were given the task of reading the teaching material and scientific articles, and students asked to submit questions in writing. Each student questions were scored separately according to the level of thinking and understanding the chemistry. The total score of the question is the sum of these two values the question level. The example such as a question analysis presented in Table 2. Distribution of skill student questions on the teaching material is presented in Figure 1 and 2.

Tables of research data show changes in levels of student skills in asking questions. Level questioning skills of students to scientific articles is more qualified than the level questioning skills of students in the teaching materials. Changes in the level of thinking skills that occur, namely: (1) knowledge question decreased by 20%, (2) a question of understanding and applications increased by 15%, and (3) higher-level thinking questions increased by 5%. Changes in the level of the chemical comprehension skills, namely the question of symbols and macro decreased by 6.89% and 10.35%, while the question of micro, macro-micro, macro-processes, micro-process, and the macro-micro-process increased by 6.89%; 1.15%; 5.75%; 2.29%; and 1.15% respectively. In this study, we analysed the quote "think-aloud" when students ask questions. Quote "think-aloud" is assumed as representations of student metacognitive process. There are three profiles of students when asked a question. First, the students asked after finding or are interested in a particular sentence. Examples of student "A" ask a question "whether there are other substances contained in the nine turns into acetic acid?" The question relates to the fact that in the article. Answers to these questions include the level of knowledge of thinking; the answer is a list of names of substances in the nira that has been written in the article. If the questions from the students "A" were analysed using the chemical comprehension level, the answer requires only one chemistry understanding level, namely the macro (the name of the substance).

Tabel 2. Analysis of the question: (a) example 1, (b) example 2, and (c) example 3

Aspect	Thinking Levels	Chemistry understanding levels
(a). Example 1: What is the chemical reaction rate?		
	Answer only requires knowledge of thinking level, which is just a definition	Answer only requires an understanding of the level of symbols: $mol/_{Ls^{-1}}$
Score	0/3	1/3
(b) Example 2: What is the relationship between the rate of decomposition of N_2O_5 and NO_2 formation on the decomposition reaction $2N_2O_5(g) \rightarrow 4NO_2(g) + O_2(g)$?		
	The answer requires understanding and application-level thinking.	Answer require two levels of understanding chemistry: • Symbol: chemical equation • Process: The decomposition of a chemical reaction rate of N_2O_5
Score	2/3	2/3
(c) Example 3: Example 3: What are the molecular collision theory and activated complex can explain the changes fast, slow, and there is no change in a chemical reaction?		

	Answer requires thinking level analysis of information, the ability to create relationships between concepts, and make inferences	The answer requires three levels of understanding chemistry: Micro: the reaction rate Micro: collisions between molecules Process: reaction changes, fast, slow, and there was no change
Score	3/3	3/3

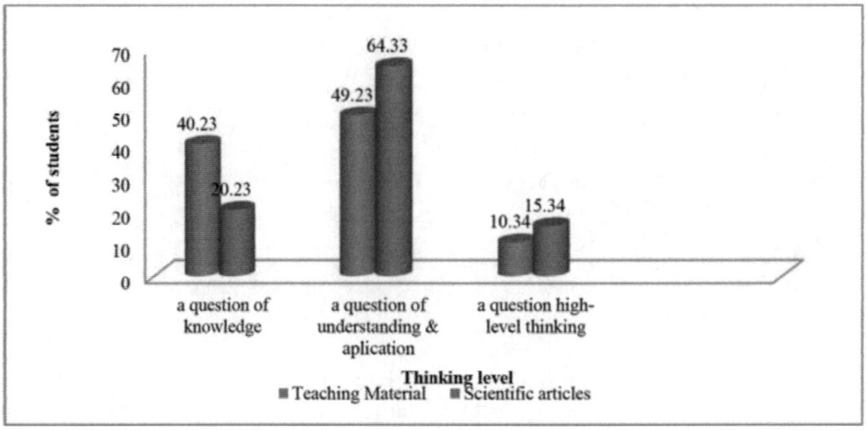

Figure 1. Distribution of skills students questions according to the thinking levels for teaching materials and scientific articles.

Second, students ask questions after they tried to know and understand the contents of some or all articles, as an example of a student "C" to ask "How did the fermentation of sucrose in the nira occur? Moreover, how mechanisms of reaction? "Through the interviews revealed that questions arose after students read thoroughly and make records that need to be studied further. It means that students recognise the need to ask questions that require detailed explanation. The answer to this question requires a level of knowledge and understanding of thought t. While the answer at the level of understanding the chemistry requires two levels, namely the symbol (equation) and processes (chemical reaction mechanisms for the formation of acetic acid by catalytic enzyme invertase). Third, the questions asked by the students based on the identification of issues referring to the central topic of the article or the result of abstraction. For example, a student "D", ask "How different the effect of temperature on the rate of reduction of sucrose and the rate of formation of acetic acid?" The student is aware of the need to ask questions of the investigation.

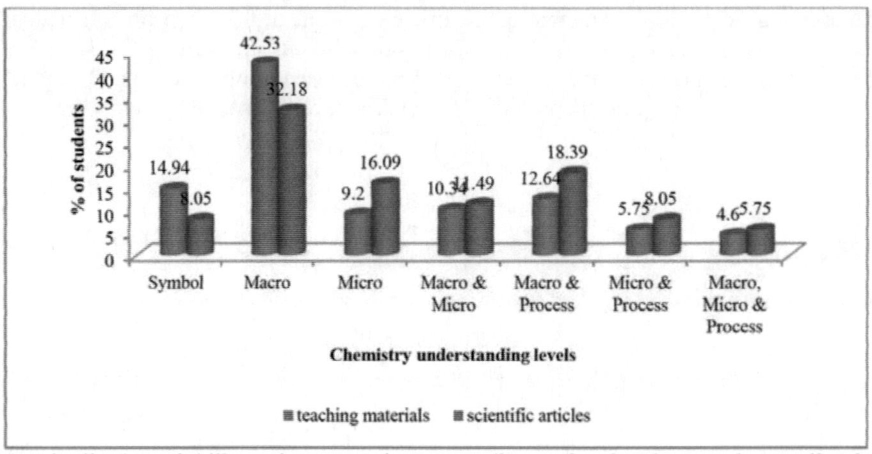

Figure 2. Distribution of skills students questions according to the chemistry understanding levels for teaching materials and scientific articles.

The answer to this question requires thinking the level of understanding, analysis and prediction; while the answer at the chemistry understanding level needed four levels. They are symbol chemical equations; macrophysical shape and taste of sucrose and acetic acid compounds; micro: the rate of sucrose reduction and acetic acid formation; process: chemical reactions mechanisms of sucrose reduction and acetic acid formation. Based on the description of the process analysis students questions mentioned above and according to Kabermen & Dori [8], the questioning student can be classified into three groups metacognitive process; (1) low-level metacognitive process, students formulate questions based on a specific sentence; (2) intermediate level metacognitive process, students formulate questions after knowing and understanding, and (3) high-level metacognitive process, students make questions based on the results to identify the central topics or abstraction the essence of scientific articles or teaching materials.

4. Conclusion

The profile of the questions formulated by chemical education students has a level of thinking, chemistry understanding, and metacognitive process. In term of thinking level, the questioning student consists of three types, i.e. knowledge, understanding and application, and higher level thinking. In the chemistry understanding levels, the questioning student has seven categories; symbol, macro, micro, macro-micro, macro-processes, micro-process, and the macro-micro-processes. While on the line with the metacognitive process level, the questioning student has three kinds; low-level metacognitive process, medium level of metacognitive process, and high-level metacognitive process. Level questioning skills of students to scientific articles is more qualified than the level questioning skills of students in the teaching materials.

References

[1] Golding C, 2011, *High. Educ. Res. Dev.*, **30(3)**, 357–370
[2] Nussbaum E and Edwards OV, 2011, *J. Learn. Sci.*, **20**, 443-488
[3] Chin C, 2004, *School Sci. Rev.* **86,** 107-112
[4] Chin C, 2007, *J. Res. Sci. Teach.*, **44,** 6, pp. 815-843
[5] Cho YH, Lee J, Jonassen DH, 2011, *Comput. Educ.*, **56**, pp. 112–126
[6] Lan YF and Lin PC, 2011. *Australas. J. Educ. Tech.*, **27(4),** 581-599
[7] Kaberman Z and Dori YJ, 2009, *Instr. Sci.*, **37,** 403-436
[8] Hofstein A, Navon O, Kipnis M, Mamlok-Naaman R, 2005, *J. Res. Sci. Teach.*, **42,** 791-806.
[9] Kaberman Z and Dori YJ, 2009, *Int. J. Sci. Math. Educ.*, **7**, 597-625
[10] Chin C and Osborne J, 2010, *J. Res. Sci. Teach.* **47(7),** 883-908
[11] Eshach H, Ziderman YD, Yefroimsky Y. 2014. J. Sci. Educ.Tech. **23**, 67-81
[12] Akyol Z and Garrison DR, 2011, *Internet High. Educ.*, **14**, 183-190.
[13] Seraphin KD, Philippoff J, Kaupp L, Vallin LM, 2012, *Sci. Educ. Int.*, **23(4)**, 366-382
[14] Schraw G, Crippen KJ, Hartley K, 2006, *Res. Sci. Educ.*, **36**, 111-139
[15] Schraw G and Moshman D, 1995, *Educ. Psychol. Rev.*, **7(4),** 351–371.
[16] King A. 1995. *Teach. Psychol., 22(1)*, 13-17
[17] Taboada A and Guthrie JT, 2006, *J. Lit. Res.* **38(1)**, 1–35.
[18] Pang K and Ross C, 2010, *J. Effect. Teach.* **10(1)**, 79-97.
[19] Schunk DH, 2012, *Learning theories an educational perspective,* Singapura, Pearson Education, Inc.
[20] Stewart PW, Cooper SS, Moulding LR. 2007. *The Researcher*, **21(1),** 32-40
[21] Almeida PA, 2011, *Proc. Soc. Behave. Sci.*, **30**, 357–362.
[22] Blonder R, Mamlok RN, Hofstein A, 2008, *Chem. Educ. Res. Pract.*, **9**, 250-258
[23] Yu FY, Tsai HC, Wu HL, 2013, *Australasian J. Educ. Technol., 29(3)*, 416-433
[24] Permendikbud, 2013, *Peraturan Menteri Pendidikan dan Kebudayaan Republik Indonesia Nomor 65 Tahun 2013 Tentang Standar Proses Pendidikan Dasar dan Menengah*, Jakarta: Kementerian Pendidikan dan Kebudayaan
[25] Watts M, Gould G, Alsop S, 1997, *School Sci. Rev.*, **79**, 57–63

Modeling Governance KB with CATPCA to Overcome Multicollinearity in the Logistic Regression

L Khikmah[1], H Wijayanto, and U D Syafitri

Department of Statistics, Faculty of Mathematics and Natural Science, Bogor Agriculture University, Bogor, Indonesia.

Corresponding author:aisyah.salsabila17@gmail.com

Abstract. The problem often encounters in logistic regression modeling are multicollinearity problems. Data that have multicollinearity between explanatory variables with the result in the estimation of parameters to be bias. Besides, the multicollinearity will result in error in the classification. In general, to overcome multicollinearity in regression used stepwise regression. They are also another method to overcome multicollinearity which involves all variable for prediction. That is Principal Component Analysis (PCA). However, classical PCA in only for numeric data. Its data are categorical, one method to solve the problems is Categorical Principal Component Analysis (CATPCA). Data were used in this research were a part of data Demographic and Population Survey Indonesia (IDHS) 2012. This research focuses on the characteristic of women of using the contraceptive methods. Classification results evaluated using Area Under Curve (AUC) values. The higher the AUC value, the better. Based on AUC values, the classification of the contraceptive method using stepwise method (58.66%) is better than the logistic regression model (57.39%) and CATPCA (57.39%). Evaluation of the results of logistic regression using sensitivity, shows the opposite where CATPCA method (99.79%) is better than logistic regression method (92.43%) and stepwise (92.05%). Therefore in this study focuses on major class classification (using a contraceptive method), then the selected model is CATPCA because it can raise the level of the major class model accuracy.

1. Introduction

In the field of modeling studies often found an association between variables, whether the relationship between the response variable with the explanatory variables, the relationship between the response variable, as well as the relationship between the explanatory variables. The relationship between the response variable will cause the relationship between remnant. This will result in inter-residual graphs cannot describe the actual state of the data, so that interpretation to be incorrect. However, the relationship between the response variable can still be controlled when the number of degrees of freedom and a lot of large remnants [2].

Relationships between more than two explanatory variables are often referred to as multicollinearity. In the case of a linear relationship between the explanatory variables in the regression parameter estimates lead to be inappropriate, although the coefficient determination generated great and significant. Multicollinearitybetween explanatory variables in the logistic regression model parameter estimation will cause to be invalid. This is because the variance covariance matrix that will be formed to be interconnected, so that parameter estimation is not appropriate [1].

Some researchers didmulticollinearitydetection before further analysis. If the explanatory variables are numeric variables, then to detect multicollinearity using correlation matrices, variance influence factor (VIF), and the eigenvalues of the correlation matrix. However, if the explanatory variables are categorical variables, then to detect multicollinearity using association analysis.

Multicollinearitytechniques for handling classified into two methods, the method of removal (discard the explanatory variables are correlated) and variable selection method. In general, the variable selection method used is the best subset, the method, and the backward stepwise regression method. But the stepwise method and VIP still not up to handle their multicollinearity, so use some method development. In the case of linear data, handling multicollinearity can be resolved by methods PLS-VIP, methods PLS-BETA, LASSO and stepwise regression [3]. Bastien [4] examined using PLS-LR and Aguilena[5] using PCLR. Both of these methods are good enough to handle multicollinearity in the logistic regression. However, if the explanatory variables in the form of nominal and ordinal data (nonlinear), Kemalbay[1] developed a method Categorical Principal Component Analysis (CATPCA).This study will compare the methods of stepwise regression and CATPCA method by using case studies of governance KB modeling with logistic regression.

2. Material and Methods

2.1. Categorial Principal Component Analysis (CATPCA)

The goal of traditional principal component analysis (PCA) is to reduce the number of m variables to a smallernumber of p uncorrelated variables called principal components which account for the variance in the data as muchas possible. Since PCA is suitable for continuous variables which are scaled at the numerical level of measurement such that interval or ratio and it also assumes linear relationship among variables, it is not an appropriate method of dimension reduction for categorical variables. Alternatively, categorical (also known as nonlinear) principalcomponents analysis (CATPCA) has been developed for the data given mixed measurement level such that nominal, ordinal or numeric which may not have a linear relationship with each other. For categorical variables, CATPCA uses optimal scaling process which transforms the category labels into numerical values while the variance accounted for among the quantified variables is maximized [6]. We refer to [7] for a historical review of CATPCA using optimal scaling. For continuous numeric variables, the optimal scaling process is as the traditional case. Suppose we have measurement of n individuals on m variables given with an $n \times m$ observed scores matrix **H** where each variable is denoted by j **X**, j=1,...,m that is the jth column of **H**. If the variable j **X** are of nominal or ordinal measurement level, then a nonlinear transformation called optimal scaling is required where each observed scores transformed into category quantification given by:

$$q_j = \varphi_j(X_j) \tag{1}$$

where**Q** is the matrix of category quantifications. Let **S**ee the $n \times p$ matrix of object scores, which are the scores of the individuals on the principal components, obtained by CATPCA. The object scores are multiplied by a set of optimal weights which are called component loadings. Let **A** be a $m \times p$ matrix of the component loadings where the jth column is denoted by j **a** . Then the loss function for minimization of difference between original data and principal components can be given as follows:

$$L(Q, A, S) = n^{-1}\sum_{j=1}^{m} tr\left(q_j a_j^T - S\right)^T \left(q_j a_j^T - S\right) \tag{2}$$

2.2. Logistic Regression Multinomial

Logistic regression analysis is an analysis that is used to model the response variable Y that is based on the categorical explanatory variables X that are numerical and categorical. Logistic regression analysis requires several things (assumptions) about the nature of the data, namely:

1. The dependent variables should be categorical.
2. There is no significant correlation between the independent variables.
3. The number of observations for each variable should be adequate, and the overall sample size is large enough.The number of samples of at least 400 units to be able to get a good fitness model [8].

The conditional probability can be written:

$$E(y|x)=\frac{\exp(\beta_0+\beta_1x_1+\beta_2x_2+\ldots+\beta_px_p)}{1+\exp(\beta_0+\beta_1x_1+\beta_2x_2+\ldots+\beta_px_p)} \quad (3)$$

or

$$\pi(x)=\frac{\exp(\beta_0+\beta_1x_1+\beta_2x_2+\ldots+\beta_px_p)}{1+\exp(\beta_0+\beta_1x_1+\beta_2x_2+\ldots+\beta_px_p)} \quad (4)$$

Equivalently, logit has a linear equation as follows:

$$\text{Logit } [\Pi (x)] = \log\left(\frac{\Pi(x)}{1-\Pi(x)}\right) = \beta_0+\beta_1x_1+\beta_2x_2+\ldots+\beta_px_p \quad (5)$$

which is a bridge function (link function) of the explanatory variables [9].

Testing parameters simultaneously performed with a likelihood ratio test (likelihood ratio test). A likelihood ratio test (likelihood ratio test) is obtained by comparing the functions of all variables explanatory Log-likelihood with Log-likelihood function without the explanatory variables. With the hypothesis that will be tested as follows:

$H_0 : \beta_1 = \beta_2 = \ldots = \beta_j = 0$ (no explanatory variables that influence the response variable)

H_1 : At least there's the one with $\beta_j \neq 0$, $j = 1,2,\ldots, p$ (at least one explanatory variables that influence the response variable).

The test statistic G formulated as follows $$G = -2\ln\frac{l_0}{l_1} \quad (6)$$

where

l_0 : likelihood without explanatory variable

l_1 : likelihood with explanatory variable

G statistic follows a *chi-square* distribution with $db = (c-1)(r-1)$. For accept or reject H_0, then compare the G value and X^2. Reject H_0 f $G > X^2_{(p,\alpha)}$ or reject H_0 if $p-value < \alpha$ [8].Partial testing is done by using the Wald test. The hypothesis tested is:

$H_0 : \beta_j = 0$ (X_j explanatory variables did not significantly influence the response variable)

$H_1 : \beta_j \neq 0$ (X_j explanatory variables did significantly influence the response variable),

for $j = 1,2,\ldots, p$.

Formula for the Wald test statistic is $$W = \frac{\hat{\beta}_1}{SE(\hat{\beta}_1)} \quad (7)$$

3. **Methods**

3.1. Data

Data were used in this research is secondary data. Data were a part of data Indonesian Demographic Health Survey (IDHS) in 2012 obtained from Demographic and Health Surveys (DHS). The unit of observation in the form of household, namely the productive age group of women aged 15-49 years as many as 29 882 people. Variables used a categorical variable by 8 variables. Variables used are shown in Table 1.

Table 1. Variables

Variable	Category
Contraseptive methods	1 = not used
	2 = used
Age in 5-year Groups (x_1)	1 = <30 years old

	2 = 30-39 years old
	3 = >39 years old
Type Place of Residence (x_2)	1 = Urban
	2 = Rural
Number of Living Children (x_3)	1 = ≤ 2 person
	2 = > 2 person
Knowledge of Any Method (x_4)	0 = No Knows
	1 = Knows
Respondent Currently Working (x_5)	0 = No
	1 = Yes
Highest Educational Level (x_6)	1 = Primary
	2 = Junior High
	3 = Senior High
	4 = University
Patner's Level Education (x_7)	1 = Primary
	2 = Junior High
	3 = Senior High
	4 = University

3.2. Analysis Method

Steps of data analysis performed in this study are as follows: (1) Explore the data to get a general overview of user IDHS data 2012. (2) Modeling tool selection without contraception with logistic regression using stepwise regression method and CATPCA. In this step will be performed with logistic regression modeling while maintaining the association between the explanatory variables. So it will be visible difference modeling results with and without maintaining the association between the explanatory variables. The model will be established by equation 6 and 7. (3) Handling associations between explanatory variables with CATPCA. (4) Logistic regression modeling tool selection method contraception stepwise regression. (5) Logistic modeling tool selection method contraception CATPCA. (6) Comparing CATPCA with stepwise regression and classification table.

4. Results and Discussion

4.1. Classification Table

After testing the suitability of the models that have been obtained, then it was performed the accuracy measurement of the classification multinomial logistic regression model that has been formed. Classification accuracy results can be seen in Table 2.

Table 2. Classification Performance with Logistic Regression Model

Performance classicification	Result
Accuracy	61.54
Sensitivity	92.43
Specificity	14.50
AUC	57.39

Based on Table 2, it can be seen that the accuracy of classification models using logistic regression models good enough for 61.54%. AUC values of 57.39% indicate that the model's accuracy is good enough. The percentage accuracy of the classification model quite well. A value large enough sensitivity that is equal to 92.43%, which means that the accuracy of classification for methods of contraception would be appropriate but still errors may occur a minority who will be classified to the majority class. Classification of a logistic regression model with a stepwise method is done by entering the variables most associated with Y is the age category, the wife of education, knowledge of contraceptives, the number of living children, living and working place today, so we get Table 3.

Table 3. Classification Performance with Stepwise Logistic Regression Model

Performance classicification	Result
Accuracy	60.80
Sensitivity	92.05
Specificity	13.20
AUC	58.66

Based on Table 3, it can be seen that the accuracy of classification models using a stepwise logistic regression model with good enough for 60.80%. AUC values of 58.66% indicate that the model's accuracy is good enough. The percentage accuracy of the classification model quite well. A value large enough sensitivity that is equal to 92.05%, which means that the accuracy of classification for methods of contraception would be appropriate but still errors may occur a minority who will be classified to the majority class. With CATPCA, we get Table 4.

Table 4. Classification Performance with CATPCA Logistic Regression Model

Performance classicification	Result
Accuracy	61.16
Sensitivity	99.79
Specificity	2.34
AUC	57.39

Based on Table 4, it can be seen that the accuracy of classification models using a stepwise logistic regression model with good enough for 61.16%. AUC values of 57.39% indicate that the model's accuracy is good enough. The percentage accuracy of the classification model quite well. A value large enough sensitivity that is equal to 99.79%, which means that the accuracy of classification for methods of contraception would be appropriate but still errors may occur a minority who will be classified to the majority class.

Comparison Model

The model that has been obtained is then compared to the level of accuracy with AUC values. AUC values on stepwise logistic regression model (57.89) are higher than the logistic regression model (57.39) and CATPCA (57.89). This shows that the stepwise logistic regression model is more accurate than the logistic regression models and models CATPCA. In addition to comparing the AUC values obtained, may be considered sensitivity and specificity of each model. AUC value was composed by sensitivity and specificity. In this study, more sensitivity focused on value, because in this case, the use of contraceptive methods is more likely than not using a contraceptive method. According to Table 4, the model can improve sensitivity CATPCA big enough though specificity decreased. Sensitivity models CATPCA (99.79) is higher than the logistic regression model (92.43) and stepwise models (92.05). Thus, based on the sensitivity of the model to predict the data CATPCA minor class with a lot better.

5. Conclusions and Recommendations

This research concluded that Based on AUC values, the classification of the contraceptive method using stepwise logistic regression model is better than the logistic regression mode and CATPCA. While based on the value of sensitivity, the model CATPCA better than the logistic regression model and stepwise. In this study focuses on major class (the use of contraceptive methods), CATPCA can raise specificity major class model. In general, in the condition data is not balanced method CATPCA less well in the accuracy of the classification, but were able to classify correctly. This research recommends that In the actual case, there are four categories in which the two categories have a small percentage so that data is not balanced. So to further research needs to be treated assessment of the data is not balanced so as to optimize the performance of classification.

References

[1] Kemalbay Gand Ozlem BK2014 *Procedia - Social and Behavioral Sciences* **109** 730 DOI: 10.1016/j.sbspro.2013.12.537

[2] Draper NR and Smith H 1998 *Applied Regression Analysis3rd Ed*ISBN: 978-0-471-17082-2(New York: John Wiley & Sons Inc)
[3] Chong I and Chi-Hyuck J 2005 *Chemometr Intell Lab***78** 103 doi:10.1016/j.chemolab.2004.12.011
[4] Bastien P, Vinzi V E and Tenenhaus M 2005 *Comput Stat Data An* **48** 17 doi:10.1016/j.csda.2004.02.005
[5] Aguilena AM, Manuel E and Mariano JV 2006 *Comput Stat Data An***50** 1905 doi:10.1016/j.csda.2005.03.011
[6] LintingM and Van der Kooij A 2012*J Pers Assess***94** 12. DOI:10.1080/00223891.2011.627965
[7] Gifi A1990 *Nonlinear multivariate analysis*ISBN: 978-0-471-92620-7 (Chichester: John Wiley and Sons)
[8] Hosmer DW and Lemeshow S2000 *Applied Logistic Regression 2nd Ed.* (New York: Wiley)DOI: 10.1002/0471722146
[9] Agresti A 2002 *Categorical Data Analysis* 2nd Ed. (New York: John Wiley & Sons)

The Algorithm of Determinant Centrosymmetric Matrix Based on Lower Hessenberg Form

N Khasanah[1], Farikhin , and B Surarso

Faculty of Science and Mathematics, Diponegoro University, Semarang, Indonesia

Corresponding author: nur_kha91@yahoo.co.id

Abstract. Centrosymmetric matrix has practical application in mathematics and engineering. Certain specialized cases of centrosymmetric matrix discuss computing determinant. Therefore, we need an algorithm to compute determinant centrosymmetric matrix efficiently on computations. Based on the structure of centrosymmetric matrix has lower Hessenberg form, so in this paper, we propose the algorithm to compute the determinant of the centrosymmetric matrix using an algorithm of determinant lower Hessenberg matrix.

1. Introduction

Centrosymmetric matrix plays a major role in some areas such as pattern recognition, antenna theory, mechanical and electrical systems, and quantum physics. Nearly 75% reduction in the multiplicative complexity is achieved for evaluation of the determinant of the centrosymmetric matrix [1,2]. Then, the algorithm of the determinant centrosymmetric matrix is needed with efficiently. On the other side, the role of Hessenberg matrix is important in numerical analysis. For example, the Hessenberg decomposition plays on matrix eigenvalues computations. A recursive algorithm for computing determinant of an n-by-n lower Hessenberg matrix is obtained [3]. This algorithm is better than studied of the matrix on the complexity based on a previous study [4-8]. Therefore, the computing of determinant of the centrosymmetric matrix with lower Hessenberg form is proposed in this paper.

2. Preliminaries

The properties and characteristics of the centrosymmetric matrix are discussed [9,10]. The following important result can be used for computing determinant of the centrosymmetric matrix.

Definition 1 [9]

Let $\mathbf{A} = (a_{ij})_{n\times n} \in R^{n\times n}$ is a centrosymmetric matrix, if

$$a_{ij} = a_{n-i+1,n-j+1},\ 1 \le i \le n,\ 1 \le j \le n \text{ or written as } \mathbf{A} = \begin{pmatrix} a_{11} & a_{12} & \cdots & a_{1,n} \\ a_{21} & a_{22} & \cdots & a_{2,n} \\ \vdots & \vdots & \vdots & \vdots \\ a_{2,n} & \cdots & a_{22} & a_{21} \\ a_{1,n} & \cdots & a_{12} & a_{11} \end{pmatrix}.$$

Theorem 2 [9]

IOP Conf. Series: Journal of Physics: Conf. Series **824** (2017) 012028 doi:10.1088/1742-6596/824/1/012028

Let $\mathbf{A} = (a_{ij})_{n\times n} \in R^{n\times n}$ equivalently $\mathbf{J_n A J_n} = \mathbf{A}$, where $\mathbf{J_n} = (\mathbf{e_n}, \mathbf{e_{n-1}}, \cdots, \mathbf{e_1})$ and $\mathbf{e_i}$ is the unit vector with the i-th elements 1 and others 0 or written as $\mathbf{J_n} = \begin{pmatrix} 0 & \cdots & 0 & 0 & 1 \\ 0 & \cdots & 0 & 1 & 0 \\ \vdots & \cdots & ⋰ & \vdots & \vdots \\ 0 & 1 & \cdots & 0 & 0 \\ 1 & 0 & \cdots & 0 & 0 \end{pmatrix}$.

Proof. Let n-by-n centrosymmetric matrix and n-by-n of the $\mathbf{J_n}$ matrix. It can be proven.■

The purpose of this section is discussed about properties square centrosymmetric matrix from the standpoint of computations. So, for the next we only discuss on n-by-n centrosymmetric matrix n is even. For next discussion, lower Hessenberg matrix is described. The n-by-n lower Hessenberg matrix form as [3] :

$$\mathbf{H} = \begin{pmatrix} h_{11} & h_{1,2} & 0 & 0 & 0 \\ h_{21} & h_{2,2} & h_{2,3} & 0 & 0 \\ \vdots & \vdots & \ddots & \ddots & 0 \\ h_{n-1,1} & h_{n-1,2} & \cdots & h_{n-1,n-1} & h_{n-1,n} \\ h_{n,1} & h_{n,2} & \cdots & h_{n,n-1} & h_{n,n} \end{pmatrix}$$

This matrix can be solved with partition, let the $(n+1)$-by-$(n+1)$ lower triangular matrix [3]

$$\widetilde{\mathbf{H}} = \left(\begin{array}{c|c} \mathbf{e}_1^{\mathrm{T}} & 0 \\ \hline \mathbf{H} & \mathbf{e_n} \end{array}\right) = \left(\begin{array}{ccccc|c} 1 & 0 & \cdots & \cdots & 0 & 0 \\ \hline h_{11} & h_{1,2} & 0 & 0 & 0 & 0 \\ h_{21} & h_{2,2} & h_{2,3} & 0 & 0 & \vdots \\ \vdots & \vdots & \ddots & \ddots & 0 & \vdots \\ h_{n-1,1} & h_{n-1,2} & \cdots & h_{n-1,n-1} & h_{n-1,n} & 0 \\ h_{n,1} & h_{n,2} & \cdots & h_{n,n-1} & h_{n,n} & 1 \end{array}\right)$$

Partition $\widetilde{\mathbf{H}}^{-1}$ into $\begin{pmatrix} \boldsymbol{\alpha} & \mathbf{L} \\ h & \boldsymbol{\beta}^{\mathrm{T}} \end{pmatrix}$, where $\boldsymbol{\alpha}, \mathbf{L}, \boldsymbol{\beta}$ are matrices of size n-by-1, n-by-n, n-by-1 respectively and h is scalar. By $\widetilde{\mathbf{H}} \cdot \widetilde{\mathbf{H}}^{-1} = \begin{pmatrix} \mathbf{e}_1^{\mathrm{T}} & 0 \\ \mathbf{H} & \mathbf{e_n} \end{pmatrix} \begin{pmatrix} \boldsymbol{\alpha} & \mathbf{L} \\ h & \boldsymbol{\beta}^{\mathrm{T}} \end{pmatrix} = \begin{pmatrix} 1 & 0 \\ 0 & \mathbf{I_n} \end{pmatrix} = \mathbf{I_{n+1}}$

then
$$\mathbf{H}\boldsymbol{\alpha} + h\mathbf{e_n} = 0 \tag{1}$$
and
$$\mathbf{HL} + \mathbf{e_n}\boldsymbol{\beta}^{\mathrm{T}} = \mathbf{I_n}. \tag{2}$$

It can be seen that $h \neq 0$. But, if $h = 0$, so $\alpha = 0$ since $\mathbf{H}\boldsymbol{\alpha} = 0$ from (1). This implies that $\widetilde{\mathbf{H}}^{-1}$ is singular, which is the contradiction. From (1), known as $\mathbf{e_n} = -h^{-1}\mathbf{H}\boldsymbol{\alpha}$. Substitution of (2) founded

$$\mathbf{H}(\mathbf{L} - h^{-1}\boldsymbol{\alpha}\boldsymbol{\beta}^{\mathrm{T}}) = \mathbf{I_n}$$

So $\mathbf{H}^{-1} = \mathbf{L} - h^{-1}\boldsymbol{\alpha}\boldsymbol{\beta}^{\mathrm{T}}$.

It is noticed that
$$\begin{pmatrix} \mathbf{I_n} & -h^{-1}\boldsymbol{\alpha} \\ 0 & 1 \end{pmatrix} \begin{pmatrix} \boldsymbol{\alpha} & \mathbf{L} \\ h & \boldsymbol{\beta}^{\mathrm{T}} \end{pmatrix} = \begin{pmatrix} 0 & \mathbf{L} - h^{-1}\boldsymbol{\alpha}\boldsymbol{\beta}^{\mathrm{T}} \\ h & \boldsymbol{\beta}^{\mathrm{T}} \end{pmatrix}. \tag{3}$$

According to the above preparation, the result can be presented as:

Lemma 3 [3]
Suppose that $\mathbf{H}$ is a lower Hessenberg matrix of order n and $\widetilde{\mathbf{H}}$ is its associated lower triangular matrix as above.
Then

$$\det(\mathbf{H}) = (-1)^n h \cdot \prod_{i=1}^{n-1} h_{i,i+1}.$$

Proof. From (3) then $\det(\widetilde{\mathbf{H}}^{-1}) = (-1)^{n+1+1} h \cdot \det(\mathbf{L} - h^{-1}\boldsymbol{\alpha}\boldsymbol{\beta}^{\mathrm{T}}) = (-1)^n h \cdot \det(\mathbf{H}^{-1})$

So $\det(\mathbf{H}) = (-1)^n h \cdot \det(\widetilde{\mathbf{H}})$.■

Theorem 4 [3]

Let $\mathbf{H}$ be a lower Hessenberg matrix and all elements of the super diagonal be non-zero, and $\widetilde{\mathbf{H}}$ be the associated matrix as above. Partition $\widetilde{\mathbf{H}}^{-1}$ as $\begin{pmatrix} \boldsymbol{\alpha} & \mathbf{L} \\ h & \boldsymbol{\beta}^{\mathrm{T}} \end{pmatrix}$, where $\boldsymbol{\alpha}, \mathbf{L}, h, \boldsymbol{\beta}$ are as above. Then

$$\det(\mathbf{H}) = (-1)^n h \cdot \prod_{i=1}^{n-1} h_{i,i+1}.$$

where $h_{i,i+1}(i = 1,2,\cdots,n-1)$ are the super diagonal elements of $\mathbf{H}$.

Proof. Using Lemma 3 and $\det(\widetilde{\mathbf{H}}) = \prod_{i=1}^{n-1} h_{i,i+1}$, it can be proven.■

3. Results and Discussion

Centrosymmetric matrix has block matrices with lower Hessenberg form, they are $\mathbf{B}$ and $\mathbf{J_m C}$. Based on the properties before, so the algorithm to compute determinant centrosymmetric matrix with algorithm determinant of lower Hessenberg efficiently is presented in this section. The algorithm is presented as follows [11] :

Input: Matrix $\mathbf{A} = \begin{pmatrix} \mathbf{B} & \mathbf{J_m C J_m} \\ \mathbf{C} & \mathbf{J_m B J_m} \end{pmatrix}$

Output : $\det(\mathbf{A})$

1. Block centrosymmetric matrix
2. Block centrosymmetric martix on Hessenberg matrix
3. Determinant of centrosymmetric matrix

Next, the following are the details of the steps of the algorithm determinant centrosymmetric matrix.

3.1. *Block Centrosymmetric Matrix*

Based on studied before [11] it seen that the determinant of centrosymmetric matrix founded with unique properties of centrosymmetric matrix, about determining block matrices centrosymmetric matrix. Based on the partition of the centrosymmetric matrix, the character of the symmetric square matrix of size n-by-n where n is even can be exploited as bellows.

Lemma 5 [11]

Let $\mathbf{A} = (a_{ij})_{n\times n} \in R^{n\times n}(n = 2m)$ is the centrosymmetric matrix, if and only if $\mathbf{A}$ has the form :

$$\mathbf{A} = \begin{pmatrix} \mathbf{B} & \mathbf{J_m C J_m} \\ \mathbf{C} & \mathbf{J_m B J_m} \end{pmatrix}, \text{ and } \mathbf{Q}^{\mathrm{T}}\mathbf{A}\mathbf{Q} = \begin{pmatrix} \mathbf{B} - \mathbf{J_m C} & 0 \\ 0 & \mathbf{B} + \mathbf{J_m C} \end{pmatrix} \tag{4}$$

where $\mathbf{B} \in R^{m\times m}$, $\mathbf{C} \in R^{m\times m}$ and $\mathbf{Q} = \frac{\sqrt{2}}{2}\begin{pmatrix} \mathbf{I_m} & \mathbf{I_m} \\ -\mathbf{J_m} & \mathbf{J_m} \end{pmatrix}$.

Proof. $\mathbf{Q}^{\mathrm{T}}\mathbf{A}\mathbf{Q} = \frac{\sqrt{2}}{2}\begin{pmatrix} \mathbf{I_m} & -\mathbf{J_m} \\ \mathbf{I_m} & \mathbf{J_m} \end{pmatrix}\begin{pmatrix} \mathbf{B} & \mathbf{J_m C J_m} \\ \mathbf{C} & \mathbf{J_m B J_m} \end{pmatrix}\frac{\sqrt{2}}{2}\begin{pmatrix} \mathbf{I_m} & \mathbf{I_m} \\ -\mathbf{J_m} & \mathbf{J_m} \end{pmatrix} = \begin{pmatrix} \mathbf{B} - \mathbf{J_m C} & 0 \\ 0 & \mathbf{B} + \mathbf{J_m C} \end{pmatrix}$.■

It is seen that determining determinant of $\mathbf{A}$, an exactly determinant of block matrices $\mathbf{B} - \mathbf{J_m C}$ and $\mathbf{B} + \mathbf{J_m C}$.

Based on structure of Hessenberg matrix, we can see that centrosymmetric matrix can be formed as

$$\mathbf{H}=\left(\begin{array}{c|c}\mathbf{B} & \mathbf{J_m C J_m}\\ \hline \mathbf{C} & \mathbf{J_m B J_m}\end{array}\right)=\left(\begin{array}{ccccc|ccccc}
b_{11} & b_{12} & & & & & & & c_{m,2} & c_{m,1}\\
b_{21} & b_{22} & b_{23} & & & & & c_{m-1,3} & c_{m-1,2} & c_{m-1,1}\\
\vdots & \vdots & \ddots & \ddots & & & \iddots & \iddots & \vdots & \vdots\\
b_{m-1,1} & b_{m-1,2} & \cdots & b_{m-1,m-1} & b_{m-1,m} & c_{2,m} & c_{2,m-1} & \cdots & c_{22} & c_{21}\\
b_{m,1} & b_{m,2} & \cdots & b_{m,m-1} & b_{m,m} & c_{1,m} & c_{1,m-1} & \cdots & c_{12} & c_{11}\\ \hline
c_{11} & c_{12} & \cdots & c_{1,m-1} & c_{1,m} & b_{m,m} & b_{m,m-1} & \cdots & b_{m,2} & b_{m,1}\\
c_{21} & c_{22} & \cdots & c_{2,m-1} & c_{2,m} & b_{m-1,m} & b_{m-1,m-1} & \cdots & b_{m-1,2} & b_{m-1,1}\\
\vdots & \vdots & \iddots & \iddots & & & \ddots & \ddots & \vdots & \vdots\\
c_{m-1,1} & c_{m-1,2} & c_{m-1,3} & & & & & b_{23} & b_{22} & b_{21}\\
c_{m,1} & c_{m,2} & & & & & & & b_{12} & b_{11}
\end{array}\right)$$

3.2. Block Centrosymmetric Martix on Hessenberg Matrix

Based on centrosymmetric matrix, it is formed block matrix as

$$\mathbf{B}=\begin{pmatrix} b_{11} & b_{12} & & & \\ b_{21} & b_{22} & b_{23} & & \\ \vdots & \vdots & \ddots & \ddots & \\ b_{m-1,1} & b_{m-1,2} & \cdots & b_{m-1,m-1} & b_{m-1,m}\\ b_{m,1} & b_{m,2} & \cdots & b_{m,m-1} & b_{m,m}\end{pmatrix} \text{ and } \mathbf{J_m C}=\begin{pmatrix} c_{11} & c_{12} & & & \\ c_{21} & c_{22} & c_{23} & & \\ \vdots & \vdots & \ddots & \ddots & \\ c_{m-1,1} & c_{m-1,2} & \cdots & c_{m-1,m-1} & c_{m-1,m}\\ c_{m,1} & c_{m,2} & \cdots & c_{m,m-1} & c_{m,m}\end{pmatrix}$$

where both of them are Hessenberg matrices.

Based on Lemma 5, we have orthogonal matrix $\mathbf{P}=\frac{\sqrt{2}}{2}\begin{pmatrix}\mathbf{I_m} & \mathbf{I_m}\\ -\mathbf{J_m} & \mathbf{J_m}\end{pmatrix}$, so

$$\mathbf{P^T H P}=\begin{pmatrix}\mathbf{B}-\mathbf{J_m C} & 0\\ 0 & \mathbf{B}+\mathbf{J_m C}\end{pmatrix}=\begin{pmatrix}\mathbf{M} & 0\\ 0 & \mathbf{N}\end{pmatrix} \tag{5}$$

where $\mathbf{M}=\mathbf{B}-\mathbf{J_m C}$, $\mathbf{N}=\mathbf{B}+\mathbf{J_m C}$ and $\mathbf{M},\mathbf{N}$ are Hessenberg matrices.

Assume $\widetilde{\mathbf{M}}^{-1}=\begin{pmatrix}\boldsymbol{\alpha}_\mathbf{M} & \mathbf{L_M}\\ h_M & \boldsymbol{\beta}_\mathbf{M}^\mathbf{T}\end{pmatrix}$, $\widetilde{\mathbf{N}}^{-1}=\begin{pmatrix}\boldsymbol{\alpha}_\mathbf{N} & \mathbf{L_N}\\ h_N & \boldsymbol{\beta}_\mathbf{N}^\mathbf{T}\end{pmatrix}$ and $\widetilde{\mathbf{M}}=\begin{pmatrix}\mathbf{e}_1^\mathbf{T} & 0\\ \mathbf{M} & \mathbf{e_m}\end{pmatrix}$, $\widetilde{\mathbf{N}}=\begin{pmatrix}\mathbf{e}_1^\mathbf{T} & 0\\ \mathbf{N} & \mathbf{e_m}\end{pmatrix}$ (6)

then $\widetilde{\mathbf{M}}\cdot\widetilde{\mathbf{M}}^{-1}=\begin{pmatrix}\mathbf{e}_1^\mathbf{T} & 0\\ \mathbf{M} & \mathbf{e_m}\end{pmatrix}\begin{pmatrix}\boldsymbol{\alpha}_\mathbf{M} & \mathbf{L_M}\\ h_M & \boldsymbol{\beta}_\mathbf{M}^\mathbf{T}\end{pmatrix}=\begin{pmatrix}\mathbf{e}_1^\mathbf{T}\boldsymbol{\alpha}_\mathbf{M} & \mathbf{e}_1^\mathbf{T}\mathbf{L_M}\\ \mathbf{M}\boldsymbol{\alpha}_\mathbf{M}+\mathbf{e_m}h_M & \mathbf{M}\mathbf{L_M}+\mathbf{e_m}\boldsymbol{\beta}_\mathbf{M}^\mathbf{T}\end{pmatrix}=\begin{pmatrix}1 & 0\\ 0 & \mathbf{I_m}\end{pmatrix}=\mathbf{I}_{m+1}$.

It seen that
$$\mathbf{M}\boldsymbol{\alpha}_\mathbf{M}+\mathbf{e_m}h_M=0 \tag{7}$$
and
$$\mathbf{M}\mathbf{L_M}+\mathbf{e_m}\boldsymbol{\beta}_\mathbf{M}^\mathbf{T}=\mathbf{I_M}.$$

It shows that $h_M\neq 0$, if $h_M=0$ than $\boldsymbol{\alpha}_\mathbf{M}=0$ because $\mathbf{M}\boldsymbol{\alpha}_\mathbf{M}=0$ from (6). It is implied that $\widetilde{\mathbf{M}}^{-1}$ is a nonsingular matrix, which is the contradiction. By the same way, we have a $\widetilde{\mathbf{N}}^{-1}$ nonsingular matrix.

3.3. Determinant Centrosymmetric Matrix

From (5), then

$$\mathbf{H}=\mathbf{P}\begin{pmatrix}\mathbf{B}-\mathbf{J_m C} & 0\\ 0 & \mathbf{B}+\mathbf{J_m C}\end{pmatrix}\mathbf{P^T}=\mathbf{P}\begin{pmatrix}\mathbf{M} & 0\\ 0 & \mathbf{N}\end{pmatrix}\mathbf{P^T},$$

so

$$\det(\mathbf{H})=\det(\mathbf{P})\cdot\det\begin{pmatrix}\mathbf{M} & 0\\ 0 & \mathbf{N}\end{pmatrix}\cdot\det(\mathbf{P^T})=\det(\mathbf{M})\cdot\det(\mathbf{N})$$

Having the above preposition, the theorem of the determinant centrosymmetric matrix is obtained as follows.

Theorem 6 [11]
Let $\mathbf{H}, \mathbf{P}, \mathbf{M}, \mathbf{N}, \tilde{\mathbf{M}}^{-1}, \tilde{\mathbf{N}}^{-1}$ are defined as aforementioned. Then

$$\det(\mathbf{H}) = h_N \cdot h_M \cdot \prod_{i=1}^{m-1} \left(g_{i,i+1} \cdot q_{i,i+1}\right).$$

Proof. Let $\mathbf{M} = \left(g_{ij}\right)_{m\times m} \in R^{m\times m}$, $\mathbf{N} = \left(q_{ij}\right)_{m\times m} \in R^{m\times m}$. From Lemma 3, it can be seen that

$$\det(\mathbf{M}) = (-1)^m h_M \cdot \det(\tilde{\mathbf{M}}) = (-1)^m h_M \cdot \prod_{i=1}^{m-1} g_{i,i+1}, \ \det(\mathbf{N}) = (-1)^m h_N \cdot \det(\tilde{\mathbf{N}}) = (-1)^m h_N \cdot \prod_{i=1}^{m-1} q_{i,i+1}$$

then $\det(\mathbf{H}) = (-1)^m h_M \cdot \prod_{i=1}^{m-1} g_{i,i+1} \cdot (-1)^m h_N \cdot \prod_{i=1}^{m-1} q_{i,i+1} = h_N \cdot h_M \cdot \prod_{i=1}^{m-1} \left(g_{i,i+1} \cdot q_{i,i+1}\right)$. ■

Numerical Example
Given the following centrosymmetric matrix

$$\mathbf{A} = \begin{pmatrix} 1 & 1 & 0 & 0 & 0 & 0 & 2 & 1 \\ 0 & 2 & 2 & 0 & 0 & 1 & 3 & 0 \\ 1 & 2 & 2 & 1 & 2 & 2 & 2 & 1 \\ 1 & 0 & 1 & 1 & 4 & 1 & 3 & 2 \\ 2 & 3 & 1 & 4 & 1 & 1 & 0 & 1 \\ 1 & 2 & 2 & 2 & 1 & 2 & 2 & 1 \\ 0 & 3 & 1 & 0 & 0 & 2 & 2 & 0 \\ 1 & 2 & 0 & 0 & 0 & 0 & 1 & 1 \end{pmatrix} \text{ then } \mathbf{B} = \begin{pmatrix} 1 & 1 & 0 & 0 \\ 0 & 2 & 2 & 0 \\ 1 & 2 & 2 & 1 \\ 1 & 0 & 1 & 1 \end{pmatrix}, \ \mathbf{C} = \begin{pmatrix} 2 & 3 & 1 & 4 \\ 1 & 2 & 2 & 2 \\ 0 & 3 & 1 & 0 \\ 1 & 2 & 0 & 0 \end{pmatrix} \text{ and } \mathbf{J}_4 = \begin{pmatrix} 0 & 0 & 0 & 1 \\ 0 & 0 & 1 & 0 \\ 0 & 1 & 0 & 0 \\ 1 & 0 & 0 & 0 \end{pmatrix}.$$

Thus

$$\mathbf{M} = \begin{pmatrix} 0 & -1 & 0 & 0 \\ 0 & -1 & 1 & 0 \\ 0 & 0 & 0 & -1 \\ -1 & -3 & 0 & -3 \end{pmatrix}, \ \mathbf{N} = \begin{pmatrix} 2 & 3 & 0 & 0 \\ 0 & 5 & 3 & 0 \\ 2 & 4 & 4 & 3 \\ 3 & 3 & 2 & 5 \end{pmatrix} \text{ and } \tilde{\mathbf{M}} = \begin{pmatrix} 1 & 0 & 0 & 0 & 0 \\ 0 & -1 & 0 & 0 & 0 \\ 0 & -1 & 1 & 0 & 0 \\ 0 & 0 & 0 & -1 & 0 \\ -1 & -3 & 0 & -3 & 1 \end{pmatrix}, \ \tilde{\mathbf{N}} = \begin{pmatrix} 1 & 0 & 0 & 0 & 0 \\ 2 & 3 & 0 & 0 & 0 \\ 0 & 5 & 3 & 0 & 0 \\ 2 & 4 & 4 & 3 & 0 \\ 3 & 3 & 2 & 5 & 1 \end{pmatrix}.$$

So, it can be founded that

$$\tilde{\mathbf{M}}^{-1} = \begin{pmatrix} 1 & 0 & 0 & 0 & 0 \\ 0 & -1 & 0 & 0 & 0 \\ 0 & -1 & 1 & 0 & 0 \\ 0 & 0 & 0 & -1 & 0 \\ -1 & -3 & 0 & -3 & 1 \end{pmatrix} \text{ and } \tilde{\mathbf{N}}^{-1} = \begin{pmatrix} 1 & 0 & 0 & 0 & 0 \\ -0.6667 & 0.3333 & 0 & 0 & 0 \\ 1.1111 & -0.5556 & 0.3333 & 0 & 0 \\ -1.2593 & 0.2963 & -0.4444 & 0.3333 & 0 \\ 3.0741 & -1.3704 & 1.5556 & -1.6667 & 1 \end{pmatrix}.$$

Therefore,

$$h_M = 1, \ h_N = 3.0741, \text{ and } \prod_{i=1}^{4} \left(g_{i,i+1}\right) = (-1)(1)(-1) = 1, \ \prod_{i=1}^{4} \left(q_{i,i+1}\right) = (3)(3)(3) = 27$$

then

$$\det(\mathbf{A}) = h_N \cdot h_M \cdot \prod_{i=1}^{m-1} \left(g_{i,i+1} \cdot q_{i,i+1}\right) = (1)(3.0741)(1)(27) = 83.0007.$$

4. Conclusion

From the previous sections, it shows that centrosymmetric matrix has a special structure which can be formed block matrices. Therefore, it can be exploited for determining determinant of centrosymmetric matrix with special block structure on lower Hessenberg matrix. So, an efficient algorithm for this case is obtained.

5. Acknowledgments
The authors would like to express heartfelt thanks to the LPDP for helping financial support. We also appreciate the Faculty of Science and Mathematics, Diponegoro University, Semarang, Indonesia for supporting our project.

References
[1] Datta L and Morgera S D 1989 *Circuits Systems and Signal Process* (*CSSP*) **8** 71 doi:10.1007/BF01598746
[2] Li H, Zhao D, Dai F and Su D 2011 *Appl. Math. Comput.* **218** 4962 doi: 10.1016/j.amc.2011.10.061
[3] Chen Y H and Yu C Y 2011 *Appl. Math. Comput.* **218** 4433 doi: 10.1016/j.amc.2011.10.022
[4] Elouafi M and Aiat Hadj A D 2009 *Appl. Math. Comput.* **214** 497 doi: 10.1016/j.amc.2009.04.017
[5] Roland A and Sweet 1969 *Comm.* ACM **12** 330 doi: 10.1145/363011.363152
[6] Tomohiro S 2008 *Appl. Math. Comput.* **196** 835 doi: 10.1016/j.amc.2007.07.015
[7] Tomohiro S 2007 *Appl. Math. Comput.* **187** 785 doi: 10.1016/j.amc.2006.08.156
[8] Tomohiro S 2008 *Appl. Math. Comput.* **201** 561 doi: 10.1016/j.amc.2007.12.047
[9] Liu Z Y 2003 *Appl. Math. Comput.* **141** 297 doi: 10.1016/S0096-3003(02)00254-0
[10] Golub G H and Van Loan C F 2013 *Matrix Computations, fourth ed p* (Baltimore : The Johns Hopkins University Press) chapter 7 pp 376-384
[11] Zhao D and Li H 2015 *Appl. Math. Comput.* **250** 721 dx.doi.org/10.1016/j.amc.2014.09.114

IOP Conf. Series: Journal of Physics: Conf. Series **824** (2017) 012029 doi:10.1088/1742-6596/824/1/012029

The Analysis of Fixed Final State Optimal Control in Bilinear System Applied to Bone Marrow by Cell-Cycle Specific (CCS) Chemotherapy

E Rainarli and K E Dewi

Universitas Komputer Indonesia, Bandung, Indonesia

Corresponding author: ednawati.rainarli@email.unikom.ac.id

Abstract. The research conducted by Fister & Panetta shown an optimal control model of bone marrow cells against Cell Cycle Specific chemotherapy drugs. The model used was a bilinear system model. Fister & Panetta research has proved existence, uniqueness, and characteristics of optimal control (the chemotherapy effect). However, by using this model, the amount of bone marrow at the final time could achieve less than 50 percent from the amount of bone marrow before given treatment. This could harm patients because the lack of bone marrow cells made the number of leukocytes declining and patients will experience leukemia. This research would examine the optimal control of a bilinear system that applied to fixed final state. It will be used to determine the length of optimal time in administering chemotherapy and kept bone marrow cells on the allowed level at the same time. Before simulation conducted, this paper shows that the system could be controlled by using a theory of Lie Algebra. Afterward, it shows the characteristics of optimal control. Based on the simulation, it indicates that strong chemotherapy drug given in a short time frame is the most optimal condition to keep bone marrow cells spine on the allowed level but still could put playing an effective treatment. It gives preference of the weight of treatment for keeping bone marrow cells. The result of chemotherapy's effect (u) is not able to reach the maximum value. On the other words, it needs to make adjustments of medicine's dosage to satisfy the final treatment condition e.g. the number of bone marrow cells should be at the allowed level.

1. Introduction

Chemotherapy is the type of medication given to patients with cancer by using medicines [1]. Chemotherapy treatment is given especially for cancer cells with high proliferation ability such as breast cancer or blood cancers [1]. Administering chemotherapy is not only killing cells of cancer but also killing normal cells in the body. One of the normal cells affected during the chemotherapy treatment is bone marrow cells. To minimize the influence of chemotherapy against bone marrow cells then it is given therapeutics that cared for cell growth phase. Type of this medication chemotherapy called chemotherapy Cell Cycle Specific (CCS). The examples of chemotherapy medicine that used are Taxol and Cyclophosphamide [2]. Model of mathematics that discussed the influence of giving chemotherapy (CSS) against to bone marrow cells was a model conducted by Fister and Panetta [2]. It was used the drug effects as a control of bone marrow cells changes. The objective function used was maximizing the number of bone marrow cells and maximizing medicine effects of cancer cells. However, both factors having influence mutual contrary. It means when we want to maximize the number of bone marrow cells then the medicine effect would be lower and vice versa. This research showed that time of treatment with a short interval time and high dosage of a drug continue to be effective in reducing bone marrow cells died that caused of cancer drugs. Ledzewicz et. al used this mathematical model and added objective of Bolza type in representing the effect of drug dosage [3]. The other research, they combined this model by adding with pharmacokinetics equation which models the time-evolution of the drug's concentration in the body [4]. Alamir M. et. al design optimal control to maximize the quantity of drug injected over treatment period while continuously restricted the value of states [5]. Skandari et. al used a new linear quadratics function to describe the effect of drug dosage in CCS chemotherapy [6].

The research uses the model of Fister & Panetta and adds a requirement at the end of treatment, some bone marrow cells at the level of allowed. The certain dosages of drugs must be able to keep the number of cells not less than 50 % of normal circumstances [1], [2]. The results would be simulated to get the optimal time in administering CCS chemotherapy drugs, so it would be known whether short time intervals of treatment and continue dosage of a drug to be effective in this case. Before simulated done, it would be shown the system controllable using Lie Algebra and analyzed the characteristics optimal control of bilinear system model using Hamilton equation.

2. Model Of Fixed Final State Optimal Control Applied To Bone Marrow By Cell-Cycle Specific (CSS) Chemotherapy

By Fister & Panetta were made a model describe the state of bone marrow cells with the influence of giving chemotherapy medicine in it [2]. Figure 1 showed the dynamic model of bone marrow by administering chemotherapy drugs. In this model, the cells divide into two phases which are P and Q. P is a proliferating phase, Q is a quiescent phase. At P phase, some of the cells died naturally or died because of administering medication, and there are cells in P who moved to quiescent phase Q. Furthermore, cells in P phase can grow due to proliferating phase. The cell on Q phase partially transformed into blood cells and get into the blood stream. Besides of that, the cell in Q can increase or reduce as a result of the cells coming from P and out of Q. Transition cells can be seen more details in Figure 1. As a control, it uses the effect of chemotherapy where compensation of less dose of medicine in the patient body resulting the number of bone marrow cells died would increase.

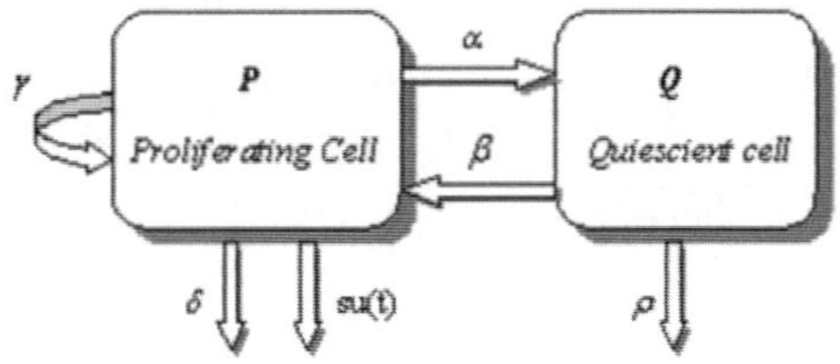

Figure 1. Model of Bone Marrow Cells with Chemotherapy CCS

Let, α = the rate of displacement cells from P to Q, β = the rate of displacement cells from Q to P, γ = cell growing rate, δ = natural death rate, ρ = the rate of cell that entering into blood stream, s = the strength of treatment/the strength of chemotherapy drugs, u = the effect of chemotherapy treatment. All the parameter applies to a unit of the day. The values of those parameters were given in Table 1 [2]. For control of u, was defined admissible space that given by equation (1)

$$U = \{u \text{ admissible function} | 0 \le u(t) \le 1, t \in [0,T]\} \quad (1)$$

$u(t)=1$ Shows maximal chemotherapy. On the other word, at time t the state of bone marrow was at the maximal influence of the drug. $u(t)=0$ Shows no chemotherapy drug in bone marrow cells. Based on Figure 1, it could be shown by equation (2) and (3).

$$\dot{P} = (\gamma - \delta - \alpha - su(t))P + \beta Q, P(0) = P_0 \quad (2)$$

$$\dot{Q} = \alpha P - (\rho + \beta)Q, \ Q(0) = Q_0 \quad (3)$$

The objective function used in this model is maximizing the effect of chemotherapy treatment without incurring excessive damage of cells. The function uses equation (4).

$$J(u)=\int_0^T\left[a\left(P(t)+Q(t)-\frac{b}{2}(1-u(t))^2\right)\right]dt \tag{4}$$

Parameter a was weight to maximize the number of bone marrow cells $P(t)+Q(t)$and parameter b was weight to maximize the chemotherapy effect $u(t)$. Part of $\frac{1}{2}(1-u(t))^2$ was the reduction of the number of bone marrow cells. $1-u(t)$ The amount of chemotherapy drug given to the patient. Weights of a and b would affect the damage of cells. If the number of patient's blood cells was below the desired level, then $\frac{a}{b}>1$ was chosen. The aim was maximizing the number of bone marrow cells. Conversely, if we want to maximize drug given, then the value of weight b would become greater, so $\frac{a}{b}<1$.

Table 1. The Value of Parameters

Value (Range)/Day	
$\gamma=1{,}47;(0{,}6667-2)$	$\delta=0$
$\alpha=5{,}643;(4{,}94-6{,}12)$	$\beta=0{,}48$
$\rho=0{,}164$	

In this research, the model would complete by adding a final state condition based on the fact that the number of bone marrow cells are not more than 50 % of the total cell before treatment, so in this model is added a limit and we have equation (5)

$$P(t)+Q(t)=0.5(P_0+Q_0), \tag{5}$$

By this equation, it would be a guarantee that total bone marrow cells at the final time would equal to a half of total cells before treatment. P_0 Indicated the number of proliferating cells at the beginning of treatment and Q_0 indicated the number of quiescent cells at the beginning of treatment.

3. Result and Discussion

Based on equation (2) and (3), a model of this system was model control of a bilinear system. The bilinear system is part of the nonlinear system [7]. One of the advantages of using bilinear system can linearize around equilibrium point as a linear system. There are some discussions about solving the optimal control of bilinear system especially using indirect method, there are Hamilton–Jacobi–Bellman (HJB) equation [8] or transform optimal control problem into a nonlinear two-point boundary value (TPBVP)[9]. In this part, we will use TPBVP to find the optimal solution, but first, it would be shown controllability of bilinear system and continue with discussing about characteristics of optimal control. At the end of this part, it would be shown the result of the simulation.

3.1. Controllability System

To ensure that systems could be controllable, it was used control theory of Lie Algebra [10], [11]. From the equation (2) and (3) it can be rewritten by equation (6).

$$A=\begin{bmatrix}a_{11} & a_{12}\\ a_{21} & -a_{22}\end{bmatrix}, B=\begin{bmatrix}-s & 0\\ 0 & 0\end{bmatrix}, x=\begin{bmatrix}x_1\\ x_2\end{bmatrix}=\begin{bmatrix}P\\ Q\end{bmatrix} \tag{6}$$

where are $a_{11}=\gamma-\delta-\alpha$, $a_{12}=\beta$, $a_{21}=\alpha$, $a_{22}=\rho+\beta$?

So, we have two vector fields $f(x)$ and $g(x)$ where are shown by equation (7),

$$f(x)=Ax=\begin{bmatrix}a_{11} & a_{12}\\ a_{21} & -a_{22}\end{bmatrix}\begin{bmatrix}x_1\\ x_2\end{bmatrix}=\begin{bmatrix}a_{11}x_1+a_{12}x_2\\ a_{21}x_1-a_{22}x_2\end{bmatrix},\ g(x)=Bx=\begin{bmatrix}-s & 0\\ 0 & 0\end{bmatrix}\begin{bmatrix}x_1\\ x_2\end{bmatrix}=\begin{bmatrix}-sx_1\\ 0\end{bmatrix} \tag{7}$$

Because of $f(x)=Ax$ and $g(x)=Bx$, it is clear that A and B are Jacobian matrices that can be written into equation (8)

$$\frac{\partial f}{\partial x}=\begin{bmatrix} a_{11} & a_{12} \\ a_{21} & -a_{22} \end{bmatrix}=A,\ \frac{\partial g}{\partial x}=\begin{bmatrix} -s & 0 \\ 0 & 0 \end{bmatrix}=B \tag{8}$$

It could be built a distribution Δ that has characteristics of the maximal integral manifold to ensure that system is weakly controllable [10]. Next, it would be shown that the controllability rank condition was fulfilled, e.g. $\dim\Delta_c(x)=n$.

Proof:

Using Lie Bracket definition, we could determine $[A,B],[A,[A,B]],[A,[A,[A,B]]],[A,[A,[A,[A,B]]]]$ in equation (9), (10), (11) and (12)

$$[A,B]=BA-AB=s\begin{bmatrix} 0 & -a_{12} \\ a_{21} & 0 \end{bmatrix} \tag{9}$$

$$[A,[A,B]]=s\begin{bmatrix} -2a_{12}a_{21} & (a_{11}+a_{22})a_{12} \\ (a_{11}+a_{22})a_{21} & 2a_{12}a_{21} \end{bmatrix} \tag{10}$$

$$[A,[A,[A,B]]]=\{(a_{11}+a_{22})^2+4a_{12}a_{21}\}[A,B] \tag{11}$$

$$[A,[A,[A,[A,B]]]]=\{(a_{11}+a_{22})^2+4a_{12}a_{21}\}[A,[A,B]] \tag{12}$$

It could be shown new vector fields that are made by vector fields $f(x)$ and $g(x)$ using Lie bracket [10],

$$[f,g](x)=\frac{\partial g}{\partial x}f(x)-\frac{\partial f}{\partial x}g(x)=\begin{bmatrix} -sa_{12}x_2 \\ sa_{21}x_1 \end{bmatrix}=[A,B]x, etc. \tag{13}$$

Then based on equation (13) it will be constructed distribution $\Delta_c=span\{[A,B]x,[A,[A,B]]x\}$, where are

$$C=\{\tau\in V(R^2):\tau(x)=Tx, T\in span\{[A,B],[A,[A,B]]\}\}. \tag{14}$$

So, using equation (14) we had matrix

$$[[f(x),g(x)][f(x),[f(x),g(x)]]]=[[A,B]x,[A,[A,B]x]]=\begin{bmatrix} -sa_{12}x_2 & -2sa_{12}a_{21}x_1+s(a_{11}+a_{22})a_{12}x_2 \\ sa_{21}x_1 & s(a_{11}+a_{22})a_{21}x_1+2sa_{12}a_{21}x_2 \end{bmatrix} \tag{15}$$

Form the equation (15) we had that $\dim\Delta_c=2$, for each x in maximal integral manifold, and it is applied equation (16)

$$U^*=\{x\in R^2:(a_{11}+a_{22})x_1+a_{12}x_2\neq 0\} \tag{16}$$

So, the system was weakly controllable for each x is defined in U^*.

3.2. Characteristics of Optimal Control

To know the characteristic of optimal control then it was used Hamiltonian equation. Lewis discusses how to solve optimal control for fixed final state [12]. Based on that, so from equation (2) and (3) we could build a Hamiltonian equation (17)

$$H=a(P+Q)-\frac{b}{2}(1-u)^2+\lambda_1(\gamma-\delta-\alpha-su)P+\lambda_1\beta Q+\lambda_2\alpha P-\lambda_2(\rho+\beta)Q+w_1(t)u+w_2(t)(1-u) \tag{17}$$

where $w_1(t)\geq 0,\ w_2(t)\geq 0$ are penalty multiplications that were the full field

$w_1(t)u=0\ \ w_2(t)(1-u)=0$ so (18)

when u^* is optimal.

$$u^*(t) = \frac{b - \lambda_1 sP + w_1(t) - w_2(t)}{b} \tag{19}$$

Using Maximum Pontryagin Principle, to get optimal control without w_1 and w_2 explicitly then the equation (19) will be divided into three cases:

1. For $\{t|0 < u^*(t) < 1\}$, to satisfy equation (18), then it should be $w_1(t) = 0 = w_2(t)$ so the equation (19) could be written into

$$u^*(t) = \frac{b - \lambda_1 sP}{b} \tag{20}$$

2. For $\{t|u^*(t) = 1\}$, $w_1(t) = 0$ and $w_2(t) \neq 0$, so it must be

$$u^*(t) = 1 \leq \frac{b - \lambda_1 sP}{b} \tag{21}$$

3. For $\{t|u^*(t) = 0\}$, it should be $w_2(t) = 0$ and $w_1(t) \neq 0$, so

$$\frac{b - \lambda_1 sP}{b} \leq 0 = u^*(t) \tag{22}$$

Based on the equation (20), (21), and (22) then the optimal control can be written as equation (23)

$$u^*(t) = \min\left(1, \left(\frac{b - \lambda_1 sP}{b}\right)^+\right), \tag{24}$$

With boundary condition $P(0) = P_0, Q(0) = Q_0$ and $P(T) + Q(T) = 0.5(P_0 + Q_0)$. Because of the initial state and final state are given then $dx_T = 0, dT \neq 0$ and to full fill, that condition is given in Lewis then it should be $H|_T = 0$ [12].

3.3. Simulation

Not only held the analysis on the model but in this research was also did a simulation of optimal control model. To the needs of simulation would be used Matlab, while the completion of model optimal was a form of the solution of boundary value problems for the ordinary differential equation. Some solution of optimal control problems is solved with variation parameter method to optimal control issues were described by Avvakumov [13]. To demonstrate the numerical solution of boundary problem in Matlab, it was used a function of *bp4c* [14]. The use of this function was also described by Wang [15]. Specifically, Wang explains how to solve the optimal control problems uses indirect method especially in the case of fixed state. However, in this simulation was used toolbox Tomlabs for optimization to Matlab [16]. In testing, the parameter value of model was used from Table 1.

From the previous part, it had shown that the system could be controlled and it could be shown the characteristics of optimal control. On this part, it would be discussed the result of model simulation obtained. For the simulations, it would be divided into three cases, e.g.:

1. The influence of changing weights *a* and *b* to the objective function and the length of time treatment.
2. The influence of changing the strength of drugs *s* through objective function and length of time treatment.
3. The influence of changing the time of treatment *T* against objective function and length of time treatment.

Figure 2a-shows the case of the treatment for 21 days with $a = b = 1, s = 1$. The obtained result was that it took waiting time for 10 days before the treatment to be done so, at the end of the treatment process, the condition of bone marrow cells still stayed above 50 percent of the total bone marrow cells at the beginning of treatment. Figure 2b showed the comparing result of simulation to a value of $a > b, a = b$ and $a < b$. When $a > b, a = b$, both of that condition showed that optimal control u^* could achieve 1. However, $a < b$ is given when we put a higher priority of the patient condition then the effect of chemotherapy drugs. In the other word, our priority was maintaining many bone marrow cells at the final state. We saw that at $a < b$, the value of u^* could not achieve 1 at the end of treatment. It means that not whole of the quantity of medicine could be given.

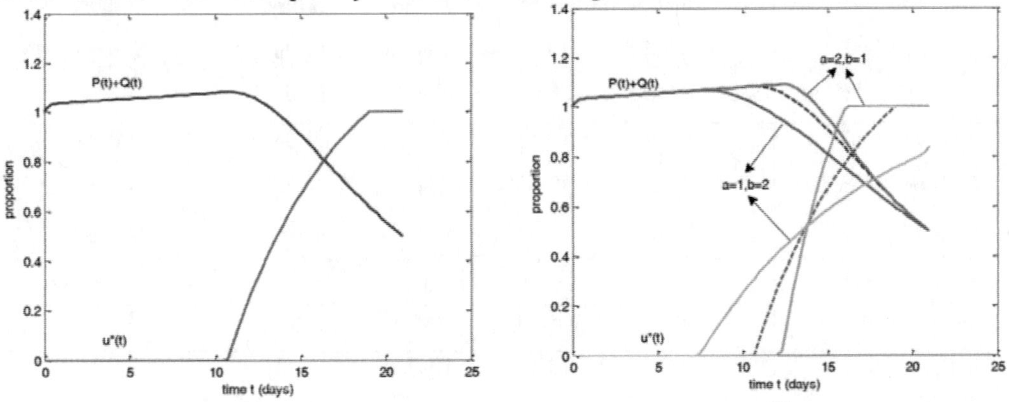

(a) (b)

Figure 2. (a) Comparison value $P(t)+Q(t)$ to $u(t)$ when $a=1, b=1, s=1$, (b) Comparison value $P(t)+Q(t)$ to $u(t)$ when $a=2\&b=1$ $a=1\&b=1$, $a=1\&b=2$

From Figure 3a it shows the effect of s change in value. We saw that when the strength of chemotherapy was increasing then waiting time of medicine given become longer. Whereas in Figure 3b describes the effect of time changing the optimal condition of $P(t)+Q(t)$ and u^*. It shows that in the short time $T = 7$ there was no waiting time in given treatment.

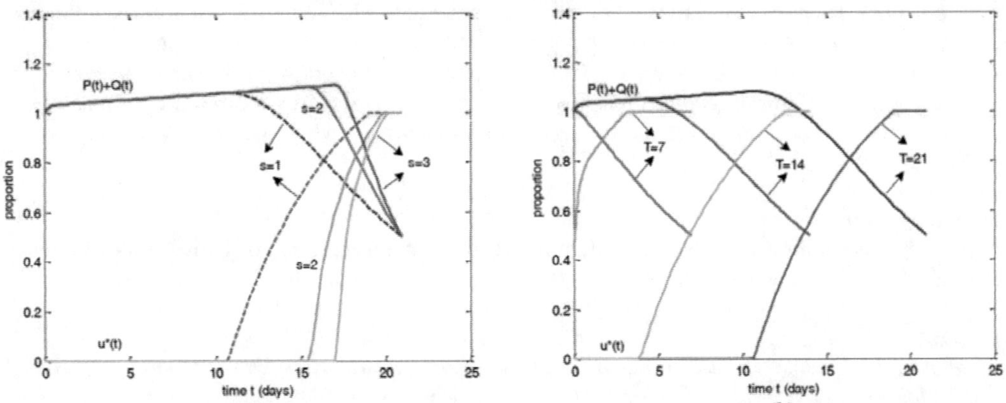

(a) (b)

Figure 3. (a) Comparison value $P(t)+Q(t)$ to $u(t)$ when $s=1, s=2\&s=3$, (b) Comparison value $P(t)+Q(t)$ to $u(t)$ when $a=2\&b=1$, $a=1\&b=1$, $a=1\&b=2$

Further, in Table 2, it shows results of a simulation of three scenarios as delivered above. Based on Table 2, when $a < b$ the value of u^* at the final time only achieved 0,8362 with the longest time treatment. By increasing the strength of chemotherapy medicine (s) with the same observation term, which is 21 days, the values of objective functions are similar. It means that giving a higher dose would be best based on normal condition. In the case of the period of observation change, Table 2 show that the largest value of J happened at time observation 21 days. Table 2 also shows that on 7 days of treatment, there is no waiting time. However, in this conditioning treatment with a short time produce the smallest values of *J*.

Table 2. The Result of Testing

T	Scenario	*u*(*T*)	*P*(*T*)+*Q*(*T*)	*J*	Waiting Time	Time Treatment	Duration of The Influence of Drug
21	*a*=2, *b*=1, *s*=1	1	0,5002	33,5580	12 days	9 days	5 days
	a=1, *b*=1, *s*=1	1	0,5002	13,5374	10 days	11 days	3 days
	a=1, *b*=2, *s*=1	0,8362	0,5001	7,5180	7 days	14 days	-
21	*a*=1, *b*=1, *s*=1	1	0,5002	13,5374	10 days	11 days	3 days
	a=1, *b*=1, *s*=2	1	0,5002	12,9306	15 days	6 days	3 days
	a=1, *b*=1, *s*=3	1	0,5004	12,7310	16 days	5 days	2 days
7	*a*=1, *b*=1, *s*=1	1	0,5001	5,1457	-	7 days	4 days
14	*a*=1, *b*=1, *s*=1	1	0,5001	9,3407	2 days	9 days	3 days
21	*a*=1, *b*=1, *s*=1	1	0,5002	13,5374	10 days	11 days	3 days

4. Conclusion and Suggestion

Based on the simulation, it could be concluded that addition of boundary condition e.g. final state of bone marrow cell must comply 50 percent of the initial conditions, then the bilinear system was made by Fister still could be used to determine optimal control of CCS chemotherapy. By using Lie Algebra, it showed that the system was controllable. It was also discussed about the characteristic of optimal control of the system. Giving a strength chemotherapy treatment would be an optimal option in maintaining patient condition. If a dose of medicine was increased then the influence of medicine absorbing not maximal, therefore it would be an indicator that to review the number of given doses.

For the next research, the model could be combined with a model of immunotherapy (immunotherapy, a process to increase the amount of bone marrow at the end of therapy) to see the dynamic interaction between chemotherapy treatment and immunotherapy treatment so then we can decide the appropriate time in giving both therapies. The model could be modified by reexamining factors that affect cell of changes besides naturally death and death caused by administering chemotherapy.

Acknowledgments

This research was supported by Research Institute and Community Service of Universitas Komputer Indonesia.

References

[1] Brunton L, Chabner B and Knollman 2011 *Goodman & Gillman's Pharmacological Basis of Therapeutics* vol 12, ed Brunton L (New York: McGraw-Hill) section 8 pp1665 - 755
[2] Fister K R and Panetta J C 2000 *J. Siam App. Math.* **60** 1059
[3] Ledzewicz U and Schattler H 2004 *J. Math. Biosci. Eng.* **1** 95
[4] Ledzewicz U and Schattler H 2007 *J. Mathematical Biosciences* **206** 320
[5] Alamir M and Chareyron S 2007 *J. Optim. Control Appl. Meth.* **28** 175
[6] Skandari M H N, Erfanian H R, Kamyad A V and Mohammadi S 2012 *Euro. J. Exp. Bio.* **2** 562
[7] Pardalos P M and YatsenkoV 2008 *Optimization and Control of Bilinear System: Theory, Algorithm, and Applications*, ed D Z Du (New York: Springer Science+Buisiness Media) pp 1-25
[8] Bellman R 1954 *Bull. Amer. Math. Soc.* **60** 503
[9] L. S. Pontryagin 1959 *Usp. Mat. Nauk* **14** 3
[10] Isidori A 1995 *NonLinear Control System* vol 3, ed Sontag E D et al (London: Springer-Verlag) pp 1-104.

IOP Conf. Series: Journal of Physics: Conf. Series **824** (2017) 012029 doi:10.1088/1742-6596/824/1/012029

[11] Jurdjevic V 1997 *Geometric Control Theory*, ed W Fulton et al (New York: Cambridge University Press) pp 125-40.
[12] Lewis F L, Vrabie D and Syrmos V L 2012 *Optimal Control* vol 3 (USA: John Wiley & Son Inc.) pp 110 - 232
[13] Avvakumov S N and Kiselev Y N 2000 *J. Spectral and Evolution Problems* **10** 147
[14] Shampine L F, Gladwell I and Thompson S 2003 *Solving ODEs with Matlab* (New York: Cambridge University Press) pp 133 - 68.
[15] Wang X 2009 Solving Optimal Control Problem with Matlab – Indirect Methods.
[16] Rutquist P E and Edval M M 2016 *PROPT – Matlab Optimal Control Software* [online].

IOP Conf. Series: Journal of Physics: Conf. Series **824** (2017) 012030 doi:10.1088/1742-6596/824/1/012030

The Bivariate Central Normal Distribution

B Pratikno, Jajang , A C C Putri, and Y W W Haryono

Department of Mathematics
Faculty Mathematics and Natural Science Jenderal Soedirman University Purwokerto, Indonesia

Corresponding author: bpratikto@gmail.com

Abstract. The research studied Probability Density Function (pdf), Cumulative Distribution Function (CDF) and graphical analysis of the bivariate central distribution. The CDF of this distribution is then used to compute power of the test Pre-Test-Test (PTT) in the testing intercept. The tables and graphs of the pdf (and CDF) of the bivariate central normal distribution (BCND) and power of the test in testing intercept are produced using R-code. The results showed that the mean and coefficient correlation have a significant affect on the curve, but the variance is not. The curves tend to be leptokurtic for small coefficient correlation, and they will be mesocurtic for large coefficient correlation. The power of the PTT increases as the CDF of the BCND increases.

1. Introduction

The normal (Gaussian) distribution is the most popular distribution. It often uses in many areas and statistical analysis [3]. Probability Density Function (PDF) of this distribution (univariate normal distribution) of X random variable with mean (μ) and variance(σ^2), $X \square N(\mu,\sigma^2)$, is given by

$$f(x)=\frac{1}{\sigma\sqrt{2\pi}}e^{-\frac{1}{2}\left(\frac{x-\mu}{\sigma}\right)^2}, -\infty < X < \infty \text{, with } -\infty < \mu < \infty \text{ and } 0 < \sigma < \infty . \quad (1)$$

The values of the pdf of the equation (1) have not seen in textbooks yet. Currently, the availability table of statistics of this distribution is a normal standard distribution table for $\mu = 0$ the end $\sigma^2 = 1$.

From the equation (1), it is clear that the computational of the probability integral of the pdf and cumulative distribution function *(CDF)* are very complicated and hard, so they should be numerically computed [8]. Due to the values of the pdf and CDF are critical values that use to accept or reject the null hypothesis (H_0), then we need to create the statistics table for both (the pdf and CDF).

Moreover, there are several problems in statistical analysis in term of two or more of random variables. In this case, all the computational of the pdf and CDF are very complicated and challenging. It is because of distribution is a joint distribution of the normal random variable X and Y. Let ρ is a coefficient correlation between them, then if $\rho \neq 0$, the bivariate central normal distribution (bivariate normal distribution) of X and Y is the joint distribution (multiple) of the marginal and conditional normal distribution [16]. Furthermore, Sahoo [6] has also studied and presented the pdf of the central bivariate normal distribution (BCND) of X and Y, that is

$$f(x,y)=\frac{1}{2\pi\sigma_1\sigma_2\sqrt{1-\rho^2}}e^{-\frac{1}{2}\left(\frac{1}{1-\rho^2}\left(\left[\frac{x-\mu_1}{\sigma_1}\right]^2-2\rho\left[\frac{x-\mu_1}{\sigma_1}\right]\left[\frac{y-\mu_2}{\sigma_2}\right]+\left[\frac{y-\mu_2}{\sigma_2}\right]^2\right)\right)},-\infty<x,y<\infty \quad (2)$$

with $\mu_1,\mu_2 \in$ □ , $\sigma_1,\sigma_2 \in (0,\infty)$, and $\rho \in (-1,1)$. The notation of the BCND is then written as

(X,Y) □ $N(\mu_1,\mu_2,\sigma_1^2,\sigma_2^2,\rho)$.

Many authors already used the BCND, such as Pratikno [2] and Khan [10] in testing intercept using *non-sample prior information* (NSPI). Here, Pratikno [2] used the BCND to compute power of the test of *pre test-test* (PTT) in testing intercept using NSPI on simple regression model, while Khan [9,10], Khan and Saleh [11,13,14], Khan and Hoque[12], Khan and Pratikno [15], Saleh [1] and Yunus [7] contributed to developing the research in estimation area. All the authors used *R-code* and *R-package* especially **mvtnorm** on the *multivariate (bivariate/BCND) normal distribution*.

Due to the importance of the BCND, we studied more detail of this distribution particularly in computational of the PDF, CDF, graphical analysis and its applied in NSPI testing. The steps of the research methodology are (1) reconstructed and figured the pdf formula, (2) computed CDF table and (3) graphical analysis. *R-code* is used to figure pdf and compute the table of the CDF of the BCND.

The research presented the introduction in Section 1. Reconstructed the pdf formula of the BCND is obtained in Section 2. The graphs of the pdf of BCND and its analysis are given in Section 3. Section 4 explained the computational of the values of the CDF of the BCND. Furthermore, computational of the power of the test is obtained in Section 5, and Section 6 described conclusion and remark.

2. Method

Following Kenny and Keeping [5], we reconstructed the formula of the equation (2) in term of two independent random variables of the univariate normal standard, Z_1 and Z_2, namely $X=\mu_1+\sigma_{11}Z_1+\sigma_{12}Z_2$ and $Y=\mu_2+\sigma_{21}Z_1+\sigma_{22}Z_2$ (3)

where X and Y are random variables of the normal distribution with mean(μ_1,μ_2), -variance $(\sigma_1^2,\sigma_2^2)=(\sigma_{11}^2+\sigma_{12}^2,\sigma_{21}^2+\sigma_{22}^2)$, and covariance$=\rho\sigma_1\sigma_2=\sigma_{11}\sigma_{21}+\sigma_{12}\sigma_{22}$. Moreover, the joint pdf of Z_1 and Z_2 are then written as $f(z_1,z_2)=f(z_1)f(z_2)=\frac{1}{\sqrt{2\pi}}e^{-\frac{1}{2}z_1^2}\frac{1}{\sqrt{2\pi}}e^{-\frac{1}{2}z_2^2}=\frac{1}{2\pi}e^{-\frac{1}{2}(z_1^2+z_2^2)}$. Using matrix properties and a simple modification, we got the equation in matrix form as below

$$\begin{pmatrix}X-\mu_1\\Y-\mu_2\end{pmatrix}=\begin{pmatrix}\sigma_{11}&\sigma_{12}\\\sigma_{21}&\sigma_{22}\end{pmatrix}\begin{pmatrix}Z_1\\Z_2\end{pmatrix} \text{ or } \begin{pmatrix}Z_1\\Z_2\end{pmatrix}=\begin{pmatrix}\sigma_{11}&\sigma_{12}\\\sigma_{21}&\sigma_{22}\end{pmatrix}^{-1}\begin{pmatrix}X-\mu_1\\Y-\mu_2\end{pmatrix}. \quad (4)$$

From the equation (4), we get the squareof Z_1 and Z_2 on equation (5) below

$$\begin{aligned}(Z_1^2+Z_2^2)(\sigma_{11}\sigma_{22}+\sigma_{12}\sigma_{21})^2&=[\sigma_{22}(X-\mu_1)-\sigma_{12}(Y-\mu_2)]^2+[-\sigma_{21}(X-\mu_1)+\sigma_{11}(Y-\mu_2)]^2\\&=\sigma_1^2\sigma_2^2\left(\left[\frac{X-\mu_1}{\sigma_1^2}\right]^2-2\rho\left[\frac{X-\mu_1}{\sigma_1}\right]\left[\frac{Y-\mu_2}{\sigma_2}\right]+\left[\frac{Y-\mu_2}{\sigma_2^2}\right]^2\right),\end{aligned} \quad (5)$$

if $\frac{1}{1-\rho^2}=\frac{\sigma_1^2\sigma_2^2}{(\sigma_{11}\sigma_{22}+\sigma_{12}\sigma_{21})^2}$, then the equation (5) is written as

$$(Z_1^2+Z_2^2)(1-\rho)^2=\left(\left[\frac{X-\mu_1}{\sigma_1^2}\right]^2-2\rho\left[\frac{X-\mu_1}{\sigma_1}\right]\left[\frac{Y-\mu_2}{\sigma_2}\right]+\left[\frac{Y-\mu_2}{\sigma_2^2}\right]^2\right).$$

IOP Conf. Series: Journal of Physics: Conf. Series **824** (2017) 012030 doi:10.1088/1742-6596/824/1/012030

Furthermore, we need Jakobian transformation for setting the joint distribution of X and Y. The Jacobian transformation is then expressed as $J = \frac{1}{\sigma_1 \sigma_2 \sqrt{1-\rho^2}}$, and the joint pdf of the X and Y is given by

$$f(x,y) = \frac{1}{2\pi\sigma_1\sigma_2\sqrt{1-\rho^2}} e^{-\frac{1}{2}(Z_1^2+Z_2^2)} = \frac{1}{2\pi\sigma_1\sigma_2\sqrt{1-\rho^2}} e^{-\frac{1}{2}Q(x,y)} \quad (6)$$

where $Q(x,y) = \frac{1}{1-\rho^2}\left[\frac{(x-\mu_1)^2}{\sigma_1^2} - \frac{2\rho(x-\mu_1)(y-\mu_2)}{\sigma_1\sigma_2} + \frac{(y-\mu_2)^2}{\sigma_2^2}\right]$. Due to the complicated formula on the equation (6), *R-code* is then used to compute the values of the pdf of the equation (6).

3. Result and Discussion

The pdf graph of the equation (2) is made by *R-code*. Here, some simulations of the graphs are presented in Figure 1. However, we did not present all the graphs in this paper. Figure 1 showed that ρ has a significant affect on the form of the curve. The curve tends to be leptokurtic for $\rho = -0.9$, and it will be symmetric (mesokurtic) on $\rho = 0$. We note that for $\rho(\rho < 0)$ the curve tends to be leptocurtic, and it will be mesokurtic for $\rho(\rho > 0)$. From several simulations, we see that the curve is concave-convex with $\mu_1 = \mu_2 = 0$, $\rho = 0$, $\sigma_1^2 = \sigma_2^2$ and tends to be mesokurtic for large σ^2 and leptocurtic for small σ^2. Conversely, the curve will be convex concave for $\mu_1 = \mu_2 = 0$; $\rho = 0$, $\sigma_1^2 \neq \sigma_2^2$ However, a parameter σ^2 is not significant in changing the form of the curve.

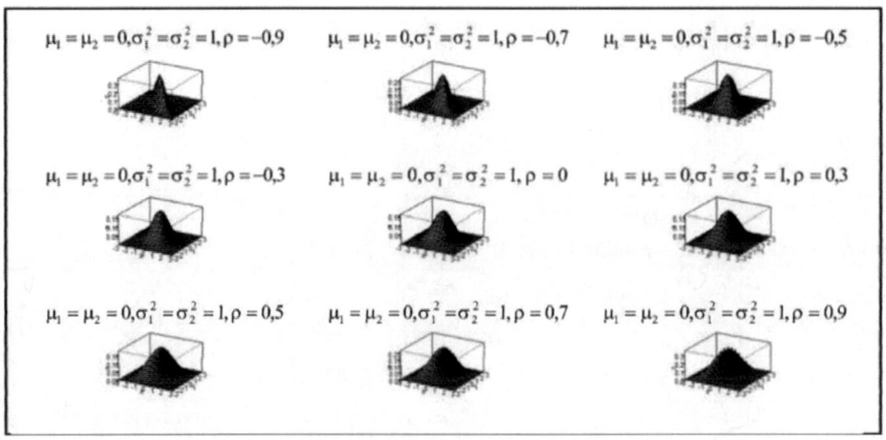

Figure 1. The pdf of the bivariate central normal distribution

To compute the values of the CDF of the bivariate central normal distribution (BCND), *R-code* is used. This is due to the process of its integral is very complicated and difficult. Here, we used *R-command adaptIntegrate* and *package cubature*, and we wrote (run): *adaptIntegrate(f,lowerLimit=c(-100,100), upper Limit=c(x,y))$integral* with lower bound (-100,-100) and up bounds (x , y) to produce the CDF as well as the power of the test of the *pre-test test* (PTT). Table 1-3 below are the values of the fCDF of the BCND for some selected ρ.

Table 1. The values of CDF of the BNCD with $\rho = 0.9$

	X						
Y	-1,5	-1,0	-0,5	0,0	0,5	1,0	1,5
-1,5	0,00000	0,00000	0,00000	0,00001	0,00048	0,00535	0,02286
-1,0	0,00000	0,00000	0,00002	0,00070	0,00898	0,04316	0,09720
-0,5	0,00000	0,00002	0,00080	0,01163	0,06321	0,15886	0,24221
0,0	0,00001	0,00070	0,01163	0,07178	0,20309	0,34205	0,43321
0,5	0,00048	0,00898	0,06321	0,20309	0,38373	0,53283	0,62466
1,0	0,00535	0,04316	0,15886	0,34205	0,53283	0,68269	0,77454
1,5	0,02286	0,09720	0,24221	0,43321	0,62466	0,77454	0,86639

Table 2. The values of CDF of the BNCD with $\rho = 0.5$

	X						
Y	-1,5	-1,0	-0,5	0,0	0,5	1,0	1,5
-1,5	0,01832	0,03242	0,04684	0,05765	0,06357	0,06594	0,06663
-1,0	0,03242	0,06251	0,09748	0,12740	0,14621	0,15487	0,15779
-0,5	0,04684	0,09748	0,16332	0,22688	0,27224	0,29609	0,30530
0,0	0,05765	0,12740	0,22688	0,33333	0,41834	0,46874	0,49084
0,5	0,06357	0,14621	0,27224	0,41834	0,54624	0,63028	0,67149
1,0	0,06594	0,15487	0,29609	0,46874	0,63028	0,74520	0,80696
1,5	0,06663	0,15779	0,30530	0,49084	0,67149	0,80696	0,88471

Table 3. The values of CDF of the BNCD with $\rho = 0.0$

	X						
Y	-1,5	-1,0	-0,5	0,0	0,5	1,0	1,5
-1,5	0,00446	0,01060	0,02061	0,03340	0,04619	0,05621	0,06234
-1,0	0,01060	0,02517	0,04895	0,07933	0,10970	0,13348	0,14806
-0,5	0,02061	0,04895	0,09520	0,15427	0,21334	0,25959	0,28793
0,0	0,03340	0,07933	0,15427	0,25000	0,34573	0,42067	0,46660
0,5	0,04619	0,10970	0,21334	0,34573	0,47812	0,58176	0,64527
1,0	0,05621	0,13348	0,25959	0,42067	0,58176	0,70786	0,78514
1,5	0,06234	0,14806	0,28793	0,46660	0,64527	0,78514	0,87085

The Table 1-3 showed that the values of the CDF is going to zero when X and Y decrease (going to be negative), and it increases to 1 as the X and Y increase (going to be positive). Here, the values of the table (CDF) are similar (the same as) for positive and negative correlation (ρ).

Following Pratikno [2] and Khan [9], there are three tests in *testing intercept* using *non-sample prior information* (NSPI), namely *unrestricted test* (UT), *restricted test* (RT) and *pre-test test* (PTT). In this case, the bivariate central normal distribution is used to compute the power of the *pre-test test* (PTT). Here, the power is a probability of reject H_0 ing *under* $H_a : \theta = \theta_a$, then it is written as $\pi(\theta_a) = P(reject\ H_0 \mid \theta = \theta_a)$ (Wackerly et al., 2008, p455). Following Pratikno (2012), the power of the PTT in testing hypothesis *one-side maximum on simple regression model* is given by

$$\pi^{PTT}(\lambda_1,\lambda_2)=\phi\left(z_{\alpha_3}-\frac{\lambda_2\sqrt{S_{xx}}}{\sigma\sqrt{n}}\right)\left(1-\phi\left(z_{\alpha_2}-\frac{\lambda_1+\lambda_2\bar{X}}{\sigma}\right)\right)$$
$$=+c_{1\rho}\left(z_{\alpha_3}-\frac{\lambda_2\sqrt{S_{xx}}}{\sigma\sqrt{n}},z_{\alpha_1}-\frac{\lambda_1}{m_1},\rho\neq 0\right),$$

Where $m_1=\sigma\left[1+\frac{n\bar{X}^2}{S_{xx}}\right]^{1/2}$, $\phi(x)$ is cumulative distribution function (CDF) of the normal distribution and $c_{1\rho}$ is probability integral of the normal bivariate distribution? The graphs of the power of the PTT are presented in Figure 2.

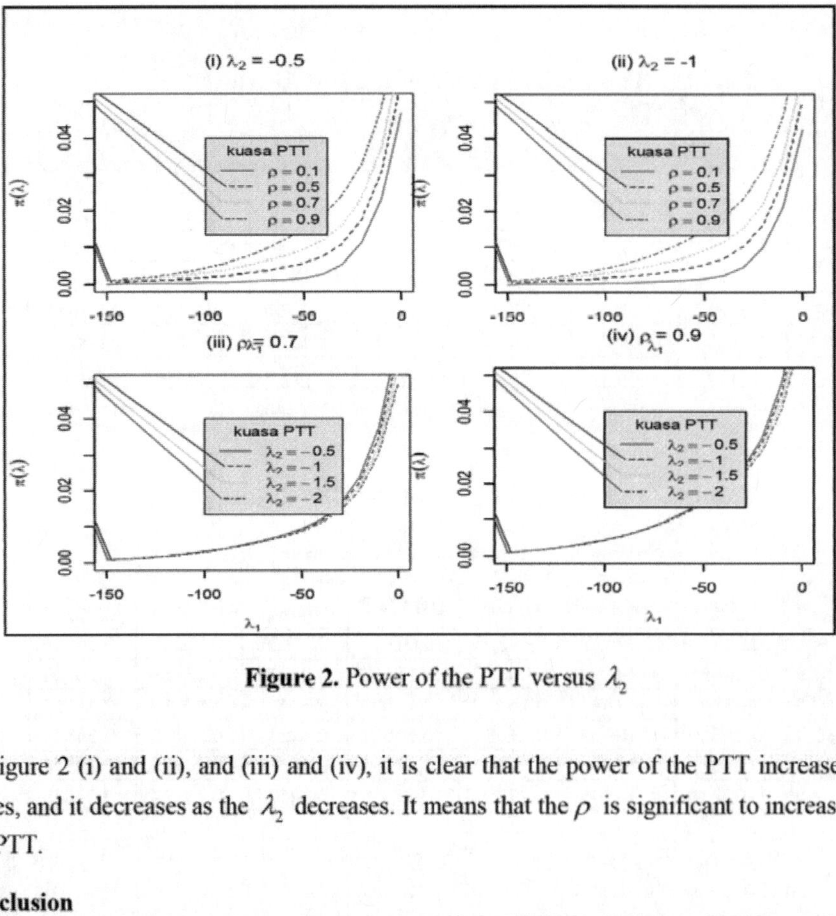

Figure 2. Power of the PTT versus λ_2

From Figure 2 (i) and (ii), and (iii) and (iv), it is clear that the power of the PTT increases as the ρ increases, and it decreases as the λ_2 decreases. It means that the ρ is significant to increase the power of the PTT.

4. Conclusion

The research studied pdf, CDF, graphical analysis of the BCND, and its application on the power of the PTT. To compute the power of the PTT in testing intercept and produce the graphs of the pdf and table CDF, *R-code* is then used. The results showed that the mean and coefficient correlation have a significant affect to the curves, but the variance is not. The curves tend to be leptocurtic for small coefficient correlation, and they will be mesokurtic for large coefficient correlation. On the other hand, the curve will be *leptocurtic* and or *mesokurtic* when the coefficient correlation decreases (and or increases). For both μ_1 and μ_2 are negative, the curve is *skew positive*, and vise verse.

In term of the power of the PTT, the BCND is used to compute the power of the PTT, and the result showed that it increases as the coefficient correlation increases.

Reference

[1] Saleh A K M E 2006 (John Wiley and Sons, Inc., New Jersey)
[2] Pratikno B 2012 Dissertation, University of Southern Queensland, Australia, 2012.
[3] Salim C O 1999 *J Kedokter Trisaktik* **18(2)** 107
[4] Wackerly D D, Mendenhall W & Scheaffer R L. 2008. *Mathematical Statistics with Application* (7th Ed. Thomson Learning, Inc., Belmont, CA, USA, p.455.
[5] Kenny J F and Keeping E S 1951. *Mathematics of Statistics Part One* (3rd Ed. Van Nostrand, Inc: New Jersey, p.92.
[6] Sahoo P, *Probability and Mathematical Statistics 2008* Departement of Mathematics: University of Louisville. p.355.
[7] Yunus R M. 2010.Unpublished Ph.D. Thesis, *University* of Southern Queensland, Australia.
[8] Hogg R V, Joseph M, and Craig A T 2012. *Introduction to Mathematical Statistics:* United States of America pp 164.
[9] Khan S 2005 *J.Stat.Res.* **39 (2)** 79
[10] Khan S 2008 *Commun.Stat.Theor. Method* **37** 247
[11] Khan S and Saleh A K Md E 1997. *Biometrical. J.* **39** 1
[12] Khan S and Hoque Z 2003 *J. Stat. Res.* Vol 37 43
[13] Khan S and Saleh A K Md E 2005.*Statistical Papers*. **46** 379
[14] Khan S and Saleh A K Md E 2008 *Commu. Stat. Theor. Method* **37**(16), 2564
[15] Khan S and Pratikno B 2013 *Statistical Papers* 54(3) 605.
[16] Turyadi,Mara M N dan Kusnandar D 2013 Bletin *Ilmiah Matematika Statistika dan Terapannya,* 2(2) 127

IOP Conf. Series: Journal of Physics: Conf. Series **824** (2017) 012031 doi:10.1088/1742-6596/824/1/012031

Analysis of Palm Oil Production, Export, and Government Consumption to Gross Domestic Product of Five Districts in West Kalimantan by Panel Regression

E Sulistianingsih[1], M Kiftiah[1] D Rosadi[2], H Wahyuni[2]

[1] Department of Mathematics, University of Tanjungpura Pontianak, West Kalimantan, Indonesia
[2] Department of Mathematics, Universitas Gadjah Mada

Corresponding author: evysulistianingsih@math.untan.ac.id

Abstract. Gross Domestic Product (GDP) is an indicator of economic growth in a region. GDP is a panel data, which consists of cross-section and time series data. Meanwhile, panel regression is a tool which can be utilised to analyse panel data. There are three models in panel regression, namely Common Effect Model (CEM), Fixed Effect Model (FEM) and Random Effect Model (REM). The models will be chosen based on results of Chow Test, Hausman Test and Lagrange Multiplier Test. This research analyses palm oil about production, export, and government consumption to five district GDP are in West Kalimantan, namely Sanggau, Sintang, Sambas, Ketapang and Bengkayang by panel regression. Based on the results of analyses, it concluded that REM, which adjusted-determination-coefficient is 0,823, is the best model in this case. Also, according to the result, only Export and Government Consumption that influence GDP of the districts.

1. Introduction

Gross domestic product (GDP) considered as a measure of economic growth in a region. So that GDP has a prominent influence to Gross National Product of a country. Central Bureau of Statistics Republic of Indonesia (BPS) argued that West Kalimantan is on the 22 place out of 33 provinces in Indonesia. Meanwhile, West Kalimantan is the second biggest region in Indonesia that has been enriched by large natural resources as well as other areas in Indonesia.

Some researchers conducted several types of research that evaluated some aspects that can contribute to GDP. The first character, palm oil plantation, was done by [1] and [2]. [1] analysed effects of palm-oil- farm development to economic growth, regarding poverty alleviation of five districts, in Riau Province as a part of central of palm oil plantation development. According to this research, it concluded that palm oil plantation in the five districts produces both positive multiplier effect score and farm welfare index.

[2] evaluated impacts of palm oil plantation which was organised by Sarawak and Land Consolidation and Development Authority (SALCRA) to the prosperity of land owner community around SALCRA, namely Iban and Bidayuh community. They summed up that the palm oil plantation contributed to the increase of community income, infrastructure improvement such as roads, clean water providing, electricity and health services that improve the live quality of the community. In another research, there is analyse expansion policy of palm oil plantation in Sambas district due to community well-being from 2006 until 2011. Based on the recent studies, it deduced that there was an increase of 3.43% in Sambas's GDP each year.

The second aspect, which has been evaluated giving influence to GDP by some researchers, is export. [3] analysed influence between infestation, labour force and export to GDP of Buleleng, a district in Bali Province by path analysis. The result of the research argued that those aspects, in this case, is export, was significantly influence Buleleng GDP.

The third issue that was considered to evaluate in this research is government consumption. The impact of government consumption to GDP has been examined by [4]. This study used panel regression and deduced that government use was significantly giving a positive contribution to district's GDP in Central Java Province.

Furthermore, panel regression has been used prominently to evaluate many cases in numerous journals such as [5], [6], [4], [7], [8], [9], [10], and [11]. Those journals will be utilised to study the methodology in analysis panel data.

In a nutshell, there is some evidence that palm oil plantation (in this case is palm oil production), export, and government consumption fosters GDP in other districts in Indonesia. Therefore, this paper aims to analyse the relationship between those factors and five-district GDP in West Kalimantan, namely Sanggau, Sintang, Sambas, Ketapang and Bengkayang by panel regression. This research attempted to analyse district-level data of those factors from 2007 to 2013 [12].

2. Panel Regression

A longitudinal or panel data set is a combination of cross section and time series data. Panel data provides multiple observations on each in the sample over time. Meanwhile, a regression model utilised to analyse panel data is called by panel regression model [13]. Besides, a panel regression model offers advantages over conventional cross section or time series data set. Some benefits of a panel regression model are [14].

1. It could be more accurate to estimate parameters of a model. It is caused by a large number of data points and increases the degrees of freedom on it. Besides, data panel also can decrease the collinearity among independent variables.
2. It can simplify computation and statistical inference. In normal condition, calculation of panel regression estimator or inference could be more complicated rather than estimator based on either cross-sectional or time series data. However, in the case of nonstationary time series analysis which the large sample approximation of the distributions of least square or maximum likelihood estimators is not longer normally distributed. Nevertheless, using panel data, researchers can invoke central limit theorem across cross section units to denote that limiting distributions of various estimators remain asymptotically normal.

[15] stated that a panel regression model (PRM) is different from both regular time series and cross section regression model.

A PRM

$$y_{it} = \alpha_{it} + \beta_{it}x_{it} + \varepsilon_{it}\,, \qquad i = 1,2,\dots,N\,;\; t = 1,2,\dots,T \tag{1}$$

Where y_{it} is the dependent variable, x_{it} is the independent variable, and ε_{it} is the error term, uncorrelated with x_{it}, with mean zero and constant variance σ_ε^2. Besides, α_{it} is a scalar and β_{it} is regression coefficient, *slope*. The subscript i denotes the cross-section dimension while t symbolises the time series dimension.

Moreover, [15] asserted that panel regression model focus on how controlling the impact of unobserved heterogeneity for getting valid inference on β_{it}. For example, in a linear regression model, suppose unobserved heterogeneity is individual specific and time invariant. So that, the individual specific effect on y_{it} could either be invariant with the x_{it} or interact with x_{it}. In this case, (1) can be written as

$$y_{it} = \alpha_i^* + \beta_i x_{it} + \varepsilon_{it}\,, \qquad i = 1,2,\dots,N\,;\; t = 1,2,\dots,T \tag{2}$$

The parameters α_i^* and β_i can be different from the various cross-sectional units, even though both of them stay constant over time. Following this assumption, a variety of sampling distribution can occur.

For example, sampling distribution can seriously mislead the least square regression of y_{it} on x_{it} when all of NT observations are utilised to estimate (3) as follows [14]

$$y_{it} = \alpha^* + \beta \ x_{it} + \varepsilon_{it}, \qquad i = 1,2,\dots,N \,;\ t = 1,2,\dots,T \tag{3}$$

Three models can be chosen to obtain estimators in panel regression, namely Common Effect Model, Fixed Effect Model, and Random Effect Model. Those types explained as follows:

1. Common Effect Model (CEM)
   CEM was written as (3) where the regression parameters estimated by Least Square. Cross section data and time series in CEM data are united and considered as one whole observation.
2. Fixed Effect Model (FEM)
   [14] stated that FEM written as
   $$y_{it} = \alpha_i^* + \boldsymbol{\beta} x_{it}' + \varepsilon_{it}, \qquad i = 1,2,\dots,N;\ t = 1,2,\dots,T \tag{4}$$
   In FEM, α_i^* vary over individuals. In this research, FEM assumes that there are no time-specific effects.
3. Random Effect Model (REM)
   REM wrote as (1), but α_{it} is considered as a random variable with the expectation α_0 so that $\alpha_i^* = \alpha_0 + \varepsilon_i$. Consequently, REM expressed as
   $$y_{it} = \alpha_0 + \beta_{it} x_{it} + w_{it} \tag{5}$$
   where $w_{it} = \varepsilon_i + \varepsilon_{it}$.

Usage panel regression to analyse independent variables and dependent variables involve some statistical tests. Those are Chow Test, Hausman Test and also Lagrange Multiplier Test. Chow test is a test that utilised to select, which one is better between CEM and FEM. Meanwhile, Hausman Test is carried out to analyse whether or not the existence of random effect in the model. On the other hand, Lagrange Multiplier Test will be conducted to choose between CEM and REM that is better to use in analyse panel regression. Those tests can be studied extensively from [9], [15], [14] and [13].

3. Analysis And Result

In this section, panel regression will be applied to analyse palm oil production, export, and government consumption, that are considered giving an influence to GDP of the five districts in West Kalimantan. For simplification reason, the variables in this research namely export, government consumption, and palm oil production successively are symbolised by $x_1, x_2,$ and x_3.

Firstly, Chow Test was done to evaluate which one is better between CEM and FEM to use in modelling panel data. Chow Test showed that F-calculated is 181.43, which is more than F-table with $\alpha = 5\%$, 2,82. So that, it can be concluded that FEM is a better model to use rather than CEM.

After Chow Test, Hausman Test is conducted to examine whether there is a random effect or not in the model. According to Hausman Test's result, REM is a better model than FEM. It is shown by chi-square statistics at 6.065 which is bigger than chi-square table, with degrees of freedom 3 and $\alpha = 5\%$, at 7.814.

Next, at the third step, to make sure the results of the first and the second phase, Lagrange Multiplier Test was executed. Lagrange Multiplier Test deduced that REM is the best model to analyse the panel data due to LM-calculated, 59.94493, that is bigger than a value of chi-square table with degrees of freedom 3 and $\alpha = 5\%$, 7.814.

According to the three test results, it concluded that REM is the best model rather than CEM and FEM. The result shows that palm oil production does not significantly influence the five district's GDP so that by backwards elimination, palm oil production variable x_3 eliminated. Moreover, the evidence for the best REM can be seen in Table 1 as follows.

Table 1. REM after Backwards Elimination

Variable	Coefficient	t-Statistic	Prob
C	574.0381	2.193829	0.0370
X1	0.849075	2.702087	0.0118
X2	6.175591	7.810495	0.0000

Random Effects (Cross)	
Sanggau-C	107.5346
Sintang-C	25.06422
C-C	770.7759
Sambas-C	-228.9822
Bengkayang-C	-674.3925
	Weighted Statistics
R-squared	0.836099
Adjusted R-Squared	0.823958

According to Table 1, REM for this case with Adjusted R-squared 0.824 written as follows

$$\hat{y}_{it} = 574.0381 + 0.8490x_1 + 6.1756x_2. \quad (6)$$

Parameter estimation using REM utilise Generalised Least Square (GLS). Consequently, the estimators are robust from heteroscedasticity and autocorrelation problems. Moreover, based on Table 1, it can be deduced that every rise of export as one billion rupiahs will increase the GDP about 849 million rupiahs. Furthermore, each increase of government consumption as one billion rupiahs, it can contribute rising of the GDP approximately 6.175 million rupiahs.

REM can distinguish effect of every district by error component, so estimation of GDP for five areas in West Kalimantan written as follows at Table 2

Table 2. REM for Five Districts in West Kalimantan

District	Estimation of GDP
Sanggau	$681,572 + 0.849075x_1 + 6.175591x_2$
Sintang	$25,064 + 0.849075x_1 + 6.175591x_2$
Sambas	$770,775 + 0.849075x_1 + 6.175591x_2$
Ketapang	$345,055 + 0.849075x_1 + 6.175591x_2$
Bengkayang	$-674,393 + 0.849075x_1 + 6.175591x_2$

The interpretation of Table 2 is analogue with the analysis in the previous table.

According to the results, palm oil production of the five districts in West Kalimantan cannot give a contribution to the GDP. Similarly, it concluded that palm oil plantation, in this case, cannot be considered as a tool to increase the community's prosperity. Some researchers reported that palm oil plantation donates adverse effects for both the environment and also the community nearby the farm areas. The negative effects of palm oil plantation were reported by [16] and [17].

4. Conclusion

According to the results, it summarised that export and government consumption indicate the positive contribution to GDP of Sanggau, Sintang, Sambas, Ketapang and Bengkayang. Meanwhile, palm oil production is not significantly giving impact to the GDP.

Acknowledgments

Authors would like to acknowledge The Ministry of Research, Technology and Higher Education of Indonesia for financial support in this research and the referees for the comments.

References

[1] Syahza A 2011 *Jurnal Ekonomi Pembangunan* Vol. 2 No. 2 pp 297-310
[2] Sanggin S E and Mersat N I 2012 *International Journal of Sustainable Development* Vol. 3: 11 (Ontario International Development Agency)
[3] Susi L I D, Kirya I K, and Yudiaatmaja F 2015 *E-Jurnal Bisma Universitas Pendidikan Ganesha* Vol 3 No.1 pp 1-10
[4] Noegroho Y S and Soelistianingsih L 2007 (Depok: Universitas Indonesia)
[5] Romain A and Potterie B V P 2003 (Japan: Hitotsubashi University)
[6] Fayissa B Et al 2008 *Tourism Economics* Vol. 14 No. 4 pp. 807-818
[7] Pujiati A 2007 *Jurnal Ekonomi Pembangunan* Vol 4 No. 2 pp 61-70
[8] Budhi M K S 2013 *Jurnal Ekonomi Kuantitatif Terapan* Vol 6 No.1 pp 1-6
[9] Ayunanda M and Zain I 2014 *Jurnal Sains dan Seni POMITS* Vol 2 No. 2 pp 237-242
[10] Ratnasari N P A, Eka I P E and Gandhiadi G K 2014 *E-Jurnal Matematika* Vol 3 No. 1 pp 1-7
[11] Prasanti T A, Wuryandari T, and Rusgiyono A 2015 *Jurnal GAUSSIAN* Vol 4 No. 3 pp 687-696
[12] BPS Provinsi Kalimantan Barat *Kalimantan Barat dalam Angka* 2009-2014 (Pontianak: BPS Provinsi Kalimantan Barat)
[13] Rosadi D 2011 (Yogyakarta: CV Andi OFFSET)
[14] Hsiao C 2014 (USA: Cambridge University Press)
[15] Baltagi B H 2005 (England: The Atrium John Wiley and Sons Ltd)
[16] Down To Earth Indonesia, 2010
[17] Jaenicke J et al. 2010 *Journal of Environmental Management* Vol.92 pp. 630-638

Models for Train Passenger Forecasting of Java and Sumatra

Sartono

Finance Education and Training Agency - Ministry of Finance the Republic of Indonesia

Corresponding author: sartono82@kemenkeu.go.id

Abstract. People tend to take public transportation to avoid high traffic, especially in Java. In Jakarta, the number of railway passengers is over than the capacity of the train at peak time. This is an opportunity as well as a challenge. If it is managed well then the company can get high profit. Otherwise, it may lead to disaster. This article discusses models for the train passengers, hence, finding the reasonable models to make a prediction overtimes. The Box-Jenkins method is occupied to develop a basic model. Then, this model is compared to models obtained using exponential smoothing method and regression method. The result shows that Holt-Winters model is better to predict for one-month, three-month, and six-month ahead for the passenger in Java. In addition, SARIMA(1,1,0)(2,0,0) is more accurate for nine-month and twelve-month oversee. On the other hand, for Sumatra passenger forecasting, SARIMA(1,1,1)(0,0,2) gives a better approximation for one-month ahead, and ARIMA model is best for three-month ahead prediction. The rest, Trend Seasonal and Liner Model has the least of RMSE to forecast for six-month, nine-month, and twelve-month ahead.

1. Introduction

The train is one of the most preferred transports in Indonesia, especially on Eid day. It is a period for people to travel from city to village to see their relatives as a part of local culture in Java and most of the regions in the country. Ministry of Transportation the Republic of Indonesia releases news that during Eid 2016 the number of passengers using a bus (and other similar public transportation) decreases 5.59% from 4,697,945 last year to 4,416,119 this year. On the other hand, the number of a passenger for the railway is 4,080,319, raised 3.78% from 3,931,712 its last year [1]. It seems that people likely prefer to take the train for some reasons. In Jakarta, passengers often need to wait for the next train since it has been full at peak time. This indicates a high demand that offers high profit for train provider.

The opportunity should be managed well to optimize revenue, otherwise, the overwhelming customers can lead to disaster. The manager may buy a new set of a train or build a new track to meet the customers need, but the cost would be a lot. Therefore, it will be useful to know how the number of passengers behaves over time. Kusakabe [2] develop a model to estimate the passengers of train choice in Japan using smart card transaction data. Changes, modifications, innovations and creative activities occur by chance and unsystematically in one of the train company in Iran [3], however, managing train capacity allocation can increase the Avenue from 2.6 to 29.3 per cent as well as 8.4 to 29 per cent in passengers carried in India [4]. Besides, modeling can help to improve the manually planned allocation of train passengers in Dutch [5]. In Indonesia, Sartono [6] shows that ARIMA+GARCH model serves a better prediction for goods delivery by train in Java than the other developed models. These results offer motivation to uncover the behavior of the passengers of the train.

Understanding the pattern of train passengers' data may provide a reasonable suggestion for the decision maker to create an accurate plan in controlling and improving the service. This paper will discuss some models which can be occupied to forecast the number of train passenger for the area of Java and Sumatra. These two regions are the first and the second dense island in Indonesia [7]. Now, business on the railway is still concentrated in those areas. This article is a preliminary study which may be advantageous as one of the references for further research on train modeling.

2. Models Building

To construct an appropriate model, a monthly data set of train passengers from January 2006 up to December 2015 is utilized. The set consists of data for Java and Sumatra region. For each region, the data is divided into two parts. One part contains the observed values from 2006 to 2014 which is used to develop some models. The rest of the data is exploited for model comparison to determine the most suitable model among methods for passengers of train data in each area.

2.1. Basic Method

There are no plenty of study which exactly talks about passengers of train prediction, but there are a huge number of researches in forecasting. Many of them use the Box-Jenkins method to build the model. Therefore, it is chosen to be the base of the method applied in this project. It will be compared to other methods to get the desired model.

Commonly, the Box-Jenkins method is known as ARIMA method. It can be written as ARIMA(*p, d, q*), in which *p, d*, and *q* represent the order of auto-regressive, differencing, and moving-average respectively (see [8] page 33-86 to figure out the background of the model). In general, if a series $\{X_t\}$ is an ARMA(*p, q*), it can be expressed in

$$X_t = \beta_1 X_{t-1} + \beta_2 X_{t-2} + \ldots + \beta_p X_{t-p} + \omega_t + \delta_1 \omega_{t-1} + \delta_2 \omega_{t-2} + \ldots + \delta_q \omega_{t-q} \quad (1)$$

where ω_t is a white noise.

Plotting the data set could be the first step to utilizing this method. It will serve a brief view of the behavior of the data, the possibility of any required transformation, and the existence of any anomaly. ACF plot helps to inspect whether differencing is needed [9]. Next step is evaluating whether the series is stationary. The plot can be used to know it, but a formal test may be more accurate. Augmented Dickey-Fuller test [10] is one of the tools to capture whether the data is non-stationary. When a series is non-stationary, the differencing process may change it becomes a stationary series. If a series is stationary after differencing once, then the order of differencing in ARIMA model is one (written as ARIMA(*p*, 1, *q*)). It is possible to have some models that fit to data, in this case, the best model can be determined by the Akaike Information Criterion (AIC) [11].

The ARIMA model is based on the assumption that the errors are independent and identically distributed to zero mean and constant variances. It is checked in the model validation step. The ACF and PACF of the residuals exhibit an indication of the independence. However, a result from formal tests such as McLeod-Li test [12] and Ljung-Box test [13] may be more reassuring. Besides, if the variances are not constant, the auto regressive conditional heteroscedasticity (ARCH) or general auto regressive conditional heteroscedasticity (GARCH) will be worthwhile to fix it.

A series may contain a seasonal effect as well as a trend. SARIMA(*p, d, q*)(P, D, Q) is the model based on ARIMA model which can capture the seasonal effect. In this model, (*p, d, q*) is the order of non-seasonal term and (P, D, Q) is the order of seasonal term.

2.2. Compared Method

The model obtained by Box-Jenkins method will be compared to models constructed by smoothing method and regression method. In this paper, they are Holt-Winters model and Trend and Seasonal Linear Model (TSLM) respectively. The Holt-Winters model can be written in

$$\left.\begin{aligned} a_t &= \alpha(X_t - S_{t-q}) + (1-\alpha)(a_{t-1} + b_{t-1}) \\ b_t &= \beta(a_t - a_{t-1}) + (1-\beta)b_{t-1} \\ s_t &= \theta(X_t - a_t) + (1-\theta)s_{t-q} \end{aligned}\right\} \quad (2)$$

IOP Conf. Series: Journal of Physics: Conf. Series **824** (2017) 012032 doi:10.1088/1742-6596/824/1/012032

where a_t and b_t are estimated the level and slope respectively, and s_t is estimated seasonal effect at the time t, and α, β, and θ are smoothing parameters. The TSLM is defined as

$$X_t = m_t + s_t + \varepsilon_t \tag{3}$$

where m_t is the trend term, s_t is the seasonal term, and ε_t is the error term at the time t. If the trend is straight line, then equation (3) can be written as

$$X_t = a_1 t + s_t + \varepsilon_t = \begin{cases} a_1 t + \beta_1 + \varepsilon_t & t = 1, 13, 25, ... \\ a_1 t + \beta_2 + \varepsilon_t & t = 2, 14, 26, ... \\ a_1 t + \beta_3 + \varepsilon_t & t = 3, 15, 27, ... \\ \vdots & \\ a_1 t + \beta_{12} + \varepsilon_t & t = 12, 24, 36, ... \end{cases} \tag{4}$$

here, $s_t = \beta_j$ if t is in the j-th season. The parameter in equation (4) can be estimated using Ordinary Least Square (OLS) method.

2.3. Comparison Tool

To find out the best model an analyst need to investigate the ability of the models in forecasting. The better model is the closer its prediction value to the real. The goodness of a model can be evaluated using in-sample and out-sample forecast. Define one-step ahead forecasting errors as

$$e_t(1) = X_t - \hat{X}_t(t-1) \tag{5}$$

Consider to equation (5), Mean Squared Error (MSE) is defined as

$$MSE = \frac{1}{n}\sum_{t=1}^{n}\left(e_t(1)^2\right) \tag{6}$$

The root of MSE is called RMSE. The more accurate a model the less its RMSE.

3. Result

The data set consist of two parts, Java and Sumatra. In each part, the first model is constructed by Box-Jenkins method. Then, the best model from this step is compared to Holt-Winters model and TSLM model. All computations, hypothesis testing, and plotting for figures are done using R-software.

3.1. Passengers of Train in Java

The plot of the original data and its decomposition can be seen at Figure1. It seems that there is a seasonal effect, so SARIMA and other models involving seasonal term could be fit the data. The first step is to build the ARIMA model. P-value = 0.9071 on Dickey-Fuller test suggests that the data is not stationary. Differencing once gives p-value = 0.01 which means the data is stationary now with d=1 in ARIMA model. Thus, the model will be in the form of ARIMA (p, 1, q).

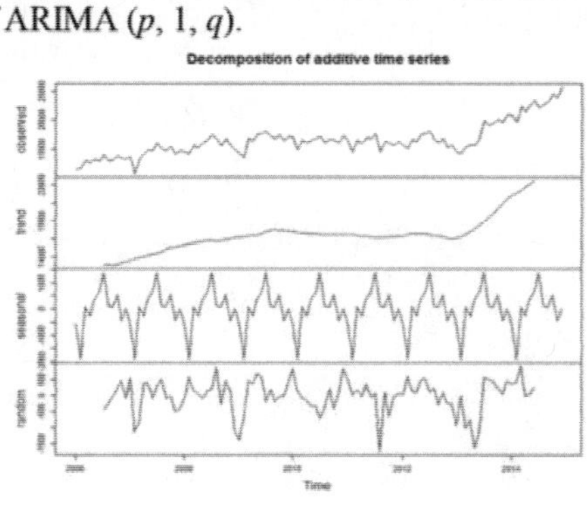

(a) (b)

Figure 1. (a) the number of passengers of train in Java 2006-2014, (b) the composition of data in (a)

Table 1. Table of AIC for Java region

No	ARIMA	AIC	No	ARIMA	AIC
1	(0, 1, 0)	1807.983	9	(2, 1, 0)	1795.162
2	(0, 1, 1)	1795.169	10	(2, 1, 1)	1795.759
3	(0, 1, 2)	1793.930	11	(2, 1, 2)	1785.396
4	(0, 1, 3)	1795.872	12	(2, 1, 3)	1785.000
5	(1, 1, 0)	1793.803	13	(3, 1, 0)	1793.785
6	(1, 1, 1)	1795.470	14	(3, 1, 1)	1795.723
7	(1, 1, 2)	1795.912	15	(3, 1, 2)	**1784.965**
8	(1, 1, 3)	1796.982	16	(3, 1, 3)	1786.962

Consider to Table 1, and the best model is ARIMA(3, 1, 2). Next, the residuals of the chosen model are investigated. Figure 2 reveals that the residuals are white noises. Moreover, Box-Ljung test provides p-value = 0.9742 which support the previous result. Therefore, no GARCH model is needed to maintain the residuals. Hence, the suitable ARIMA model is ARIMA(3,1,2). Using the AIC criterion too, the best model for SARIMA is SARIMA(1,1,0)(2,0,0) with AIC = 1738.395.

(a) (b)

Figure 2. (a) ACF of the square of ARIMA(3, 1, 2) residuals,
(b) PACF of the square of ARIMA(3, 1, 2) residuals

3.2. Passengers of Train in Sumatra

Figure 3 shows the number of passengers of the train in Sumatra from 2006 to 2014 and its decomposition. It is likely seasonal effect exists. Just as in Java, models which include seasonal term will be useful. SARIMA, Holt-Winters, and TSLM are some of them. They will be compared each other and so will the ARIMA model. To begin with ARIMA model, the series must be analyzed whether it is stationary. Dickey-Fuller test gives p-value = 0.4485, and this means that the series is not stationary. By first order differencing the p-value now is 0.01, so the process can be continued for the ARIMA(p, 1, q).

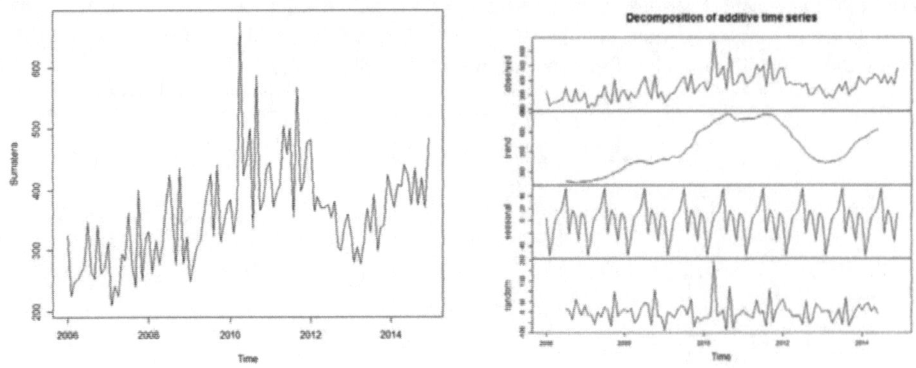

(a) (b)

Figure 3. (a) the number of passengers of train in Sumatra 2006-2014,
(b) The composition of data in (a)

Table 2. Table of AIC for Sumatra region

No	ARIMA	AIC	No	ARIMA	AIC
1	(0, 1, 0)	1255.976	9	(2, 1, 0)	1198.158
2	(0, 1, 1)	1197.117	10	(2, 1, 1)	1197.165
3	(0, 1, 2)	1195.574	11	(2, 1, 2)	1198.253
4	(0, 1, 3)	1197.014	12	(2, 1, 3)	**1191.712**
5	(1, 1, 0)	1213.509	13	(3, 1, 0)	1198.755
6	(1, 1, 1)	1195.383	14	(3, 1, 1)	1197.873
7	(1, 1, 2)	1197.319	15	(3, 1, 2)	1199.800
8	(1, 1, 3)	1198.463	16	(3, 1, 3)	1201.665

Based on AIC criterion as shown on Table 2, ARIMA(2, 1, 3) is the best among the others. The residuals are white noises (see fig. 4). Furthermore, p-value for Box-Ljung test is 0.9782 which strengthen this result, hence, GARCH process is not required. On the other hand, by the AIC criterion, the best SARIMA for Sumatra series is SARIMA(1,1,1)(0,0,2) with AIC = 1171.225.

Figure 4. (a) ACF of the square of ARIMA(2, 1, 3) residuals,
(b) PACF of the square of ARIMA(2, 1, 3) residuals

3.3. Models Comparison

From the previous work, the ARIMA(3,1,2) and SARIMA(1,1,0)(2,0,0) are the best on their class for modeling passengers of the train in Java. For Sumatra, they are ARIMA(2,1,3) and SARIMA(1,1,1)(0,0,2). In this section, those models are compared to the Holt-Winters model and TSLM model. For this purpose, the observed values in the year of 2015 are exploited. The MAE and RMSE are used in this article to select the best model, thus, the model which provides the least RMSE.

Table 3. Table of RMSE value in forecasting for Java region

Time Forecast (Month)	ARIMA	SARIMA	Holt-Winters	TSLM
1	708.9387	685.2047	**560.6887**	4,707.7290
3	1,889.8508	1,427.5119	**1,145.4301**	5,127.0130
6	1,788.9301	1,181.6453	**1,133.6436**	5,649.3640
9	1,830.3599	**980.7162**	1,016.8504	5,730.9110
12	2,213.6562	**873.8687**	1,195.6856	6,060.6570

Table 4. Table of RMSE value in forecasting for Sumatra region

Time Forecast (Month)	ARIMA	SARIMA	Holt-Winters	TSLM
1	4.7707	**1.7723**	6.7918	9.1319
3	**11.6466**	24.0475	21.1706	21.7184
6	22.6385	22.3371	19.9976	**19.2721**
9	41.9038	37.9631	39.1960	**28.7422**
12	43.4062	34.2374	36.3579	**28.8428**

Table 3 shows that the Holt-Winters model is more accurate than the others to estimate the number of passengers of the train for one, three, and six-month ahead, while, SARIMA is more suitable for nine-month and twelve-month estimation. On the other hand, SARIMA model only best for one-month ahead prediction for the region of Sumatra, and to predict in three-month ahead the ARIMA model makes the least error. Besides, Trend and Seasonal Linear Model is more promising to forecast for six, nine, and twelve-month in the future (see Table 4).

4. Summary

Some models have been developed and tested according to get the most suitable one for passengers of train data set. There is no model which dominates all area for all time forecast. For Java, the model obtained by Holt-Winters technique is better than ARIMA, SARIMA, and Trend and Seasonal Linear Model to predict in one, three, and six-month ahead. Nevertheless, it is no more accurate than SARIMA model as the best to foresee in nine and twelve-month in the future. However, it does not mean that SARIMA is better for long-time prediction in general. In fact, for the area of Sumatra SARIMA is more appropriate for one-month time forecast, and less accurate when used for three months or more. Besides, ARIMA is good for three-month ahead forecasting. The rest, Trend and Seasonal Linear Model is the fittest model to forecast for next six, nine, and twelve-month than other models for Sumatra region.

Holt-Winters, SARIMA, and TSLM are models which include the seasonal effect on their formula. It seems that the passengers of train follow a trend or seasonal pattern since those models coming up as the best model in almost all time forecast for both region. Further research may be executed to reveal the trend and seasonal specifically. Other methods considering seasonal effect may give the different view or even strengthen the result in this project.

References

[1] http://www.dephub.go.id/welcome/readPost/menhub-jonan--penumpang-angkutan-umum-lebaran-ini-naik-4.30
[2] Kusakabe, T., Iryo, T. and Asakura, Y. 2010 Estimation method for railway passengers' train choice behavior with smart card transaction data. *Transportation*, 37(5), pp 731-749.
[3] Feyzbakhsh, S. A., Sadeghi, R. and Shoraka, S. 2008 A case study of intrapreneurship obstacles: The RAJA passenger train company. *Journal of Small Business and Entrepreneurship*, 21(2), pp 171-180,251.
[4] Gopalakrishnan, R. and Rangaraj, N. 2010 Capacity management on long-distance passenger trains of indian railways. *Interfaces*, 40(4), pp 291-336.
[5] Abbink, E., Berg, B. v. d., Kroon, L. and Salomon, M. 2004 Allocation of railway rolling stock for passenger trains. *Transportation Science*, 38(1), pp 33-41.
[6] Sartono 2016 *Proc. Nat. Conf. Modelling for Goods Delivery by Railway in Java and Sumatra* (Surakarta: Muhammadiyah University of Surakarta, Mar 12, 2016) pp 790–799.
[7] Agency, I. S. C. 2015 *Percentage distribution of population by province*, 2000-2014, Retrieved Aug 5, 2016, from www.dephub.go.id
[8] Wei, W. W. S. 2006 *Time Series Analysis: Univariate and Multivariate Methods, 2nd ed.* (New York: Pearson)
[9] Sumway, R. H. and D. S. Stoffer, *Time Series Analysis and Its Application - with R Examples, 2nd ed.*(New York: Springer) pp 141–142.
[10] Dickey, D. and W. Fuller, 1979 *J. Am. Stat. Assoc* 74, pp 427–431.
[11] Akaike, H. 1973 *Biometrika* 60, pp 255–265.
[12] McLeod, A. I. and W. K. Li 1983 *Journal of Time Series Analysis* 4, pp 269–273.
[13] Ljung, G. and G. Box 1978 *Biometrika* 65(2), pp 297–303.
[14] Ardebili, A. S. M., M. Hashemi and M. Barough 2015 *European Online Journal of Natural and Social Sciences* Vol 4 No 1 (s), pp 831–844.
[15] Box, G. E. P., G. M. Jenkins and G. C. Reinsel 1994 *Time Series Analysis: Forecasting and Control, 3rd ed.* (New Jersey: Prentice Hall).
[16] Chatfield, C. 2000 *Time-Series Forecasting* (New York: Chapman and Hall).
[17] Cowpertwait, P. and A. Metcalfe 2009 *Introductory Time Series with R* (New York: Springer).

IOP Conf. Series: Journal of Physics: Conf. Series **824** (2017) 012033 doi:10.1088/1742-6596/824/1/012033

Simultaneous Co-Clustering and Classification in Customers Insight

M Anggistia , A Saefuddin, and B Sartono

Department of Statistics, Bogor Agricultural University, Indonesia

Corresponding author: meta.anggist@gmail.com

Abstract. Building predictive model based on the heterogeneous dataset may yield many problems, such as less precise in parameter and prediction accuracy. Such problem can be solved by segmenting the data into relatively homogeneous groups and then build a predictive model for each cluster. The advantage of using this strategy usually gives result in simpler models, more interpretable, and more actionable without any loss in accuracy and reliability. This work concerns on marketing data set which recorded a customer behaviour across products. There are some variables describing customer and product as attributes. The basic idea of this approach is to combine co-clustering and classification simultaneously. The objective of this research is to analyse the customer across product characteristics, so the marketing strategy implemented precisely.

1. Introduction

Competition in current business is tighter than a few years ago. High profit always became the priority indicator for the success of the company. Campaign product is important for customers to buy products. However general campaign model usually does not fit particular customer's interest. Customer diversity may be addressed by the different type of campaign strategy [1].

Market segmentation is one of the popular marketing strategies to face customer diversity. Smith in [2]were mentioned that "Market segmentation is to divide a market into smaller groups of a buyer with distinct needs, characteristics, or behaviours who might require separate products or marketing mixes". In other words, segmentation is a strategy to divide customers become homogenous groups. Sun [2] explained six roles of market segmentation as the following: 1) to choose customers interest and desire appropriately, 2) to seek market advantages, 3) to determine features in market aim, 4) to make strategy for marketing mixes, 5) helping small company to develop business and survive among another group, and 6) easier to get feedback and arrange marketing strategy. Based on these phenomena, statistical analysis plays a crucial role. Study of marketing segmentation classified in two, these are supervised and unsupervised technique. Clustering is one of the unsupervised methods. According to [3], clustering may reduce the data dimension by forming groups with a particular characteristic in each cluster.

Clustering by combined with grouping form two ways is called co-clustering. Cheng [4] were described co-clustering as simultaneous column and row grouping from matrix data set. Each co-cluster will have homogeneous observations. Co-clustering is more favourable than clustering in some viewpoints, i.e. more informative, use fewer parameters, and time reduces in computation [5]. The objective of the research is to apply simultaneously co-clustering and logistic regression approach in transaction data set. So that analysing influence factors in each co-cluster can be done by logistic regression. Therefore offered campaign is more efficient because attained by specific targeted customers.

2. Related Work

One of the popular marketing strategies is segmentation, targeting, and positioning (STP). Segmentation is grouping the homogenous clients so that each cluster will have distinct needs and desire while targeting is direct marketing activities at each cluster which customers believe that the company can satisfy better the competitors. Moreover, positioning is a position the product to offering the customers to buy. Therefore, making the segment is important in the marketing strategy [6]. In [2], it mentioned that there are four bases in segmentation, i.e. geographic, psychographic, demographic, and behaviour segmentation. Among all segmentations, demographic is the most popular one. Partly because of the customer desire, preference, and usage rate are related to demographic variables. Segmentation in statistical analysis can be done by using supervised or unsupervised learning. The supervised learning uses the known corresponding target training data set to give feedback after the procedure finished. The supervised learning involves building a regression or a classification model. The unsupervised learning is data categorization without any corresponding target values. The unattended is grouping the data by finding the correlation from the raw dataset, where it is called clustering. Singh [7] was used the clustering method to segmenting the customer using RFM (recency, frequency, and monetary) variable from transaction data set. The group of clients, which have high RFM value will be chosen as a marketing target because they give great profit for the company while the small one needs another market strategy.

Haider [8] were improved the segmentation technique using clustering and classification method simultaneously. This method divides into two main steps. The first step is clustering the observation, while the second phase is model building. The objective is put the new customer to the proper group. The algorithm uses iteration to find the best members of the cluster, but it does not guarantee the objective function will increase monotonically. So the highest objective value is the best choice.

Clustering technique improved from years to years. The researchers want to group not only from one way but also from two ways. In 2000, Cheng [4] were proposed co-clustering using expression data set. The expression dataset builds from gene as a row across condition as a column. The co-clustering in an expression dataset will generate a homogeneous group of genes and conditions by minimising the mean square error. Each co-cluster will have different characteristics of genes and conditions. However, the observations inside the co-cluster will have a high similarity to the interpretation is easier. Cho [9]were developed the Cheng and Church's method. They proposed sum square error as the objective function, while Cheng and Church used mean square error. The best co-clustering determines by minimised the objective function. When the Cheng and Church's co-clustering method founds one co-cluster each time, the Cho *et al.*'s method could find $k \times l$ co-clustering each time. So this new method is more efficient in computational time. Co-clustering can be written down as a mapping from m rows data to k rows cluster that symbolise with ρ and mapping from n columns data to l columns group that express with γ. The mapping is written down by:

$$\begin{aligned} \rho &: \{1,2,3,\ldots,\mathrm{m}\} \rightarrow \{1,2,3,\ldots,\mathrm{k}\} \\ \gamma &: \{1,2,3,\ldots,\mathrm{n}\} \rightarrow \{1,2,3,\ldots,\mathrm{l}\} \end{aligned} \quad (1)$$

With $\rho(i) = g$ interpreted as a row i is assigned to row cluster g and $\gamma(j) = h$ construed as the column j assigned to the column grou h p. Cho [9] used iteration to find the best objective value. The repetition guarantees the objective function monotonically decreased. So, compared with the previous method, this method is more efficient in computational time.

The co-clustering method also can be used in text mining problem. Dhillon [3] were did the co-clustering using the data of words across documents. This approach uses information theoretic as the objective function. The aim in the previous co-clustering method is to minimising the objective function, but this research's aim is to maximising the objective function. The iterations in this approach ensure the objective function monotonically increased. The result from this method concludes that a high dimension and heterogeneous data handled. Another co-clustering technique is collaborative filtering, which proposed by George [10]. By using the movie rating dataset, this method can predict the missing data. The missing data from each co-cluster predicts with the average from non-missing data in their co-cluster. The better method of missing data prediction in co-clustering was developed by Banerjee [5]. This research uses the Bregman divergence, and it gives a better result than collaborative filtering in predict the missing data.

In general, the supervised approach captures global structure but disregards any local structure in response measurement. On the other hands, the unsupervised approach focuses exclusively on achieving a local structure in response measurement. Therefore Agarwal [11] were proposed predictive discrete latent factor models. The idea combines the benefits of both supervised and unsupervised learning approach. Expected-Maximization(EM) algorithm used in this method. The E-step does renewal the members in each co-cluster while M-step does renewal the prior function in each co-cluster. The dyadic data structure is the data with response value that can describe in two (or more) interconnected group at the matching point. This data consists of two data matrix. The first data is an array of response value. While the second data is a set of explanatory variables called covariate vector. The covariate vector forms from row attributes, column attributes, and another particular feature of column and row pair. In 2010, Dheodar [12] were introduced another co-clustering algorithm using the dyadic data structure as a covariate. The method is simultaneous co-clustering and classification. The algorithm explained in the next chapter. This algorithm applied in this research by using transaction data set from cosmetic and body care company.

3. Simultaneous Co-Clustering and Classification

The response variable **Z** is sized $m \times n$. With m is the number of rows and n is the number of columns. The value of reply deals with a binary class, which coded -1 if a customer buys a product and 1 if a customer does not purchase a product. The covariate vector $\mathbf{x}_{ij}$ forms from row attribute $\mathbf{c}_i^T$, column attributes $\mathbf{p}_j^T$, and another particular feature of row and column pair $\mathbf{a}_{ij}^T$. Logistic regression is used to model building in classification method. Because response values are binary, the transformation needed in logistic regression. The conversion makes the model into linear. The change is called logic. The linear equation for a linear model written by:

$$\ln\left[\frac{P(z_{ij}=1\mid x_{ij})}{1-P(z_{ij}=1\mid x_{ij})}=\beta^T x_{ij}\right] \tag{2}$$

with $\boldsymbol{x}_{ij}^{\;T}=\left[1,\boldsymbol{c}_i^T,\boldsymbol{p}_i^T,\boldsymbol{a}_{ij}^T\right]$and $\beta^T=\left[\beta_0,\beta_{\boldsymbol{c}}^T,\beta_{\boldsymbol{p}}^T,\beta_{\boldsymbol{a}}^T\right]$.

The aim of this research is simultaneously clustering row and column. Which are called co-clustering? So will be formed k row clusters and l column clusters or $k \times l$ co-cluster (block). Each co-cluster will have different model and characteristic. To get the best co-cluster (ρ,γ), minimising the loss function is needed. The loss function created from both co-clustering and logistic regression. The loss function defined as

$$\sum_{g=1}^{k}\sum_{h=1}^{l}\sum_{u:\rho(u)=g}\;\sum_{v:\gamma(v)=h}\ln\left(1+\exp\left(-z_{uv}\beta^{gh^T}x_{uv}\right)\right) \tag{3}$$

With $\mathbf{z}_{uv}$ is the response value in a row u and column v, and $\boldsymbol{\beta}^{gh}$ is the vector coefficient in each model from sub-matrix **Z**. In simultaneous co-clustering and classification, initialization for co-clustering (ρ,γ) in matrix **Z** takes first. After initialization, the next is model building using logistic regression that is changing the row in the best row cluster by calculating row error. Row error is formulated by

$$E_u(g)=\sum_{h=1}^{l}\sum_{v:\gamma(v)=h}\ln\left(1+\exp\left(-z_{uv}\beta^{gh^T}x_{uv}\right)\right) \tag{4}$$

The row goes into a row cluster if the row error value in that row group is the smallest one. In other words row, u is added in row group g if $\rho^{baru}(u)=\arg\min_g E_u(g)$. After all the rows added in the best row cluster, the next changes the column to the best column group by using the same step in the row changing. The column error can be defined by

$$E_v(h)=\sum_{g=1}^{k}\sum_{u:\rho(u)=g}\ln\left(1+\exp\left(-z_{uv}\beta^{gh^T}x_{uv}\right)\right) \tag{5}$$

Input: $Z_{m\times n}$, covariate
Output: Co-clustering (ρ,γ) and co-clustering model βs

```
Begin with a random co-clustering (ρ,γ)
Repeat
Step 1
    Update co-cluster models
    For g=1 to k do
        For h=1 to l do
            Build logistic regression model each co-cluster (g,h), to obtain update β^gh
        End for
    End for
Step 2a
    Update ρ – assign each row to the row cluster that minimises the row error
    For u=1 to m do
```

$$\rho(u) = \arg\min_g \sum_{h=1}^{l} \sum_{v:\gamma(v)=h} \ln\left(1+\exp\left(-z_{uv}\beta^{gh^T} x_{uv}\right)\right)$$

```
    End for
Step 2b
    Update γ – assign each column to the column cluster that minimises the row error
     For v=1 to n do
```

$$\gamma(v) = \arg\min_h \sum_{g=1}^{k} \sum_{u:\rho(u)=g} \ln\left(1+\exp\left(-z_{uv}\beta^{gh^T} x_{uv}\right)\right)$$

```
    End for
    Optional: Repeat 2a and 2b until convergence
Until convergence
Print (ρ,γ) and β
```

Source: Dheodar [12]

Figure 1. Algorithm of simultaneous co-clustering and classification

Column moving will make the best row cluster arrangement different. Likewise, row moving will make the best column group method different. Therefore iteration is needed until the column and row assignment not changing. The sign of that is the convergence of loss function. Finding the best co-cluster does not mean that finding the best model. So the iteration is also needed in the parameter estimation and co-clustering. The iteration held until the loss function is convergence in one value. Note that iteration ensures the loss function is monotonically decrease. But the iteration in this approach just convergence in a local minimum. A simple algorithm was proposed by Dheodar [12]. As can be seen in Figure 1 there is two first step of this method. The first step is minimising objective function using logistic regression, while the second phase is exchanging row and column to obtain optimal co-cluster. The algorithm needs some initialization on co-cluster membership which and then optimise during the iteration. As often most optimisation algorithm, various initialization may produce a different final result. So the authors suggest to runs the algorithm several times and select the best result with smallest loss value.

Working with a huge number of observations may lead very long computation time in estimating logistic regression model. Instead of using the complete set of observation, we propose to modify the algorithm by using a reasonably small subset of view. In this paper, we randomly select 10% remarks from each co-cluster to obtain the logistic model. The reader may use the other portion depend on the size of the data set.

4. Experimental Results

4.1. Data Description

A beauty and body care company produce many types of products, i.e. makeup, body care, and perfume. A lot of number branches in the different point of sale area make the diverse customer characteristics can't be avoided to facilitate the marketing product, and the company was applied the membership card for the regular customer. Fortunately, 90% of the customers have a membership card. Therefore the company can store the client data such as age, address, marital status, gender, and so on.

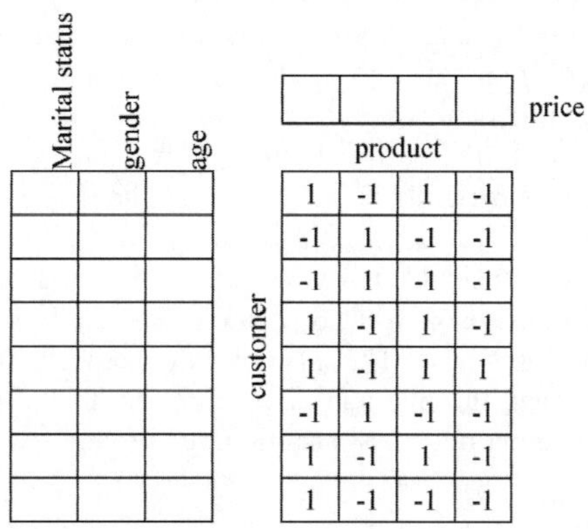

Figure 2. Data illustration

The various customers make the company do the segmentation the customer. Along with many of products, segmentation of customer across product is a need. The client segment information used for making marketing strategy, such as campaign held in the right place for the right customer while products segment used for product development like pricing strategy. The company always records the product transaction. The documents are called transaction data set, will be employed in this paper. The transaction data set records 198,537 members who purchased the product in four months. Furthermore, 66 selected products used in this analysis. The data of purchased product by the customers presented in matrix **Z** with the binary value. The value 1 in matrix **Z** means the customer buys the product and -1 means the customer did not purchase the product.

This paper was not only using transaction as primary data set but also the auxiliary data set. The auxiliary consists of the customer attribute and the product attribute. The variables that can explain the characteristics of the client are age, gender, and marital status. That three variables were called customer attributes $\mathbf{c}_i$ while the product quality $\mathbf{p}_j$ is just a price.

On the data analysis, the customers denoted as row and products as a column, so $198{,}537 \times 66$ is the size of matrix **Z**. The customer attributes called as matrix **C** has size $198{,}537 \times 3$ and product quality will be matrix P with size in 66×1. Figure 2 displays the data illustration. The number of customer clusters and product clusters was objectively chosen as 5 and 4 by authors. As mentioned above, running data analysis in several times can't be avoided, so this study decides to use 5 repeats. The result from the repetition is the fourth one has smallest loss value. So the fourth repetition is chosen to clustering and logistic regression for this paper. Figure 3 display the line plot of objective value in the fourth repetition. By looking at Figure 3, there are 11 iterations of this algorithm to makes the real value convergence. The smallest loss value is 1,983,035.

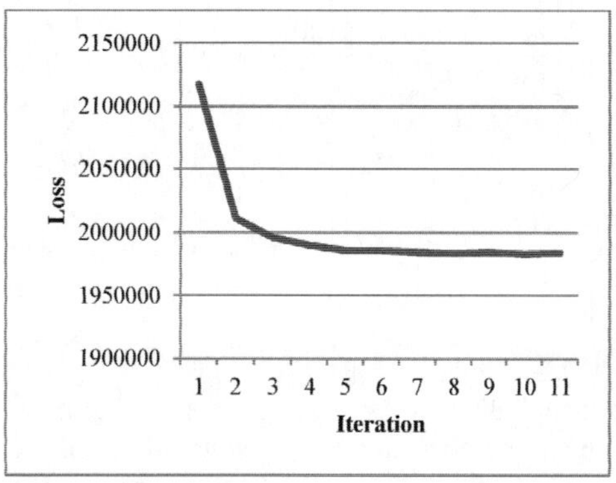

Figure 3. Line plot of loss value in 4th replication

4.2. *Result*

In each customer cluster, the number of females always has a higher percentage than male. This result fits in this problem because the data taken from a transaction of cosmetic and body care company. Furthermore, the customer cluster number three and five have more male customers than the other cluster. However, in customer number three, the married customers are dominated, but the single customer in customer group number five is dominating. That means young clients, both female and male were gather in customer group number 5 and the old ones were gather in customer cluster number 3. Product group number three collects 23 products with slightly low price than others. There is an outlier in product category number three, which the outlier has the lowest price for the other product. The product group number five only has 8 products, but the box plot shows that the product's price is diverse and has positive skewness. This skewness concluded that there are products in this cluster that have a higher price than the average price. Same with the product price in the product group number 4, the product cluster number 1 is also gathered products with the high price.

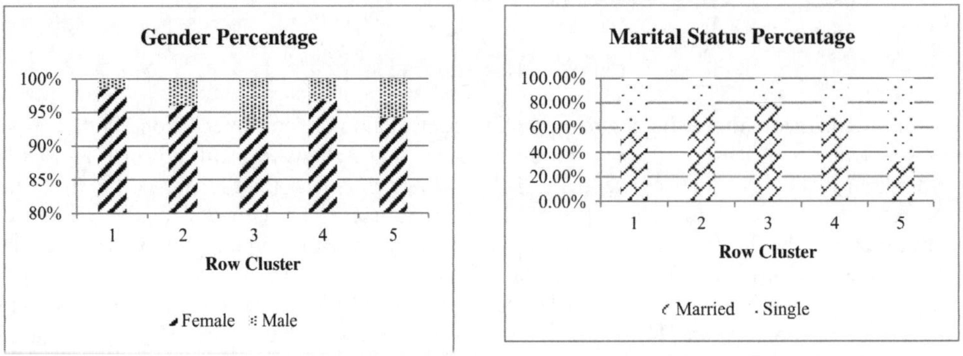

(a) (b)

Figure 4. Row cluster characteristics (a) Number of gender each row group in percentage (b) Number of marital status each row cluster in percentage

The percentage of buyers in each co-cluster displayed in Table 1 and the plot of customers across consumer show in Figure 6. The black dot means the customer purchased the product. Product cluster number three almost has full of dots. So in product group number three, the products are the favourite one. The price in product category three is low enough maybe this is one of the reasons. But the cheap one does not mean that the product will takes customer interest because the product cluster number two has the low price but the small number of buyers.

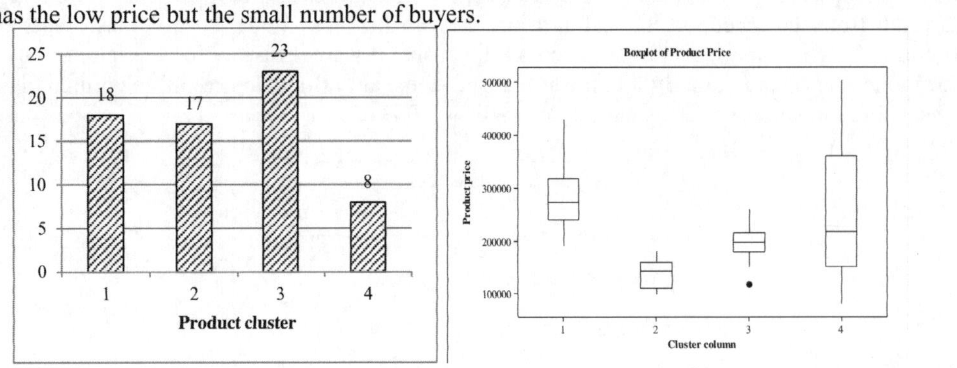

(a) (b)

Figure 5. Product group characteristic (a) Number of product (b) Box plot of product price

The second favourite products gather in the product category number four. The products price is quite high than the products cluster number three. So, it concluded that the product group number four gives superior profit for a company. Otherwise, the product category number one and two have a small number of buyers. It is shown in Figure 6 that the black dot just a little. The product cluster with a low percentage of consumers indicates that the goods do not catch the customer interest, so this group of the goods need some development.

Table 1. Percentage of buyers in each co-cluster

Customer cluster	Product cluster 1	2	3	4
5	1.42%	1.09%	9.18%	9.48%
4	0.25%	1.59%	8.88%	4.57%
3	1.37%	1.79%	13.58%	9.78%
2	1.68%	0.22%	3.67%	12.37%
1	0.02%	0.69%	4.26%	1.54%

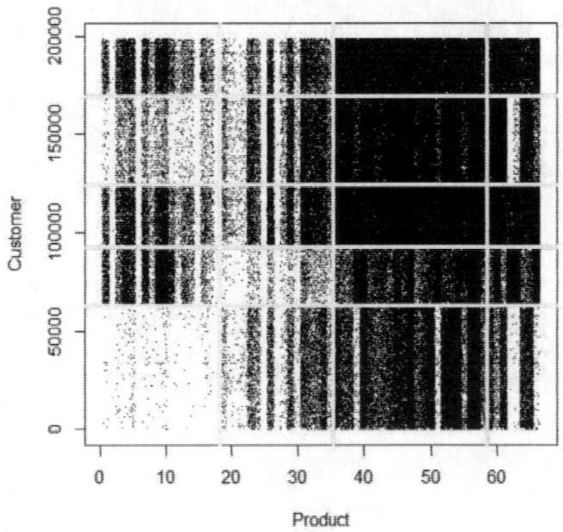

Figure 6. Plot of the product purchase by customer

The customers can be chosen as a market target if they like to purchases many products. That the clients are on the co-cluster number (3,3), (2,4), (5,4), (5,3) and (4,3). The other side, the co-cluster with a small number of buyers such as co-cluster number (1,1), (2,2), (4,1), (5,1), and (1,4) need a better marketing strategy to increase the number of buyer and frequency of products purchase. Product cluster number one and two show that there are around 5 white lines. The White line in Figure 6 means that the goods are rarely bought by the customer. The company should take action from this case. The better market strategy can save the product with catch the client's interest to buy the products. Figure 6 shows in co-cluster (4,4) some products rarely purchased. While in co-cluster (3,4) and (5,4) that products were many purchased by customers. So, it concluded that the clients in co-cluster (3,4) and (5,4) interest in this products, while the customers in co-cluster (4,4) not take interest for this products. This products campaign for the clients in co-cluster (3,4) and (5,4) may lead to a high frequency of purchasing the products. The coefficient of logistic model for each co-cluster displayed in Table 2. All intercept coefficients are significant in alpha 5%; this indicates that there are another variables that would make the customer purchase the product. Beside intercept factor, almost all the price coefficients are significant in all models, so it concluded the product price changing gives influence for the customer decision to buy a product. If the price decreases then the probability of the client buys product higher because the cost coefficient has a negative sign. The high value of fixed price ratio means the customers are sensitive in price changing.

Table 2. Logistic regression coefficient

Customer cluster	Variable	Product cluster 1	2	3	4
1	Intercept	0.62*	-1.53*	-1.75*	-3.16*
	#Customer				
	Gender	0.04	0.17	0.14*	0.02
	Age	0.00*	0	0.00*	0.00*
	Marital status	-0.03*	0.04	0.04*	-0.01
	#Product				
	Price	-12.87*	-21.16*	0.97*	-0.17
2	Intercept	-0.93*	-4.34*	-2.28*	-7.38*

	#Customer				
	Gender	-0.10*	-0.48	0.02	-0.11
	Age	0	0.01	0.00*	0.01*
	Marital status	-0.09*	-0.33	0.05*	-0.12*
	#Product				
	Price	-13.21*	-31.77*	-3.09*	8.03*
3	Intercept	2.49*	-1.36*	-1.15*	-0.37*
	#Customer				
	Gender	-0.07*	0.21*	0.60*	0.03
	Age	-0.01*	0.01*	0.01*	0.01*
	Marital status	-0.02	-0.04	0.11*	-0.02
	#Product				
	Price	-28.01*	-34.33*	-20.05*	-24.29*
4	Intercept	-3.78*	-5.58*	-5.37*	-8.92*
	#Customer				
	Gender	0.48*	1	0.48*	-0.13*
	Age	0.02*	0.01	0.01*	0
	Marital status	-0.12*	-0.19	0.04*	0.02
	#Product				
	Price	-3.02*	-23.8	8.59*	20.86*
5	Intercept	0.83*	-1.14*	-1.30*	-2.92*
	#Customer				
	Gender	0.03	0.05	0.13*	-0.16*
	Age	0.00*	0	0.00*	0.01*
	Marital status	-0.07*	-0.08	0.05*	-0.07*
	#Product				
	Price	-16.83*	-30.98*	-4.66*	-4.81*

*significant in alpha 5%

All marital status coefficient in product cluster number three has a positive value and great. That indicated that the marital status gives influence for customer decision, such as the probability of a married customer to buy the product in product cluster three is higher than a single client. Not significantly of the gender coefficient in co-cluster means that the goods usually purchase both female and male customer. Likes in the co-cluster (1,1), (1,2), and (1,4), the probability of the product purchase for male and female is same. That supported by the percentage of buyer that displays in Table 1. The not significant of age coefficient means the product can use for all ages.

5. Summary

Simultaneous co-clustering and classification is the right method for grouping observations and building models in a dyadic data structure with a covariate. Note that in this research assume that there are no missing data. If there are missing data, weight included in the loss function calculation. The repetition in this approach is needed because initialization takes place in the first step. So the number of repetitions may lead to the better result. While the huge number of observation will lead a long computational time, sampling used as the one of a solution.

Based on the result, the simultaneous co-clustering and classification give the significant result in transaction data set with covariates. The product cluster with small numbers of buyer needs product improvement or attracts the customer interest for that product. The products with high consumer interest gather in the client cluster number three. Not high product price may be one of the reasons the product became popular. For the company, use at least one-year transaction data set may give a better result for the customer and product segmentation than use four months data set because each month has a different event.

References

[1] Goyat S 2011 EJBM **3** 45 (http://www.iiste.org/Journals/index.php/EJBM/article/view/647)
[2] Sun S 2009 *IJBM* **4** 63 doi: 10.5539/ijbm.v4n2p63.

[3] Dhillon I, Mallela S and Modha D 2003 *Proc. of the Ninth ACM SIGKDD Int. Conf. on Know. Disc. and Data Min* p 89 doi:10.1145/956750.956764.
[4] Cheng Y and Church G 2000 *Proc. Int. Conf. Intell. Syst. Mol. Bio.* (*ICMB*) **8** *p* 93 (https://www.ncbi.nlm.nih.gov/pubmed/10977070)
[5] Banerjee A, Dhillon I, Ghosh J, Merugu S and Modha D S 2007 *JMLR* **8** 1919 (http://www.jmlr.org/papers/v8/banerjee07a.html)
[6] Lynn M 2011 *Segmenting and targeting your market: Strategies and limitations* (In Sturman M C, Corgel J B and Verma R (Eds.)) The Cornell School of Hotel Administration on Hospitality: Cutting edge thinking and practice (pp. 353-369). Hoboken, NJ: Wiley. (http://scholarship.sha.cornell.edu/articles/243)
[7] Singh A, Rumantir G, South A and Bethwaite B 2014 *Proc. of the 2014 Int. Conf. on Big Data Sci. and Comput.* doi: 10.1145/2640087.2644161.
[8] Haider P, Chiarandini L and Brefeld U 2012 *Proc. of the 18th ACM SIGKDD Int. Conf. on Know. Disc. and Data Min. p* 417 doi: 10.1145/2339530.2339600.
[9] Cho H, Dhillon I, Guan Y and Sra S 2004 *SIAM Int. Conf. on Data Min* (SDM) *p* 114 (http://bigdata.ices.utexas.edu/publication/minimum-sum-squared-residue-co-clustering-of-gene-expression-data)
[10] George T and Merugu S 2005 *Proc. of the Fifth IEEE Int. Conf. on Data Min.* (*ICDM*) *p* 625 doi: 10.1109/ICDM.2005.14.
[11] Agarwal D, Merugu S. 2007. *Proc. of the 13th ACM SIGKDD Int. Conf. on Know. Disc. and Data Min. p* 26 doi: 10.1145/1281192.1281199.
[12] Dheodar M and Ghosh J 2010 ACM *Transactions on Knowledge Discovery from Data* (*ACM TKDD*) **4** doi:10.1145/1839490.1839492.

Mathematical Model of Seasonal Influenza with Treatment in Constant Population

M Kharis [1] and R Arifudin [2]

[1] Mathematics Department, Universitas Negeri Semarang, Indonesia.
[2] Computer Science Department, Universitas Negeri Semarang, Indonesia

Corresponding author: kharis.mat@mail.unnes.ac.id

Abstract. Seasonal Influenza is one of disease that outbreaks periodically at least once every year. This disease caused many people hospitalized. Many hospitalized people as employers would infect production quantities, distribution time, and some economic aspects. It will infect economic growth. Infected people need treatments to reduce infection period and cure the infection. In this paper, we discussed about a mathematical model of seasonal influenza with treatment. Factually, the disease was held in short period, less than one year. Hence, we can assume that the population is constant at the disease outbreak time. In this paper, we analyzed the existence of the equilibrium points of the model and their stability. We also give some simulation to give a geometric image about the results of the analysis process.

1. Introduction

In 2009 avian influenza was become epidemic and followed by swine flu epidemic [1]. Those epidemics made some death cases and many people hospitalization. The strain of H5N1 virus was identified as a cause of avian influenza epidemic, and Strain of H1N1 virus was identified as a cause of swine influenza while the cause of seasonal influenza is H3N2 virus. The symptoms of all influenza disease is similar. Influenza viruses were responsible for many death cases and many people hospitalization. Seasonal influenza usually doesn't cause death, and it has some effects economically i.e. productivity of employee will decrease, some production process needs longer time, some distribution of goods was suspended [2].

Factually, the infected persons need some treatments to cure infection like drugs assimilation and hospitalization. In [3], it was stated extending oseltamivir to infected children (1 – 3 years old) which don't exceed 24 hours after the symptoms give the significant result. Oseltamivir is known as the only one exact drug for the children under 15 years old. Committee on Infectious Diseases ([4]) also stated treatment should be offered as early as possible. In [5], it was stated that early treatment with antiviral medications may reduce the severity and duration of symptoms, hospitalizations, and complications (otitis media, bronchitis, pneumonia), and may reduce the use of outpatient services and antibiotics, extent, and quantity of viral shedding, and possibly mortality in certain populations. In [6], it was stated that giving the drug in earlier time will be reduced the infection period. In [6], it was also stated diagnosis and treatment would be faster the result of oseltamivir therapy. Urashima [7] suggested that one of the actions to reduce the incidence of influenza A, especially in specific subgroups of schoolchildren was vitamin D3 supplementation. In [8], it was stated that using oseltamivir prophylaxis can shorten the duration of influenza outbreaks. In [9,10], there were mutated viruses that have immunities of oseltamivir.

2. Methods

The first step to do this research was literacy study. In this step, we study the fact and some assumptions from various scientific literacy. The facts were also got from our previously research (see

[11]). After that, we complete the facts with some assumptions to build the model. The second step was building and analyzing mathematics model. In this step, we build the mathematics model and then analyze it to determine the equilibrium points and their stability. The third step was making a simulation with parameters value got from our previously research in 2015.

3. Mathematics Model

[11] stated that was no difference periods date between age group except average infection period with treatment between child and elderly and also treatments were effective to all of the human age criteria. Almost all of period date had no significant difference so that we can neglect the human age criteria. We also got the fact that seasonal influenza doesn't cause death (almost any). In this paper, we assume that the population is constant. Transfer diagram of a seasonal epidemic is given in Figure 1.

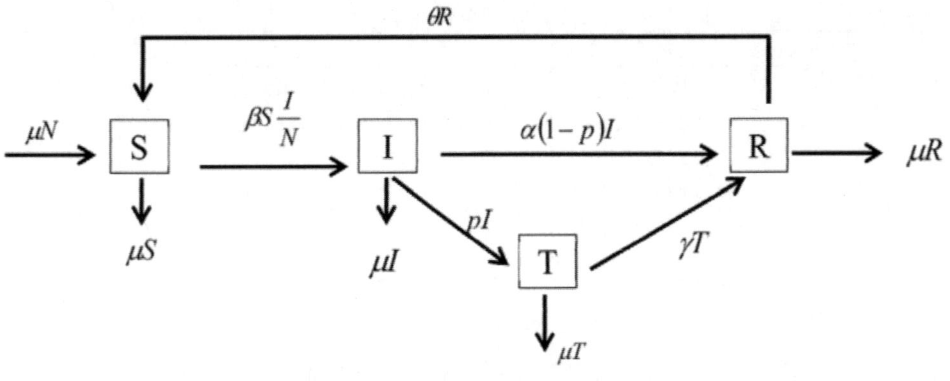

Figure 1. Transfer diagram of seasonal epidemic with treatment

Where N is the total population, S is a total number of the susceptible person, I is a total number of an infected person, T is a total number of treated person, and R is the total number of recovered person. The meaning of parameters in the model was given in Table 1. From Figure 1 we construct the system of an ordinary differential equation as System 1.

$$\begin{aligned}
\frac{dS}{dt} &= \mu N + \theta R - \left(\beta S\frac{I}{N} + \mu S\right) \\
\frac{dI}{dt} &= \beta S\frac{I}{N} - [\alpha(1-p) + p + \mu]I \\
\frac{dT}{dt} &= pI - (\mu + \gamma)T \\
\frac{dR}{dt} &= \alpha(1-p)I + \gamma T - (\mu + \theta)R \\
S + I &+ T + R = N
\end{aligned} \tag{1}$$

Because the population is constant so we can define new variables, i.e. $s = \frac{S}{N}$, $i = \frac{I}{N}$, $\tau = \frac{T}{N}$, and $r = \frac{R}{N}$. Hence, System 1 changes to System 2.

IOP Conf. Series: Journal of Physics: Conf. Series **824** (2017) 012034 doi:10.1088/1742-6596/824/1/012034

$$\begin{aligned}
\frac{ds}{dt} &= \mu + \theta r - (\beta si + \mu s) \\
\frac{di}{dt} &= \beta si - [\alpha(1-p) + p + \mu]i \\
\frac{d\tau}{dt} &= pi - (\mu + \gamma)\tau \\
\frac{dr}{dt} &= \alpha(1-p)i + \gamma\tau - (\mu + \theta)r \\
s + i &+ \tau + r = 1
\end{aligned} \tag{2}$$

Table 1. The meaning of parameters

Parameter	The Meaning
μ	: Birth rate is assumed same with death rate
β	: The probability of infectious contact was happen
α	: Recovery rate without treatments
γ	: Recovery rate with treatments
p	: the proportion of infected person get treatments
θ	: Immunity loss rate

The existence of equilibrium points of System (2) is given in Theorem 1.

Theorem 1.

Let $R_0 = \dfrac{\beta}{\alpha(1-p) + p + \mu}$.

1. If $R_0 \le 1$ then System (2) has only one equilibrium point i.e. nonendemic equilibrium point $Q_0 = (s,i,\tau,r) = (1,0,0,0)$.
2. Jika $R_0 > 1$ make then System (2) has two equilibrium points i.e. nonendemic equilibrium point $Q_0 = (s,i,\tau,r) = (1,0,0,0)$ and endemic equilibrium point $Q_1 = (s^*, i^*, \tau^*, r^*)$ where

$s^* = \dfrac{1}{R_0}$, $i^* = \dfrac{(\mu+\theta)(\mu+\gamma)(R_0 - 1)}{R_0[(\mu+\gamma)[\alpha(1-p)+p+\mu+\theta]+\theta p]}$, $\tau^* = \dfrac{p(\mu+\theta)(R_0-1)}{R_0[(\mu+\gamma)[\alpha(1-p)+p+\mu+\theta]+\theta p]}$,

$r^* = \dfrac{[\alpha(1-p)(\mu+\gamma)+\gamma p](R_0-1)}{R_0[(\mu+\gamma)[\alpha(1-p)+p+\mu+\theta]+\theta p]}$.

Proof.

The equilibrium points were a solution of System 3.

$$\begin{aligned}
\mu + \theta r^* - (\beta s^* i^* + \mu s^*) &= 0 \\
\beta s^* i^* - [\alpha(1-p) + p + \mu]i^* &= 0 \\
pi^* - (\mu + \gamma)\tau^* &= 0 \\
\alpha(1-p)i + \gamma\tau^* - (\mu + \theta)r^* &= 0 \\
s^* + i^* + \tau^* + r^* - 1
\end{aligned} \tag{3}$$

From the second equation, we get $i^* = 0 \vee s^* = \dfrac{\alpha(1-p)+p+\mu}{\beta} = \dfrac{1}{R_0}$.

The case of $i^* = 0$:

From the third and fourth equation, we get $\tau^* = 0$ and $r^* = 0$. From the fifth equation, we get $s = 1$.

Hence, we get a non-endemic equilibrium point i.e. $Q_0 = (s^*, i^*, \tau^*, r^*) = (1,0,0,0)$.

The case of $i^* \neq 0$:

Clear that $s^* = \frac{\alpha(1-p)+p+\mu}{\beta} = \frac{1}{R_0}$. From the third equation, we get $\tau^* = \frac{pi^*}{\mu+\gamma}$. From the fourth equation, we get $r^* = \frac{\alpha(1-p)i^* + \gamma\tau^*}{\mu+\theta} = \frac{[\alpha(1-p)(\mu+\gamma)+\gamma p]i^*}{(\mu+\theta)(\mu+\gamma)}$. Substitute τ and r to the first equation of System (3) and we obtain $i^* = \frac{(\mu+\theta)(\mu+\gamma)\left(1-\frac{\alpha(1-p)+p+\mu}{\beta}\right)}{(\mu+\gamma)[\alpha(1-p)+p+\mu+\theta]+\theta p}$. From the value of R_0, we get $i^* = \frac{(\mu+\theta)(\mu+\gamma)(R_0-1)}{R_0[(\mu+\gamma)[\alpha(1-p)+p+\mu+\theta]+\theta p]}$. Clear that $i^* > 0$ if $R_0 > 0$. Because $i^* > 0$ of then $\tau^* > 0$ and $r^* > 0$. Hence we get an endemic equilibrium point $Q_1 = (s^*, i^*, \tau^*, r^*)$ where all values were defined at Theorem 1. The Stability of equilibrium points of System 2 is given in Theorem 2.

Theorem 2.

Let $R_0 = \frac{\beta}{\alpha(1-p)+p+\mu}$.

1. If $R_0 < 1$ then the non endemic equilibrium point Q_0 was locally asymptotically stable.
2. If $R_0 > 1$ then the non-endemic equilibrium point Q_0 was unstable and the endemic equilibrium point Q_1 was locally asymptotically stable.

Proof.

The Jacobian Matrix of System 2 was given below.

$$J(Q) = \begin{bmatrix} -(\beta i+\mu) & -\beta s & 0 & 0 \\ \beta i & \beta s - [\alpha(1-p)+p+\mu] & 0 & 0 \\ 0 & p & -(\mu+\gamma) & 0 \\ 0 & \alpha(1-p) & \gamma & -(\mu+\theta) \end{bmatrix}. \quad (4)$$

where $Q_1 = (s, i, \tau, r)$.

The stability analysis of Q_0

The eigenvalues of matrix $J(Q_0)$ were $\lambda_1 = -\mu$, $\lambda_2 = -(\mu+\gamma)$, $\lambda_3 = -(\mu+\theta)$, and

$$\lambda_4 = [\alpha(1-p)+p+\mu](R_0-1)$$

Clear that λ_1, λ_2, and λ_3 are negative. Clear that λ_4 is negative when $R_0 < 1$, λ_4 is zero when $R_0 = 1$, and λ_4 is positive when $R_0 > 1$. If λ_4 is zero then the equilibrium point Q_0 is non-hyperbolic point. Hence we neglect the condition of $R_0 = 1$. Hence, if $R_0 < 1$ then the non-endemic equilibrium point Q_0 was locally asymptotically stable and if $R_0 > 1$ then the non-endemic equilibrium point Q_0 was unstable.

The stability analysis of Q_1

The point Q_1 is exist only when $R_0 > 1$. The characteristic polynomial of the Jacobian matrix of Q_1 $(J(Q_1))$ is $(\lambda+\mu)(a_0\lambda^3 + a_1\lambda^2 + a_2\lambda + a_3) = 0$

where $a_0 = 1$, $a_1 = 2\mu + \gamma + \theta + \beta i^*$, $a_2 = \beta i^*[2\mu + \alpha(1-p) + p + \gamma] + \mu(\mu + \gamma + \theta)$

$a_3 = \beta i^*[(\mu+\gamma)(\alpha(1-p) + p + \mu] + \theta(\mu + p + \gamma)]$

From $\lambda + \mu = 0$ we get $\lambda = -\mu$. Clear that a_1, a_2, and a_3 are positive.

Clear that $a_1a_2 - a_3 = \beta i^*\left[\mu[2\mu + \alpha(1-p) + p] + \mu(2\mu+\gamma) + \theta[\mu + \alpha(1-p)]\right] + \gamma\left(2\mu + \gamma + \theta + \beta i^*\right)]$

From the value above, it is true that $a_1a_2 - a_3 > 0$. Hence, the equation $a_0\lambda^3 + a_1\lambda^2 + a_2\lambda + a_3 = 0$ satisfies Routh-Hurwitz criteria. It means all roots of the equation have negative real part. From the result of analysis, we get that all of roots of characteristic polynomial of $J(Q_1)$ have negative real part. Hence, the endemic equilibrium point Q_1 is locally asymptotically stable.

3.1. Minimum Value of p

The minimum value of p such that the influenza doesn't spread widely can be determined from the value of R_0. We must bound the value of $R_0 < 1$ and we get $p > \frac{\beta - (\mu + \alpha)}{1-\alpha}$. Clear that $0 \le \alpha \le 1$, so if $\beta \le \mu + \alpha$ then every $p > 0$ will prevent the epidemic to spread widely.

4. Simulation

We will give the simulations for $R_0 < 1$ and $R_0 > 1$ using the parameters value which is got from our previously research. The parameters value is given in Table 2.

Table 2. The parameters value of seasonal influenza in Central Java province

Parameter	Average Value (day/days)
Incubation period	3.2
Infection period without treatment	10.2
Infection period with treatment	5.5
Reinfection period	26.8

We also get the proportion of person taking treatments i.e. 96% or 0.96. From Table 2, we get the paramaters value of mathematics model i.e. $\mu = 0.00004$ from the expectation of living period (65 years), β has range from 0 until 1, $\alpha = 0.098$ from Infection period without treatments, $\gamma = 0.182$ from Infection period with treatments, p = 0.96 from proportion of person taking treatments, and $\theta = 0.037$ from Reinfection period

4.1. Simulation for $R_0 < 1$

We use the value of $\beta = 0.5$. The value of $R_0 = 0.52 < 1$. The simulations under this condition are given in Figure 2.

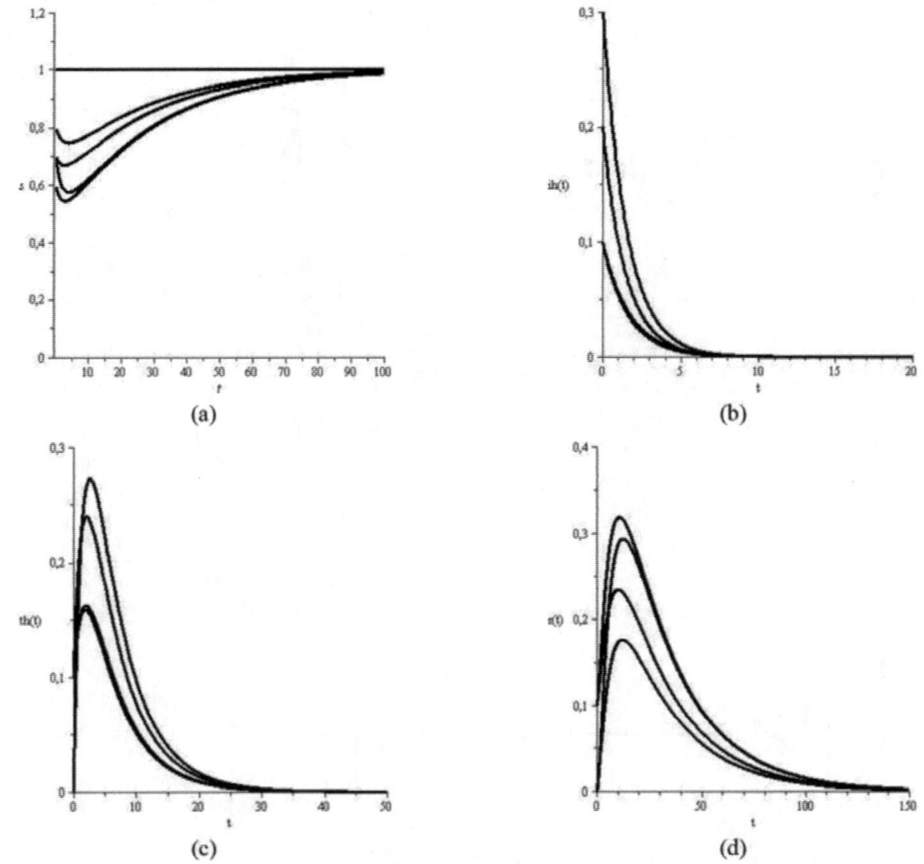

Figure 2. Graphics of variables in model (a) $s(t)$ (b) $i(t)$ (c) $\tau(t)$, and (d) $r(t)$

For $\beta = 0.5$ and $p = 0.96$ we can see that the value of $i(t)$ decreased faster. Because of that, the value of $\tau(t)$ and $r(t)$ also decrease but slower than $i(t)$. From the parameters value in this condition, we can determine that minimum value of p is 45%.

4.2. Simulation for $R_0 > 1$

We use the value of $\beta = 1$ and $p = 0.5$. The value of $R_0 = 1.82 > 1$. The simulations under this condition are given in Figure 3. From the simulation in Figure 3, all of the graphics have to decrease sinusoidal graph. It was caused because the dynamic of every variable affects between one and other. It can also indicate that the eigen values of the Jacobian matrix for the endemic point have complex value.

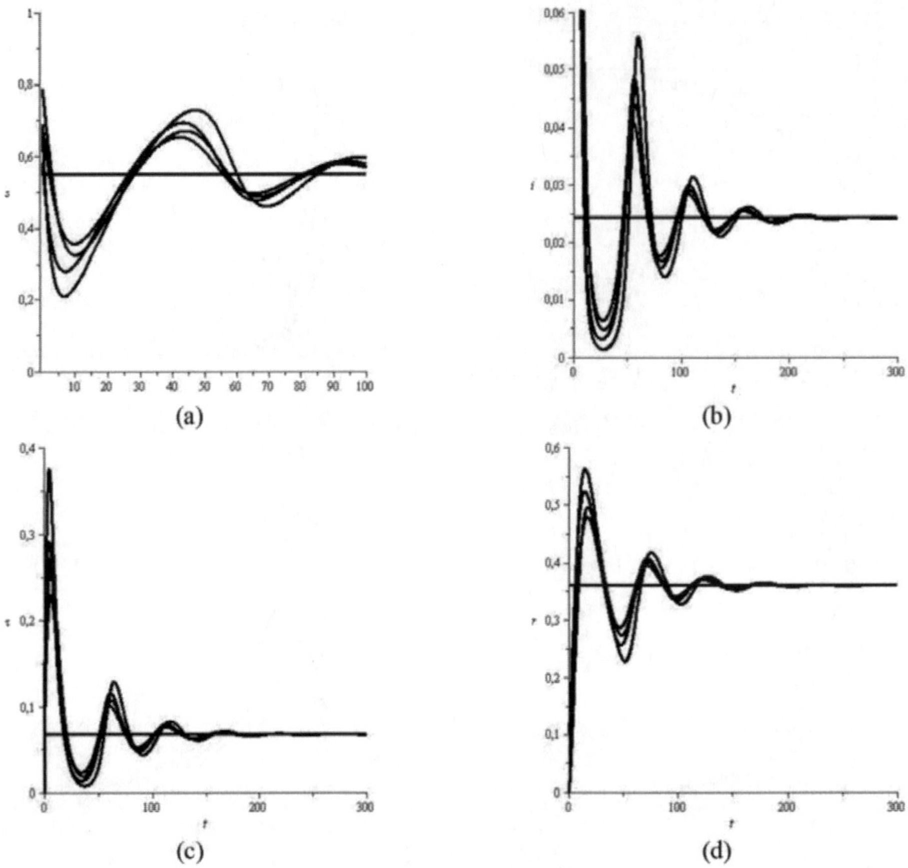

Figure 3. Graphics of variables in model (a) $s(t)$ (b) $i(t)$ (c) $\tau(t)$, and (d) $r(t)$

5. Conclusion and Discussion

From analysis and simulation above, we get the dynamic of mathematics model of seasonal epidemic with treatment, especially for constant population. We also got the reproduction number which can be used to determine whether the epidemic spread widely or vanish. From that number, we get the minimum proportion of treated person. For the next research, we propose to make the mathematics model for non-constant population.

6. Acknowledgments

This research was funded by Kementerian Riset Teknologi dan Pendidikan Tinggi (Kemristekdikti) especially in schema of HIBAH BERSAING since 2015.

References

[1] Yang Y, Sugimoto J D, Halloran M E, Basta N E, Chao D L, Matrajt L, Potter G, Kenah E and Longini Jr I M 2009 *Science* **326** 729 doi: 10.1126/science.1177373
[2] Jansen A G S C, Sanders E A M, Hoes A W, van Loon A M and Hak E 2007 *Eur Respir J.* **30** 1158 DOI: 10.1183/09031936.00034407
[3] Heinonen S, Silvennoinen H, Lehtinen P, Vainionpa R, Vahlberg T, Ziegler T, Ikonen N, Puhakka T and Heikkinen T 2010 *Clin Infect Dis.* **51** 887 doi: 10.1086/656408
[4] Committee on Infectious Diseases 2015 *Pediatrics* **136** doi: 10.1542/peds.2015-2920

[5] Harper S A, Bradley J S, Englund J A, File T M, Gravenstein S, Hayden F G, McGeer A J, Neuzil K M, Pavia A T, Tapper M L, Uyeki T M and Zimmerman R K 2009 *Clin Infect Dis.* **48** 1003 doi: 10.1086/598513

[6] Aoki F Y, Macleod M D, Paggiaro P, Carewicz O, El Sawy A, Wat C, Griffiths M, Waalberg E and Ward P 2003 *J. Antimicrob. Chemother.* **51** 123 doi: 10.1093/jac/dkg007

[7] Urashima M, Segawa T, Okazaki M, Kurihara M, Wada Y and Ida H 2010 *Am J Clin Nutr* **91** 1255 doi: 10.3945/ajcn.2009.29094

[8] Ye M, Jacobs A, Khan M N, Jaipaul J, Oda J, Johnson M and Doroshenko A 2016 *BMJ Open* **6** 1 doi:10.1136/bmjopen-2016-011686

[9] de Jong M D, Thanh T T, Khanh T H, Hien V M, Smith G J D, Chau N V, Cam B V, Qui P T, Quang Ha D, Guan Y, Peiris J S M, Hien T T and Farrar J 2005 *N Engl J Med* **353** 2667 DOI: 10.1056/NEJMoa054512

[10] Gooskens J, Jonges M, Claas E C J, Meijer A, van den Broek P J and Kroes A C M 2009 *JAMA* **301** 1042 doi:10.1001/jama.2009.297

[11] Kharis M and Arifudin R 2015 *Proc. of Int. Conf. on Math, Sci, and Edu. (ICMSE)*, *p* M-1 (http://icmseunnes.com/2015/?page_id=563)

Self-Regulation and Problem Solving Ability in 7E-Learning Cycle Based Goal Orientation

Mulyono[1] and N L Noor

Mathematics Department, Universitas Negeri Semarang, Semarang, Indonesia

[1]Corresponding author: mulyono.mat@mail.unnes.ac.id

Abstract. Goal orientation differences between mastery goals and performance goals can be a cause of high and low self-regulation and problem-solving abilities. To overcome these problems applied 7E-learning cycle in which students learn and develop ways to optimise the power of reason through the learning phase elicit, engage, explore, explain, elaborate, evaluate, and extend. This study aimed to test the effectiveness of learning by 7E-learning cycle and describe self-regulation and mathematics problem solving based on goal-orientation after the implementation 7E-learning cycle. This study used mix method design with research subject is graders XII sciences MA NU Nurul Ulum Jekulo Kudus which divided into goal orientation is mastery goal and performance goal. The independent variable of this research is learning model, while the dependent variable is problem solving and self-regulation. Then, collecting data using scale, interviews and tests. The data processed with the proportion of test, t-test, paired samples t-test, and Normality-gain. The results show problem-solving abilities of students through 7E-learning cycle the average of mathematical problem-solving capability class, self-regulation at 7E-learning cycle is better than the traditional model study. The problem-solving skills at 7E-learning cycle are better than the traditional model study, there is an increase in self-regulation through 7E-learning cycle of 0.4 (medium), and there is an increased problem-solving ability through 7E-learning cycle by 0.79 (high). Based on the qualitative analysis, self-regulation and problem-solving ability after the implementation of 7E-learning cycle students of a mastery goal group are better than the performance goal team. It is suggested to implement 7E-learning cycle to improve self-regulation and problem-solving ability as well as directing and fostering mastery goal on the student in the learning process.

1. Introduction

Mathematics concerning the ideas, structures and relationships governed by mathematical logic so that it is concerned with abstract concepts. The process of the good learning mathematics can make students understand and learn math well too so easy to apply it to next the new situation, namely resolve both in mathematics problems and other science in a daily life.

The problems of mathematics education in Indonesia seen from several indicators, one of is the small achievement of Indonesia in the international ratings. International Survey Program for International Student Assessment (PISA) in 2012 shows that Indonesia in second-bottom ratings of the 65 countries in mathematics [1].

The scope of the subjects of mathematics in senior high school that must be mastered students of grade XI one of them is a differential. In this study takes the material differential. Differential material retrieval is because the material often found difficulty in solving the problem requires an understanding of concepts, reasoning and accuracy. There are variations of the problem and the formula so that students should be good at planning strategies and analysing it. Most are unable to determine the first step that should be used to solve the differentials problem. It shows the low of problem-solving abilities of students. As a result, the core competencies associated with the differential material can not be achieved to the maximally. Based on data obtained from one of the teachers of mathematics in grade XI MA NU Nurul Ulum Jekulo Kudus, it is known that value of differential material students' daily tests in the academic year 2014/2015 there are still many students who did not complete or did not reach the minimum completeness criteria.

One of the goals of mathematics learning is to develop problem-solving abilities. According to [2], stressed the importance of solving the problem because problem-solving is a part in mathematics, so it should not be released from the learning of mathematics. A problem-solving process in which students find the combination of rules that have been studied before, which can be applied to obtain the solution of a problem or a particular situation [3].

Some of the reasons for the need to teach problem-solving among students according to [4] is (1) Problem solving develops general cognitive skills. Besides, (2) problem-solving fosters creativity and (3) problem solving is a part of the scientific application process, and (4) problem solving motivates pupils to learn mathematics. According to [5] mathematic test that measures aspects of problem-solving can be solved by using the steps (1) understanding the problem, (2) devising a plan, (3) carry out the plan, and (4) looking back.

Steps in the problem solving one of which is to determine a strategy to resolve the problem. The ability to plan learning strategies and targets to be achieved in education is one of the characteristics possessed by students who have self-regulation. Individuals who do self-regulation in learning is an individual who has a knowledge and strategic objectives as well as have the independence to direct their resources effectively in education [6].

Table 1. Dimensions of Self-regulation by Zimmerman [7]

Learning Issues	Self-Regulation Sub-Process
Why	*Self-efficacy* and self-goals
How	Strategy use or routinized performance
When	Time management
What	Self-observation, self-judgment, self-reaction
Where	Environmental structuring
With Whom	Selective help seeking

Zimmerman [7, 8] defines self-regulation as the degree of metacognitively, motivationally, and behaviorally active participants in the learning process to achieve the learning goal. Kauffman [9] has also described self-regulation as the student's effort to control and conduct complex learning activities. The dimensions of self-regulation described in Table 1.

Goal orientation concerning the purpose or goal to be achieved individual in an assignment. The characteristics goal orientation is mastery goal and performance goal [10]. Mastery goal is a motivational orientation of the individual, emphasising acquired new skills or mastery of the material. While the performance goal is a student's orientation to get good results.

Characteristics of goal orientation according to [11, 12] is divided into two: the mastery orientation and performance orientation. Pintrich explains mastery orientation and performance orientation in approach and avoidance states. Detailed descriptions are shown in Table 2.

Based on the problem of a little problem-solving ability of students, it's time to fix the mathematics learning process, especially regarding the strategy and learning model used in the study. 7E-learning cycle developed by Eisencraft is a learning model that provides opportunities for students to optimise learning and students develop reasoning power. The students try to construct their thoughts, so the 7E-learning cycle is a learning model centred on the learner. Steps 7E-learning cycle according to [13]: (1) Elicit, (2) Engage, (3) Explore, (4) Explain, (5) Elaborate, (6) Evaluate, and (7) Extend.

Table 2. Characteristics of goal orientation

Goal Orientation	Approach state	Avoidance state
Mastery Orientation	Focus on mastering task, learning, understanding	Focus on avoiding misunderstanding, avoiding not learning or not mastering task

	Use of standards of self-improvement, progress, thorough understanding of task	Use of standards of not being wrong, not doing it incorrectly about task
Performance Orientation	Focus on being superior, besting others, being the smartest, best at task in comparison to others	Focus on avoiding inferiority, not looking stupid or dumb compared to others
	Use of normative standards such as getting best or highest grades, being top or best performer in class	Use of normative standards of not getting the worst grades, being lowest performer in class

Based on the introduction described, the formulation of research problems in this study are (1) is the learning cycle 7E-learning useful? (2) How students' self-regulation and mathematics problem solving after implementation 7E-learning cycle based on goal orientation?

Based on the formulation of the problem, the aims of this study are: testing the effectiveness of learning by 7E-learning cycle, include (1) the ability of students' mathematical problem solving who acquire learning by using 7E-learning cycle can achieve mastery learning. The meaning of mastery learning in this research that if the experimental class of students who scored greater than or equal to 70 reached more than or equal to 75% of the class. Besides, (2) students' mathematical problem-solving ability who acquire learning by using 7E-learning cycle of learning is better than conventional models. Then, (3) students' self-regulation who acquire learning by using 7E-learning cycle of learning is better than traditional models. Also, (4) an increase in students' problem-solving ability before and after the implementation of learning by using 7E-learning cycle. Latter, (5) an increase in students' self-regulation before and after the implementation of learning by using 7E-learning cycle, and describing the self-regulation and solving math problems based on the student's goal orientation after the implementation of the 7E-learning cycle.

2. Method

This study design is a mixed method of a sample were students of grade XI MA NU Nurul Ulum Jekulo Kudus grouped based on goal orientation namely the mastery goals and performance goals. The study conducted in five meetings, both in the experimental group: the group obtained using the learning-7E learning cycle and the control group that received learning using conventional learning models. Data collection was performed employing the scale, test and documentation. In the experimental group, students were given a range of goal orientation for grouping into groups mastery goal or a performance goal. At the start of research activities, the pretest tested about problem-solving ability and scale of self-regulation. At the end of the research activities, given problem-solving ability posttest and range of self-regulation. The independent variable of this research is learning model, while the dependent variable is the problem solving and self-regulation. Data processed by the proportion of test, t-test, paired test, and the gain normalised.

3. Results and Discussion

Data retrieval goal orientation through a scale for classifying categories of students' goal orientation. Grouping using standard deviation method which limits the group by a particular standard deviation. The results of data processing which is goal orientation scale of 34 students experimental group obtained results in Table 3.

Table 3. Results Grouping Students' Goal Orientation.

No	Group	Number of Students
1.	Mastery goal	11
2.	Performance goal	8
3.	Indistinguishable	15

The effectiveness of learning to 7E-learning cycle shown in (1) Based on the results of the problem-solving ability post-test, the proportions test, shows that the students' problem-solving ability experimental group reached the classical completeness, i.e., students who scored more than equal to 70 entering more than equal to 75%. Besides, (2) based problem-solving ability post-test experimental and control groups, t-test result students' mathematical problem-solving ability who acquire learning by using 7E-learning cycle is better than conventional models of learning. Then, (3) Based on data from the final score of self-regulation scale experimental and control groups, t-test results obtained self-regulation students acquire learning by using 7E-learning cycle is better than conventional models of learning. (4) Based on data from the problem-solving abilities pre-test and post-test experimental group, the pair tested obtained results students' problem-solving abilities after subjected to learning by using 7E-learning cycle is better than problem-solving skills before subjected to learning by using 7E-learning cycle, then the normalised gain test score of 0.79 was obtained with a high category. (5) Based on data from an initial score and the final score of the scale of experimental self-regulation group, the pair tested obtained by students' self-regulation after subjected to learn by using 7E-learning cycle. That is better than self-regulation before subjected to learning by using 7E-learning cycle, then the normalised gain test obtained a score of 0.4 with a medium category.

The results of data processing condition of self-regulation based on the goal orientation of students after the implementation of the 7E-learning cycle consisting of 11 students of mastery goals groups and eight students of target performance group obtained results in Table 4.

Table 4. Distribution of Students, Self-Regulation Based Goal Orientation after 7E-Learning Implementation Cycle.

No	Mastery Goal		Performance Goal	
	Frequency	Percentage	Frequency	Percentage
High	4	36.4 %	0	0 %
Medium	7	63.6 %	5	62.5 %
Low	0	0 %	3	37.5 %
Total	11	100 %	8	100 %

Based on Table 4 can be concluded that after the implementation of 7E-Learning Cycle students MA NU Nurul Ulum Jekulo Kudus mastery goal group had high criteria of self-regulation more than the students ' performance goal team. Both groups are on the average score of the medium standards of self-regulation, the common measures of the mastery goal group there were seven students (63.6%) it is more than the performance goal unit that is five students (62.5%). At small criteria of performance goal group, there are three students (37.5%) whereas no mastery goal group that has low standards. According [14] there is a difference self-regulated learning between students' mastery goal with students' performance goal.

The goal orientation based groupings have two students of mastery goal groups, and two students of performance goal groups analysed more depth through the method of interviews about self-regulation and problem-solving abilities. Summary results of interviews of self-regulation based on goal orientation group consisting mastery goals and performance goals after the implementation of the 7E-learning cycle in the matrix presented in Table 5.

Table 5. Condition Self-Regulation Based on Goal Orientation After Implementation 7E-Learning Cycle

Self-Regulation Indicators	Goal Orientation	
	Mastery Goal	Performance Goal
Self-efficacy and self-goals	Frequently target value during the test, even predicting a problem checked.	At the time of the test do not determine a target value.
Strategy use	Have a particular approach to learning to increase understanding of the material.	Do not have a specific strategy in learning.
Time management	Studious, but still feel the time reserved for work on the test is not enough.	Studious, but still feel the time reserved for work on the trial is not sufficient.
Self-observation, self-judgment, self-reaction	Often taking the time to re-examine the answers to quiz before being collected. Always listen to the discussion of the test questions and feel challenged if it gets an awkward question.	Not taking the time to re-examine the answers to quiz before being collected. Sometimes not listen to the discussion of the test questions and quickly discouraged when gets difficult questions.
Environmental structuring	If the discussion has expressed their opinions, feel comfortable with the learning environment in class.	If the discussion has expressed their opinions, feel uncomfortable with the learning environment in class.
Selective help seeking	If difficulties understanding the material, ask to the cleverer, maximising other educational resources such as books and the internet.	If difficulties understanding the material, ask to the more intelligent, the learning resources only textbook from the teacher. So less to maximise help or educational resources.

Consistent with a study done by [15] shown students who held high mastery goal had a higher ability level than other students holding other types of goals orientation (performance goal or combined mastery goal -performance goal). Based on analysis of interviews students' problem-solving abilities after learning implementation 7E-learning cycle is obtained an increase in the problem-solving ability presented in Table 6.

Table 6. Students' Problem Solving Ability Based on Goal Orientation After 7E-Learning Implementation Cycle.

No	Problem Solving Indicators	Goal Orientation	
		Mastery Goal	Performance Goal
1.	Understanding the problem	Students wrote down what is known and asked of problems efficiently.	Students wrote down what is known and asked of the problem, but incomplete
2.	Devising a plan	Students can write a problem-solving plan, is already complete.	Students can write a problem-solving plan, but incomplete.
3.	Carry out the plan	Students can implement problem solutions based plan	Still, there is a mistake in the application of solutions based on the strategy.
4.	Looking back	Students check answers that have been done.	Students did not check back the answers that have been done

4. Conclusion

The conclusions of this study were (1) the learning 7E-learning cycle efficiently (2) self-regulation after implementation 7E-learning cycle group mastery goal has high self-regulation criteria. That is more than the student's performance goals group and the learning problem-solving abilities with 7E-learning cycle students of mastery goal group. It increased can understand the problem to be more precise, planning and problem solving write more systematically, and checks back to the answer by reading the back problems then examine each of these problems solving steps or to conduct a recount only.

Besides, there are some things to advise. There are (1) 7E-learning cycle used as a reference for teachers to improve students' self-regulation and problem-solving abilities. Then, (2) the increase in students' self-regulation and problem-solving skills can be done by directing and fostering mastery goal on the students in the learning process, and (3) Need for further research as the development of different grade level or material.

References

[1] PISA O. Results in Focus: What 15-year-olds know and what they can do with what they know. 2014-12-03].
[2] National Council of Teachers of Mathematics, editor. Principles and standards for school mathematics. National Council of Teachers of Mathematics; 2000.
[3] Orton A 2004 *Learning Mathematics: Issues, theory and classroom practice*. (London: Bloomsbury Publishing)
[4] Pehkonen E 1997 The state-of-art in mathematical creativity *ZDM* **29** 63-7.
[5] Polya G 2014 How to solve it: A new aspect of mathematical method. (Princeton: Princeton University Press)
[6] Peverly ST, Brobst KE, Graham M, Shaw R 2003 College adults are not good at self-regulation: A study on the relationship of self-regulation, note taking, and test taking *Journal of Educational Psychology* **95** 335.
[7] Schunk DH, Meece JR, Pintrich PR 2012 Motivation in education: Theory, research, and applications. (London: Pearson Higher Ed)
[8] Zimmerman BJ 1990 Self-regulated learning and academic achievement: An overview. *Educational Psychologist* **25** 3-17
[9] Usta G, Bozpolat E 2014 An Assessment of Education Faculty Students' Views on their Self-regulated Learning Skills *Procedia-Social and Behavioral Sciences* **116** 1620-5.
[10] Ames C, Archer J 1988 Achievement goals in the classroom: Students' learning strategies and motivation processes. *Journal of educational psychology* **80** 260.
[11] Pintrich PR 2000 An achievement goal theory perspective on issues in motivation terminology, theory, and research *Contemporary educational psychology* **25** 92-104.
[12] Treasure DC, Roberts GC 1995 Applications of achievement goal theory to physical education: Implications for enhancing motivation *Quest* **47** 475-89.
[13] Eisenkraft A 2003 Expanding the 5 E Model A purposed 7 E model emphasizes "transfer of learning" and the importance of eliciting prior understanding *The Science Teacher* **70** 56–59
[14] Puspitasari A 2013 Self Regulated Learning Ditinjau dari Goal Orientation. *Educational Psychology Journal* **2** 1-6
[15] Mattern RA 2005 College students' goal orientations and achievement. *International Journal of Teaching and Learning in Higher Education* **17** 27-32.

Impact of Implementation of Direct Cash Transfer Program 2008/2009 on Household Consumption in Central Java Province

S Subanti[1,2,5], A R Hakim[2,4,6], and I M Hakim[3]

[1]*Department of Statistics Faculty of Mathematics and Natural Science, Universitas Sebelas Maret*
[2]*Institute for Research and Community Services, Universitas Sebelas Maret*
[3]*Department of Industrial Engineering, Faculty of Engineering, Universitas Indonesia*
[4] *Faculty of Economics and Business, Universitas Indonesia*
[5]*Research Group of Applied Statistics, Universitas Sebelas Maret*
[6]*Department of Economic Development, Universitas Ahmad Dahlan*

Corresponding author: sri_subanti@yahoo.co.id

Abstract. This study aims to see the impact of direct cash transfer program for 2008/2009 on household consumption of food, nonfood, education, and health in Central Java Province. The study is expected to provide important findings for the improvement of a similar program in the future.This study findings that (1) the increasing in food and non-food consumption for direct cash transfer recipients than non direct cash transfer recipients; (2) the impact of households expenditure on education for direct cash transfer recipients is higher than non direct cash transfer recipients; (3) the impact of households expenditure on health for direct cash transfer recipients is lower than non direct cash transfer recipients. This study recommended that (1) implementation of direct cash transfer program 2008/2009 must be managed to be better because this program can defend household welfare. It shows from several indicators of well-being such as consumption spending, education, and health; (2) data targets for poor households (very poor, poor, nearly poor) must be updated.

1. Introduction

The increase in international crude oil prices pushed the government to raise fuel prices and reduced subsidies. As compensation for the rise in fuel prices, the government provides direct cash transfer program for poor households to maintain purchasing power. During the years 2005-2011, the government has raised fuel prices twice in 2005 and 2008. In 2005, the fuel was increased by 24 percent to 2,400 rupiahs per liter from the previous 1,810 rupiahs per liter. In October 2005, the government raised fuel prices by 87.5 percent to 4,500 rupiahs per liter. In May 2008, the government raised fuel prices by 25 percent to 6,000 rupiahs per liter [6,7,8,9].

Direct cash transfer program awarded to several reasons, namely the initial budget allocation of energy subsidies in the Budget (October 2007) is greater than the budget of the poverty reduction program. Based on data from the Government budget for the poor program only about Rp70, 6 trillion, or 7.5 billion U.S. dollars.While the budget for energy subsidies amounted to approximately Rp77, 2 billion (8.2 billion dollars). Where is the energy subsidy is 70 percent mostly enjoyed by high-income people? Therefore, if the policy of fuel price increases is not taken to be accommodated by the direct cash transfer program feared would happen in fiscal instability, the implementation of the Government's development program funds are ultimately political and security instability. Direct cash transfer program implemented in 2008-2009 to compensate for the rise in fuel prices is expected to have an impact on the livelihoods of the poor. Households who are targeted poor households and almost destitute poverty based census conducted by known as Central Bureau of Statistics [6,7,8,9,10].

From background above, the purpose of this study is to estimate the impact of direct cash transfer program for 2008/2009 on household consumption on food, nonfood, education, and health in Central Java Province.

2. Data And Methodology

2.1. Data

The data used in this paper come from Susenas (National Economy Social Survey) in February 2008, and 2009 are utilized in the present evaluation. We can identify most households across all two survey rounds allowing for both a short- and medium-term evaluation of the program. Although implemented before the design of the Direct Cash Transfer (DCT) program, we used the February 2008 Susenas provides a good baseline for impact. The baseline survey contains 2,422 households, while the follow-up in February 2009 contains only 2,454 households. After matching on geographic and household identifiers, we have a balanced panel data for 2008 and 2009. It contains 2,422 households; 897 of which received BLT and 1,525 of which not received BLT.

2.2. Mean Difference Approach

The basic principle of this method is compared differences between the treatment and control groups before and after the start of the DCT program. It is measured by calculating the mean difference (MD) in these two groups.This calculation uses the following formula [1]:

$$MD = \left(\mu_{i,2009} - \mu_{i,2008}\right)^{BLT} - \left(\mu_{i,2009} - \mu_{i,2008}\right)^{NonBLT} \quad (1)$$

where,

μ2009BLT	: the average value of consumption per month for DCT household in 2009
μ2008BLT	: the average value of consumption per month for DCT household in 2008
μ2009NonBLT	: the average value of consumption per month for Non DCT household in 2009
μ2008NonBLT	: the average value of consumption per month for Non DCT household in 2008

3. Result And Discussions

We will show the households demographic profile in Central Java Province in Table 1. This table shows that the proportion of head of household if male (84.39%) is higher than female (15.61%). The marital status of respondents with married status represent 84.06%. It is greater than another status, represent 15.94%. Age groups were also relatively distributed, except for the head of household with age between 35 or below. Majority respondents have not attended junior high school. Majority head of household has finished primary school or less. Respondents have attended junior high school, represented by 10.32%, whereas an amount of 9.25% from a head of household have finished senior high school education. Only 2.64%, head of household have finished university education. From main job status, the majority of respondents work as entrepreneur represent 35.55% then followed by employee and self employee represented by 27.99% and 15.77%. Based on origin, majority head of household cames from rural with 65.11% and urban with 34.89%. Majority head of household has a monthly income below 1.5 million rupiahs. For the head of household with monthly income 1.51 – 2.25 and 2.25 – 3 million rupiah accounted for 2.19% and 1.20%. Only 0.58%, head of household have a monthly income between 3.01 million or above. Then, direct cash transfer recipient in 08/09 accounted for 37.04% compared with the not recipient (62.96%). This table shows that direct cash transfer recipient in 05/06 and 08/09 accounted for 33.86% compared with the not recipient (66.14%).

Table 1. Household Demographic Profile In Central Java Province

Characteristics		Freq.	Percentage	Characteristics	Freq.	Percentage
Sex				Head of Household Origin		
	Male	2044	84.39	Urban	845	34.89
	Female	378	15.61	Rural	1577	65.11

Age			Education		
14 – 24	21	0.87	Primary School or Less	1518	62.68
25 – 35	314	12.96	Junior High School	250	10.32
36 – 46	716	29.56	Senior High School	224	9.25
47 – 57	664	27.42	University	64	2.64
>=58	707	29.19	Not Answered	366	15.11
Marital Status			Status in Main Job		
Other	386	15.94	Self Employee	382	15.77
Married	2036	84.06	Entrepreneur	861	35.55
Monthly Income			Employee	678	27.99
(in Rupiah) 1US$ = Rp 13000			Other	240	9.91
<= 1,5 million	1144	48.31	Not Answered	261	10.78
1,51 - 2,25 million	53	2.19	Direct Cash Transfer Recipient Year 08/09		
2,25 - 3 million	29	1.20	Yes	897	37.04
3,01 - 4 million	4	0.17	No	1525	62.96
>4 million	10	0.41	Direct Cash Transfer Recipient Year 05/06 & 08/09		
Not Answered	1156	47.73	Yes	820	33.86
			No	1602	66.14

Table 2. Summary Of Estimate Result

Variable Dependent	Impact	Direction	Significance
Food Consumption	3,4 until 3,9	Positive	Significance
Non Food Consumption	8,8 until 10,1	Positive	Significance
Education Expenditure	6,9 until 9,9	Positive	Not Significance
Health Expenditure	62,8 until 67,4	Negative	Significance

From Table 2, it was found that direct cash transfer 2008/2009 increase household consumption expenditure growth (food and non-food). Increasing of household consumption expenditure for direct cash transfer recipients are smaller than non-recipients of direct cash transfer. While the effects of direct cash transfer in 2008/2009 in the field of education have increased although not significantly. While the provision of direct cash transfer in 2008/2009, there was no increase in health expenditures for households receiving direct cash transfer. From this result, we can explain that the recent social assistance programs, direct cash transfer program, for example, it is administratively more efficient that other social assistance programs, likes better child nutrition for example. Because this program incurred per unit value of the benefit for each household and this program also used for the consumption of the commodity subsidized by the government [3,4]. Moreover, the main purpose of the program is to provide compensation as a result of rising fuel prices are always followed by the increase in prices of other needs. Supposedly, the increase in the value of BLT funds was higher than the value of the price of staples such as rice [8].

The impact of the program on food consumption and non food consumption were positive and significance. This occurs because the beneficiary households prioritize the needs for food, especially staples, to be fulfilled in the long term [8]. Also, non-food expenses are sometimes allocated for other things such as cigarettes [5,6,7]. The impact of the program on health consumption was negative and significance. This program can indicate that it can improve member of household health care because frequency and duration of illness have found lower. Then, it can also indicate that improved nutrition and preventive health care have made a member of household more robust against illness [2,4]. The impact of the program on education was positive and not significance. It is indicated that the difference of priority for the allocation of funds. The order of priority may include the consumption of food and non-food, pay off debt, buy clothes, and venture capital. Others priorities, the consumption of food and non-food, pay off debt, transportation, and school fees [7,8].

4. Conclusion

According to the analysis and findings, this study was found that first, the increasing in food and non-food consumption for direct cash transfer recipients than non direct cash transfer recipients; (2) the impact of households expenditure on education for direct cash transfer recipients is higher than non direct cash transfer recipients; (3) the impact of households expenditure on health for direct cash transfer recipients is lower than non direct cash transfer recipients.

Recommendation from this study, (1) implementation of direct cash transfer program 2008/2009 must be better because this program can defend household welfare. It shows from several indicators of well-being such as consumption spending and education; (2) data targets for poor households (verypoor, poor, nearlypoor) must be updated. Because targeting is very important to achieve cost effectiveness in the social assistance program, social protection program, and poverty reduction program; even though it is not easy is often suggested. Not only administrative costs, but also additional costs like disincentive costs, stigma costs, and political economy costs [10]. So, it must be possible that data targeted becomes more costly, but it is necessary and is a must, for the similar program in the future to be better again. Also, the presence of local initiative needs to be fostered, because it can improve the direct cash transfer program's management.

References

[1] D.A. Lind, W.G. Marchal, and S.A Wathen, Statistical Techniques in Business and Economics 15th Edition (McGraw – Hill, America, 2012), pp. 371 – 408.

[2] E. Skoufias. "PROGRESA and Its Impacts on the Welfare of Rural Households in Mexico", IFPRI Research Report No 139, 1-83 (2005).

[3] E. Skoufias, M. Unar, & T. Gonzalez-Cossio. "The Impact of Cash Transfer and In-Kind Transfers on Consumption and Labor Supply", World Bank Policy Research Working Paper No 4778, 1 – 48 (2008).

[4] J. Hoddinott, & E. Skoufias, Economic Development and Cultural Change 53 (1), pp. 37 – 61 (2004).

[5] O. Noyorono, "An Analysis of Implementation of BLT 2008/09 on Consumption, Education, Health, and Household Working Pattern," *National Seminar in Poverty held by TNP2K Office*, 2012.

[6] O. Noyorono, "The Impact of BLT 2008/09 on Consumption, Education, Health, and Household Working Pattern", Ph.D. Thesis, Universitas Indonesia, 2012.

[7] Hastuti, N. Toyamah, B. Sulaksono, S. Budiyati, W.D Widyanti, M. Rosfadhila, H. Sadaly, S. Erlita, R. J. Sodo, S. Bazzi, and S. Sumarto, "Quick Study on Implementation of BLT Year 2005 in Indonesia : Case Study in 5 Regency and Municipalities", SMERU Research Report, 1 – 37 (2006).

[8] M. Rosfadhila, N. Toyamah, B. Sulaksono, S. Devina, R. J. Sodo, and M. Syukri, "Study Implementation of Direct Cash Transfer Program in 2008 and Evaluation of BLT Program Recipients in Indonesia, SMERU Research Report, 1 – 125 (2013).

[9] S, Subanti, Respatiwulan, W. Sulandari, & A. R. Hakim. (2016) "The Impact of Direct Cash Transfer Program (BLT) on Household Consumption in Central Java Province", AIP Conference Proceeding 1746, 020054.

[10] S. Sumarto, and A. Suryahadi. (2001) "Principles and Approaches to Targeting : with Preference to Indonesian Social Safety Net Programs", Smeru Working Paper.

Economic Valuation on Change of Tourism Quality in Rawapening, Indonesia: An Application of Random Utility Method

S Subanti[1,8,9a)], B R M B Irawan[2,8,9], G Sasongko[3], and A R Hakim[3,4,5,6,7,9]

[1]*Department of Statistics Faculty of Mathematics and Natural Science, Universitas Sebelas Maret*
[2]*Department of Economic Development Faculty of Economics and Business, Universitas Sebelas Maret*
[3]*Department of Economics Faculty of Economics and Business, Universitas Kristen SatyaWacana*
[4]*Department of Economic Development, Universitas Ahmad Dahlan*
[5] *Faculty of Economics and Business, Universitas Darul Ulum Islamic Centre Sudirman GUPPI*
[6]*Department of Economic Development, Universitas Terbuka*
[7]*Faculty of Economics and Business, Universitas Indonesia*
[8]*Center for Research and Development for Tourism (PUSPARI), Universitas Sebelas Maret*
[9]*Institute for Research and Social Services, Universitas Sebelas Maret*

Corresponding author: sri_subanti@yahoo.co.id

Abstract. This study aims to determine the profit (loss) earned economic actors tourism activities if the condition or quality of tourism in Rawapeningbeimproved(deteriorated). Change condition or quality can be seen by traveling expenses, natural environment, Japanese cultural performances, and traditional markets.The method used to measure changes in the economic benefits or economic loss with a random utility approach.The study was found that travel cost, natural environment, Japanese cultural performances, and traditional markets have significant factors about respondent preferences to choose the change of tourism condition. The value of compensation received by visitors as a result of changes in conditions improved by 2,932 billion, while the change in the condition worsens by 2,628 billion. Recommendation of this study is the local government should consider environmental factors in the formulation of tourism development in Rawapening.

1. Introduction

Rawapening, administratively, located in Semarang Regency, Central Java Province, Indonesia. It located in four districts, including Banyubiru, Tuntang, Bawen, and Ambarawa. Rawapening represents one of developing tourism destination. Rawapening development should pay attention to visitors preferences. It is important that changes in the condition or quality of tourism can provide multiple benefits for both the tourism (visitors) and tourism management (local government) [4].

This study analyzed changes occurring conditions associated a in the condition or quality of tourism Rawapening. Change condition or quality of tourism could be expected to affect the preferences of visitors and potential visitors to Rawapening. The combination of changes in conditions or the quality of tourism expressed regarding attributes. There are four attributes to formulate changes in the condition. The four attributes are (a) travel cost, (b) the natural environment; (c) Javanese cultural arts performances and (d) traditional market.

Based on this background, the objective of this study is determinant of respondent choose the changing of tourism quality in Rawapening and how much economic value if the Rawapening condition (quality) improved or deteriorated.

2. Data And Methodology

2.1. Data

In this study, the survey concentrated on the visitors in Rawapening. We choose respondents as well as household head for an interview. Respondents defined who use the lake for recreation, and they must pay the entrance ticket. The visitors were randomly chosen for the interviews that completed the questionnaire [2,4,5].

In total, 150 copies of the questionnaire were prepared and distributed by interviewers. Of the 135 questionnaires returned, 120 could be used for analysis. The amount of 5 invalid replies because the respondents were not the household head, 8 did not complete the questions, and 2 were considered inconsistent in answer the questionnaires.

The questionnaire consisted of four sections, covering (1) information objects of study and research purposes, (2) demographic characteristics of respondents; (3) respondents perception to Rawapening, and (4) respondents rating for tourism quality. For section fourth, questionnaire consisted of multiple choice questions and dichotomous yes or no question.

2.2. Random Utility Approach

Random utility method is a method of decomposing the overall utility into two components, namely, deterministic (observed utility) and random (utilities are not observed). Utilities were not observed in the form of attributes that are not observed, unobservable taste variation, error and imperfect information, and the instrumental variables [1,2,3,6]. The random utility model is used to overcome the disadvantages of Contingency Valuation Method (Contingent Valuation Method) which based on hypothetical behavior directly. In the contingency valuation method, willingness to pay (WTP) or willingness to accept (WTA) asked directly, so it is possible that more assessment (overvalued) or even less (undervalued). This can happen because the respondents could not correctly convert their judgment in monetary units. Therefore the WTP or WTA should be estimated, not observed. Random utility models is one way to measure the WTP and WTA indirectly [1,6].

The random utility method is assuming that site characteristics are an important determinant of recreational choice decisions [8]. The method models try to determine how a representative consumer chooses from a set of discrete sites each of which embodies a vector of attributes (qualities). Following McFadden, some of the researchers make a basic model of RUM. This model consists of a deterministic core and a random component [7,8,9,10]. If known, there is a group of individuals n: i = 1, ..., i, ..., n. For every opportunity to recreation, such individuals may choose an alternative m which is part of a group of alternative preference (called AP): j = 1, ..., j, ..., m. Each has income (Mi) vectors, and then each AP has an attribute of tourism quality (Qj). Individual travel costs are Cij. In this study, j and k represent alternative development (changing conditions) in Rawapening. For individual i who chooses APj, utilities can be formulated as follows:

$$U_{ij} = V_{ij}(M_i - C_{ij}, Q_j) + \epsilon_{ij} \qquad (1)$$

Where ϵji is the unobservable component of utility and assumed to be random. The price of goods or services market (other consumption) omitted to simplify notation. From equation (1), we can derive the model for estimation as follows:

$$\text{Choices}_{ij} = ASC_{ij} + \beta_2 TC_{ij} + NATENV_j + JAVACUL_j + TRDMRKT_j + \epsilon_{ij} \qquad (2)$$

Equation (2) estimates with conditional logit. It is using the method of maximum likelihood (ML). The next procedure, we must test the model, and there are three procedure, (1) individual test, this test compares p value with alpha (α), and this test uses two way hypothesis, (2) overall test, this compares likelihood ratio statistics with chi squared, and (3) goodness of fit test, this test uses Pseudo R-Squared.

Last, we must estimate the welfare. It estimates in the RUM method is expected measures that depend on how the choice set is defined and how the distribution of the random error in the model. Because the observation in RUM function is ordinal and not cardinal, we can calculate welfare changes based on linear attributes assumption. So, the deterministic utility change for the representative consumer using the linear attributes as follows [7,8]:

$$\Delta Z_i = \Delta U(Z_i^1 - Z_i^0)/(-\lambda) \quad (3)$$

Then, the description of the variables used in equation (2) will be explained in Table 1.

Table 1. Description of Variables

Variable	Description
CHOICE	Dependent variable, represent conditional probability where 1 if respondent choose a choice, 0 if respondent do not choose a choice
ASC	Alternative spesific constant, used to divide choice alternative. In this model, 1 represent change condition and 0 current condition
TC	Travel Cost
NATENV	Tourism quality for natural Environment
JAVACUL	Tourism quality for Japanese cultural performance art
TRDMRKT	Tourism quality for traditional market

3. Results And Discussions

Table 2. Demographic Profile of Respondents in Rawapening

Characteristics	Freq.	Percentage	Characteristics	Freq.	Percentage
Sex			Respondent Origin		
Male	92	76.67	Semarang Regency	83	69.17
Female	28	23.33	Others	37	30.83
Age			Education		
18 – 20	14	11.67	Junior High School or Less	14	11.66
21 - 30	64	53.33	Senior High School	72	68.33
31 – 40	27	22.50	Graduate	30	25.00
41 – 50	10	8.33	Master	4	3.33
51 – 55	5	4.17	Occupation		
Marital Status			Student	10	8.34
Not Married	56	46.67	Government Staff	72	60.00
Married	61	50.83	Private Staff	30	25.00
Not Answer	3	2.50	Pension	4	3.33
Monthly Income			Entrepreneur	4	3.33
(in Rupiah) 1US$ = Rp 13000					
<= 1,5 million	49	40.83			
1,51 - 2,25 million	39	32.50			
2,25 - 3 million	27	22.50			
3,01 - 4 million	5	4.17			

Based on table 2, the proportion of male respondents (76.67%) is higher than female respondents (23.33%). The marital status of respondents with married status represent 50.83%. It's greater than single status, represent 46.67% of respondents. Age groups were also relatively distributed, except for people age 41 – 50 or older. Majority respondents have attended senior high school, even though many respondents have attended colleges or university graduates. Respondents that have attended junior high school or less represent 11.66% of respondents, whereas 3.33% of the respondents had master degree. From another job, the majority of respondents work as government staff represent 60% of respondents. Based form respondent origin, majority respondents cames from Semarang regency with 69.17% and others only 30.83%. Persons with a monthly household income of 1.51 – 2.25 million rupiah and 2.25 – 3 million rupiah accounted for 32.5% and 22.5%. While those, respondents with incomes less than 1.5 million rupiahs represented 40.83%. Only 4.17% of respondents had incomes over 3 million rupiahs.

Tabel 3 Estimation Results

Variable	Coef.	Std.Error	Sign.
ASC	0.274	0.428	
TC	-0.657	0.107	***
NATENV	0.734	0.110	***
JAVACUL	0.240	0.107	**
TRDMRKT	-0.473	0.091	***
Num of Obs		120.000	
LR chi2		125.440	
Prob> chi2		0.000	
Pseudo R2		0.040	

*** : sign α = 1%; **; sign α = 5%; * sign α = 10%

Based on Table 3, the attributes that affect the choice of respondents in Region Rawapening is travel cost, natural environment, Javanese cultural art performances and traditional markets. The four attributes significantly influence respondents' choice. ASC variable indicates the average individual utility when other variables unchanged. ASC value is positive for Rawapening Region, and it suggests the possibility of respondents would vote for a new scenario (scenario improved) higher than the scenario worse. Variable travel cost has a negative value. Travel expenses such as admission price, consumption while in tourist attractions, souvenirs, hand pieces, travel place transformation, and other costs (including toilets and parking). This variable can represent with a price. The sign of this variable indicates that it follows the demand functions. It can happen because visitors from different origins face different prices [5,7]. Then, natural environment variable is positive in Rawapening. This value indicates that respondents appear increasingly aware of the importance of natural environment when they are conducting activities traveled.Variable Javanese cultural art performance is positive in Rawapening. This value indicates the possibility of change scenarios selected respondents will increase along with the rise of Javanese cultural art performances held in Rawapening.

Then, traditional market variables are negative in Rawapening. This value indicates the possibility of change in the selected scenarios of respondents would decline with traditional market located in Rawapening. Based on these findings, the respondents did not pay much attention souvenir or handmade. Visitors choose a tourist attraction that has a beautiful natural environment, can interact with the local community, Javanese cultural art performance, and affordable with income respondents. In line with this opinion, the experience a visitors or recreationist is a must because they will have in particular activities at given sites [8].

Table 4 Economic Value (Rp)

Quality	Rawapening
Deteriorated	-2,628,557,743
Improved	2,932,693,095

Based on Table 4, the total economic value at Rawapening amounted to Rp 2.932 billion per year for improved scenario compared with worsened scenario amounted to Rp 2.628 billion per year. This calculation obtained by the attributes used in the study (the natural environment, Javanese cultural art performances, and traditional markets). Attributes are determined by comparing the change in the conditions of new scenarios to the status quo scenario. From this value, we can conclude that the sites operator and policy makers would like to keep measures of environmental quality, shopping facilities, and cultural performance which visitors or recreationists perceive as important for their recreational enjoyment [9].

4. Conclusion

According to the analysis and findings, this study found that (1) all of the attributes affect the preferences of respondents in choosing the changing conditions in Rawapening. There are travel cost, natural environment, Javanese cultural performances, and traditional markets; (2) the economic per year due to changing scenario of the region improved Rawapening higher than the scenario worse. Value scenarios improved compensation Rp 2.93 billion per year, while the worsening scenario of Rp 2.62 billion per year. Recommendation from this study, (1) the government needs to formulate policies Rawapening tourism development areas better. Javanese cultural performances and traditional markets as the main visitor areas need to be updated as conditions change Rawapening more attention to the natural environment; (2) the development of cultural tourism and tourism trips can be applied to the development of Rawapening. Development of natural environment combined with an increased frequency of Javanese cultural performances earned positive response.

References

[1] A.A. Patunru, "Three Essays on Choice Based Estimations of The Economic Benefits of Contamined Site Cleanup",Ph.D. Thesis, University of Illinois at Urbana-Champaign, 2004.

[2] A.Fauzi, Economic Valuation and Assessment of Damage to Natural Resources and Environment(IPB Press, Bogor, 2015), pp. 151 – 164.

[3] A. M. Freeman III, J. A. Herriges, and C. L. Kling, *The Measurement of Environmental and Resources Values: Theory and Methods*(Resources for the Future, Washington DC, 1993), pp. 386 – 410.

[4] A. R. Hakim, S. Subanti, and M. Tambunan, Journal of Sustainable Development **4 (2)**, pp. 91-101 (2011).

[5] J. Englin, and R. Mendelsohn, Journal of Enviromental Economics and Management **21**, pp. 275 – 290 (1991).

[6] L.F. Martha, "Measuring Economic Value of Change of Tourism Environment in Tanjung Pinang : An Application of Random Utility Model". National Seminar in Economics held by Universitas Indonesia, (2007).

[7] L.Pendleton, and R. Mendelsohn, Enviromental and Resource Economics **17**, pp. 89 -108 (2000).

[8] P.W. Schuhmann, and K.A. Schwabe, Enviromental and Resource Economics **27**, pp. 429 – 450 (2004).

[9] Y. Kaoru, Resource and Energy Economics **17**, pp. 119 – 136 (1995).

[10] Y. Kaoru, V. Smith, and J.Liu, American Journal of Agricultural Economics **77**, pp. 141 – 151 (1995).

Recurrent Neural Network For Forecasting Time Series With Long Memory Pattern

Walid and Alamsyah

Department of Mathematics, Universitas Negeri Semarang

Corresponding author: alikuwalid87@yahoo.com

Abstract. Recurrent Neural Network as one of the hybrid models are often used to predict and estimate the issues related to electricity, can be used to describe the cause of the swelling of electrical load which experienced by PLN. In this research will be developed RNN forecasting procedures at the time series with long memory patterns. Considering the application is the national electrical load which of course has a different trend with the condition of the electrical load in any country. This research produces the algorithm of time series forecasting which has long memory pattern using E-RNN after this referred to the algorithm of integrated fractional recurrent neural networks (FIRNN).The prediction results of long memory time series using models Fractional Integrated Recurrent Neural Network (FIRNN) showed that the model with the selection of data difference in the range of [-1,1] and the model of Fractional Integrated Recurrent Neural Network (FIRNN) (24,6,1) provides the smallest MSE value, which is 0.00149684.

1. Introduction

Electricity plays a significant role in realising the strategic objectives and national development, and therefore the power supply business controlled by the state and provision should be increased in line with the elaboration of the construction that provided electricity in sufficient quantities, equitable and quality. Supply of electric power is capital intensive and technology and in line with the principle of decentralisation and democratisation in the order of society, nation, and state the role of local governments and communities in the provision of electric power needs to be improved.

The importance of energy for society can be indicated by the amount of electricity used by the public both for household consumption and for industrial and business activity in local and national scale. Obviously, this will significantly affect the quantity and quality of goods or services that exist. Another sector that is not less necessary for the functioning of electricity are the advances in communications technology and informatics who took part in expanding the space for the transportation of goods and services in all areas of life today.

According to Law No. 30 of 2009 on electricity, PT. PLN (State Electricity Company) as the holder of the power of electricity business is required to provide continuous electrical power (constant) with high quality and reliability, are also needed to provide excellent service to customers/consumers of electricity. It turns out the circumstances encountered much different now to what has been established by the Electricity Act. As a result of the power supply crisis experienced in almost all corners of the country, PT. PLN is not able to provide continuous power to its customers.

Estimate or forecast the amount of electricity used to optimise the use of power in society, so that there will be no wastage or power outage. According to F. Chan and C. Lim (2011) forecasting is an important tool in planning the effective and efficient.So if it based on a count of the Central Bureau of Statistics which is abbreviated by BPS, there are 50 billion SMEs, mean total losses reached 2.5 billion unavoidable due to blackouts by PLN. It is not surprising because the electrical energy is indeed a prime mover in industrial activity and a full range of human needs. Electricity has become a necessity indispensable to man as closely linked to the overall economic base. This ranging from the smallest to the greatest extent.Specifically, in urban areas, dependence on electrical energy is very high, and it concerns the core activities within the scope of the industry that allows subdued overall economic security.

Other aspects that also suffered a setback as a result of planned blackouts by PLN is the effect on the decrease in the interest of investors to invest in Indonesia because he felt there was no guarantee of the power supply becomes the driving force of industrial production processes. According to [1], there are several stages in forecasting, among others, model identification, parameter estimation, verification of models and forecasting. At the time series prediction models forecasting the future state based on the past. Time series approach can use several methods, may include analysis using autocorrelation function and partial function autocorrelation to study changes with time series parametric models known as ARIMA time domain analysis.

Besides, there is an alternative approach that used as a method of spectral analysis. The spectral method is one form of the Fourier transform.To be able to forecast correctly then one of the first steps is to identify the data model to know whether the data included in the data group which has a long memory categories or short memory. Identification of this initial data will provide valuable information relating to the next step of estimating the parameters towards-existing parameters, then verify and new forecasting. Information good start would be a significant contribution in providing a level of accuracy of prediction.

Based on the above background, this paper will focus on a study to identify the model of extended memory using the rescaled range. The data is the data used to simulate the electrical load is calculated every hour throughout the year. Issues to be studied is how does the identification of long memory models using the rescale range?

2. Time Series Analysis And Long Memory Model

Time series analysis by [2] has several purposes, namely forecasting, modelling, and control. Forecasting issues related to the establishment of models and methods that can be used to produce an accurate estimate. Modelling aims to obtain an appropriate statistical model to represent the behaviour of a long-term time series data. Difference modelling with forecasting is predicting more likely on a model of the "black-box" to get the forecast, while modelling tends to models that can be interpreted to explain what is going on concerning the relationship between variables in a time series data. While the purpose of the control widely used in engineering, especially signal processing.

Statistical modelling of time series analysis if traced back, preceded by Yule (1927) which introduced autoregressive linear models (AR) to predict yearly sunspot numbers. Since the publication concerned with the analysis of time series is growing rapidly. Until 1980, most of the research has focused on the model of linear time series, mainly linear model class Autoregressive Integrated Moving Average (ARIMA). [1] developed a complete procedure for ARIMA model methodology hitherto used as a standard procedure in the establishment linear time series models. Some literature which discusses many ARIMA model seen in [3] and [4]. Also, the properties related to statistical theory for the ARIMA model has also been extensively analysed and developed by several researchers, among others, have been conducted by [5].

Time series is a series of observations Y_t on a variable Y, which each observation recorded at a particular time $t \in T$. In this case, T is the set of time in which the comments made. If T is a discrete set, then $\{Y_t,\ t \in T\}$ is a separate time series. $\{Y_t,\ t \in T\}$ is a notation whole time series, where Yt is the observation of $\{Y_t,\ t \in T\}$ in time to t. In the case of discrete time series, views usually taken at the same time interval.

A statistical approach to the analysis of time series conducted using statistical models to describe the dynamic behaviour of a time series. It is assumed that a time series generated by a mechanism or a stochastic model, which is usually defined by a stochastic difference equation. It is stochastic difference equation that consists of an equation and some initial conditions. Results or a solution of this model is a stochastic process, which is a sequence of random variables $\{Y_t\}$ defined on the probability space (Ω, $\mathfrak{K}$, P). For certain ω□Ω, $Y_t(\omega)$ is called a realisation (sample path or trajectory) of $\{Y_t\}$ Each observation

$Y_t(\omega)$ is a realisable value of the random variable Y_t whose values obtained in d-dimensional Euclidean space $\mathbb{R}^d$.

The most significant assumptions in the analysis of time series are stationary data. Nonstationarity inspection data done with the help of time series plots using scatter plots. Nonstationarity in the mean can be resolved by the process of differentiating (differencing). According to [1] stationary data invariance can be seen with a value of λ (lambda) estimate the Box-Cox transformation

[5] expressed a process $\{Y_t\}$ is said to be stationary stronger if $(y_{t_1}, y_{t_2}, \ldots, y_{t_k})'$ and $(y_{t_{1+l}}, y_{t_{2+l}}, \ldots, y_{t_{k+1}})'$ have the same joint distribution function for all integers k≥1dan for all $t_1, t_2, \ldots, t_k, k \in \mathbb{N}$. Processes with the first and second moments that are independent of time is also a concern in time series analysis. The following definitions relating to the concept of a weak fixed or stationary until the second order.

If given $\{Y_t\}$ is a process with $E|Y_t|^2 < \infty$ for every $t \in \mathbb{N}$, then $\{Y_t\}$ is said to be fixed weak (weakly stationary) if $E(Y_t) = \mu$ for all $t \in \mathbb{N}$ and $Cov(Y_r, Y_s) = Cov(Y_{r+h}, Y_{s+h})$ for all $r, s, h \in \mathbb{N}$. If $\{Y_t\}$ is a process that is stationary weak (weakly stationary), then the function autocovariance $\gamma_y(.)$ of $\{Y_t\}$ defined $\gamma_y(h) = Cov(Y_t, Y_{t+h})$ to all $t, h \in \mathbb{N}$ [5].

Time series that have a long-term dependence, or better known as the long memory is a time series in which observations are far apart still have a high correlation. Autocorrelation of a process of long memory fell hyperbolic. A sequence $\{Y_t\}$ is said to follow the model Autoregressive Integrated Moving Average if the distinction to d namely $W_t = \nabla^d Y_t$ is a stationary ARMA process. If W_t is ARMA (p, q), then Y_t is ARIMA (p, d, q). In practice, the value of d used is 1 or at most 3 [3]. Model ARFIMA able to model the dependence of short-term and long-term. Observations generated by the structure of ARMA shows the relationship of short, while the distinction fractional parameter d, which caused values to fall hyperbolic ACF showed long-term dependence.

According to Wilfredo Palma (2007), a definition of long memory understood as follows. Let $\gamma(h) = \langle y_t, y_{t+h} \rangle$ is the covariance function of the lag h of a stationary process $\{y_t : t \epsilon \mathbb{Z}\}$. The usual definition of long memory is

$$\sum_{h=-\infty}^{\infty} |\gamma(h)| = \infty \tag{1}$$

However, there is an alternative explanation. In particular, long memory can be defined by specifying the hyperbolic decline in this autocovariance. Recalling that a function of measuring active defined in some environments [a, ∞) infinite is said to change slowly in a sense because if and only if for every $c > 0, l(cx)/l(x)$ converges to 1 of x tends Infinity. Examples slowly changing function is $l(x) = \log(x)$ and $l(x) = b$, where b is a positive constant. Furthermore, notation $x_n \sim y_n$ which means that $\frac{x_n}{y_n} \to 1$ of $n \to \infty$, unless otherwise stated.

$$\gamma(h) \sim h^{2d-1} l_1(h) \tag{2}$$

For $h \to \infty$, where d is called long memory parameters and $l_1(.)$ is a function that is changing slowly.

Another widely used definition of a high dependence on the spectral domain is

$$f(\lambda) \sim |\lambda|^{-2d} l_2(1/|\lambda|) \tag{3}$$

To λ in zero environments and $l_2(.)$ is a slowly changing function. Furthermore, an alternative definition of long memory on the rules that are based directly on the decomposition Wold

$$\psi_j \sim j^{d-1} l_3(j) \tag{4}$$

For j> 0, where $l_3(.)$ is a slowly changing function. Model ARFIMA (p, d, q) developed [6] is written as:

$$\phi(B)(1-B)^d(Y_t - \mu) = \theta(B)a_t \tag{5}$$

With t: index of observation

d: distinguishing parameters (fractions)

IOP Conf. Series: Journal of Physics: Conf. Series **824** (2017) 012038 doi:10.1088/1742-6596/824/1/012038

μ: average observation

$a_t \sim \text{IIDN}\ (0, \sigma_a^2)$,

$\phi(B) = 1 - \phi_1 B - \phi_2 B^2 - \cdots - \phi_p B^p$ is a polynomial AR (p),

$\theta(B) = 1 - \theta_1 B - \theta_2 B^2 - \cdots - \theta_q B^q$ q are polynomials MA (q),

$(1 - B)^d = \nabla^d = \sum_{k=0}^{\infty} \binom{d}{k} (-1)^k B^k$ differentiator fractional operator,

$$\sum_{k=0}^{\infty} \binom{d}{k} (-1)^k B^k = F(-d, 1; 1; B)$$

$$= \sum_{k=0}^{\infty} \frac{\Gamma(k-d)}{\Gamma(k+1)\Gamma(-d)} B^k$$

F is the hypergeometric function is defined as follows.

$$F(a, b; c; d) = 1 + \frac{a.b}{1.c} B + \frac{a(a+1)b(b+1)}{1.2.c(c+1)} B^2 + \cdots \tag{6}$$

a, b and c are real numbers and B is the operator Backshift.

For d is value fraction, fractional differencing operator $(1 - B)^d$

Defined as

$$(1 - B)^d = 1 + \sum_{k=1}^{\infty} \frac{\Gamma(-d+k)}{\Gamma(-d)k!} B^k \tag{7}$$

If the equation $\lambda_k(d) = \frac{\Gamma(-d+k)}{\Gamma(-d)k!}$ in equation (7) described for various values of k then:

For k = 1, is obtained $\frac{\Gamma(-d+1)}{\Gamma(-d)1!} = \frac{(-d)!}{(-d-1)!1!} = -d$,

For k=2, diperoleh $\frac{\Gamma(-d+2)}{\Gamma(-d)2!} = \frac{(-d+1)!}{(-d-1)!2!} = \frac{-d(1-d)}{2}$,

k=3, diperoleh $\frac{\Gamma(-d+3)}{\Gamma(-d)3!} = \frac{(-d+2)!}{(-d-1)!3!} = \frac{-d(1-d)(2-d)}{6}$ and so on so that the equation (3.3) can be written back into

$$(1 - B)^d = 1 + \sum_{k=1}^{\infty} \lambda_k B^k \tag{8}$$

with $\lambda_0(d) = 1, \lambda_1(d) = -d\ \lambda_2(d) = -\frac{1}{2}d(1-d), \lambda_3(d) = -\frac{1}{6}d(1-d)(2-d)$ and so on. So that equation (8) above can be written back into

$$(1 - B)^d = 1 - dB - \tfrac{1}{2}d(1-d)B^2 - \tfrac{1}{6}d(1-d)(2-d)B^3 - \cdots \tag{9}$$

In general model of seasonal autoregressive integrated moving average (ARIMA) (P, D, Q) S is as follows.

$$\Phi_p(B^S)(1 - B^S)^D Z_t = \Theta_Q(B^S) a_t \tag{10}$$

with $\Phi_p(B^S) = 1 - \Phi_1 B^S - \Phi_2 B^{2S} \ldots - \Phi_P B^{PS}$

$\Theta_Q(B^S) = 1 - \Theta_1 B^S - \Theta_2 B^{2S} \ldots - \Theta_Q B^{QS}$

Multiplicative ARIMA model written in the form

$$\phi_p(B)\Phi_p(B^S)(1 - B)^d(1 - B^S)^D Z_t = \theta_q(B)\Theta_Q(B^S) a_t \tag{11}$$

This model is often called by the model SARIMA (p, d, q) (P, D, Q)^{S}S

This model reduces to the ARIMA model (p, d, q) when no seasonal effects as well as being ARMA (p, q) when the stationary time series.

Additive SARIMA generalized models can be written in the form

$$\left(1-\phi_1 B-\phi_2 B^2-\cdots-\phi_p B^p-\Phi_1 B^S-\Phi_2 B^{2S}\ldots-\Phi_P B^{PS}\right)(1-B)^d(1-B^S)^D Z_t$$
$$=\left(1-\theta_1 B-\theta_2 B^2-\cdots-\theta_q B^q-\Theta_1 B^S-\Theta_2 B^{2S}\ldots-\Theta_Q B^{QS}\right)a_t \qquad (12)$$

3. Research Methods

The data used in this study is the simulation data generated from the values of Hurst, including Hurst value: H = 0.12345; H = 0.1567; H = 0.2389; H = 0.3456; H = 0.4535; H =0.4995; H = 0.5000; H = 0.5235; H = 0.5853; H = 0.6534; H = 0.7654; H = 0.8745; H = 0.90345; H = 0.9999.

In addition to the above simulation data, this study also uses data the electricity consumption data which is secondary data from the PLN in Java and Bali. Data is data consumption every half-hour heat load Se Bali-Java, taken from January 1 to the date of the end of December 2010. Through the program Minitab and Matlab searchable form plot (scatter) data. Furthermore, the data in the analysis to determine the magnitude of the value difference d (difference) method rescale range (R/S). This process begins with determining the average value of the data, the adjusted mean and standard deviation.

Furthermore, the determination of the value of the cumulative deviation and range of the cumulative difference, and the final stage is to determine the statistical value of Hurst (H) with logarithmic statistics (R/S) and assess the value of H through Ordinary Least Square method (OLS).

4. Result

Results of analysis with Matlab program for Hurst value seen in figure 1.

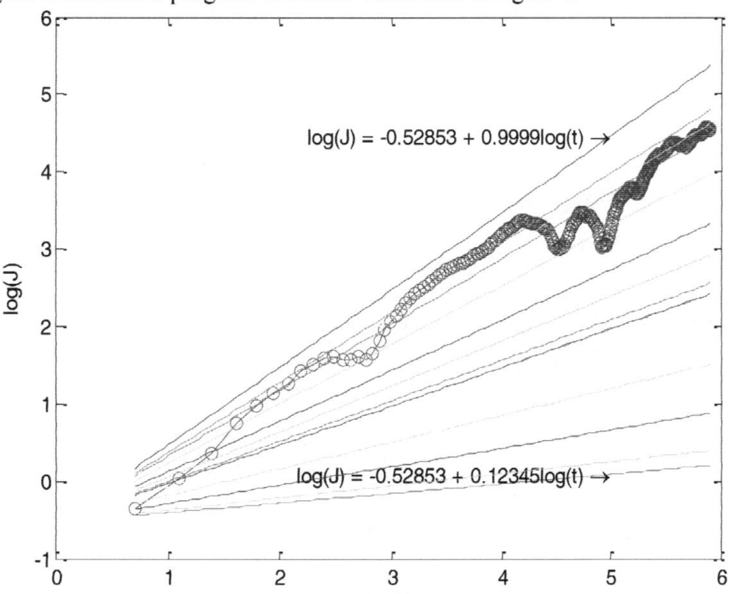

Figure 1 Hurst Value

Preprocessing done by using transformation, in this study conducted transformations on the data difference in the range [1,1] and conversion of data difference in the range [-1,1]. While the first conversion function used is function prestd (targtrain) to obtain the target data normally distributed, namely the achievement of a zero mean value and variance value is one. While the role of the second transformation in the preprocessing is by using the conversion prestd (inptrain) with the aim to obtain the input data were normally distributed, namely the achievement of a zero mean value and variance value is one. The result of the preprocessing stage using data difference through two models namely the type function net = newff (minmax (in), [1,1], { 'tansig', 'purelin'}) and the model function of net = train (net, in, tn).

The next step is to provide training using difference NN data on the range [1,1], the two models in NN used that model RNN. Training in RNN model by modifying the number of hidden layers to obtain optimal results in forecasting, optimisation methods used in training are Levenberg Marquardt. The optimal lead to forecasting this will be demonstrated by evaluating the value of MSE on each model of training used.

Based on the above steps, obtained several models for use in forecasting time series patterned long memory. The results of model identification, among others model FIRNN (24,1,1), FIRNN (24,2,1), FIRNN (24,3,1), FIRNN (24,4,1), FIRNN (24,5,1), FIRNN (24,6,1), and model FIRNN (24, 7.1).

Also, the results of training iterations difference 24 RNN with the data in the range [1,1] each model found in the appendix. FIRNN training results using data difference in the range [-1,1]. Some models RNN also developed, with the type identification obtained several models. Among others: the model RNN (24,1,1) using 1 unit hidden and 1 unit of output and optimisation methods used in training are Levenberg Marquardt, the results obtained in the epoch to 21 model provides MSE value of 0.00570881 with an amount of gradient $0{,}23376\text{x}10^{-5}$. It appears that at this period to 23 models reach towards convergence with absolute values. The results of the iteration shown in figure 2.

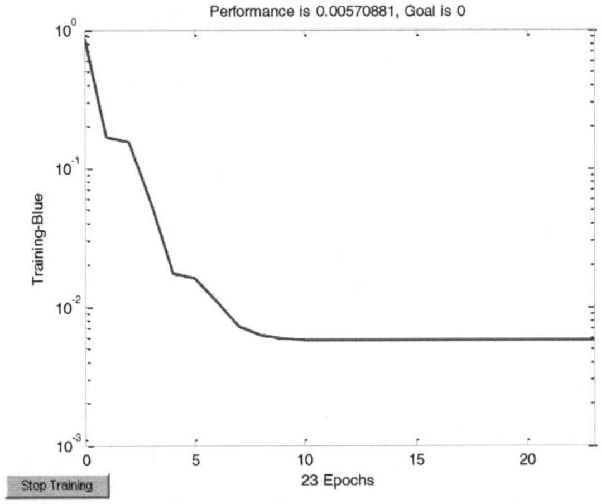

Figure 2 The Results Of Iteration

Training with the model FIRNN (24,2,1) using 2 hidden units and 1 unit of output and optimisation methods used in training are Levenberg Marquardt, the results obtained in the epoch to the 100 model provides MSE value to the value of 0.0392602 with gradient $2,\ 11941\text{x}10^{-5}$. This model produces period to 100 to reach convergent. The results of the iteration shown in figure 3.

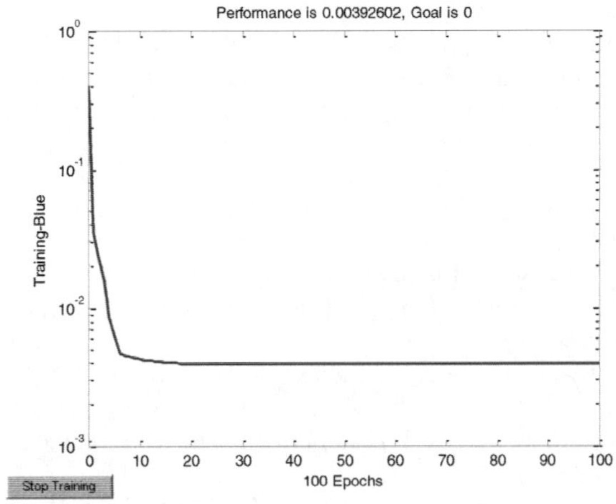

Figure 3 The Results Of Iteration

Training models FIRNN (24,3,1) with 3 hidden units and 1 unit of output and optimisation methods used in training are Levenberg Marquardt, the performance in the epoch of the 100 model provides MSE value of 0.0301456 with gradient value 0, 48871x10^{-5}. This model produces period to 100 to reach convergent. The results of the iteration shown in figure 4.

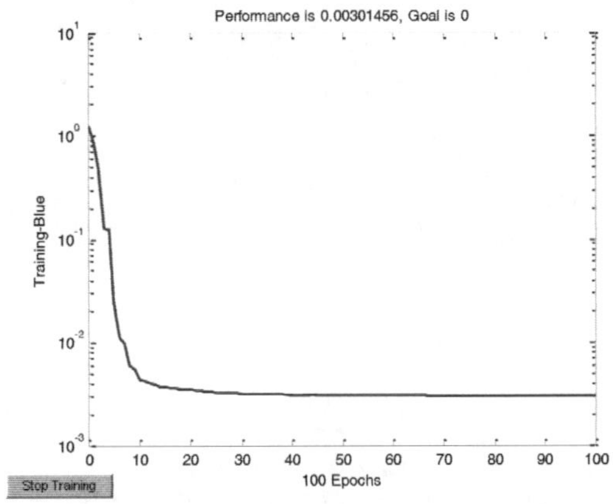

Figure 4 The Results Of Iteration

Training models FIRNN (24,4,1) using 4 units hidden and 1 unit of output and optimisation methods used in training are Levenberg Marquardt; the result converges on epoch to 100, this model provides MSE value of 0.00240606 and value of gradient by 1,31178x10^{-5}. Training models FIRNN (24,5,1) using 5 hidden units and 1 unit of output and optimisation methods used in training are Levenberg Marquardt; the result converges on epoch to 100, this model provides MSE value of 0.00202306 value gradient 0,34653x10^{-5}. Training models FIRNN (24,6,1) using 6 hidden units and 1 unit of output and optimisation methods used in training are Levenberg Marquardt, obtained results converge to the epoch to 100 model provides MSE value of 0.00177697 with value amounted gradient 0,37509x10^{-5}.

Training models FIRNN (24,7,1) using 7 hidden units and 1 unit of output and optimisation methods used in training are Levenberg Marquardt, obtained results converge to the epoch to 100 model provides MSE value of 0.00170185 and value gradient 0,54 130 x10^{-5}.

5. Conclusions

The results of long memory time series forecasting using models Fractional Integrated Recurrent Neural Network (FIRNN) showed that the model Fractional Integrated Recurrent Neural Network (FIRNN) (24,7,1) gives the smallest MSE value, which is 0.00177697. Besides, using simulated data, the result that for the simulated data with H <0.5 forecasting results obtained by using FIRNN shows the level of accuracy of 80%, while for H> 0.5 provides a 90% accuracy rate. Use of the activation function in the E-RNN needs to be modified with other functions to get the variance results in line with expectations so that it will be obtained an activation function appropriately in forecasting.

References

[1] Box G E P & Jenkins G M 1976 *Time Series Analysis, Forecasting and Control, 3rd Edition* (Englewood Cliffs: Prentice Hall)
[2] Chatfield C 2001 *Time Series Forecasting* (Chapman)
[3] Wei W W S 1999 *Time Series Analysis: Univariate and Multivariate Methods* (USA: Addison-Wesley Publishing Co)
[4] Box E.P George *et al.* 1994 *Time Series Analysis, Forecasting and Contro* (New Jersey: Prentice Hall Inc)
[5] Brockwell P J and Davis RA 1991 *Time Series Theory and Methods, 2nd Edition* (New York: Springer-Verlag)
[6] Joyeuk 1980 An Introduction to Long Memory Time Series Models and Fractional Differencing, Journal of Time Series Analysis, Vol. 1, pp. 15-29.

The Development of Probability Material using Edmodo

I Sujadi,R Kurniasih, and S Subanti

Postgraduate Mathematics Education FKIP, Universitas Sebelas Maret, Indonesia

Corresponding author: imamsujadi@ymail.com

Abstract. This research is aimed to describe the process and to get product development of learning material using Edmodo. Learning material is developed in the form of interactive learning material using Edmodo. This research belongs to Research and Development (R&D). The procedure includes the steps of Borg and Gall such as conducting research, collecting information, planning and developing preliminary form product, preliminary field testing and main product revision, main field testing and operational product revision, operational field testing and final product revision, dissemination and implementation. Expert judgment conducts data validity with the average score of media expert 3.57 and 3.67 from the material expert with a maximum score of 4 on a Likert scale. Practicability is obtained by the implementation of learning materials in each meeting which has exceeded 75%. The effectiveness of the response questionnaire of thestudent has achieved 80%. The students have received instructional materials well. The result of focus group discussion (FGD) can be concluded that learning material can be used in theteaching-learning process in curriculum 2013.

1. Introduction

Probability is a material in amathematic subject that has existed in elementary school to university. Probability is essential in daily activity in Indonesia. Based on thecurriculum in Indonesia, probability material starts to be taught in primary to middle school [9]. As noted at National Conference of Teachers in Mathematics (NCTM), theprobabilityis taught in curriculum from kindergarten [6]. Gal stated that probability is an essential material to help students in life [11]. On the other hand, Jones et al. Ministry of Education, Pratt, Schield, stated that the "importance of probability in everyday life has led to calls for increased attention to this strand in the mathematics curriculum"[11]. Therefore, probability material is essential to be mastered. However, students' ability on probability is low. It can be seen in national examination of senior high school in 2015. It shows thelow result with only 8.09%.

Some experts have examined Students' understanding of probability concept with various perspectives. Jones, et. al. described how students think probabilistic. It means how the students solve problems in probability (see [4] and [5]). Probabilistic matter contains uncertainty. According to Sujadistated that uncertainty matter tends to be a random activity that can deliver probable result [12]. However, thecertain result cannot be previously determined accurately. Students who think in an uncertain situation are supposed to think probabilistic. To know how students think probabilistic, the researcher conducts pre-survey by using theinstrument of Jun [6]. The result shows that many students are in subjective to quantitative informal level (see [4] and [5]).

Now, Indonesia applies curriculum 2013 that emphasises the development of students in some aspects, such as knowledge, behaviour and skill. Curriculum 2013 is also expected to be able to implement all integrated subjects in communication and information technology. Minister of culture and education, Hestated that nowadays, developing technology becomes important information in knowledge. In an international seminar on "21th Century Learning" in January 20nd, 2016, Minister of culture and education stated that main resource of information is a way to reach all sides in giving information in education and teaching-learning process. One of the implementations that havebeen conducted is national examination with Computer Based Test (CBT). Furthermore, he conveyed that the use of technology is not only used for theexamination but also in teaching-learning process.

In thedigital era, most of thestudents have used various social media. Most of them prefer to use technology from smartphone, tablet, gadget and computer then anopen book or repeat material in school. It is also probable that bad information can turn up as well. Therefore, teachers must also use technology in theteaching-learning process.

Fast development of information technology can be used in theteaching-learning process as aprovider of online media. Online media is usually named as e-learning, a kind of teaching-learning process that uses electronic technology and the internet. Teaching-learning process with the internet can use Edmodo. According to Pitoy [10], Edmodo is a platform of social media for teachers and students to share idea, file, agenda, activity and task. Edmodo is aimed to help teachers in using social media in theteaching-learning process. Edmodo is almost same as Facebook. However, it is not as free as Facebook. Edmodo feature is a special design in education to make an interaction among teachers, students and parents.

The government expects that developed technology can be used well in education. Utomo has developed e-learning material basis Edmodo in lithospherematerial of geography subject. Based on the above explanation, theresearcher will develop interactive material through social media of Edmodo [14]. Development of Edmodo material is expected to create high thought skill of probability. It can also facilitate students to use acellphone or tablet and internet service as learning media. Tomlinson added material development is what the writers, teachers, and students conduct to give input resource as a designed experience to improve learning [13]. Therefore, amaterial using Edmodo is expected to give easy way when the students want to repeat material and facilitate them to be independent in learning.

2. Method

This research uses research and development method. This research modifies the Borg & Gall model of research and development [1]. Education research and development is a process that is used to improve and validate an education product such as teaching-learning model, instrument, textbook, module, and media[1]. The following are research procedure of this research: 1) Collecting research and information: in this stage, researcher conducts a literature review about level probabilistic thought, analyses students' level probabilistic thought, observation and interview with the teachers. 2) Planning; it is conducted by defining, arranging interactive material design with aworksheet, determining material arrangement. 3) Developing preliminary form product; it includes preparing material or content design and producing amaterial draft in which product form that is developed is an interactive material with theworksheet. Afterwards, it is well kept by using Edmodo.4) Testing preliminary field and revising main product; it is conducted to state that the previous product has been valid with average valid criteria more than 2.5 with thescale of Likert (1-4). 5) Testing themain field and revising operational product; researcher conducts limited experiment towards developed material. The limited experiment is conducted by teaching ten students using Edmodo material. After conducting alimited experiment, revision is conducted based on the result of thelimited experiment through response questionnaire that has been validated by the experts. 6) Testing operational field and revising final product; it is conducted by applying theproduct in theteaching-learning process. Validated lesson plan must be preparedfor this testing. The operational testing field is conducted in a class that is chosen randomly to get practical and effective data.

The used instrument is material observation sheet for practice and response questionnaire for effectiveness. Criteria for practice is obtained if the percentage is more than 75% [8]. It is scored by two observers with statistical analysis "percentage of agreement" Grinnel [8]. An effective data product is obtained from theresult of response questionnaire using Edmodo in Poll feature. Effective criteria product is observed by students' response towards product implementation if no more than 80% of students give positive average result toward the developed product [12]. Furthermore, to declare valid product whether it is practical or effective, Focus Group Discussion (FGD) is conducted through assessment of practitioner or teachers.

3. Results and Discussion

Development process starts from research with data that is obtained probabilistic thought level. Collecting information is obtained from observation and teachers' interview. Data of probabilistic thinking level is mostly in subjective and transitional level. Observation shows that teachers use material from official government. The same result is also obtained from teachers' interview in which the material is from compulsory book and worksheet. Therefore, it is needed to develop interactive material with thetechnological element.The idea will be supported because school has adequate facility and students has smartphone that can be used to study

Steps of product arrangement are planning and developing preliminary form product by designing material content and teaching-learning and producing material. The resulting product is interactive material. The form is interactive multimedia that can be accessed by smartphone or computer. The material will be completed with theworksheet. According to Depdiknas [2], theInteractive material is arranged based on theprincipal of material development, such as 1) starting from easy to difficult, and from concrete to abstract. 2) repetition will strengthen understanding, 3) positive feedback will give strength towards students understanding, 4) high learning motivation is a factor of successful learning, 5) achieving success seems like rising stairs, step by step and finally it will reach the top, 6) knowing result that will be achieved encourages students to continuously achieve the aim. Probability in theinteractive material is arranged based on main competence of Junior high school in curriculum 2013. Display of interactive material is shown in Figure 1.

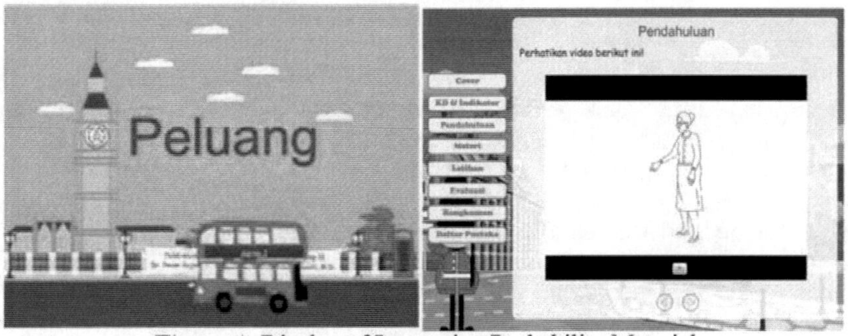

Figure 1. Display of Interactive Probability Material

Next step is preliminary field testing and main product revision. Experts conduct preliminary field testing by scoring related product that is developed from media and material aspect. The experts are lecturers of mathematics education of Sebelas Maret University (UNS), informatics technique of UNS, thetechnique of information and computer education of UNS and outstanding teachers. The experts have previously ever made teaching material and media. Data for avalid product with anaverage percentage of three validators is shown in Table 1.

Table 1. Result of Interactive Valid Material by Experts

Scoring	Validator			Statement
	1	2	3	
Media experts	3.75	3.52	3.46	very valid
Material experts	3.67	3.74	3.59	very valid

Main product revision is conducted by improving thecontent and material display (interactive material and worksheet) as the experts suggest. After conducting revision, there will result in interactive material as below:

Figure 2. Display of Interactive Probability Material

The interactive material contains probability material that has been arranged orderly. It starts withrecognising experiment and statistic experiment, differentiating experiment and statistic experiment, sample point and sample space, recognising possible, impossible, and certain event, empirical and theoretical probability of an experiment, differentiating empirical and theoretical probability and probability complement of an event. The valid interactive material is completed with theworksheet. The interactive material can be accessed through asmartphone. However, the display adjusts with the smartphone screen of each user.

Next step is main field testing and operational product revision. It needs to prepare response questionnaire to know thereadable product that is developed. Main field testing is conducted by using interactive material and worksheet. It is taught to 10 students who are chosen randomly. Before testing, students have had Edmodo account because it has been able to be downloaded. The limited test result is obtained from thevote of students through Edmodo or response questionnaire. The result of response questionnaire obtains apercentage of 89.10 %. It means that students can accept material well and use it as alearning source. Operational product revision is conducted based on aquestionnaire, students' suggestion and researcher's note as material for revision.

The sixth step is operational field testing and final product revision. An instrument that must be prepared is arranging lesson plan that adopts the use of developed material and observation sheet of conducting material. All teaching-learning process that includes introduction, content, and closing is conducted by using interactive material and worksheet. The teaching- learning process is conducted in agroup. Each group use one laptop, and each student is given a worksheet. Before conducting teaching-learning process, lesson plan and observation sheet are validated by three experts. The result of teaching-learning includes planning of lesson plan, observation sheet and response questionnaire that stated lesson plan couldbe conducted, observation sheet can be used to measure the material conducted, and response questionnaire can be used to measure the acceptance of students' material.

Operational field testing is conducted in a class to know practicability and effectiveness. Practicability is based on observation result with material observation sheet in theteaching-learning process. Observation result of 2 observers in the first meeting is 93.75% and in the second meeting is 100%. According to Borich [8], criteria of practicability is more than 75%. Effectiveness based on response questionnaire is 93.25%. It means that students can accept material well. Besides that, notes from observers and students can be used as aremedial and teaching-learning process.

Final product revision is the last step that is conducted in this development process. This last step is aimed to reveal aproduct that is developed. It is a valid, practical and effective final product. It is discussedin Focus Group Discussion (FGD) with mathematic teachers. The process of FGD uses assessment sheet filled by five teachers. The average of assessment result and comment is 3.82. It means that the developed product is very valid. Comments can also be used as amaterial alternative which can be used in theteaching-learning process. The interactive material can help students to understand theconcept of probability because theabstract material can be represented in animation as mentioned in research result of Yung and Pass [17] related to the influence of visual representation and computer on mathematic learning. The result of teaching-learning mathematic with visual representation by using acomputer can support teaching-learning process, transform abstract mathematic concept and give influence on cognitive skill. The following is the display of interactive material and worksheet as the last product of development process.

Figure 3. Final Product of Probability Material

The use of Edmodo has not been effective yet since many students have not been accustomed to asmartphone to study. According to Hankins, teaching-learning result using Edmodo does not give significant influence towards students' achievements [16]. However, Edmodo helps students to study. Besides that, students are taught to use technology. Research of Wardono also supports this case; he found that e-learning using Edmodo gives good result in developing theliterate skill of junior high school students [15]. Furthermore, research conducted by Jun shows that only 4 % of students do not understand probability concept [6]. It can be concluded that the step conducted can improve concept understanding of probability.

The development process of material using Edmodo is an introduction. It includes research and collecting information. Development step includes planning and developing an early product, testing field and revising, testing the main field and revising, testing operational field and revising. The result of introduction step shows that probabilistic level of students is on the subjective and transitional level. Collecting information is obtained through interview and observation. In the step of product development results validity with average 3.57 from media experts and 3.6 from material experts. Validator experts state that Edmodo can be used in the teaching-learning process. It is also suitable with the basic competence of probability. Practicability means that percentage of material using Edmodo has achieved an average of more than 75%. Effectiveness also shows that students well accept Edmodo with response percentage of 93.25%. The result of FGD shows that material using Edmodo can be used and expanded in the teaching-learning process.

Learning material of probability using Edmodo can be used to the learning process of mathematic either inside the class or outside the class. Learning material of probability can help students to understand the concept of probability and to increase the level of students' probabilistic thinking.

4. Conclusions

Conclusions of this study are: (1) the ability to think critically mathematical junior high school students are still, (2) in terms of cognitive styles critical thinking skills mathematical junior high school students with a higher FI cognitive styles of the students FD, (3) From the aspect of students' critical thinking aspect with FI cognitive style is better than the FD on the viability inference, assumptions, deduction, and argument evaluation.

IOP Conf. Series: Journal of Physics: Conf. Series **824** (2017) 012039 doi:10.1088/1742-6596/824/1/012039

References

[1] Borg W R and Gall M D 1979 *Educational Research An Introduction Third Edition* Longman, New York

[2] Depdiknas 2008*Panduan Pengembangan Bahan Ajar* Jakarta Depdiknas Direktorat Jendral Manajemen Pendidikan Dasar dan Menengah Direktorat Pembinaan SMA

[3] Hankins N S. St. Thomas University, Miami Gardens, Florida

[4] Jones G A and Tarr J E1997 *Mathematics Education Research Journal***9**, No 1, pp 39-59

[5] Jones G A, Langrall W C, Thornton A C, Mogill T 1999 *Journal for Research in Mathematics Education***30**, pp 487-519

[6] Jun 2000*Chinese Student's Understanding of Probability* Singapura: Nanyang Technological University, In fulfillment of the requirement for the degree of Doctor of Philosophy. 2000.

[7] Kemp J E 1994*Proses Perancangan Pengajaran*Edited by Asril Marjohan, ITB, Bandung

[8] Kuswardi Y, Kurniawati I and Chrisnawati H E 2007 *Laporan Pengembangan dan Peningkatan Kualitas Pembelajaran di LPTK (PPKP). Universitas Sebelas Maret*

[9] Permendikbud No. 20 Tahun 2016

[10] Pitoy Y P 2014 *Making Mathematics Learning Excited: New Hope, New Spirit with Edmodo*

[11] Sharma S 2012 *Journal of Mathematics Research*, Vol **4** No 5

[12] Sujadi I 2010 *Tingkat-Tingkat Berpikir Probabilistik Siswa Sekolah Menengah Pertama. Disertasi*. Universitas Negeri Surabaya

[13] Tomlinson, B 1998 *Material Development in Language Teaching Cambridge* Cambridge University

[14] Utomo D S 2015 *Pengembangan Bahan Ajar E-learning Berbasis Edmodo pada Materi Litosfer Kelas X Sekolah Menengah Atas (Online) Abstrak* Universitas Malang

[15] Wardono S. B. Waluya, Scolastika M, S Candra D 2016 *Journal of Physics: Conference Series***693**, 012014

[16] Watson M J and Kelly A B 2009 *International Journal of Science and Mathematics Education* 7:25-54

[17] Yung H I and Paas F 2015 *Educational Technology & Society*, **18 (4)**, pp 70-77

IOP Conf. Series: Journal of Physics: Conf. Series **824** (2017) 012040 doi:10.1088/1742-6596/824/1/012040

Undergraduate Students' Initial Ability in Understanding Phylogenetic Tree

S Sa'adah[1], T Hidayat[2], Fransisca Sudargo[3]

[1] Postgraduate Science Education Student Universitas Pendidikan Indonesia (UPI), Jalan Setiabudi 229 Bandung 40154 Indonesia, UIN SGD Bandung, Jalan AH Nasution 105 Bandung Indonesia
[2, 3]Universitas Pendidikan Indonesia (UPI), Jalan Setiabudi 229 Bandung 40154 Indonesia

Corresponding author: topikhidayat@upi.edu

Abstract. The Phylogenetic tree is a visual representation depicts a hypothesis about the evolutionary relationship among taxa. Evolutionary experts use this representation to evaluate the evidence for evolution. The phylogenetic tree is currently growing for many disciplines in biology. Consequently, learning about the phylogenetic tree has become an important part of biological education and an interesting area of biology education research. Skill to understanding and reasoning of the phylogenetic tree, (called tree thinking) is an important skill for biology students. However, research showed many students have difficulty in interpreting, constructing, and comparing among the phylogenetic tree, as well as experiencing a misconception in the understanding of the phylogenetic tree. Students are often not taught how to reason about evolutionary relationship depicted in the diagram. Students are also not provided with information about the underlying theory and process of phylogenetic. This study aims to investigate the initial ability of undergraduate students in understanding and reasoning of the phylogenetic tree. The research method is the descriptive method. Students are given multiple choice questions and an essay that representative by tree thinking elements. Each correct answer made percentages. Each student is also given questionnaires. The results showed that the undergraduate students' initial ability in understanding and reasoning phylogenetic tree is low. Many students are not able to answer questions about the phylogenetic tree. Only 19 % undergraduate student who answered correctly on indicator evaluate the evolutionary relationship among taxa, 25% undergraduate student who answered correctly on indicator applying concepts of the clade, 17% undergraduate student who answered correctly on indicator determines the character evolution, and only a few undergraduate student who can construct the phylogenetic tree.

1. Introduction

The phylogenetic tree is a diagram that depicts evolutionary relationships among a group of the organism (Taxa) that are believed to have a common ancestor [1][2][3]. A phylogenetic tree is an important tool for organising knowledge about the diversity of organism and a tool for structuring classifications, and for providing insight into events that occurred during evolution [4].

Phylogenetic trees are currently growing in nearly all disciplines of biology [5]. Phylogenetic used by researchers to answer fundamental questions about the history and diversity of life on earth and applied by the researchers, such as in the field of human epidemiology, antibiotic resistance, artificial selection for the domestication of animals and plants [6][7]. Therefore, studying about phylogenetic trees has become an important component in the biology education and an interesting area for biology education research [8]. The ability to understand and make sense of phylogenetic trees (referred to as cladistics thinking/tree thinking) is an important skill for biology students [9]. Unfortunately, students

IOP Conf. Series: Journal of Physics: Conf. Series **824** (2017) 012040 doi:10.1088/1742-6596/824/1/012040

often are not taught how to make sense of evolutionary relationship that is depicted in the diagram. Students are also not provided with information regarding the underlying theory and process of the phylogenetic tree [9]. Therefore, no wonder many undergraduate students have misconceptions in understanding phylogenetic tree [9][10] and students have difficulty in interpreting, constructing, and comparing among the phylogenetic tree [2][10][11]. The researcher has an opinion that understanding phylogenetic tree is a complex cognitive task and without proper scaffolding, many students cannot transfer the empirical data into visual structure [12]. The purpose of this study was to investigate the initial ability of undergraduate students in understanding and reasoning the phylogenetic tree. Information obtained from this study can give an idea to educators in developing undergraduate students' skill to understanding and make sense of the phylogenetic trees.

2. Research methodology

2.1. Research subjects

The research has been conducted on Program Studi Pendidikan Biologi UIN SGD in Bandung West Java. The subjects were 87 undergraduate students enrolled in vertebrate zoology course. All students were a sophomore.

2.2. Methods, data collection and analysis techniques

The method of this study is descriptive research. Data collection technique used was tests and questionnaire. Tests conducted to reveal undergraduate students' initial ability in understanding and reasoning phylogenetic tree. Subject received multiple choice questions and essay. Each question is adjusted with tree thinking skill modified from Novic and Catley[13]. It is describing the character of a taxon from the phylogenetic tree, determining the character of evolution, comparing the type of the phylogenetic tree, applying the concept of clade, evaluating relative evolutionary relatedness, determine the common ancestor between two or more species (MRCA/Most Recent Common Ancestor) and built a phylogenetic tree. Furthermore, the correct answer for every question made a percentage. Sample question can be seen in Figure 1.

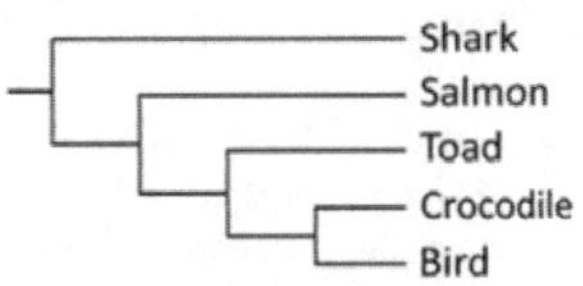

Observe the phylogenetic tree above what animals have the closest evolutionary relationship? Why, tell your reason!

(a)

	vertebrate	Bone	Four limb	Amnion	Hair	Two post-orbital fenestra
Shark	Yes	No	No	No	No	No
Mackerel	Yes	Yes	No	No	No	No
Frog	Yes	Yes	Yes	No	No	No
Rabbit	Yes	Yes	Yes	Yes	Yes	No
Monkey	Yes	Yes	Yes	Yes	Yes	No
Crocodile	Yes	Yes	Yes	Yes	No	Yes
Peacock	Yes	Yes	Yes	Yes	No	Yes

Observe the table that contains character for some animals above, create a phylogenetic tree base on data in the table!

(b[17])

Figure 1. Sample question to investigate undergraduate students' initial ability in understanding and reasoning phylogenetic tree

IOP Conf. Series: Journal of Physics: Conf. Series **824** (2017) 012040 doi:10.1088/1742-6596/824/1/012040

3. Results and Discussion

Undergraduate students' initial ability in understanding and reasoning phylogenetic tree shown in Figure 2.

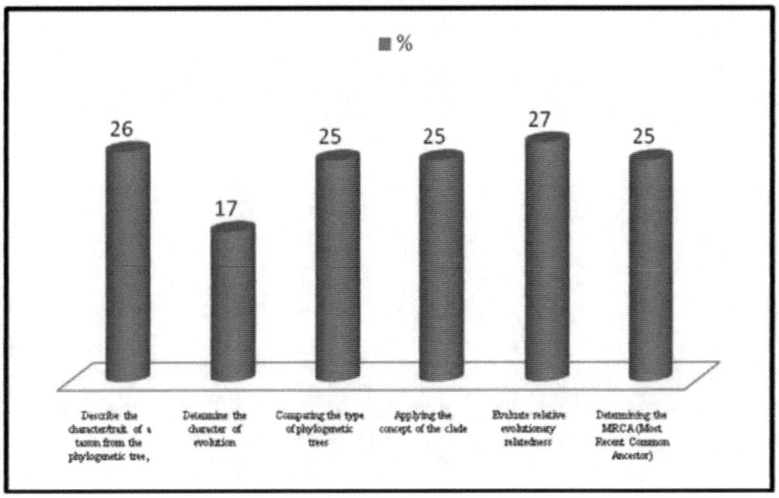

Figure 2. The percentage of undergraduate student' initial ability in understanding and reasoning phylogenetic tree

Based on Figure 2, undergraduate students' initial ability in understanding and reasoning phylogenetic trees are low. Students had difficulty answering the questions that we provided. Undergraduate students who answered correctly on each tree thinking skill, none up to 30%. The lowest percentage (17%), obtained from the indicator tree thinking skill namely determine the character of evolution. Students have difficulties in determining character evolution such as synapomorphy and autapomorphy from a phylogenetic tree provided because they do not understand the definition of each character evolution and often confuse one character with another character.

Base on Figure 2, the highest percentage (27%), obtained from the indicator tree thinking skill namely to evaluate the evolutionary relationship among taxa. Further, we ask about the indicators to evaluate the evolutionary relationships between taxa (question shown in Figure 1a), and most students answered incorrectly. According to them, an animal that has the closest evolutionary relationship is a shark and salmon, because of sharks and salmon both belong to a group of fish (Pisces), both live in the water, and their position in the phylogenetic tree is near. Some students answered the question correctly, but not with the reasons. Some students answered correctly, but less precise reasons, such as crocodiles and birds have a close evolutionary relationship because their position in the tip a phylogenetic tree is adjacent. Only a few students were answered correctly and give a good reason. Previous studies show that college students have misconceptions about reading the phylogenetic tree, among these students thought similarity indicates relatedness [14]. The species are drawn closer together at the tips of the tree were more closely related to each other than those drawn farther apart [11,][15], taxa relatedness is determined by a prior knowledge and not provided by phylogenetic trees]16][13].

Undergraduate students' initial ability in constructing the phylogenetic tree was not much different from reading the phylogenetic tree. Only a few students were able to build the phylogenetic tree based on the data provided. Figure 3, 4, and 5 below presents the answers of students in constructing the phylogenetic tree based on data provided (Figure 1b).

IOP Conf. Series: Journal of Physics: Conf. Series **824** (2017) 012040 doi:10.1088/1742-6596/824/1/012040

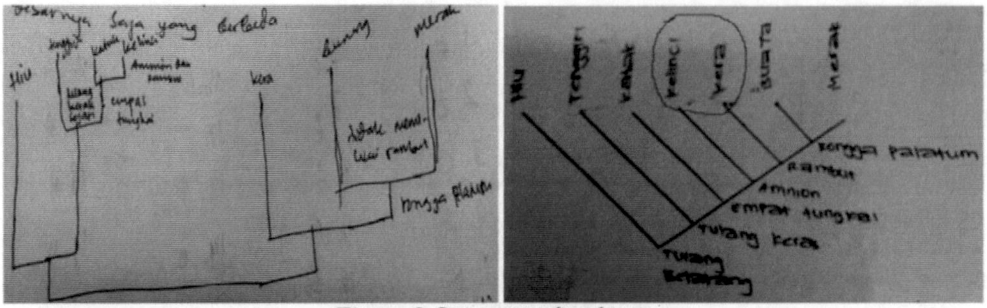

Figure 3. Incorrect relatedness

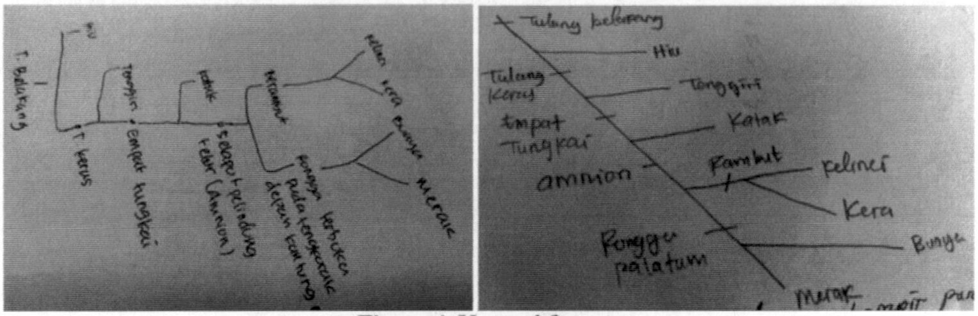

Figure 4. Unusual feature

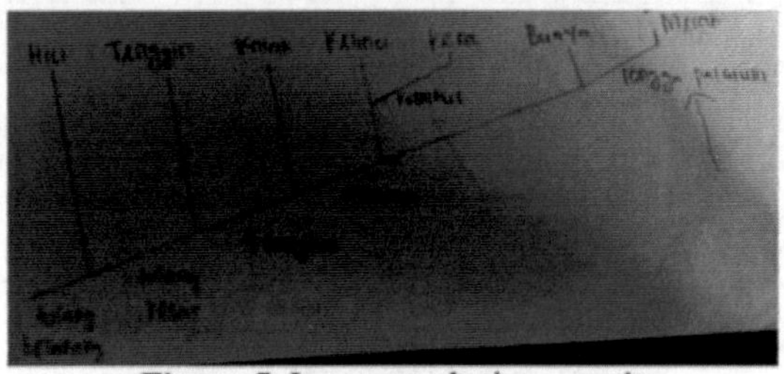

Figure 5. Incorrect placing a trait

Phylogenetic trees can be presented in a many ways, as shown in Figure 6, but some students draw the phylogenetic tree with an unusual figure (Figure 4). Based on Figure 3. many students draw phylogenetic trees incorrectly especially placing taxa in the tip phylogenetic trees, so phylogenetic trees that they draw doesn't describe the actual evolutionary relationship (incorrect relatedness). For example, in Figure 3a. Incorrect relatedness showed among mackerel (*tenggiri*), frog (*katak*) and rabbit (*kelinci*), and in Figure 3b. Incorrect relatedness showed among rabbit (*kelinci*) and monkey (*kera*). Based on Figure 5. The student was incorrect in placing trait (fur = rambut); therefore, the fur only belong monkey (kera), and rabbit (kelinci) don't have fur.

Results of student answer as shown in the image above, not much different from previous studies, that common major construction error was contemporary descent (extant taxa are descended from other extant taxa), empty branches, extra nodes, incorrect relatedness, incorrect traits[18].

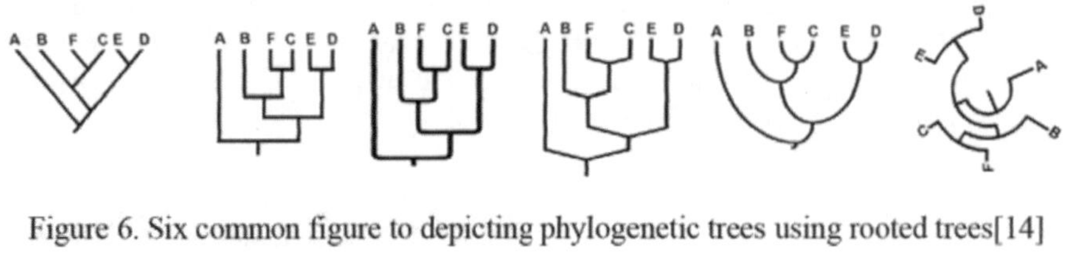

Figure 6. Six common figure to depicting phylogenetic trees using rooted trees[14]

These results were supported by the results of the questionnaire given to the students that 62% students stated no to the question "*Are you able to read the phylogenetic tree to see the evolutionary relationships among organisms?*" and 86% stated "no" to the question "*Are you able to build the phylogenetic tree base on data provided?*". Base on information from the questionnaire indicates that students lack knowledge of the phylogenetic tree, although 95% of the students admitted that they had previously been taught about the phylogenetic tree in Invertebrate Zoology and Botany Cryptogamae courses.

4. Conclusion

Based on the results and discussion, it can conclude that generally, undergraduate students' initial ability in understanding phylogenetic is low. Many students are not able to answer questions about the phylogenetic tree. Undergraduate students have to experience difficulties, especially in constructing a phylogenetic tree. Thus, it is necessary to plan great strategy in solving the problems. The results provide insight to researchers and educators to develop learning or lab for fostering undergraduate students in understanding the phylogenetic tree.

References

[1] Zachos F E 2016 *Mammalian Biology* 81, p.185–188.
[2] Dees J, Momsen J L, Niemi J, and Montplaisir L 2014 *CBE—Life Sciences Education*, 13, 666–676.
[3] Baum D A and Smith S D 2012 *Tree Thinking an Introduction to Phylogenetic Biology*. (Roberts And Company Publishers, Greenwood Village Colorado) p. 35.
[4] Baum D A 2008 *Nature Education* 1(1), 190
[5] Omland K E, Cook L G, and Crisp M D 2008 *BioEssays*, 30 (9) 854–867.
[6] American Association For The Advancement Of Science (AAAS). 2011 *Vision And Change In Undergraduate Biology Education: A Call To Action*. (Washington Dc), p 1-11.
[7] Dharmayanti I. 2011 *Wartazoa* 21 (1)
[8] Dees, J. and Momsen, J L. 2016 Student Construction of Phylogenetic Trees in an Introductory Biology Course *Evolution Education Outreach* 9: 3, p 1-9.
[9] Phillips B P, Novick L R, Catley K M, and Funk, D J 2012 *Evo Edu Outreach* 5, 595–602
[10] Novick L R, Shade C K, and Catley K M 2011 *Topic in Cognitive Science* 3, 536-559.
[11] Novick L R, Stull A T, and Catley K M 2012 *Bioscience* 62, 757-764.
[12] Halverson K L, Pires C J, and Abell S K 2011 *Science Education* 95,794–823.
[13] Novick L R and Catley K M. 2013 *American Educational Research Journal*, 50: 1, 138–177.
[14] Gregory T R M 2008 *Evoutionl Education Outreach*, 1,121–37.
[15] Meir E, Perry J, Herron J C, and Kingsolver J 2007 *The American Biology Teacher*, 71-76
[16] Smith J J, Cheruvelil K C, and Auvenshine S 2013 *CBE-Life Science Education* 12, 542–552.
[17] Modification from http://evolution.berkeley.edu/evolibrary/article/0_0_0/phylogenetics_07

IOP Conf. Series: Journal of Physics: Conf. Series **824** (2017) 012041 doi:10.1088/1742-6596/824/1/012041

Calculus Problem Solving Behavior of Mathematic Education Students

M Rizal and J Mansyur

University of Tadulako, Indonesia

Corresponding author: rizaltberu97@yahoo.com

Abstract. The purpose of this study is to obtain a description of the problem-solving behaviour of mathematics education students. The attainment of the purpose consisted of several stages: (1) to gain the subject from the mathematic education of first semester students, each of them who has a high, medium, and low competence of mathematic case. (2) To give two mathematical problems with different characteristics. The first problem (M1), the statement does not lead to a resolution. The second problem (M2), a statement leads to problem-solving. (3) To explore the behaviour of problem-solving based on the step of Polya (Rizal, 2011) by way of thinking aloud and in-depth interviews. The obtained data are analysed as suggested by Miles and Huberman (1994) but at first, time triangulation is done or data's credibility by providing equivalent problem contexts and at different times. The results show that the behavioral problem solvers (mathematic education students) who are capable of high mathematic competency (ST). In understanding M1, ST is more likely to pay attention to an image first, read the texts piecemeal and repeatedly, then as a whole and more focus to the sentences that contain equations, numbers or symbols. As a result, not all information can be received well. When understanding the M2, ST can link the information from a problem that is stored in the working memory to the information on the long-term memory. ST makes planning to the solution of M1 and M2 by using a formula based on similar experiences which have been ever received before. Another case when implementing the troubleshooting plans, ST complete the M1 according to the plan, but not all can be resolved correctly. In contrast to the implementation of the solving plan of M2, ST can solve the problem according to plan quickly and correctly. According to the solving result of M1 and M2, ST conducts by reading the job based on an algorithm and reasonability. Furthermore, when SS and SR understand the problem of M1 and M2 similar to the ST's, but both of the problem solvers read the questions with not complete so that they cannot pay attention to the questions of the problems. SS and SR create and execute M2 plan same as ST, but for M1, SS and SR cannot do it, but only active on reading the statement of the problem. On the checking of the M2 task, SS and SR retrace the task according to the used formula.

1. Introduction

Every people have ever been confronted a problem, which is not they hoped and naturally it opposites with the facts. Schunk [1] states that a problem occurs only when a man met a difficulty in which its solution is not soon available, but the difficulty is not the intrinsic character. A problem is a duty that the solving procedure is not known yet to emphasise this idea. Thus the procedure should be pursued to solve it [2]. A problem always depends on knowledge and experience of the problem solver. Hayes [3]posits a problem for someone could be training for others; all of it are depends on personal knowledge and skills. It can be said that a problem probably becomes a problem for someone but not for others.

Someone who has a problem should solve it. If someone finds a way to change a problem become becomes a wished purpose, then it means someone has to solve it [4]. Problem-solving "means a precise way to cover the gap" [3]. Whereas, a problem resolving is an instructed idea to find a solution for a specific issue directly [5]. A problem-solving means an instructed behaviour to a goal which needs a precise mental representation from trouble and the implication of particular method or a strategy to gain a required purpose [6].

According to Polya [7], there are four steps in problem-solving: understand the problem, devising a plan, carrying out the plan, and looking back. Those four steps are elaborated as follows. (1) *Understand the Problem*. Understand a problem is an important step in answering it. Without understanding the problem well, someone is not able to solve his problem. This step is started by knowing what is unknown and what will be gained. It is followed by a way to observe things were already understood and existed data then perceive whether the existed data and occurred condition adequate to decide things which are achieved. (2) *Devise a Plan*. This step needs a capability to see a relation between data and occurred condition with data and things, which are not known. If the relation is not found, observing other way is probably as an alternative. The next procedure is making a resolve plan by concerning these following aspects: did the students ever face the problem, do the students can use the theorem to answer the problem. For a wider problem, it can be solved in piecemeal then students can arrange a plan by devising systematic steps. (3) *Carry Out the Plan*. The previous plan made before should be done cautiously in every step. In undertaking the previous plan or fixing mathematic questions, students are expected to heed the principle (formula) of procedure to get the result of the correct mathematic model. Making the wrong answer can cause a fault in answering the problem. Therefore, checking each of the solution steps must be done to ensure the truth of obtaining the expected purposes. (4) *Looking Back*. A step is an effort to look back and ensure whether the solution was done is appropriate with the expected solution. If the result is not appropriate for the purpose, then looking back the each solving steps, analyse the result to the problem, and grasp the possibilities to solve the problems. The process above can observe the improper strategy, and by this, the improper strategy can be re-fixed.

Problem-solving is becoming a fundamental issue and a particularly interesting issue for Pribyl and Bodner [8] who affirm that the first step in problem-solving includes disembedding (discharging) relevant information from the problem and re-arrange or problems transformation to individual understanding. Those first steps are a crucial part in deciding the success or failure process for problem-solving. The purpose of first steps in problem-solving is an experiment to understand problems or to find the problem. In working out the problem, a problem solver must convert the series of words in which represents an internal mental which is manipulated into the effort to solve the problem [9].

Table 1. Behaviour characteristic in solving problem-based on the Polya

The steps of Problem Solving	Characteristic
Understand Problems	1. Understand the existed data or information of the given problem 2. Question the issues of the given problem 3. Identify needed requirement to solve the given problem 4. Analyse the previous identic questions
Devise A Plan	1. Strategy used to solve the problem 2. Probable steps to solve the problem
Carry Out Plan	1. Strategy used to solve the problem 2. Probable steps to solve the problem
Look Back	1. Re-investigate the conducted solution 2. Apply other solution/strategy

This research is aimed to disclose a behaviour in working out the calculus problems done by students based on the indicators/steps of Polya's problem-solving steps. The details behaviour characteristic in solving the problem based on the Polya can be seen in Table 1.

2. Method

The subject of this research is students of the Mathematic Department who take a calculus course. The research subject is one student of each mathematic thinking level; high, moderate, and low. The behaviours of subjects in solving the problem are observed based on the Polya's [7], understand the problem, devise a plan, carry out the plan and look back.

The instrument used in this research is not only the researcher himself as the main instrument but also the supported instrument: mathematic test and equal calculus problems code by P1 and P2 as follows.

P1	M1	Decide the line equation of graph below. 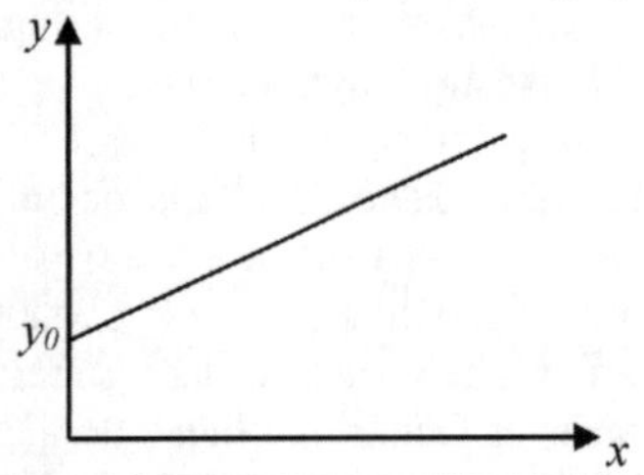 What is the graph meaning? Make the interpretation! Find the equation $g = y_0 x + \frac{1}{2} m x^2$ from the line equation above.
	M2	The relation between y and x displayed below: 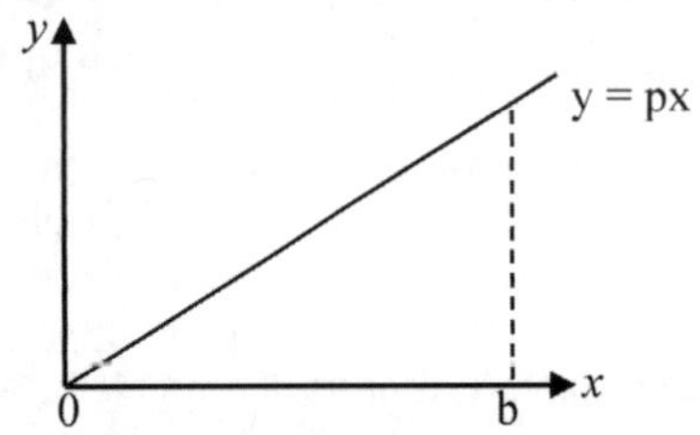 Use the integral to decide the area width limited by a curve $y = px$ (p constant) and x-axis (in interval $x = 0$ and $x = b$)
P2	N1	Decide the motion equations of a substance represent by the following graph: 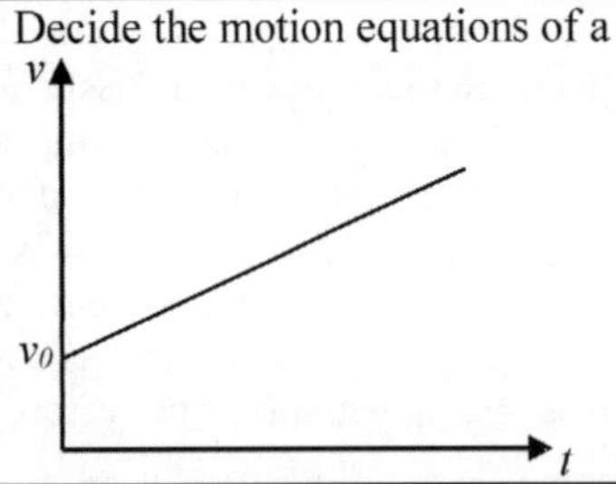 What is the meaning of the motion equations? Make the interpretation! Find the equation $s = v_0 t + \frac{1}{2} a t^2$ from the equation above.
	N2	Use integral concept (based on the graph) to decide the effort of the elasticity if the elasticity is pressed then one of the edge move from points $x = 0$ to $x = a$ (relation between force and length change is $F = kx$). 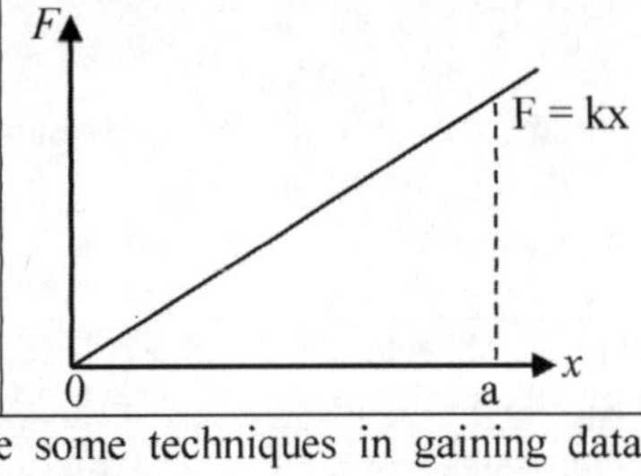

There are some techniques in gaining data in a matter of subjects' behaviours in solving the problems. The techniques used are observation technique, thinking aloud and in-depth interview, the techniques are applied to identify subject behaviour in understanding the problem, devise a plan, carry out plan, and look back.

After the data collected, credibility test is conducted by using triangulation of time. It aims to compare gathered data to a different time. If the data is collected at different times provide a consistent result, then the data can be decided credible.

The consistent research data is analysed based on; data reduction, data display, and conclusion [10].

3. Result and Discussion

The result of triangulation data ST, SS, and SR clarifies that it is consistent in solving P1 and P2 in a different time. Thus it can be concluded that those three data are credible. Therefore, the data of problem solver behaviour is analysed when they do P1.

3.1. The Result of Data Analysis and Discussion of ST as Problem Solver Behavior

3.1.1. The Steps in Understanding P1 (M1 and M2)

The interview transcript of ST in understanding P1 (M1 and M2) is written clear as follows.
P: Before you do the question, try to read the problem
ST: Directly see the graph
P: Read the problem loudly
ST: Read P1 loud, in piecemeal and repeatedly
P: Do you understand?
ST: read thoroughly, and read slowly when seeing the picture and gaining equations
P: Do you understand?
ST: Yes, Sir!

Based on the observation result and interview transcript above that ST in understanding M1 and M2 tend to concern the picture first, reading the texts piecemeal and repeatedly, then thoroughly and more focus on the equation sentences, numbers or symbol. However, for M1 for ST is not all information from the problem can be accepted well. Thus the subject cannot relate information which is gained and the questioned information. However, for M2, ST can assimilate information from the problems directly, because of it is suitable for the information which is existed in cognitive structure, because the information which is gained fit with the previous experience. This phenomenon is in line with Ausubel [11] that the information can be assimilated meaningfully if a man can relate it to the previous knowledge.

3.1.2. The Step in Make a Planning Solving Problem P1 (M1 and M2)

Transcript interview with ST in the making a plan of problem-solving of P1 explained in detail:

M1 Problem Solving Plan

P: Based on your understanding, how is the way to solve this problem?
ST: Graph on (0, y)
P: So, how is the way to solving the problem?
ST : use known formula, by line equations through points (a,b) by gradient m: y-b = m(x-a)
P: is there any other way?
ST: Silent, while thinking and seeing figure, then they say "nothing."

M2 Problem Solving Plan

P: how is the way to solve the second problem (M2)
ST: by looking back the result and focus to see the picture
P: so, how is the way to solve it?
ST: The result by pointing" $y = mx + y_0$" substitute to $g: y_0x + \frac{1}{2}mx^2$

Based on the interview transcript is above known that ST makes a plan to solve problem M1 and M2 by using formula. To make a plan of M1, ST used a formula of line equations through points (a,b) with gradient m, y – b = m (x-a), but several questions for M1 is not made the plan yet because the accepted information is not appropriate to the scheme thus it cannot be assimilated by ST directly. Therefore, for M2, ST make a plan to solve the problem by using integral pattern. It can be planned as that because the accepted of the problem based on the same experience gained formerly. This way is based on the Gestalt [12] state that if the previous aspect is unique, then it appear again. Thus the aspect cannot emerge by imbalance. Those aspects can be easily assimilated into the cognitive structure.

Repetition can increase memorising memory caused by the enhancement of activities relation between information [13]. Information saved in mind in the form of information. Thus it is getting frequent in using one-way information. Information in that way is strengthened and can simply be accessed.

3.1.3. The Steps in Carrying Out the Plan Problem Solving P1 (M1 and M2)

The details of observation result transcripts and *thinking-aloud* ST when executing the problem-solving plan of P1 are as follows.

Implementation of M1 Plan

P : "Now, please finish the problem", it should be said with loud voice

ST : There is a graph on coordinate axis

: **First**, ask about line equations on the graph

: Second, deduce the graph meaning. Interpret it!

: Third, deduce the equation $g: y_0x + \frac{1}{2}mx^2$

: To finish the first part, use straight-line equation through point (a, b) with a gradient *m* namely*: y-b = m(x-a)* as Figure 1:

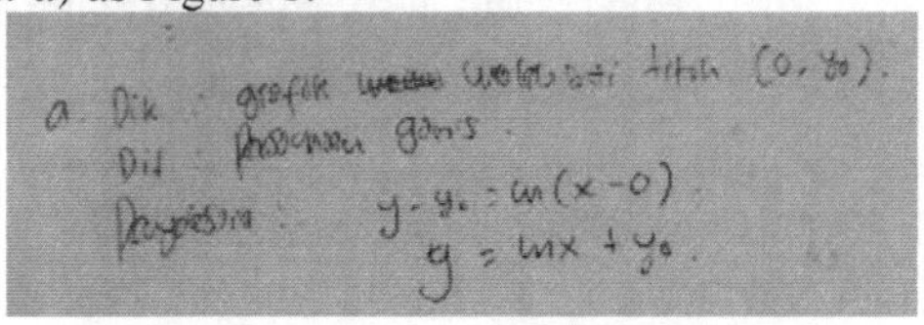

Figure 1. Graph on coordinate axis

: **part two,** the value of x≥0

x value is directly proportional to y, the greater of value y will make value x greater also with x ≥0, Figure 2.

Figure 2. The value of x≥0

: **part three**, the value is substituted as Figure 3

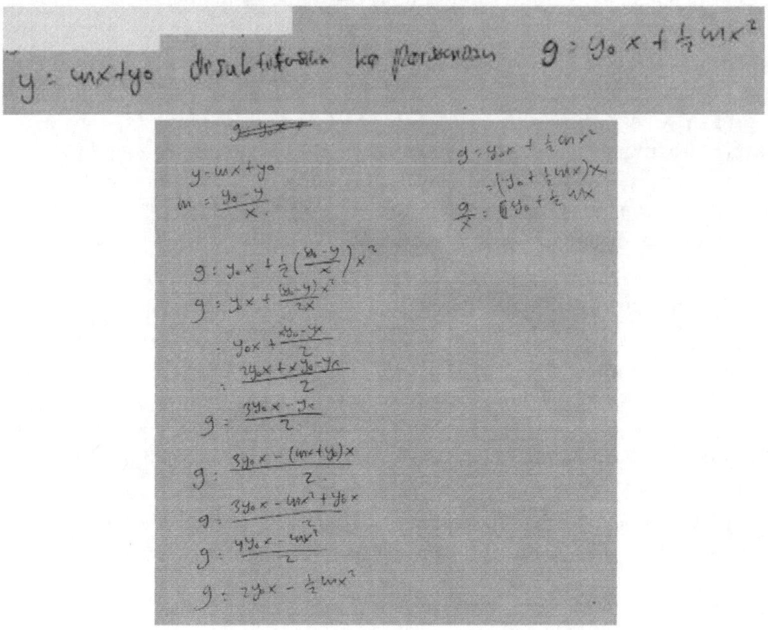

Figure 3. The value is substituted

The results of observations and *thinking-aloud* showed that ST was able to interpret the graph on M1. Therefore, ST could complete the first and second parts of M1. Nevertheless, the information contained in M1 of part three could not be assimilated by ST because it did not comply with their scheme. The aforementioned is in line with Piaget statements [14] assimilation is the incorporation of new events into intelligence as a scheme or concept. During assimilation, the stimulus is interpreted according to the scheme owned by the subject. When the given stimulus is matched to the existing scheme, then the subject can respond directly to it. The assimilation allows the present situation to be responded by the subject by its previous knowledge [12].

Implementation of M2 Plan

Subject ST can interpret the graph to make easier in finishing it. It can resolve th problem M2 quicker because ST has had a similar experience in finishing th problem. The subject's work in finishing the M2, Figure 4:

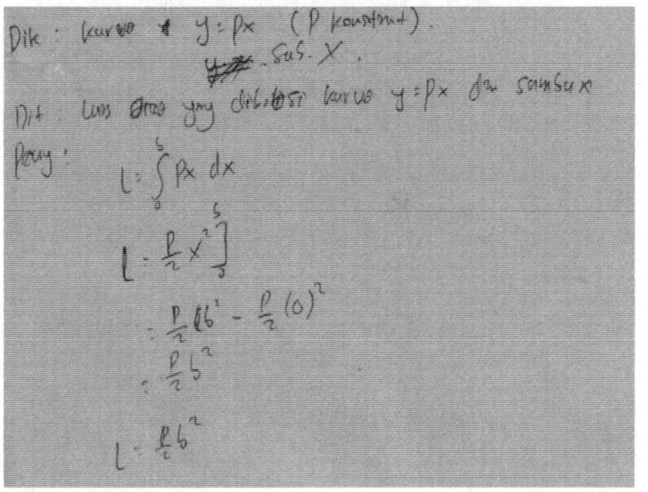

Figure 4. The subject's work in finishing the M2

The results of *thinking-aloud* showed that ST's behaviour in completing the M2 could finish it quicker because the received information corresponded to its previous experience. ST could assimilate all of the available information on the M2 problem because the received information was by the scheme.

3.1.4. Steps of looking back the completed work, P1 (M1 and M2)

The observation and *thinking-aloud* ST showed that ST examined the solved problem of M1 and M2 by re-read the work based on an algorithm and logic. ST examined only the working process that has been done, and not paying attention to the truth on the process of how it is done.

3.2. Results of Data Analysis and Discussion on Behaviour of SS and SR Problem Solving

The result of observation and *thinking-aloud* showed that SS and SR had almost a same way with ST in understanding the M1 namely reading piecemeal and over and over to the text. It is to interpret the meaning of the graph. However the both subjects could not connect the information received from the problems that were stored in *working memory* with information on *long-term memory*. As a result, both subjects were unable to resolve the problem of M1.

To understand M2, although with the readings as in understanding the M1, and experienced a similar problem before, both subjects still can find out and write down all of the information contained on the problems. Based on the understanding of both subjects in M2, the problem-solving plan on M2 used definite integrals. The transcription on the results of written test and *thinking-aloud* of SS and SR while implementing problem-solving plans of M2 are described in detail as follows.

Implementation Plan of M2

The Subjects SS and SR can interpret the graph, by working it with easy. It can quickly resolve the problem of M2 because both subjects have had the experience of working the similar. The work results in both subjects in resolving the problem of M2

Subject SS

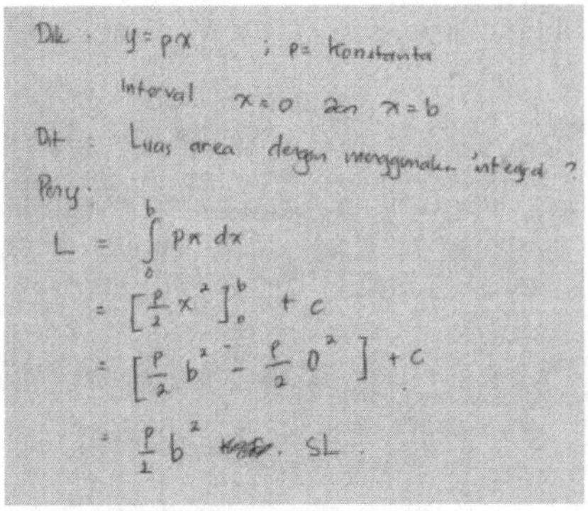

Subject SR

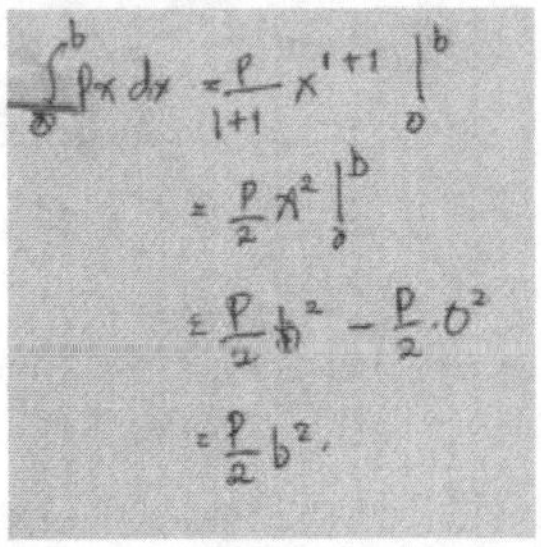

The results of the written test and *thinking-aloud* showed that the behaviour of SS and SR in resolving the M2 problem could be resolved smoothly because the received information was by the previous experience. ST can assimilate all of the available information on the M2 problem.

4. Conclusion

Based on the description above, the writer concluded that the behavioural of problem-solving from mathematics education students were as follows. (1) In understanding the problems, they tend to put their first attention to an image, reading the texts piecemeal and repeatedly, then in a whole and more focused on the sentence with equations, numbers or symbols. Nevertheless, the students could not interpret problems containing the statement that did not lead to a resolution. (2) In making problem-solving plans, they tend to use a similar formula to the previous problems they met. (3) Resolving the problems as planned. (4) Examining the truth of problems solving by re-reading and paying attention to the algorithms and logic from the previously completed work.

References

[1] Gok T 2010 *Eurasian Journal of Physics and Chemistry Education***2**pp 110-122 (
[2] Anderson W and Krathawohl R. 2001 *A Taxonomy For Learning, Teaching, and Assessing* (New York: David McKey Company, Inc)ISBN-13: 978-0801319037 ISBN-10:080131903X
[3] BouJaoude S and Barakat H 2000 *School Science Review***81** 91
[4] Weisberg R W 2006 *Creativity, Understanding Innovation In Problem Solving, Science, Invention, and the Arts*(Canada: John Wiley & Sons, Inc)
[5] Solso R L, Maclin O H and Maclin M K 2008 *Psikologi Kognitif, Edisi kedelapan* (Jakarta: Penerbit Erlangga) ISBN 9789790337404
[6] Metallidou P 2009 *Teaching and Teacher Education***25** pp 76-82
[7] Polya G 1973*How to Solve It: A New Aspect of Mathematical Method, 2nd ed.*(Princeton, *New Jersey:* Princeton University Press) ISBN: 9781400828678

IOP Conf. Series: Journal of Physics: Conf. Series **824** (2017) 012041 doi:10.1088/1742-6596/824/1/012041

[8] Solaz-Portolés J J and Lopez V S 2007 *Asia-Pacific Forum on Science Learning and Teaching*8Article 4
[9] Larkin J H 1985 Understanding, Problem Representations, and Skill in Physics. *Thinking and Learning Skills***2** pp 141-159
[10] Miles M B and Huberman A M 2009*Analisis Data Kualitatif*(Jakarta: UI Press)ISBN:979-456-103-7
[11] NovakD 2011 *Aprenzagen Significativa em Revista/Meaningful Learning Review***6** pp 1-4
[12] Hergenhahn B R and Olson M H 2009 *Theories of Learning, Edisi ketujuh*(Jakarta: Kencana Prenada Media) ISBN-10 : 0136057721, ISBN-13: 9780136057727
[13] Yovan P P 2008 *Memori dan Pembelajaran Efektif* (Bandung: CV. Yrama Widya) ISBN 979-543-723-2
[14] Brooks J G and Brooks M G 1993*In Search of Understanding: The Case for Constructivist Classroom*(Alexandria, VA: The Association for Supervision and Curriculum Development)

IOP Conf. Series: Journal of Physics: Conf. Series **824** (2017) 012042 doi:10.1088/1742-6596/824/1/012042

Teacher's Stimulus Helps Students Achieve Mathematics Reasoning and Problem Solving Competences

Isti Hidayah[1,2], Emi Pujiastuti[1], and Jeanet Eva Chrisna [1]

[1] Mathematics and Natural Sciences Faculty, Universitas Negeri Semarang, Indonesia

Corresponding author: isti.hidayah@mail.unnes.ac.id

Abstract. The students' problem-solving ability in mathematics learning still becomes a challenge for teachers, especially in primary education. The scientific approach, with its activities including observing, asking, collecting information/experimenting/trying, associating/analysing information/reasoning, communicating/presenting/ networking is expected to be able to help students to achieve their competence of reasoning and problem-solving. The Missouri Mathematics Project learning by using student worksheet and manipulative (classical and group) have helped students achieved problem-solving competence. The implementation of scientific approach in the activities of observing, experimenting, and communicating are good. However, the questioning and associating activities are still less promoted. The result of observation towards four meetings of learning by using teaching aids shows that the expected activity which did not emerge during the learning is "students ask questions from the factual thing to hypothetical thing, starting with guidance from teacher until they can do by themselves". The result of analysis towards theoretical background and research result conclude that the students' asking and thinking abilities can be developed gradually by delivering stimuli in the form of tasks which have been designed by the teacher. The task could be a problem or a clue; then the students determine things such as: "what the question?", "facts from pictures/text/graphs/tables", "find the hidden question", what's extra?", "what's missing?", "what's wrong?", alternatively, "make up the problem.

1. Introduction

Problem-solving ability still becomes a problem in mathematics learning both in primary and secondary education. Mathematics problem-solving ability is related to the students' thinking ability in the process of mathematics learning. Various studies about problem-solving ability and reasoning have been conducted, but they have not affected significantly by the practical level of mathematics learning at school. The scientific approach with its activities including observing, asking, collecting information/experimenting/trying, associating/analyzing information/reasoning, communicating/ presenting/networking is expected to be able to develop students' knowledge, thinking ability, and skill to use the knowledge through direct interaction with learning sources which have been designed by teacher [5]. Krulik and Rudnick [3] also suggest that teacher will become a choreographer who designs activities in which students could get necessary experiences to develop their mathematics power. Problem solving and reasoning are the main skills which students have to possess when they leave the school world into the real world. The students' reasoning ability will help them to solve problems and to create decisions. The current curriculum demands scientific approach be learning to achieve problem-solving competence. Meanwhile, in the meantime, the problem-solving ability or it is known in primary education as the ability to solve word problem still becomes a challenge for the teacher. The learning guide as narrated in the appendix of Minister of Education and Culture Regulation number 103/2014 states that students are subject who possess the ability to search actively, process, construct, and use knowledge. Therefore, learning has to give a chance for them to construct knowledge within their cognitive system. Missouri Mathematics Project (MMP) is a learning model with the syntax: review, development, cooperative work, seat work, and homework. The step of review is an apperception stage. It is continued by step of development which allows the teacher to guide students such that they can construct knowledge. The use of worksheet and manipulative will help students to find concepts. In the cooperative work step, teacher guides students to discuss in a group. Then, the students will work independently in seat work step, and it ends with homework step. According to Dwiningrat [1], one of the advantages of MMP is the availability of opportunity to do exercise so that the students can be skilled in solving various types of problem. Does the MMP learning model assist with students' worksheet and manipulative help students to achieve the problem-solving competence? How does the scientific approach work?

2. Material and Methods

This research was conducted at VIIIB grade students of SMP Negeri 1 Ungaran. It was an empirical testing of manipulative in a scientific approach based learning. This testing is part of a multiyear research and development. The learning implemented in this research was scientific approach learning aided with student worksheet, task, classical manipulative for teacher's demonstration, and manipulative for individual or group of students to work in group activities. The research design is a one-shot case study. According to Sugiyono [8], this experiment design uses treatment as the independent variable and the result of the treatment as the dependent variable (problem-solving ability). In this testing, the effectiveness of manipulative use is not only measured by the results, but also the implementation of scientific approach activities shown by the description emergence of the students' scientific activities in response to the stimulus of teacher that has been designed previously. The effectiveness of learning was analysed with the completeness, and the effectiveness of the process is measured by adherence to students' scientific activities.

3. Result and Discussion

The problem-solving ability test resulted that there are three students among 35 students who did not pass the test. The result of proportion hypothesis testing analysis suggested that the null hypothesis that the percentage of students passed the test individually is less than 75% was rejected, which conclude that the alternative hypothesis that the percentage of students pass the test individually is greater than or equal to 75% was accepted with a significance level of 5%. This suggests that Missouri Mathematics Project learning with a scientific approach, aided with student worksheet, and manipulative can help students achieve the classical completeness of problem-solving competence.

The result of similarity test towards the students' responses (scientific activities indicator) from the observer 1 and two at each meeting did not differ significantly with $\alpha = 5\%$. Likewise, the Anova test gave Sig = 0.052> 0.05; then we accepted the null hypothesis which means there are no significant observed differences among the four meetings. Based on multiple comparison tables, we can see that all Sig> 0.05 then the four meetings did not have significant observed differences each other. Description of the observation of the student responses (scientific activities) at the four meetings is presented in Table 1.

In Table 1 above it can be seen that the activities which get not enough percentage (maximum score of 4) are asking activity and reasoning/associating activity, with the lowest percentage, is the asking activity. It commonly happens that Indonesian students are reluctant to ask questions during learning. The teachers are also not accustomed to delivering stimuli that require or provide opportunities for students to respond, to questions, or to express opinions. The stimulus that requires students asking questions is when the teacher asks students to create a question in written, as students do not leave questions orally. While in the activity of "reasoning/associating", teachers have been aided by the presence of media, namely the manipulative (classical and group/individual) and the student worksheet. It is suitable with the result of research conducted by Pujiastuti and Madhuri [6] which states that the use of manipulative could assist teachers in teaching mathematics with a scientific approach. The scientific activities (observing, questioning, experimenting, associating, and communicating) are set of activities which might appear during learning and enable students to find or to construct mathematics concepts and principles. Meanwhile, the problem-solving ability is developed as the ability to apply the concepts/principles earned. Reasoning ability is used to find the concept/principle as well as in problem-solving. Similar research conducted by Siregar & Margit [7] about the influence of discovery learning approach with analogy aspect towards the student's reasoning ability. The presence of student worksheet helps students to solve the problem in a good structure and supports the development of students reasoning ability. The problem which requires reasoning is a non-routine problem (problem-solving).

Table 1. Percentage of Observed Score Earned against Maximum Score of the Implementation of Scientific Approach Learning aided with Manipulative

Activities	Percentage of Maximum Score for Each Meeting			
	1	2	3	4
Observing	93.8	100	87.5	93.8
Questioning	56.3	62.5	43.8	43.8
Experimenting	75.0	87.5	90.6	96.9
Associating	54.2	75	69.4	70.8
Communicating	65.6	96.9	93.8	84.4

Results of studies have shown that learning mathematics in materials of cuboid and cube, assisted with manipulative (classical and group) and students' worksheet is effective towards problem-solving ability, shown by the achievement of classical completeness. If the teacher can increase the percentage of activity particularly the "asking" and "reasoning", it can be expected to provide more optimal results. Reasoning ability of students contributes to the smooth running of students in problem-solving. It is supported by research of Tambychik and Subahan [9] they suggest that cognitive aspects of learning include the ability to do the perceptual thinking, the ability to use logic thinking, ability to memorise, and ability to recall. The main cognitive ability in learning which causes the difficulty for students is the ability to memorise and ability to recall the facts which support the making of connection in students' mind. Reasoning includes basic thinking, critical thinking, and creative thinking. Problem-solving is a means of an individual to use prior acquired knowledge, skills and understanding to meet the demand of unfamiliar situations. Problem-solving can and should be taught, problem-solving is a process that has been analysed and can be represented as a series of steps, from now on called heuristic [3]. Problem-solving pattern that is often used at school or in the study is the pattern of Polya, that in teaching students in problem-solving, the students are guided to carry out the steps of (1) understanding the problem; (2) planning; performing the plan; (4) confirmation of the answer [9]. In this study, we also integrated Polya patterns at the stage of training with a scientific approach aided by manipulative to improve students' problem-solving ability. Overview of the results of applying Polya strategy is illustrated by the conditions on the field today; while heuristic of Krulik & Rudnick [3] with five stages can be described as follows.

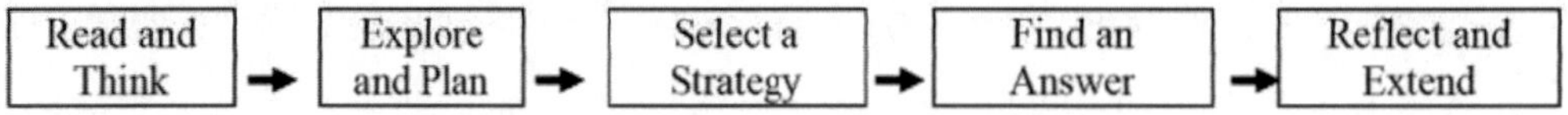

Figure 1. Steps of problem-solving [3]

Among the above five steps, we can see that it is no different with the stages on Polya pattern, but Krulik and Rudnick stated that improving the competence of reasoning and problem solving will also emphasise the change in the ways teachers teach. It is no longer relevant that a teacher demonstrates in front of the classroom. Instead, the teacher would be a choreographer who designed an activity in which students gain experiences which are necessary for the development of mathematical power. Novotna et. all [4] in his paper illustrates how to develop a creative approach to students problem-solving. It is part of the long experiment that is focused on promoting a culture of problem-solving by students. The problem solving used heuristic strategies: analogy - guess - check - revise - systematic.

Higher order thinking skill is important in society. Therefore, students should be equipped with knowledge and skills so that they can solve problems in everyday life. Teachers must be able to stimulate students to think; teachers should make sure to design and implement successful learning, teachers should be capable of setting up a map-bank of i-think map. Similar learning is done continuously so that the cultural thinking becomes a habit for students at all time [2]. In his research, he presents a visual tool of graphical representation as strategic thinking that will help students to see the appropriate thinking skills used to solve the problem. Besides, it also uses a group activity that requires students to give a reason of the thinking process that will never end. Missouri Mathematics Project learning with student worksheets and manipulative (classical and group) have successfully helped students achieve problem-solving competence, and the implementation of the scientific approach has been well performed on the activity of observing, experimenting, and communicating, but not yet on the activity of questioning and associating (reasoning).

Krulik and Rudnick present activities that are designed to assist teachers in creating classroom experiences that will help students achieve competence in reasoning and problem-solving. Here are some of the stimuli activities presented in mathematics learning refers to Krulik and Rudnick [3]. Stimulus in the form of the problem at each stage to train (habituation) students to think. The following questions are taken from the book of Krulik & Rudnick [3].

Step of Read and Think (facts from pictures/text/graphs/tables)

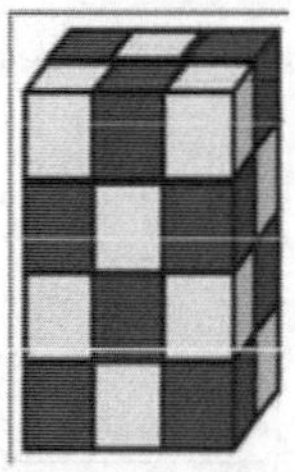

a) What is the shape of this model? (cuboid)
b) If the base of the cuboid is the lower face, then:
   - What is the length? (3 unit)
   - What is the width? (2 unit)
   - What is the height? (4 unit)
c) If the base of the cuboid is the front face, then:
   - What is the length? (3 unit)
   - What is the width? (4 unit)
   - What is the height? (2 unit)
d). What is the volume? (24 unit)
e) How is the appropriate way to calculate the volume of this cuboid? (4 unit x 3 unit x 2 unit)

[10]

The step of Explore and Plan (What is the hidden question?)

Some problems have more than one part. You have to find the answer to a "hidden question" before you can finish the problem. The problem of the following, pick out the hidden question. Last week Amanda sold 18 boxes of Girl Scout cookies. This week she sold 15 boxes. If each box cost $2, how much did she receive for all the boxes of cookies she sold? *What is the hidden question*? (a) How many boxes did of-of Girl Scout cookies Amanda sell last week? (b) How many boxes did of-of Girl Scout cookies Amanda sell during both weeks? (c) How many money did Amanda receive for the cookies she sells?

4. Conclusion

The result of analysis and discussion lead to the conclusion that the Missouri Mathematics Project aided with students worksheet and manipulative could help the students achieve classical completeness in the learning result. The implementation of scientific approach in the activities of observing, experimenting, and communicating is good. However, it did not perform well in the activities of asking and associating, which are low. If the scientific activities could be optimised, then the problem solving and reasoning could also be optimised. Furthermore, it is recommended that teacher optimise the asking and associating activities by delivering stimuli to make a habituation of students to ask and to reason in the form of tasks which have been designed by the teacher. The task could be a problem or a clue; then the students determine things such as: "what the question?", "facts from pictures/text/graphs/tables", "find the hidden question", what's extra?", "what's missing?", "what's wrong?", alternatively, "make up the problem, ".

References

[1] Dwiningrat, I. G. A. A., N. W. Suniasih, & I.B. S. Manuaba. 2014. Pengaruh.Model Missouri Mathematics Project terhadap Kemampuan Pemecahan Masalah Matematika Siswa. E-Journal MIMBAR PGSD Universitas Pendidikan Ganesha Jurusan PGSD. 2(1).

[2] Hasan, S.R, Rosli, R, Zakaria, E. 2016. The Use of i-Think Map dan Questioning to Promote HOTS in Mathematics. Creative Education. (1069-1078). http://creativecommons.org / licenses/by/4.0/

[3] Krulik, S and Rudnick, J.A. 1995. The New Sourcebook for Teaching Reasoning and Problem-Solving in Elementary School. Boston: Allyn and Bacon

[4] Novotna, J, Ersenmann, P, Pribyl, J, Ondrosora, J, Brehovsky, J. 2014. Problem-Solving in School Mathematics Based on Heuristic strategies. ERIES Journal. Vol 7 no 1.

[5] Permendikbud. 2014. Lampiran Permendikbud Nomor 103 tahun 2014 tentang proses pembelajaran.

[6] Pujiastuti, E dan Mashuri. 2015. Tracing of SMP Mathematics Teacher's Skills in Utilisation of Teaching Aids through Scientific Approuch. International Journal of Contemporary Applied Sciences. Vol 2 No. 11. Nov 2015. ISSN: 2308-1365. (www.IJCAS.net),

[7] Siregar, N.C & Marsigit. 2015. Pengaruh Pendekatan Discovery yang Menekankan Aspek Analogi terhadap Prestasi Belajar, Kemampuan Penalaran, Kecerdasan Emosional Spiritual. Jurnal Ristek Pendidikan Matematika, Vol 2-No. 2, Nov 2015 (224-234).

[8] Sugiyono. 2010. Metode Penelitian Pendidikan, Pendekatan Kuantitatif, Kualitatif, dan R & D. Bandung: Alfabeta

[9] Tambychik, T & Subahan, T. 2010. Students' Difficulties in Mathematics Problem- Solving: What do they Say? Procedia Social and Behavioral Sciences. International Conference on Mathematics Education Research (ICMER). (142 – 151)

[10] Hidayah, I, Sugiarto, and Dwijanto. 2016. The Role of Manipulatives Usage in Mathematics Learning of Primary Education with Scientific Approach towards the Students' Thinking Ability. International Journal of Applied Business and Economic Research. Vol. 14, No. 10(2016): 7215-7228

Visualisation Ability of Senior High School Students with Using GeoGebra and Transparent Mica

M Thohirudin[1], TK Maryati[1], and G Dwirahayu[1]

[1]Mathematics Education Department, Faculty of Tarbiyah and Teachers Training, UIN Syarif Hidayatullah Jakarta, Indonesia

Corresponding author: miftahthohirudin@gmail.com

Abstract. Visualisation ability is an ability to process, inform, and transform object which suitable for geometry topic in math. This research aims to describe the influence of using software GeoGebra and transparent mica for student's visualisation ability. GeoGebra is shortness of geometry and algebra. GeoGebra is an open source program that is created for math. Transparent mica is a tool that is created by the author to transform a geometry object. This research is a quantitative experiment model. The subject of this research were students in grade XII of science program in Annajah Senior High School Rumpin with two classes which one as an experiment class (science one) and another one as a control class (science two). Experiment class use GeoGebra and transparent mica in the study, and control class use powerpoint in the study. Data of student's visualisation ability is collected from posttest with visual questions which are gifted at the end of the research to both classes with topic "transformation geometry". This research resulted that studying with GeoGebra and transparent mica had a better influence than studying with powerpoint to student's visualisation ability. The time of study in class and the habit of the students to use software and tool affected the result of research. Although, GeoGebra and transparent mica can give help to students in transformation geometry topic.

1. Introduction

In mathematics, there are so many topics such as algebra, logic, calculus, and geometry. This geometry topic is as same as with some ideas of National Council of Teachers of Mathematics (NCTM). The standard of geometry topic is to analyze the characteristic and feature D-2 and D-3 and to develop mathematics argument about geometry connection, establish location, and describe the spatial connection with using geometry coordinates and other representation system, doing transformation and using symmetrical to analyze the mathematics situation, and using visualization, spatial thinking, and geometry model to solve problem. Based on that standards, visualisation and spatial thinking are a primary skill in geometry topic [1].

In case, teachers and parents still think that visual thinking or visual ability is just limited in arts. In geometry topic at school, start with mathematics conceptual comprehension until problem-solving was almost used counting concept. However, if we saw NCTM's standards, there was another alternative way to solve geometry problems that are exercise student's visualisation ability to understand lessons in a different way.

Howard Gardner divided intelligence into some kinds [2], one of them is visual/spatial intelligence. This visual intelligence can be trained and developed in geometry topic to solve geometry problems by using visualisation ability. Mental Imagery and Human-Computer Interaction Laboratory or called

"Imagery Lab" is a lab which is built by Kozhevnikov [3] and her friends did the research and developed many things about education and education media. One of their research is about visualisation ability. They divide visualisation ability into two sections, spatial visualisation and object visualisation. Spatial visualisation is an ability to process visual information about spatial relations between objects or their parts and perform mental spatial transformations and manipulations. Object visualisation is an ability to process visual information about appearances of objects and scenes regarding their pictorial properties (e.g.,.shape, colour and textures).

To develop the using of this visualisation ability, one of the indicator is the using assist tools in study geometry topic [1]. In this era, the technology especially computer is very needed to help people in many activities[14]. One of the software which can be used to help teachers and students in geometry topic is GeoGebra. GeoGebra is an open source and dynamic mathematics software in one easy-to-use package which can be used for learning and teaching at all levels of education [6][7]. It joins interactive 2D and 3D geometry, algebra, tablets, graphing, calculus, and statistics [4][5]. GeoGebra facilitates the creation of mathematical constructions and models by students. It allows interactive exploration by dragging objects and changing parameters that can be useful to develop visualisation ability in geometry topic. Later, NCTM said about direct using of geometry tools, so in this research, we add transparent mica that can be used by students to understand the process while transformed geometry objects. This transparent mica has excess in price, cheap, light, and easy to use [9][10] to give an image process of geometry transformation by the students directly.

Furthermore, the research problems of the study were as followed, how influence GeoGebra and transparent mica was, how students activities were, how test result of experimental class and control class compared. Then, considered on the research resulted that there is a positive influence of spatial ability towards student's achievement in math [11]. So, the researcher tries to more exercise this visual/spatial ability with using various tools. So we research the influence of using GeoGebra and Transparent Mica to visualisation ability of Senior High School students.

2. Material and Methods

This research is a quantitative experiment model. The subject of this research were students in grade XII of science program in Annajah Senior High School Rumpin-Bogor with two classes which one as an experiment class (science one) and another one as a control class (science two). The students of experiment class were 20 students, and the students of control class were 30 students. Experiment class used GeoGebra and transparent mica in the study, and control class used powerpoint in the study.

The researcher arranged study plan, chose a school, asked permission letter, looked for a teacher to give some advice and guidance, and prepared to learn instruments. The learning instruments have validated by author's lecturers. Data of student's visualisation ability is collected from posttest with visual questions which are gifted at the end of the research to both classes with topic "transformation geometry". All question of visualisation ability posttest have validated by five math teachers with content validity ratio (CVR), with CVR formula [12]:

$$CVR = \frac{n_e - \frac{N}{2}}{\frac{N}{2}}$$

With on$_e$ = expert who said "essential", and N = how many experts

The experimental stage did 8 times for learning and 1 times for doing a visual test. The students in experiment class and control class divided into some groups to make all of the students can participate in active learning and the students can help other students to understand the lesson. Both classes were divided into some group with 3-4 students/group. In experiment class, each group used 1 notebook to study with GeoGebra and transparent mica.

3. Result and Discussion

Based on research by Hohenwarter resulted that GeoGebra can be used in classrooms to explore basic calculus concepts, research by Widyaningrum and Murwanintyas resulted that GeoGebra made a better influence on student's motivation and score [8] and research by Tambunan resulted that spatial ability made a better influence to student's score [11]. So in this research, researcher combined GeoGebra and transparent mica to find out this tools influence to student's visual/spatial ability.

The experimental stage began with the using of GeoGebra and transparent mica in experiment control while study and using powerpoint in control class. The experimental stage did 8 times for learning and 1 time for doing a visual test. The topic for this research was "Transformation Geometry" in twelve classes. While studying, the students in experiment class and control class divided into some groups to make all of the students can participate in active learning and the students can help other students to understand the lesson.

Both classes were divided into some group with 3-4 students/groups. In experiment class, each group used 1 notebook to studied with GeoGebra and transparent mica. At the first time, the researcher gave them module how to use GeoGebra step by step to solve transformation geometry problems like translation, rotation, reflection, and dilatation. They confused to used GeoGebra because they never used the software in math, but step by step they tried to learn it and got better to used it. They used GeoGebra to made dot, polygon, and line. Then they tried to transform it with translation, reflection, rotation, and dilatation.

Beside of using GeoGebra, they used transparent mica to transformed object geometry directly. They could transform object geometry step by step so they could understand what the main purpose of transformation geometry. They could see how dot, polygon, or line is moving from the first position to transformed position. This activity with used GeoGebra and transparent mica could train their visualisation ability. They could try to solve transformation geometry problems without used counting formula.

After the students had been given treatment, at the end the students were given a test with twelve questions about visualisation ability, and they must answer it with pictures like dots, shapes, or lines. Some answers were chosen for the most answers from each class which appropriated with kinds of transformation geometry.

1-Translation

Question: "ΔPQR is a translation from ΔABC by $T = \begin{pmatrix}-5\\4\end{pmatrix}$, draw ΔPQR!"

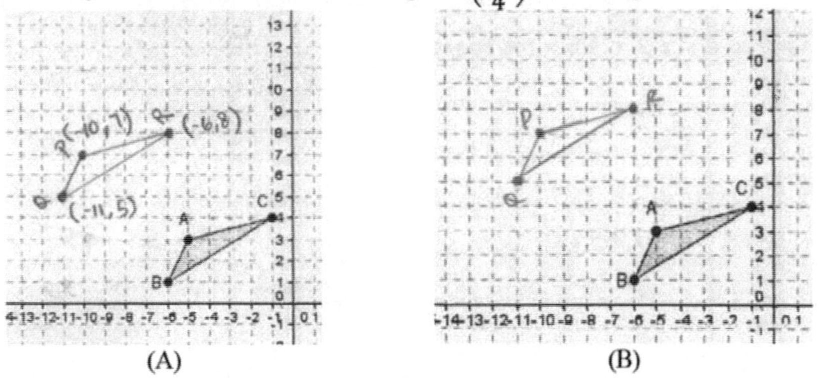

(A) (B)

Figure 1. The Most Answers in Experiment Class (A), Control Class (B)

2-Reflection

Question: "Draw a reflection of line-a toward line y=x as a line-b!"

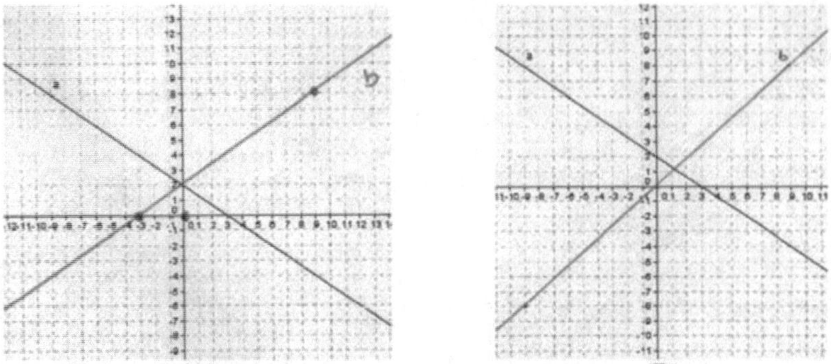

(A) (B)

Figure 2. The Most Answers in Experiment Class (A), Control Class (B)

3-Rotation

Question: "If line-a was rotated by 180° towards central point (4,2) results line-b, draw the line-b!"

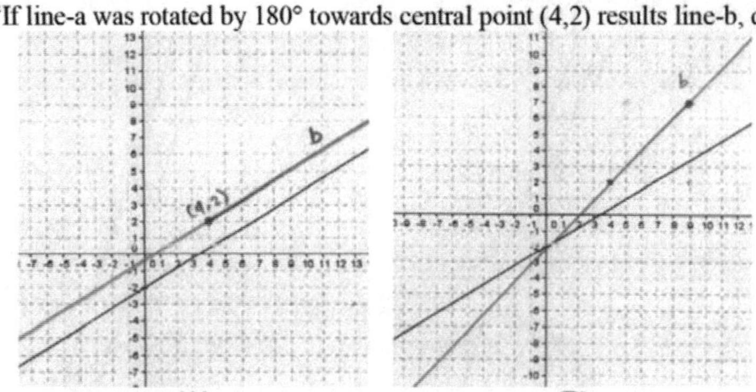

(A) (B)

Figure 3. The Most Answers in Experiment Class (A), Control Class (B)

4-Dilatation

Question: "If dilatation polygon ABCD toward a central point (0, -2) with scale 2 results polygon WXYZ, draw the polygon WXYZ!"

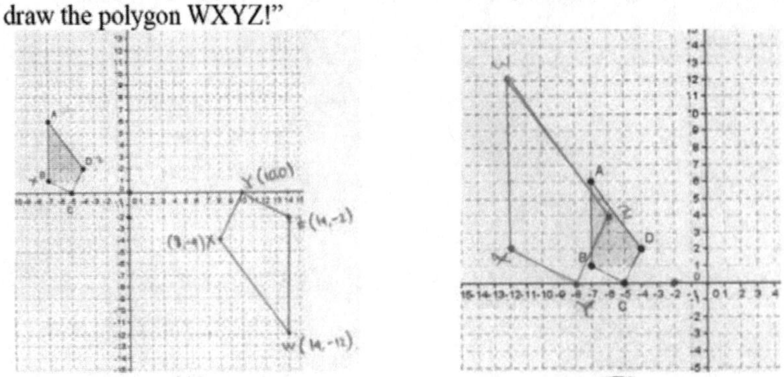

(A) (B)

Figure 4. The Most Answers in Experiment Class (A), Control Class (B)

Based on all answers of the students, Figure 5 was the data recapitulation of visualisation ability scores by experimental class and control class.

IOP Conf. Series: Journal of Physics: Conf. Series **824** (2017) 012043 doi:10.1088/1742-6596/824/1/012043

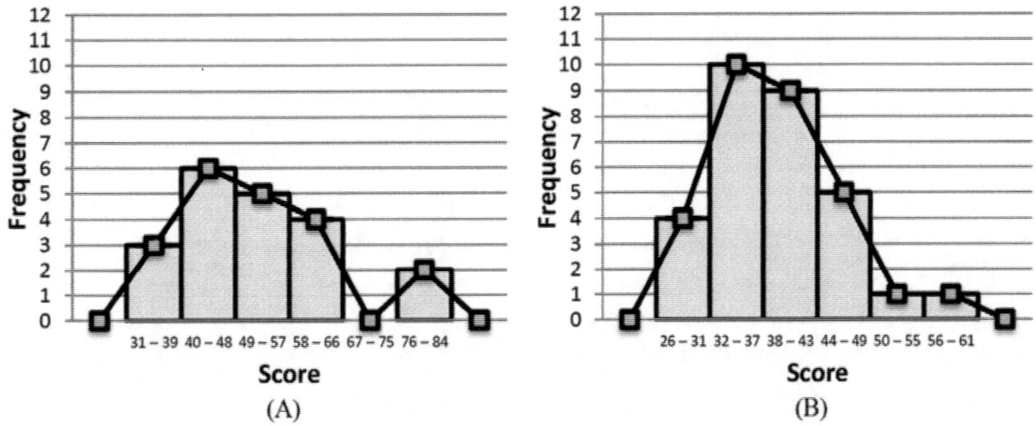

Figure 5. Visualisation Ability Score of Experimental Class (A), Control Class (B)

Figure 5A showed that coefficient slope 0.41 (slightly right slope), and sharpness 2.95 (platykurtic curve). Because the coefficient slope was positive, it means the student's score were still many below the average. Figure 5B showed that coefficient slope 0.31 (slightly right slope), and sharpness 3.29 (leptokurtic curve). Because the coefficient slope was positive, it means the student's score were still many below the average. Then the difference between experiment class and control class can be seen in Table 1.

Table 1. The Difference Between Experiment Class and Control Class

	Class	
	Experiment	Control
Amount of students	20	30
Maximum (X_{max})	82,61	58,70
Minimum (X_{min})	34,78	26,09
Average ($\bar{x}$)	52,10	38,90
Median (Me)	50,30	38,17
Modus (Mo)	46,25	36,64
Varians	169,67	51,97
Standard deviation	13,03	7,21
Slope	0,41	0,31
Sharpness	2,95	3,17

Table 1 showed that the highest score was in experiment class and the lowest score was in control class. The experiment class had range 47.83 from the maximum score to minimum score. The control class had range 32.61 from the maximum score to minimum score. As we can see, because the range score of experiment class was longer than control class, it made variance and standard deviation of experiment class higher than control class. The spreading scores of both classes can be seen in Figure 6.

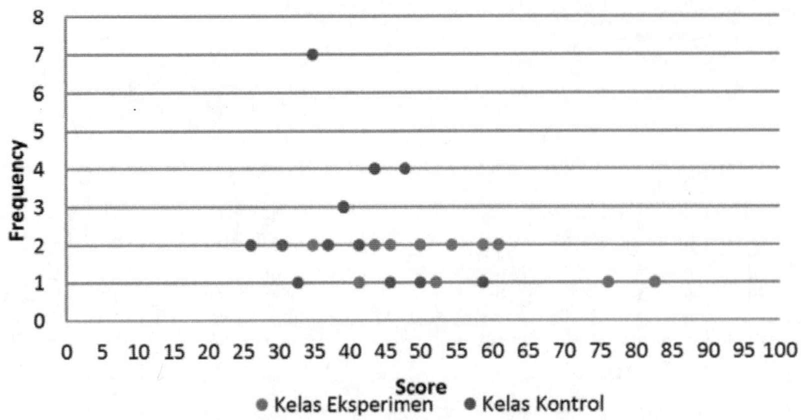

Figure 6. The Spreading Scores of Both Class

Figure 6 showed that many students of experiment class had scores between 30 – 60, whereas many students of control class had scores between 30 – 45. Figure 1, 2, 3 and Table 1 showed that visualisation ability score of experiment class which studied with GeoGebra and transparent mica were better than control class which studied with powerpoint. However, with just this data, we cannot conclude that visualisation ability of students who used GeoGebra and transparent mica were better than students who used Powerpoint. So researcher did a statistical test like normality test, homogeneity test, and hypothesis test [13]. The normality test result was presented in Table 2.

Table 2. Normality Test Result

Class	n	χ^2_{count}	χ^2_{table}	Conclusion
Experiment	20	6,225	7,815	Normal Distribution
Control	30	2,645	7,815	

Table 2 showed that normality test with chi-square α= 0.05 and df= 3 resulted both classes had normal distribution because of both $\chi^2_{counts} < \chi^2_{table}$. Because both classes had a normal distribution, then researcher did next test. The next test is homogeneity test, and the result was presented in Table 3.

Table 3. Homogeneity Test Result

Class	n	Varians	F_{count}	F_{table} (α=0,05)	Conclusion
Experiment	20	169.674	3.26	1,96	Not homogeny
Control	30	51.972			

Table 3 showed that homogeneity test with Fisher α= 0.05, df numerator = 19 and df denominator = 29 resulted that this class was not homogeny because of $F_{count} > F_{table}$. Because it was not homogeny, then researcher did next test. The next test is hypothesis test with normal and not homogeny type, and the result was presented in Table 4.

Table 4. Hypothesis Test Result

count	table (α=0,05)	**Conclusion**
4,13	2,09	Average of Experiment class in visualisation ability was better than average for control class

Table 4 showed that hypothesis test with normal and not homogeny type α= 0.05 and df = 19 resulted that average of experiment class which studied with GeoGebra and transparent mica was better than average for control class which studied with powerpoint. It means GeoGebra and transparent mica can make positive influence toward student's visualisation ability.

4. Conclusion

Considered on the purpose of the study that has been explained, the using of tools such as a computer, visual model must be maximised to help various students, especially in geometry. The teachers are supposed to create learning activity and use tools that is adjusted with student's ability or learning style. So, the schools and teachers can try to use this GeoGebra and transparent mica as alternative tools to create learning activity that more variation. According to these whole research, One of the tools that can be used in geometry to help or develop visualisation ability was GeoGebra or transparent mica.

References

[1] NCTM 2000 *Principles and Standards for School Mathematics*. (Reston, VA: National Council of Teachers of Mathematics)
[2] Armstrong T 2009 *Multiple Intelligences in the Classroom 3rd Edition* (Alexandria, VA USA: ASCD) ISBN 978-1-4166-0789-2
[3] Kozhevnikov M *Development of Spatial Ability Tests* (Imagery Lab, Massachusetts) in
[4] Hohenwarter M, Hohenwarter J, Kreis Y and Lavicza Z 2008 *ICME 11* (Monterey, Mexico) p 1.
[5] Hohenwarter M and Fuchs K 2004 *Combination of Dynamic Geometry, Algebra, and Calculus in the Software System GeoGebra*
[6] Hohenwarter M and Lavizca Z 2010 *GeoGebra its comunity and future*
[7] Lavizca Z, Hohenwarter M and Lu A *Establishing a Professional Development Network with an Open-Source Dynamic Mathematics Software* (NCETM)
[8] Widyaningrum Y T, Murwanintyas C E 2006 *Proc. Nat. Conf. on Mathematics and Mathematics Education* (Yogyakarta: UNY) 103
[9] European Specialty Minerals Association *Mica* (Belgium, IMA-Europe)
[10] IMERYS 2010 *Mica Uses and Applications* (USA, IMERYS)
[11] Tambunan S M 2006 *MAKARA, SOSIAL HUMANIORA* **10** 27-32
[12] Lawshe C H 1975 *PERSONNEL PSYCHOLOGY, INC* **28** 563 DOI: 10.1111/j.1744-6570.1975.tb01393.x
[13] Kadir 2010 *Statistika Untuk Penelitian Ilmu-Ilmu Sosial* (Jakarta: PT. Rosemata Sampurna) ISBN 978-979-15571-1-5
[14] Ibad S (Ed) and IKAPI 2012 *ICT dan Pembelajaran* (Jakarta: Referensi) ISBN 978-979-915135-3

Implementation of Statistics Textbook Support with ICT and Portfolio Assessment Approach to Improve Students Teacher Mathematical Connection Skills

P Hendikawati and N R Dewi

Mathematics Department, Faculty of Mathematics and Natural Science
Universitas Negeri Semarang

Corresponding author: p.aji.unnes@gmail.com; putriaji.mat@mail.unnes.ac.id

Abstract. Statistics needed for use in the data analysis process and had a comprehensive implementation in daily life so that students must master the well statistical material. The use of Statistics textbook support with ICT and portfolio assessment approach was expected to help the students to improve mathematical connection skills. The subject of this research was 30 student teachers who take Statistics courses. The results of this research are the use of Statistics textbook support with ICT and portfolio assessment approach can improve students mathematical connection skills.

1. Introduction

Data analysis is an important part of the research process that is often regarded as a difficult thing. Data analysis is very important in research process because after processing and analysing data, the researchers can draw conclusions and make generalisations framework uses concepts or theory. Statistics subjects discussed various statistical methods that to obtain a conclusion that will be useful for decision making. Statistics help the process of data analysis in real life and has a broad implementation. By understanding the concept of Statistics, students are expected to implement the knowledge into various real-life cases related to the application of statistics and can solve complex problems related to the analysis of statistical data.

During this time many students are still experiencing problems and made mistakes in analysing cases related to statistical data. Difficulties and mistakes that occur include inaccuracy in calculating the data analysis and the precision and accuracy of the resulting solutions [1]. Information technology is growing today, especially the development of the computer, software, and various applications that can be used to help speed up and simplify the process of understanding and analysis of statistical data, which is expected to improve the competence of students in Statistics lectures. The process of data analysis for decision-making has a fairly broad implementation in real life, so this research is important to carry out.

Nowadays, many criticisms of the low quality of education in the various educational unit. One of the highlights is an instrument or tool of evaluation used considered not valid and reliable, as well as the aspects that are less complete. It will eventually come to a conclusion, that the assessment approach now needs to be updated with other approaches. Portfolio-based assessment is an assessment model which is expected to reveal and assess students with more accurate and more complete based on the evidence possessed by students.

In this study, the Statistics textbook design that already developed will provide wider opportunities for students to go directly into the field to learn the source of the data and issues surrounding, then analysed to obtain solutions using various materials have been studied in the lecture. In this case, students are trained to be able to connect the knowledge obtained with real cases encountered in life and then provide a solution that is accurate and communicate appropriately and effectively. Mathematical connection capability is the ability of high-level mathematical thinking that either directly or indirectly have contributed greatly to the success of a student's study. The design of this course collaborated with a team of lecturers/computing laboratories so that the data can be simulated and analysed more systematically. The work of the students subsequently compiled into a set of tasks that form a portfolio.

Mathematical connection ability is an ability relating mathematical concepts either among mathematical concepts themselves (in mathematics) or relating mathematical concepts with other scopes beyond mathematics [2]. Because the material of mathematics is an interconnected unity, the mathematical connection ability is required in the learning process. Connection questions are important to be studied by the students because the concept in mathematics related to each other not only in mathematics theory but also among mathematics branch [3]. According to [4] the indicators of mathematical connection ability are (1) searching for the relationship of any concept and procedure representation; (2) understanding the relationship among mathematics topics; (3) applying mathematics in other fields or in daily life; (4) understanding the equivalent representation of a concept; (5) searching for the relationship between one procedure to another procedure in an equivalent representation; (6) applying the relationship among mathematics topics and between mathematics topics and topics beyond mathematics. There are three kinds of connections i.e. the connection between new information and current knowledge, the connection between mathematical concepts, and connection with everyday experience [5].

Nowadays, ICTs that used digital technologies such as computers and the internet are potential and powerful tools for education change and reform [6]. The use of ICTs in education help raise educational quality, helping teaching in learning process into an engaging and active that connected to real life.

The abstract object of mathematics can study easier with ICT so that the occurrence of students' misinterpretations can be avoided. [7] State that abstract mathematics is easier to understand and more interesting to learn by using ICT. Lack of lecturers' creativity that makes monotonous learning usually makes students bored and not interested in the mathematical subject. ICTs potential to innovate, accelerate, enrich, and deepen skills, to motivate and engage the student, to help related school experience to work practices, and strengthening teaching [8].

Learning mathematics in college is not just memorise or only apply a simple mathematical formula, but requires a High-Level Thinking Skills Mathematically that would be beneficial to the students themselves [9] [10]. This research will develop mathematical connection abilities that are part of the high-level mathematical thinking skills that important regarding data analysis in the course of Statistics. Statistics Textbook development of ICT-supported approaches designed a portfolio that is expected to improve the ability of students' mathematical connections in addition to improving learning outcomes.

2. Method

This research uses a quasi-experimental design with pre-test post-test control group design. This study used time series design which does not use a control group and only one group using all subjects in the group for treatment. Research conducted at the Department of Mathematics Faculty of Mathematics and Natural Sciences, Universitas Negeri Semarang. The population were all students of Mathematics Education in the second semester of 2015/2016 at the Department of Mathematics, Universitas Negeri Semarang in Statistics course. The samples in this research were selected using simple random sampling technique, and select one group of students learning group were given treatment learning using Statistics textbook supported with ICT and portfolio assessment approach. The variables observed in this study is students' mathematical connection ability and Statistics learning outcomes. These variables were measured before and after implementation of using Statistics textbook supported with ICT and portfolio assessment approach in the learning process.

2.1. The Steps of Research

The first steps in this research were determined to sample and trial class from a population of the study. Develop research instruments include syllabi, lesson plans, pre-test and post-test. After that, perform the test questions in trial class that have received Statistics material and then analyse the results of trials test to determine validity, reliability, the level of difficulty, and index of differentiated power. Next, set items that will be used in the pre-test and post-test in the experimental class. Items that do not meet the criteria are not used in the test. Before treatment, a pre-test was given to measure students' mathematical connection ability. Then, conduct learning in experiments class using Statistics textbook supported with ICT and portfolio assessment approach and give post-test to measure students' mathematical connection ability after treatment. After all, data was collected then analyse the data from the pre-test and post-test to verify the hypothesis.

2.2. Data Collection Technique and Research Instrument
The data collected in this research consisted of quantitative data. The students GPA in odd semester 2015/2016 was the data that collected with documentation method. Tests method conducted to test student mathematical connection ability. Tests carried out before the students acquire treatment (pre-test) and after the students treated using Statistics textbook supported with ICT and portfolio assessment approach (post-test). Data were then used to test for normality, homogeneity, and the average similarity.

Instruments in this research were tested item to measure students' mathematical connections ability, and lesson plan. Mathematical connections ability test consist of pre-test and post-test. The pre-test was used to determine students' ability before the treatment is given. Post-test was used to noticing an increase in students' mathematical connection ability after getting treatment. The material contained in the test question is all of the material in the Statistics course for one semester. The lesson plan was made for one semester in 11 meetings. Lesson plan made accordance with the curriculum used in Statistics.

3. Result

3.1. Preliminary Data Analysis Result
The initial analysis conducted before students were given treatment. The sample consisted of 30 students. The analysis carried out in the early stages of grade point average (GPA) data in the odd semester of the 2015/2016 academic year. Initial data taken from the odd semester GPA is used to determine the category of students' initial ability. The division of students initial categories and distribution of the sample based on students initial ability are summarised in Table 1.

Table 1. Criteria of Students Initial Ability

Interval	Category	Quantity
$\mathbf{GPA \geq 3.00}$	High	27
$\mathbf{2.20 \leq GPA < 3.00}$	Medium	3
$\mathbf{GPA < 2.20}$	Low	0

Modified from [11]

From 30 students subject of the research, based on early data analysis from GPA score, obtained students grouping according to student initially ability. There were 27 (90%) of students who fit high capability, 3 (10%) of students in moderate ability category, and there was no student in low ability category.

3.2. Data Analysis Mathematical Connections Test
After doing the research and the data obtained from the results of tests that measure the ability of mathematical connections, then tested the hypothesis that has been proposed. Analysis of test data is performed to determine whether the results of the pre-test and post-test mathematical connection satisfy the hypothesis formulated or not. Before the test, the hypothesis of differences average of pre-test and post-test first tested the prerequisites normality test.

3.2.1. Test Difference Average Pre-Test and Post-Test Students Mathematical Connection Ability

Normality Test of pre-test and post-test mathematical connection ability scores aims was to determine whether the results of the pre-test and post-test mathematical connection ability have a normal distribution. Table 2 shows the results of Kolmogorov-Smirnov test with sig 0.004 worth less than 0.05 so that H_0 is rejected. While the value of sig to post-test mathematical connection ability was 0.096 greater than 0.05, then H_0 is accepted. It shows the value of data pre-test of students mathematical connections were not normally distributed, while the post-test scores mathematical connection ability were normally distributed.

Table 2. Normality Test of Pre-test and Post-test Student Mathematical Connections Ability

Statistic	Mathematical Connections	
	Pre-test	Post-test
N	30	30
Kolmogorov- Smirnov Z	1.770	1.232
Sig.	0.004	0.096
H_0	Reject	Accept

The hypothesis of students' mathematical connection difference testing students' mathematical connection ability before and after treatment using Statistics textbook supported with ICT and portfolio assessment. This test was performed to determine the average difference in the improvement of student mathematical connection abilities before given treatment using Statistics textbook supported with ICT and portfolio assessment (μ_1) and after treatment (μ_2). Based on normality test, the hypothesis of difference used nonparametric Mann-Whitney.

Table 3 shows that Mann-Whitney U (Z) value -6.720 with sig 0.000, so accept H_1. It means that there are differences in student mathematical connection ability average using Statistics textbook supported with ICT and portfolio assessment approach. It shows that the implementation of Statistics textbook supported with ICT and portfolio assessment approach significantly proven make a difference and increase the student's mathematical connections ability. Improved connection ability, the increase in the average value of the pre-test and post-test which initially 0.67 in the pre-test score increased to 8.93 on a post-test. The use of ICT in learning has also been widely perceived benefits on several previous studies which have been carried out [12], [13], [14], [15] and [16].

Moreover, regarding the category of post-test connection capabilities mathematically students, there were 21 students (70%) belongs to a high category, 6 students (20%) included in medium category and 3 students (10%) belongs to a low category.

Table 3. Mann-Whitney Test of Mathematical Connections

Statistic	Mathematical Connections	
	Pre-test	**Post-test**
N	30	30
Mean	0.67	8.93
Mann-Whitney U	-6.720	
Sig.	0.000	
H_0	Reject	

3.2.2. Test of Mathematical Connection Ability Achievement Based on Initial Ability

Normality test of post-test mathematical connection ability based on students' initial ability aims to determine whether the results of the post-test scores have a normal distribution.

Table 4. Normality Test of Post-test Student Mathematical Connections by Initial Abilities

Statistic	Mathematical Connections	
	High	Medium
N	27	3
Kolmogorov- Smirnov Z	1.368	0.583
Sig.	0.047	0.886
H_0	Reject	Accept

Based on the result in Table 4, Kolmogorov-Smirnov value for mathematical connection ability with higher ability category was 1.368 with sig value 0.047 so that H_0 rejected. Moreover, Kolmogorov-Smirnov value for mathematical connection ability with medium ability category was 0.583 with sig value 0,886 so that H_0 accepted. It shows that post-test of students' mathematical connection abilities with high ability category were not normally distributed, while post-test for medium ability category students were normally distributed.

Hypothesis testing of students mathematical connection ability differences based on initial ability GPA scores performed to determine the difference in student mathematical connection ability achievement of high ability category (μ_1) and medium ability category (μ_2). The achievement of students' mathematical connection ability was indicated by the value of post-test mathematical connections ability. A different test average of students' mathematical connection ability calculated using Mann-Whitney test because post-test scores were not normally distributed. Based on the result of the difference in mathematical connection ability achievement in Tabel 5, it obtained Mann-Whitney U (Z) -1.816 and sig value 0.086, so can conclude to accept H_0. It means that there was no difference in the average of students mathematical connection ability achievement using Statistics textbook supported with ICT and portfolio assessment approach in students with high and medium ability categories.

It shows that the implementation of Statistics textbook supported with ICT and portfolio assessment approach significantly proven increase the ability of students' mathematical connection ability. These results correspond to the results of the research [17] that computer-based can increase students scores in low and average achievers. Achievement of the mathematical connection ability on high and medium ability category have post-test mean score consecutive 9.22 and 6.33. It shows that with Statistics textbook supported with ICT and portfolio assessment approach, students' mathematical connection ability achievement give high ability.

Table 5. Mann-Whitney Test of Achievement Mathematical Connection

Statistic	Initial Ability	
	High	Medium
Mean	9.22	6.33
Mann-Whitney U	-1.816	
Sig.	0.086	
H_0	Accept	

4. Result
Based on the results of this research can be concluded that the Statistics textbook supported with ICT and portfolio assessment approach can improve students mathematical connections ability in Statistics lectures of Mathematics Education students that seen with the increased scores of the mathematical connection abilities.

References
[1] Hendikawati P and Arini F Y 2016 The Development of Statistics Textbook Supported with ICT and Portfolio-Based Assessment *Journal of Physics: Conference Series* 693 (1) 012020 (IOP Publisher)
[2] Ruspiani 2000 *Kemampuan Mahasiswa dalam Melakukan Koneksi Matematika* Thesis (Bandung: PPS UPI)
[3] Ruseffendi E T 2006 *Pengantar Kepada Membantu Dosen Mengembangkan Kompetensinya dalam Pengajaran Matematika Untuk Meningkatkan CBSA* (Bandung: Tarsito)
[4] Sumarmo U and Nishitani I 2010 High Level Mathematical Thinking: Experiment with High School and Under Graduate Student using Various Approaches and Strategies, Gunma University, Maebashi. Vol 58, 14 pages, (http://www.des.emory.edu/mfp/Bandura 1989.pdf)
[5] Mousley J 2004 An Aspect of Mathematical Understanding: The Notion of "Connected Knowing" Proc. 28*th* *Conference of the International Group for the Psychology of Mathematics Education* ***3*** p 377-384
[6] Tinio, V.L 2002 *ICT in Education: UN Development Programme* (Retrieved from http:www.epimers.org on September 2015)
[7] Wintarti I A Yamasari, Y Johan A 2009 Pengembangan Media Pembelajaran Matematika Berbasis ICT Untuk Menumbuhkan Minat dan Motivasi Siswa dalam Memahami Konsep Matematika *Jurnal Pendidikan Matematika dan Sains* **16(2)**
[8] Amin S N U 2013 An Effective Use of ICT for Education and Learning by Drawing on Worldwide Knowledge, Research, and Experience: ICT as a Change Agent for Education *Scholarly Journal of Education* **2(4)** pp. 38-45 Online at http://www.scholarly-journal.com/SJE
[9] Dwijanto 2007 Pengaruh Pembelajaran Berbasis Masalah Berbantuan Komputer Terhadap Pencapaian Kemampuan Pemecahan Masalah dan Berpikir Kreatif Matematik Mahasiswa Ph.D. thesis, (Bandung: UPI)
[10] Sumarmo U 2005 Pengembangan Berfikir Matematik Tingkat Tinggi Mahasiswa SLTP dan SMU serta Mahamahasiswa Strata Satu (S1) Melalui Berbagai Pendekatan Pembelajaran Report Penelitian Hibah Penelitian Tim Pascasarjana-HTPT Tahun Ketiga (Bandung: UPI)
[11] Tandililing E 2011 The Enhancement of Mathematical Communication and Self Regulated Learning of Senior High School Students Through PQ4R Strategy Accompanied by Refutation Text Reading *PROCEEDINGS International Seminar and the Fourth National Conference on Mathematics Education* ISSN 978-979-16353-7-0
[12] Oviawe J I and Oshio L E 2011 Impact of Information and Communication Technology on Teaching and learning Ability of Education Students in Universities in Edo State Nigeria *International Review of Social Science and Humanities* ***2(1)*** pp.126-133 www.irssh.com ISSN 2248-9010 (Online) ISSN 2250-0715 (Print)
[13] Eisenberg M et all 2010 Information, Communications, and Technology (ICT) Skills Curriculum Based on the Big6 Skills Approach to Information Problem-Solving *The Big6 Library Media Connection* May/June pp. 24-27

[14] Shade O K et all 2012 ICT: An Effective Tool in Human Development International *Journal of Humanities and Social Science* ***2(7)*** pp.157-162 April www.ijhssnet.com

[15] Fuglestad A B 2004 ICT Tools and Students' Competence Development *Proceedings of the 28th Conference of the International Group for the Psychology of Mathematics Education* ***2*** pp.429-446

[16] Marshall S and Taylor W 2016 Editorial: Adoption, Integration and Use of ICT in Teaching *International Journal of Education and Development using Information and Communication Technology (IJEDICT)* ***12(1)*** pp.2-3

[17] Afzal M T et all 2014 The Effect of Computer Based Instructional Technique for the Learning of Elementary Level mathematics Among High, Average and Low Achievers *International Journal of Education and Development using Information and Communication Technology (IJEDICT)* ***10(4)*** pp.47-59

The Refinement of The Preliminary Genetic Decomposition of Group

K Wijayanti[1]

Department of Mathematics Universitas Negeri Semarang
Gd. D-7 Kampus Unnes Sekaran Gunungpati, Semarang, Central Java - Indonesia

[1]Corresponding author: kristina.mat@mail.unnes.ac.id

Abstract. Mathematics proof is one of the characteristics of advanced mathematics thinking, and proof plays an important role in learning the abstract algebra included group theory. The depth and complexity of individual's understanding of a concept depend on his/her ability to establish connections among the mental structure that constitute it. One of the cognitive styles is field dependent/independent. Field independent (FI) and field dependent (FD) learners have different characteristics. Our research question is (1)How is the proposed refinement of preliminary genetic decomposition of group that is designed with a preliminary study of the learning with APOS works; (2) What understanding about group that is generated by student (Field Independent, Field Neutral, and Field Dependent) when learning through designed material. This study was a descriptive qualitative. The participants of this study were nine (9) undergraduate students who were taking Introduction of Algebraic Structure 1, which included group, in the even semester of academic year 2015/2016 at Universitas Negeri Semarang. Each of type of cognitive styles consisted of 3 participants. There were two instruments used to gather data: written examination in the course and a set of the interview. The FD and FN participants generated Action for a binary operation. The FI participant generated Action and Process for a binary operation. The FD, FN and FI participants generated Action, Process, Object, and Scheme for the set. The FD and FN participant did not generate mental structure for axiom. The FI participant generated Scheme for axiom. The FD, FN and FI participants tend to have no Coherence of Scheme of the group. Not all mental structure on the refinement of the preliminary genetic decomposition can be constructed by participants so well that there are still obstacles in the process of proving.

1. Introduction

Some of the studies about the learning of group theory (algebraic structure) state that confusing a theorem and its converse [1], difficulty in managing the distinction set and its element [2]. Mathematics proof is one of the characteristics of advanced mathematics thinking, and proof plays an important role in learning the algebraic structure. According to [3], Proof construction is a mathematical task in which the prover is provided with some initial information (e.g. assumptions, axioms, definitions) and is asked to apply rules of inferences (e.g. recall previously established facts, apply theorems) until the desired conclusion is deduced.

Tabitha and Richard defined proof as "a collection of true statements linked together in a logical manner that serves as a convincing argument for the truth of a mathematical statement" [4]. This definition point to the idea, that a proof is a logical, deductive argument. Hersh in [4] addressed the different roles of proofs as both convincing and explaining. "Mathematical proof can convince, and it can explain. In mathematical research, its primary role is convincing. At the high-school and undergraduate level, its primary role is explaining". According to [5], that the majority of future teachers were not able to execute formal proofs requires only lower secondary mathematical content, in an adequate and mathematically correct way or to recognise and satisfactorily generalise a given mathematical proof.

Ed stated that there are five types of reflective abstractions or mental mechanism (interiorization, coordination, reversal, encapsulation, and generalisation), that lead to the construction of mental structure: Action, Process, Object, and Schema (APOS). The depth and complexity of individual's understanding of a concept depend on his/her ability to establish connections among the mental structure that constitute it.

According to Piaget and adopted by APOS Theory [6], a concept is first understood as an Action, that is an externally directed of a previously conceived object (objects). An Action is external regarding each step of the transformation need to be performed explicitly and guided by external instruction; each step of action cannot be imagined and skipped.

Some APOS-based studies report that encapsulation mechanism is the most difficult. In their studies about mental construction function of two variable that is only one student has constructed Object conception [7]. While [8] reported that to see something familiar in a totally new way is never easy to achieve. The difficulties arising when a Process is changed into an Object are like those experienced during the transition from one scientific paradigm to another.

Once a Process has been encapsulated into a mental Object, it can be de-encapsulated if needed, back to its underlying Process. In another word, by applying de-encapsulation mechanism, one can go back to the Process that is resulting in the Object. A coordination mechanism is very needed in the construction some Objects. Two Objects can be de-encapsulated, their Processes coordinated, and the coordinated Processes are encapsulated to form a new Object [6].

According to [9], Schema is characterised by the dynamism and the continuity reconstruction as determined by one's mathematical activities of subjects in a certain mathematical situation. The coherence of Schema is determined by one's ability to ascertain whether it can be applied in a particular mathematical situation. Once Schema is constructed as a coherent collection structure (Actions, Processes, Objects, and other Scheme) and connections are established among the structure, Schema can be transformed into the static structure (Object) and can be used as a dynamic structure that assimilates Object or other related Objects or Schemas.

The genetic decomposition is a hypothetic model that describe mental structures and mechanisms that students might need to learn the mathematical concepts. The genetic decomposition may include a description of how these structures are related and organised into a larger mental structure that is called Schema. In the description of Schema, it may be explained of how Schema is thematized into Object. The genetic decomposition also explains whatever that is known about student's expected performance that indicates the differences in the development of student's construction [6].

Based on the literature review about group theory, experienced as a student and as a teacher, and also students' difficulties in learning the concept of the group, it was designed a preliminary genetic decomposition of the group. According to [6] a genetic decomposition not only a sequence of steps or a list of conceptions students may have. Rather, the genetic decomposition is a description of mental construction may needed students to make in learning of a mathematical concept.

To find the genetic decomposition that reflects the cognition of a concept that close to many individual and can be used in the design of teaching and learning that is positively affecting student's learning, then it needs the implementation of preliminary genetic decomposition in learning.

The refinement of genetic decomposition of a group resulting from the implementation of the preliminary genetic decomposition is as follows.

Table 1. Refinement of Preliminary Genetic Decomposition.

Concept	Preliminary Genetic Decomposition	Refinement of Preliminary Genetic Decomposition
Binary Operation	1. The action is given binary operation rules on a non-empty set A, individuals applying the rules of the binary operation to	1. The action is given binary operation rules on a non-empty set A, individuals applying the rules of the binary operation to

Concept	Preliminary Genetic Decomposition	Refinement of Preliminary Genetic Decomposition
	two elements in A.	two elements in A. 2. The action using the rules above, for two pairs of specific elements of the set, defined in the binary operation.
	2. Interiorized Actions at step 1 consisting of a. Taking two elements. b. Acting on 2 objects of these elements according to the rules that have been defined. c. Generate new object (i.e. map of a pair of elements on 2.a)	3. Interiorized Actions at step 2 consisting of a. Taking two elements. b. Acting on 2 objects of these elements according to the rules that have been defined. c. Generate new object (i.e., map of a pair of elements on 3.a) 4. Interiorized Actions at step 2 consisting of a. Taking two pairs of elements which are the same. b. Acting on 2 objects of these pairs of elements according to the rules that have been defined. c. Generate new object (i.e., map of a pair of elements on 4.a)
	3.Encapsulates the process in step 2 so that a binary operation becomes Objects	5. Encapsulates the process in step 3 and 4 so that a binary operation becomes Objects with indicators can check binary operation satisfies an axiom. 6. Thematization Objects in step 5 into a scheme with indicators can define a binary operation and can check whether a rule imposed on a set is a binary operation
Set	1. Action takes elements in a set	1. The action takes any element of the set expressed by signing up members. 2. Action takes any element of the set expressed by the set membership conditions
	2. Interiorized Action, collect and put Objects in the collection based on a condition.	3. Interiorized Action, collect and put Objects in the collection is based on a condition with the indicator to check condition of membership or identify a set of elements taken from a set
	3. Encapsulates the process in step 2 so that the set becomes Object.	4. Encapsulates the process in step 3 so the set becomes Objects with indicators operate a set or states relations of two sets.
Axiom	1. Process: Checking axiom, involving the coordination of common sense checking its properties and processes defined by the specific nature of those	1. Process: Checking axiom, involving the coordination of common sense checking its properties and processes defined by the specific nature of those

Concept	Preliminary Genetic Decomposition	Refinement of Preliminary Genetic Decomposition
	checked 2. Schema of axiom contains a general sense and to check whether the pair of set and binary operation satisfies a nature.	checked 2. Schema of axiom contains a general sense and to check whether the pair of set and binary operation satisfies a nature

Scheme of the group was mentally constructed in coordination with the Scheme of Axiom, Scheme of sets and binary operations.
Scheme of Axiom includes two main components:
1. check the properties of binary operations defined on a set,
2. The three axioms of the concept of group constructed as an object.
Coherence Scheme of the group was supported by
1. The ability to construct examples and not an example,
2. The ability to recognise relationships that exist in the scheme when the group faced a problem if the characteristics of the problem are within the scope of the group scheme,
3. The ability to ensure that the relationships that exist in the scheme when the group faced a problem if the characteristics of the problem are within the scope of the group scheme.

Witkin& Goodenough defined Cognitive styles as preferred ways of selecting, perceiving, and processing new information [10]. There are various recognised cognitive styles available in the literature, among which are visual/haptic, visualise / verbalizer, levelling/sharpening, serialist /holist, and field dependent/independent.
Witkin, Oltman, Raskin, and Karp in [11] suggested that there is three field related cognitive styles: field independent (FI), field dependent (FD), and field neutral (FN).

The Group Embedded Figures Test (GEFT) developed by Witkin *et al.* [11] was designed to measure individuals' levels of field independence by tracing simple forms in the larger complex figures. Some studies have proved the reliability and validity of the test instrument over the years. GEFT had a reliability coefficient of 0.82 and was a standardised paper-pencil test, which measured visual perceptiveness [12].

Research Question

(1) What understanding about a group that is generated by student Field Independent, Field Neutral, and Field Dependent when learning through designed material?

(2) How is the proposed refinement of preliminary genetic decomposition of a group that is designed with a preliminary study of the learning with APOS works?

2. Method

This study was a descriptive qualitative. We wanted to know how and why did student response the question as they did. We discuss our findings based on the refinement of the preliminary genetic decomposition of the group. Data was collected by a written test, videotape, and interviews.

The participants of this study were nine (9) undergraduate students who were taking Introduction of Algebraic Structure 1, which included group, in the even semester of academic year 2015/2016 at Universitas Negeri Semarang. Firstly, we conducted GEFT (Group Embedded Figure Test) to all of the students who were taking Introduction to Algebraic Structure 1 (48 students) to determine their cognitive style. Based on this test, we choose 3 students from each type of cognitive style.

The class for the study was Introduction to Algebraic Structure 1 intended to be taken by students in semester 3. The class met for 3x50 minutes, once a week, for 16 weeks. There were two midterm tests each of 60 minutes and one final exam for 120 minutes. All the exams were a closed book. The standard textbook (Fraleigh, 1989) was used as a reference for examples, problems, and explanations. The course focused on group theory which includes the concepts of groups, subgroups, cyclic groups, closet and Lagrange's Theorem, homomorphism, and quotient group.

The material of this study focused on the concepts of the group. Learning implemented by presenting the material with frequently asked questions about the concept of the material presented and the underlying previous concepts. Examples and non-example were taken from things known to students. Problems presented in class discussions by the needs of the students. Homework was given to reinforce students' understanding of the material studied and discussed in class when students need it.

There were two instruments used to gather data: one written examination in the course and a set of the interview. The interview was conducted after the written closed book examination.

In the data collection, the participants working on a written test on the concepts of the group was then interviewed recorded by videotape. By comparing the recorded interviews and written work, each transcript annotated to clarify the spoken and written. Additional information can be inserted to the events and statements of interest due to his typical. A brief description of the interesting events was developed with guidance research questions.

3. Result and Discussion

We discuss our findings based on the written test, recorded the interview and the refinement of the preliminary genetic decomposition of the group (Table 1).

3.1. Binary Operation

In a binary operation, FI participants tended to perform Action 1 and Action 2, whereas FN and FD participants tended to perform Action 1 but were not able to do Action 2. The FN 2 participant was able to do the Action 1 for binary operation indicated by applying the rules of the binary operation to two elements. The FN 2 participant was not able to do the Action 2 for binary operation indicated by using the rules, for two pairs of specific elements of the set defined in the binary operation. All FD participants did not do anything in this section so that we concluded that the FD participants were not able to do Action 2 for a binary operation.

FI participants tended to interiorize Action in step 3a, 3b, and 3c. However, they were not able to interiorize Action in step 4a, 4b and 4c. Only one of the three of FI participants who was able to do that. He was FI 3 participant. FN participants were not able to perform the above steps yet except FN 3 although not yet at every step. Participants FD were not able to perform each step above yet. The FI 3 participant was able to take two elements and act on 2 Objects of these elements according to the rules that were defined, but he was not able to generate new Object. This fact showed that FI 3 participant was able to do the Process 3a and 3b but was not able to do the Process 3c. Process 4 was able to be made by FI 3 participant. The FN 3 participant was able to do Process 4 but was not able to do Process 3. The FN 3 participant was able to take two elements and act on 2 Objects of these elements according to the rules that were defined, but he was not able to generate new Object.

Participants FI, FN and FD, were not able to encapsulate the Process to Object except FI 3 and FN 3. It was seen from the indicators that individual was able to check binary operation satisfies axioms, in this case, they were associative, the existence of identity element, and the existence of inverse element.

The FI 1 participant was able to check associative properties but failed to check the existence of identity element and the existence of inverse element. The FN 2 participant was able to check associative properties but failed to check the existence of identity element and the existence of inverse element. All FD participants did not do anything in this section so that we concluded that the FD participants were not able to encapsulate Process to Object for a binary operation.

Based on the results of interviews showed that each FN participant transformation step needs to be done explicitly and guided by external instructions. According to [13] which is in line with the characteristics of the FN, participants tend to be similar to the characteristics of FD participants, namely

(1) Prefer externally defined goals and reinforcements, and clear definitions of desired outcomes
(2) Extrinsically motivated
(3) Less structured, less autonomous

3.2. Set

On the concept of set, participants FI, FN, and FD tended to do the Action, interiorize Action becomes Process, encapsulate the Process become Objects and thematize Scheme of the set, except FI 1 was able only to do Action, and FD 1 was able only to do Action, and interiorized Action becomes Process.

The participants were able to take any element of the set which indicates that participants were able to do Action. FD 1 participant was able to express condition of a set membership, but FI 1 participant was not able to do that. Thus, FI 1 participant was able to do Action, while FD 1 participant was able to perform Action and interiorize Action into a Process. The FI 1 and FD 1 participants were not able to express the relationship between 2 sets which showed that participants did not encapsulate the Process become Object.

3.3. Axiom

Participants FI tended to show that they understood the axioms in group scheme, but were not able to check the axioms except associative. Only FI 3 participant who showed that he understood the axiom of the group and was able to check in the case group. In the case of sub-group, FI participants showed that they understood and were able to check the axioms.

FN participants tended to understand and able to check the associative axiom, but not so for the other axioms of the group. The FD participants tended not to demonstrate that they understood the axiom and were not able to check the group axioms. All participants FI, FN, and FD had no Coherence of Scheme of the group, but FI 3 and FN 1 participants. Based on interviews, it is by the findings of [14] seven (7) major difficulties for students in doing proofs:

(1) The students did not know the definitions, that is, they were unable to state the definitions.
(2) The students had a little intuitive understanding of the concepts.
(3) The students' concept images were inadequate for doing the proofs.
(4) The students were unable, or unwilling, to generate and use their examples.
(5) The students did not know how to use definitions to obtain the overall structure of proofs.
(6) The students were unable to understand and use mathematical language and notation.
(7) The students did not know how to begin proofs.

While [15] found most of the mathematics students have difficulty in constructing, understanding and validating proofs.

4. Conclusion and Remark

4.1. Conclusion

Conclusions of this study were: (1)binary operation: (a) The FD and FN participants generated Action,(b) The FI participant generated Action and Process; (2) The FD, FN, and FI participant generated Action, Process, Object, and Scheme for set; (3) axiom: (a) The FD and FN participant did not generate mental structure, (b)The FI participant generated Scheme; (4)The FD, FN, and FI participant tend to have no Coherence of Scheme of group, and (5) Not all mental structure on the refinement of the preliminary genetic decomposition can be constructed by participants so well that there are still obstacles in the process of proving.

4.2. Remark

It needed a transitional class for group theory to provide mental structure they need to learn mathematical concepts. In this class, we proposed to stabilise concepts in logic, mapping and its property, partition and equivalence relation, and to prove.

References

[1] Orit H and Uri L 1996 *For the Learning of Mathematics* Students' Use and Misuse of Mathematical Theorem: The Case of Lagrange's Theorem **16(1)** 23-26

[2] Orit H and Uri L 1996 *For the Learning of Mathematics* Students' Use and Misuse of Mathematical Theorem: The Case of Lagrange's Theorem **16(1)** 23-26

[3] Keith W 2001 *Educ Stud Math* Student Difficulty In Constructing Proofs: The Need For Strategic Knowledge **48** 101–19

[4] TabithaT Y M and Richard M G 1999 *School Science and Mathematics* Preservice Teacher Beliefs About Proofs **99(8)** 438-44

[5] Bjorn S and Gabriele K 2009 *Proc. of the ICMI Study 19 Conference: Prove and Proving in Mathematics Education* Professional Competence of Future Mathematics Teachers on Argumentation and Proof and How to Evaluate It 'ed Fou-Lai Lin *et al.* (The Department of Mathematics, National Taiwan Normal University, Taipei, Taiwan) pp 190-95

[6] Ilana A *et al.* 2014 *Apos Theory A Framework for Research and Curriculum Development in*

Mathematics Education (New York: Springer)

[7] Maria T and Rafael M P 2010 *Educ Stud Math* Geometrical representations in the learning of two-variable functions **73** 3–19

[8] Anna S 1991 *Educ Stud Math* On The Dual Nature of Mathematical Conceptions: Reflection on Processes and Objects as Different Sides of the Same Coin **22** 1-36

[9] Ed D *et al.* 2005 *Educ Stud Math* Some Historical Issues And Paradoxes Regarding The Concept Of Infinity: An Apos-Based Analysis: Part 1 **58**: 335–59

[10] Erdat C and Salih A 2013 *Taiwan Int. J. Sci. Math. Edu.* The Effects of Cognitive Styles on Naive Impetus Theory Application Degrees of Pre-Service Science Teachers

[11] Eunjoo O and Doohun L 2005 *J. of Interactive Online Learning* Cross Relationships between Cognitive Styles and Learner Variables in Online Learning Environment **4** 53-66

[12] Sibel S *et al.* 2008 *The Turkish Online Journal of Educational Technology – TOJET* The Effects Of Individual Differences On Learner's Navigation In A Courseware **7** 32-40

[13] Witkin, Moore, Goodenough and Cox 1977 *Review of Educational Research* Field-Dependent and Field-Independent Cognitive Styles and Their Educational Implications **47** (1) pp 1-64

[14] Robert C M 1994 *Educ Stud Math* Making The Transition to Formal Proof **27** 249-66

[15] Susanna S E 2003 *The Mathematical Association Of America* The Role of Logic in Teaching Proof [Monthly 110 December 2003] pp 886-99

Geometry Problem Solving Ability and Tolerance Character of Students 8th Grade with Assessment Project

Masrukan[1] and N A Mufidah

Mathematics Department, Faculty of Mathematics and Natural Sciences
Universitas Negeri Semarang, Semarang, Central Java, Indonesia

[1]Corresponding author: masrukan.mat@mail.unnes.ac.id

Abstract. The 4K model is a learning model that is characterized by character, performance, creative and conservative. The use of a 4K model in this research is to improve the character of tolerance and Geometry problem-solving ability of students. Assessment of learning done thoroughly and during the learning process, using project assessment that includes assessment at the time of student do the project and assessment of students' project report. The result that the ability of geometry problem solving and tolerance character of students with project assessment achieve mastery and improved by project assessment.

1. Introduction

The school is a social system in which students learn to interact with each other, learn to understand social norms, learn to cooperate, learn to appreciate and learn various aspects of life in the community as appropriate. Within school students have very diverse backgrounds, whether religious, ethnic, socio-economic background, educational background of parents, areas, including customs and culture. Under conditions of this diversity are certainly going to find much of a difference in the attitudes and behavior of students. This difference must be understood and respected each other, thus allowing the growth of solidarity and unity among fellow students. Therefore, the value of tolerance become one of the pillars that need to be developed in the school.

The responsibility to develop the values of tolerance in students is a shared responsibility of all educators in the schools. However, such efforts more systematic and planned should become an integral part of the learning activities. Character education is a deliberate attempt to help people understand and care about the values of ethical/moral values [1]. So, teacher intentionally needs to design learning models to develop tolerance values of students. Character education can be integrated into the classroom [2]. Research conducted by a research team CDP also mentions a class that uses a character-based curriculum show a positive attitude in the student's character [2]. The integration of tolerance character through learning also does not give a negative impact on students' academic achievement. In this study observe at the tolerance character of 8thgrade students, represented by three samples of the experimental class 1, experimental class 2 and control class. Tolerance is an attitude and action that respects differences of religion, race, ethnicity, opinions, attitudes, and actions of others who are different from themselves. Indicators of achievement character of tolerance, namely: (1) work in groups with friends of the opposite sex, religion, race, and level of ability, (2) does not impose any opinion or the will to others, (3) mutual respect, and (4) speaks with polite and courteous. Fourth indicators are then developed into eight faceted tolerance observed, namely (1) the attitude of students to cooperate in a group with friends of the opposite sex, religion, race, and level of ability, (2) the attitude of the students want to help friends who have difficulty in learning, (3) the attitude of students willing to give a chance someone else expression, (4) the attitude of students to appreciate others who have different views, (5) the attitude of students to respect other people who are talking, (6) the attitude of students never mock / blame others, (7) the attitude of students speaking politely to others, (8) the polite demeanor of students to teachers to answer when the teacher asked [3]. In addition to observing the tolerance character of students, this research will raise the problem-solving ability as a measure of academic achievement of students. According to Permendiknas 22 of 2006 on the Content Standards (SI) Subjects, one of the junior high school mathematics courses goal is that the students were able to solve problems that include the ability to understand the problem, devised a mathematical model, solve the model, and interpret the obtained solution [4].

The geometry is a branch of mathematics that is taught with the aim that students can understand the properties and relationships between geometric elements and can be good problem solvers [5]. A branch of mathematics geometry is very precise in measuring the students' problem -solving abilities. This is by Bobango [6] which says that one of the objectives of learning geometry is for students to become good problem solvers. However, geometry is considered to be elusive and tend to be hated by most students. Mathematics materials that are considered difficult and dreaded students in math is a matter of geometry [7]. This resulted in reluctant students learn geometry, and ultimately learning objectives geometry to develop problem-solving skills can not be achieved. Problem solving is usually defined as formulating new answers, going beyond the simple application of previously learned rules to achieve a goal [8]. This means the problem is defined as formulating a new answer, beyond the simple application of a previous learning process to achieve the goal. The solution about problem solving includes a four-step phase of completion, which is to understand the problem, obtain eventually a plan of the solution, carry out the plan, and examine the solution obtained [9].

The learning model is a plan or pattern that can be used by teachers in implementing the learning in the classroom, ranging from preparing a learning device, media, and tools to evaluation tools that lead to the achievement of learning objectives [10]. The learning model used in this study is the 4K model. 4K model is a learning model that charged character, creative, performance, and conservative. This learning model integrates character education learning activities. This learning model nuanced conservative, this means that in learning activities using teaching aids from recycled materials. Teaching aids is a learning media is best used in the material geometry. The 4K model syntax is (1) character development illustration, (2) investigation, (3) collaborative exploration, (4) creative activity, (5) communication, and (6) appreciation. In addition to the use of a model of learning, assessment of learners also need to be considered as one important aspect of the learning process. As Zainul said in [11] that the assessment of learning outcomes is not only connected with the result but more importantly it is an important part of the learning process. The assessment used in this study is the project assessment that includes assessment at the time of student projects and assessment tasks student project report. Assessment of the project is an assessment of the tasks to be completed within a period/time, the task in the form of an investigation into something that includes planning, data collection, organizing, processing, and presentation. During the project, students prepare a plan, implement the plan, and discuss. During the project implementation process, students are expected to develop problem-solving skills. The most obvious benefit obtained by students from the project works their increasing interest towards science and technology class [12]. The most significant change seen by students regarding project preparation is their increasing grades in exams during and following the project works.

The use of the 4K model and project assessment in this research is expected to improve the character of tolerance and geometry students' problem-solving abilities. Aim of the study are (1) to test the completeness of the geometry problem solving ability of students and to determine the completeness of the tolerance character of students through 4K learning with project assessment, 4K model, and cooperative model; (2) to examine differences in geometry problem solving ability and tolerance character of students through 4K learning with project assessment, 4K model , and cooperative model; (3) to test the highest geometry problem solving ability and tolerance character of students through 4K learning with project assessment, 4K model, and cooperative model.

2. Methods

The method used in this study is quantitative. The population in this study is all students 8th grade in State Junior High School 22 Semarang and the sample is a student in class VIII E (experimental class 1), VIII B (experimental class 2), and VIII F (control class).The variables in this study are the independent variable and the dependent variable. The independent variables, namely: 4K learning model with project assessment, 4K learning model, and cooperative learning model. The dependent variable, namely: tolerance character and problem-solving ability of students in the geometry after obtaining learning by using 4K learning model with project assessment, 4K learning model, and cooperative learning model. The data collection techniques are used documentation, testing, and observation.The data used in this study is a data of problem-solving ability in geometry and tolerance character of students. The data of geometry problem-solving ability of students gathered with the test method. Meanwhile, students' tolerance character data collected by observation. The observations were made using the observation sheet to reveal the tolerance character of students. In this study the documents used in the form of video recordings of learning. The results of analysis of observations and tests conducted by researchers. Recorded video can view the event and listen to as a whole and can repeat events seen or heard repeatedly.

In this study, the test developed by the researchers own customized teaching materials studied. The tests will be used to obtain geometry problem-solving ability data of students has met the criteria of valid and reliable. Grains contained in the test also meets the criteria of validity, the level of difficulty, and a good of distinguishing.The research design used in this study is Posttest-Only Control Design. The first group was given a treatment called experimental class 1 is the class by learning the 4K model with project assessment. The second group is the experimental class 2 by learning 4K models. While the third group is called a control class is the class by cooperative learning model. Test steps are performed in this study are (1) test for normality, (2) test of homogeneity, (3) test of mastery learning, (4) the three means equality test (parametric and non-parametric), (5) advanced test (parametric and non-parametric).

3. Results and Discussion

Before the study conducted, researchers tested the initial data (students' final mathematics examination score in the 1st semester) of the experimental class 1, experimental class 2, and control class. Hypothesis testing results indicate that the initial data were normally distributed and homogeneous. One Way Analysis of Varian's test for equality of three means showed no difference in the initial mean ability of students. So that the experimental class 1, experimental class 2, and control class departs from the same knowledge. Once the researchers conducted learning as much as four meetings, then the students of three classes were given a test. The test results provide data on the geometry problem-solving ability scores of students experimental class 1 (E1), experiment class 2 (E2), and control class (K). The graph of the data is presented in the form of a diagram in Figure 1.

Figure 1 shows that the geometry problem-solving ability of the three classes is different. The highest score obtained by the experimental class 1 with a score of 58. The lowest score obtained by a control class with a score of 45. The data showed that all students in experimental class 1 could write down what is known and what is asked of all questions correctly and complete. Students can understand the problem (P1) solving by 81.25%, which means most of the students have been able to write a formula or strategy to be used in solving the problem. This is because students at the time of project tasks type 1, which is making props cubes and blocks using second-hand goods, helped students to find the concept of cubes and cuboid, ranging from the elements, surface area, and volume. This is in line with the opinion of Hung-His Wu that said to prepare students better to learn how to solve problems must begin by teaching purely mathematical problem solving, in this case about the concept of geometric surface area and volume of a cube and cuboid [13].

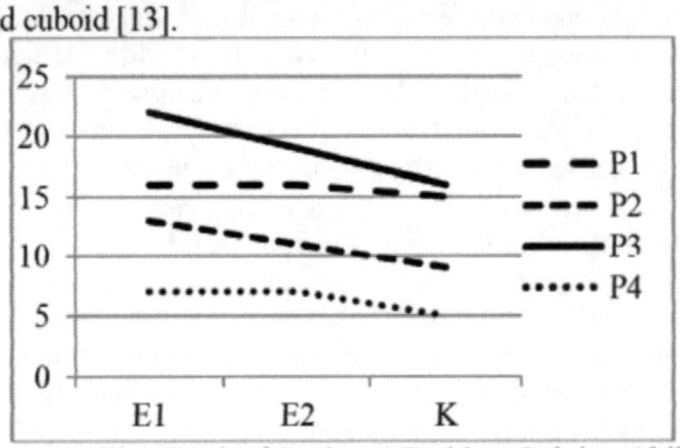

Figure 1. The Graph of Students' Problem Solving Ability

The ability to plan the completion (P2) of all three classes showed a degree that says that the experimental class 1 regarding planning will be more than an experimental class 2, and the experimental class 2 regarding planning will be more than the control class. From the data obtained, the experimental class 1 got a score of 13; the experimental class 2 got a score of 11, and the control class got a score of 9. Students in the control class lack to remember the formula that caused the students wrong to determine the formula that will be used to solve the problem. Step resolve the problems correlated with completion planned steps. The accuracy of the students in solving the problem (P2) greatly influenced by the ability to plan resolution (P3). If the planning is already better then the solution is also good, and vice versa. From the data obtained, the experimental class 1 got a score of 13 in P2 and got a score of 22 in P3. The experimental class 2 got a score of 11 in P2 and got a score of 19 in P3. Moreover, the control class got a score of 9 in P2, so that got a score of 16 in P3. The ability to re-examine the completion of the steps (P4) that have been carried out experimental class 1 and experimental 2 is good as it gets a score of 7 out of a total of 8 scores. It is better than the control class that received a score of 5. Thus, the ability to understand and re-examine of the experimental class 1 and the experimental class 2 higher than the control class. While the ability to plan and solve problems of the experimental class 1 higher than the experimental class 2 and class control.

Higher overall mathematics achievement was found among students who conceived that assessment makes students themselves accountable for learning and who conceived of assessment as good for them. In mathematics, students who conceived of assessment as a means of taking responsibility for learning, who did not ignore assessment, and who used it to improve their learning, tended to get higher achievement scores [14]. This is in line with the result of this experiment that the problem-solving abilities experimental class 1 higher than the experimental class 2 and experimental class 3 or the problem-solving ability of the experimental class 1 is the best one. This is reinforced by the test results of problem-solving ability data. The data testing aims to answer the hypothesis in this research. The hypothesis test result of students' problem-solving ability data of the experimental class 1, the experimental class 2, and the control classes contained in Table 1.

During the learning process, the character of tolerance was observed by the observation sheet that's been made. The observations provide tolerance character data of the students then analyzed to determine tolerance character of the experimental class 1, the experimental class 2, and the control class. Table 2 below show the tolerance character score improvement of the experimental class 1, the experimental class 2, and the control class.

The character tolerance score of the experimental class 1 and the experimental class 2 shows the high and very high criteria. Characters of tolerance in this study (in the experimental class 1 and the experimental class 2) were grown through habituation and reinforcement strategies. This is done by showing students a video about the character of tolerance displayed by teachers and scantlings of the teachers by providing advice to the students to always have the character of tolerance. In contrast, the tolerance character of control class indicates the criteria of the medium. This is because 4 of 8 tolerance character aspects of the control class indicate the medium criteria. Students in the control group are less regarding tolerance when working together in groups and discussion. They promiscuous friends during the discussion and did not cooperate with anyone. They are also lacking regarding helping a friend with learning difficulties. Also, they are lacking regarding accepting the opinions of others and tend to regard their opinions most correct. However, they demonstrate a positive attitude regarding respecting the teacher who was teaching the class. This happens because there is no habituation strategy and the strengthening of tolerance character in the control class exercised by the teacher. Tests were also conducted on the data of tolerance character to answer the hypothesis in this research. Data were analyzed using non-parametric tests. The result of the test then showed in Table 3.

Table 1 The Hypothesis Test Result of Students' Problem Solving Ability Data

No.	Examination	Result
1	Hypothesis 1: Mastery Test of Experimental Class 1	Percentage of students' problem-solving ability in geometry through the 4K model with project assessment more than 74.5% (complete).
	Hypothesis 1: Mastery Test of Experimental Class 2	Percentage of students' problem-solving ability in geometry through 4K model more than 74.5% (complete).

	Hypothesis 1: Mastery Test of Control Class	Percentage of students' problem-solving ability in geometry through cooperative learning model is less than or equal to 74.5% (not complete).
2	Hypothesis 2: Three Mean Similarity Test *One-Way Anova*	Students' problem-solving ability in geometry through the 4K model with project assessment, 4K model, and cooperative learning are the difference.
3	Hypothesis 3: Advanced Post Hoc Test	Geometry problem-solving ability of students through the 4K model with project assessment better than geometry problem-solving ability of students through the 4K model.
4	Hypothesis 4: Advanced Post Hoc Test	The problem-solving ability of students through 4K model better than geometry problem-solving ability of students through cooperative learning model.

Table 2 Score of Tolerance Characters

Indicator	1st Meeting			2nd Meeting			3rd Meeting			4th Meeting		
	E1	**E2**	**C**	**E1**	**E2**	**C**	**E1**	**E2**	**C**	**E1**	**E2**	**C**
T1.1	2.9	2	2	3.2	2.6	2.6	3.3	2.7	2.9	3.8	3	2.6
T1.2	2.9	2.3	2.3	3.1	2.5	2.5	3.5	2.7	2.7	3.5	3.2	2.7
T2.1	3	2.3	2.4	3.1	2.8	2.5	3.3	3	2.6	3.8	3.2	2.6
T2.2	3.1	2.4	2.5	3.3	2.8	2.7	3.6	3.2	2.7	3.8	3.3	2.8
T3.1	2.6	2.4	2.8	2.8	2.6	2.9	3	3	3.2	3.4	3.3	3.2
T3.2	2.6	2.2	2.6	2.6	2.6	2.7	3	3	3	3.2	3.3	3.2
T3.4	2.6	2.2	2.6	2.9	2.6	2.7	3.2	2.7	2.8	3.6	3.1	2.9
T3.5	2.5	2.4	2.5	2.8	2.8	2.5	3.3	3	2.7	3.4	3.2	2.9

Table 3 The Hypothesis Test Result of Students' Tolerance Character Data

No.	Examination	Result
1	Hypothesis 5: *Kruskal-Wallis* Test	There are differences in the tolerance character of students through the 4K model with project assessment, 4K model, and cooperative learning model.
2	Hypothesis 6: Mann-Whitney U Test	There is no difference in the tolerance character of students through the 4K model with project assessment and 4K model.
3	Hypothesis 7: Mann-Whitney U Test	Tolerance characters of students through 4K model are better than the character of tolerance of students in cooperative learning model.

4. Conclusion

Based on the results of research and discussion, several conclusions can be drawn. Geometry problem-solving ability and tolerance character of students through the 4K model with project assessment achieve mastery, through 4K model achieve mastery, and through the cooperative learning model not achieve mastery. Geometry problem-solving ability and tolerance character of students through the 4K model with project assessment, 4K model learning, and cooperative learning is different. Geometry problem-solving ability and tolerance character of students through the 4K model with project assessment is highest of the 4K model learning and cooperative learning. The integration of tolerance character education in learning activities provides a good influence on the students' tolerance character. Also, the learning-laden character is also not allowed to leave the cognitive aspects of the students so that learning is designed will give good character without compromising students' academic. Learning strategies exemplary, habituation, and strengthening can foster the tolerance character so well. Thus, teachers can use those strategies in the learning process to instill students' tolerance character.

References

[1] Elkind D H and Sweet F 2004 *How to do character education*. (Mill Valley, CA USA: Live Wire Media)

[2] Suzanne N 2014 *Proceeding of the Third International Seminar on Languages and Arts Padang* (Padang: Universitas Negeri Padang) ISBN: 978-602-17017-4-4

[3] Lidyasari A T 2014 *Procedia - Social and Behavioral Sciences* **123** 189

[4] Depdiknas 2006 *Permendiknas No 22 Tahun 2006* (Jakarta: Depdiknas)
[5] Safrina K, Ikhsan M and Ahmad A 2014 *Jurnal Didaktik Matematika* **1** 9
[6] Oktorizal, Elniati S, and Suherman 2012 *Jurnal Pendidikan Matematika* **1** 60
[7] Adolphus T 2011 *International Journal of Emerging Sciences* **1** 143
[8] Woolfolk-Hoy A 2005 *Educational Psychology* (Boston: Pearson Allyn & Bacon) ISBN: 0-205-43529-7
[9] Polya G 1973 *How to Solve It* (New Jersey: Princeton University Press)
[10] Joyce B and Weil M 2003 *Models of Teaching* (New Jersey: Prentice-Hall, Inc.)
[11] Masrukan 2015 *International Conference on Mathematics, Science, and Education* p M-328 (Semarang: Universitas Negeri Semarang)
[12] Akinoglu O 2008 *International Journal of Instruction* **1** 1
[13] Brown G, Hirschfeld G 2007 *Australian Journal of Educational & Developmental* **7** 63
[14] Wu H 2012 *Journal of Mathematics Education at Teachers College* **3** 14

Manipulatives Implementation For Supporting Learning Of Mathematics For Prospective Teachers

D Sulistyaningsih[1], V D Mawarsari[1], I Hidayah[2] ,and Dwijanto[2]

[1]Departement of Mathematics Education, Universitas Muhammadiyah Semarang
[2]Departement of Mathematics Education, Universitas Negeri Semarang

Corresponding author: dwisulis@unimus.ac.id

Abstract. Manipulatives are needed by teachers to facilitate students understand of mathematics which is abstract. As a prospective mathematics teacher, the student must have good skills in making manipulatives. Aims of this study is to describe the implementation of learning courses of manipulative workshop in mathematics education courses by lecturer at Universitas Muhammadiyah Semarang which includes the preparation of learning, general professional ability, the professional capacity specifically, ability of self-development, development class managing, planning and implementation of learning, a way of delivering the material, and evaluation of learning outcomes. Data collection techniques used were questionnaires, interviews, and observation. The research instrument consisted of a questionnaire sheet, sheet observation and interview guides. Validity is determined using data triangulation and triangulation methods. Data were analyzed using an interactive model. The results showed that the average value of activities in preparation for learning, fosters capabilities of general professional, specialized professional, self-development, manage the classroom, implementing the learning, how to deliver the material, and how to evaluate learning outcomes are 79%, 73%, 67%, 75%, 83%, 72%, 64%, and 54%, respectively

1. Introduction

Advances in mathematics education characterized by increased learning tools for students in understanding the mathematical sciences. Accordingly, teachers are constantly find ways to enhance their teaching to help students understand mathematics. In recent years, there has been an increase in the use of manipulatives in the mathematics teaching [1-5]. Research in educational [6-8] had shown that when manipulatives were used effectively, students able to construct their mathematical understanding which in turn can be a valuable learning experience. They can connect their ideas and then integrate their mathematics knowledge so that they gain a deep understanding of mathematical concepts by many different ways.

The National Council of Teachers of Mathematics (NCTM) strongly encourages the use of manipulatives as a tool for students to create an understanding of mathematical concepts, and students actively construct new knowledge from experience and prior knowledge [9]. Many studies [2,7,10-12,] support the use of manipulatives for teachers in the teaching of mathematics. They believe that manipulatives are very important in helping students improve their ability to provide a cognitive model for not only the abstract mathematical ideas but also the abstract mathematical processes. In fact, other researchers [13-17] emphasized the importance of using manipulatives in the long-term use because it is more effective than short-term use.

The student in university as preservice teachers should develop teachers' knowledge and skills in the implementation and development of appropriate resources, materials and technological approach. The aim of this study is to describe the implementation of the course of manipulatives workshop. Student's perception was discussed. This study was focused on the implementation of the undergraduate course of manipulatives workshop at Universitas Muhammadiyah Semarang.

2. Method

This is a descriptive analytic research with the sample population were all students at the Department of Mathematics Education semesters 4 and 6 totaling 93 students. Samples were taken at random by using Purposive Random Sampling in students who have taken courses of math manipulatives workshop, totaling 36 students. Observation, interviews, and questionnaires are used as data collection techniques. Observations carried out to obtain a description in detail about the condition of the implementation of learning mathematics manipulatives workshop. Interviews with students and a lecturer of course in mathematics manipulative workshop conducted by referring to the lattice [18] to obtain information about the implementation of learning mathematics manipulative workshop.

There are eight components that are measured in the questionnaire, namely the preparation of lecture, general professional competence, specific professional competence, the ability of self-development, ability to manage the class, planning, and implementation of learning, delivery of subject material, and the way to evaluate learning outcomes. The questionnaire design refers to the guidelines provided Schwarz [19], Lucas and Baird [20], Crocker and Algina [21], Krosnick [22], and Braun et.al [23]. Four answers choices were provided in the questionnaire namely very often, often, rarely, and very rarely with the scoring for each answer is 4, 3, 2 and 1, respectively. The questionnaire results for each component is analyzed using the formula below [24]:

$$Percentage\ of\ answers = \frac{Amount\ of\ respondent\ answers}{Ideal\ score\ answer}$$

where ideal score = maximum score x amount of questions x amount of respondents.

Criteria for assessment questionnaire refers to the criterion of Riduwan [24]. Data were analyzed using triangulation techniques which combine and generalize the results of the data into the form of descriptive sentences detailed and candid.According to Riduwan [24], the criteria for the percentage range of 100-80, 80-61, 60-41, 40-21, and 20-0 is very good, good, fair, poor, and very poor, respectively.

3. Result And Discussion

Three students did not return the questionnaire. Perception of the 35 students who return the questionnaire is presented in Table 1.

TABLE 1. The students' perceptions of the implementation of the lectures

Assessment component	Percentage (%)	Criteria
Preparing for learning	79	Good
General professional competence	73	Good
Specific professional competence	67	Good
The ability of self-development	75	Good
Ability to manage the class	83	very good
Planning and implementation of learning	72	Good
Delivery of subject material	64	Good
The way to evaluate learning outcomes	54	Fair

Component 1. Learning plan

Results shown in Table 1 was confirmed through interviews and observations. Learning plan with a percentage of 79% and good criteria indicate that lecturers always deliver the learning objectives in each meeting. However, from the questionnaire detail, it can be shown that learning plans were always taken but not implemented in detail, learning to use the main reference book without the support of additional references. It should be fixed on the preparation component is additional supporting books so that students can have a wider knowledge.

Component 2. General professional competence

The total scores obtained from four question is 376 with the ideal score is 512. In other words, the percentage of 73% with good criterion. This result describes that the lecturer has always encouraged students to understand mathematical concepts and cultivate an attitude of reasoning that can design appropriate manipulatives material. Students suggest that the lecturer can further grow fighting spirit and competitiveness among students.

Component 3, Specific professional competence

The data in Table 1 on components for general professional skills can also be described that the total score is 346 with an ideal score of 512. This result was confirmed in interviews and observations include the habit of thinking objectively and openly, exemplary attitudes, habits appreciate the aesthetic and religious values. Feedback from students to lecturer is, even more, emphasis on the teaching of habits in logical abstract thinking, systematic and creative.

Component 4, The ability of self-development

Components of an ability of self-development shown in Table 1 obtained a score of 388 with an ideal score of 512. Based on results of this questionnaire, interview and observation can be described that lecturers always develop themselves regarding knowledge and attitude of learning, develop appropriate skills of science and technology. Things need to be improved is the ability to develop knowledge and attitudes for further studies not yet visible in learning.

Component 5, The ability to manage the class

Based on the results shown in Table 1 questionnaires, interviews and observation, the ability of the faculty in classroom management can be described that the very good academic atmosphere is created at each meeting, and the spirit of the students to ask successfully developed. Lecturers also to train students to be responsible and respectful to others. Student feedback given to the lecturers for improvement is the ability of lecturers to train students to critical thinking skills need to be pursued more intensively.

Component 6. Planning and implementation of learning

Table 1 shows that the planning and implementing learning was done well. These results have been confirmed by interviews and observations. Planning and implementation of learning can be described that the learning plan is created for each meeting, learning executed sequentially from the material of manipulative design, lesson plan of manipulative and the practice of manipulative making.

Component 7. Delivery of subject material.

Good criterion with the percentage of 64% as shown in Table 1 indicate that the material is delivered with an efficient, clear, and students actively participate in the practice of manipulative making. However, from interviews and observations, there is an important note to be addressed in the limited time of the meeting led to some of the material cannot be delivered, the time for discussion in completing the task students are not provided with sufficient while students are always given the task group or individually. Regarding improving the delivery of content, students asked lecturers to create media that can assist students in developing the skills of manipulatives making.

Component 8. The way to evaluate learning outcomes

Lecturer considered sufficient (54%) regarding learning outcomes evaluation design that includes assessment, reporting and practice the use of manipulative. Group work given by lecturer led to an objective assessment was difficult to carry. Assessment of learning outcomes for the task presented that can be represented by one member, causing some students are not able to assess the results of their own learning.

The very good value in classroom management component indicated that the lecturer could create a pleasant lecturing atmosphere so that students do not get bored with the learning process. In other words, In the learning process, the lecturer always foster the student's spirit to understand the material well. While fair in the evaluation component indicates that only some groups are evaluated by lecturers so that students are not evaluated could not find out his mistake in judging themselves. In addition to assessments by the lecturer does not involve students so less objective assessment. Overall assessment component has reached a good category, with an average acquisition value for the eight components amounted to 70.87% and included a good category.

Through the evaluation of the work on the effectiveness of using math manipulatives, the author found many works that supported the use of virtual [25,26,27], pictorial [28,29], and concrete [30,31] manipulatives. For decades, many authors have been demonstrating the good impact of using pictorial and concrete math manipulatives for their students. The works that are more coeval have extended these findings to virtual manipulatives. Although some works have reported unimportant when using math manipulatives, it appears that these results are related with the method used of instructional. The lecturer who adheres the best practice recommendations for manipulative used tends to undergo positive outcomes.

4. Conclusions

As a conclusion, this study showed that the learning process for a workshop of mathematics manipulatives course had done well by the lecturer. This creates a great opportunity for students at the University as a prospective teacher develop their implicit conceptions about the teaching and learning of mathematics. This learning process can lead to increased confidence by obtaining a new learning experience.This learning process is the mechanism driving change belief-systems for students who use manipulatives to reveal the conception of teaching and learning mathematics.It can also help teacher monitor the process of reflection and to help students understand abstract concepts into concrete.

Acknowledgments

The authors would like to thank with much appreciation to The Directorate for *Research and Community Service* Indonesian Directorate General of Higher Education for financial support in this study.

References

[1] G L Musser, B E Peterson, W F Burger 2014 *Mathematics for Elementary Teachers: A Contemporary Approach, 10th Edition* (Wiley)
[2] J L Bellonio, J L 2012 *Multi-Sensory Manipulatives in Mathematics: Linking the Abstract to the Concrete* Retrieved March 5 (Yale-New Haven Teaching Institute: http://www.cis.yale.edu/ynhti/curriculum/units/2001/6/01.06.12.x.html)
[3] M Boggan, S Harper, and A Whitmire 2010 Using manipulatives to teach elementary mathematics *Journal of Instructional Pedagogies* (3) (Retrieved from http://www.aabri.com/manuscripts/10451.pdf)
[4] L G Marsh and L N Cooke 1996 *Learning Disabilities Research & Practice* 11(1) 58–65
[5] E Marzola 1987 *Journal of Reading, Writing, and Learning Disabilities International* 3(1) 9-20
[6] N M McNeil and J Jarvin 2007 *Theory into Practice* 46(4) 309-316
[7] C A Kelly 2006 *The Montana Mathematics Enthusiast* 3(2) 184-193
[8] A Cockett and P Kilgour 2015 *TEACH Collection of Christian Education* 1(1) 47-54
[9] The National Council of Teachers of Mathematics (NCTM) 2000 *Principles and Standards for School Mathematics* (Reston VA: NCTM)
[10] P Swan and L Marshall 2010 *Australian Primary Mathematics Classroom 15*(2) 13-19
[11] D Raphael and M Wahlstrom 1989 *Journal for Research in Mathematics Education.* 20(2) 173-190
[12] E Sowell 1989 *Journal for Research in Mathematics Education,* 20 498–505
[13] H Sandir 2016 *Eurasia Journal of Mathematics, Science & Technology Education* 12(8) 2103-2114
[14] P S Moyer 2001 *Educational Studies in Mathematics,* 47 175–197
[15] M Green, C Flowers, J Piel 2008 *Journal of Educational Research,* 101(4) 234-232

[16] R Ruzic and K O'Connell 2001 *"Manipulatives" Enhancement Literature Review* (http://www.cast.org/ncac/Manipulatives1666.cfm.(21.08.2016))
[17] S Ikun and Z Toluk 2004 *Issues in the Undergraduate Mathematics Preparation of School Teachers* 2 11
[18] M A Goracke 2009 *Action Research Projects.* Paper 71 http://digitalcommons.unl.edu/mathmidactionresearch/71
[19] N Schwarz 199 *American Psychologist* 54 93–105
[20] R Lucas and B Baird 2006 "Global self assessment," In *Handbook of psychological measurement: A multimethod perspective*, M. Eid & E. Diener (Eds.), Washington, DC: American Psychological Association 29–42
[21] L Crocker and J Algina 1986 Introduction to classical and modern test theory. Fort Worth, TX: Harcourt Brace
[22] J Krosnick 1999 *Annual Review of Psychology* 50 537–567
[23] E Braun, A Woodley, J T E Richardson, B Leidner 2012 *Educational Research Review* 7 1–18
[24] Riduwan 2009 *Metode dan Teknik Penyususnan Proposal Penelitian* (Bandung: Alfabeta)
[25] P S Moyer, J J Bolyard, M A Spikell 2000 *Teaching Children Mathematics* 8(6) 372-377
[26] P S Moyer, D Niezgoda, J Stanley 2005 Young children's use of virtual manipulatives and other forms of mathematical representations. In W. J. Masalaski & P. C. Elliot (Eds.), *Technology-Supported Mathematics Learning Environments* 17–34. Reston, VA: National Council of Teachers of Mathematics.
[27] P S Moyer-Packenham and J M Suh 2012 *Journal of Computers in Mathematics and Science Teaching 31*(1) 39-59
[28] P S Moyer-Packenham and L A 2012 Ulmer, K. L. Anderson, *Journal of Interactive Online Learning* 11(3) 103-120
[29] A Arcavi 2003 *Educational Studies in Mathematics 52*(3) 215-241
[30] C Kamii, B Lewis, L Kirkland 2001 *Journal of Mathematical Behavior 20*(1) 21-31
[31] T Martin, V Svihla, C P Smith 2012 *Journal of Education and Human Development 5*(1) 1-17

Development of "OQALE" Based Reference Module for School Geometry Subject and Analysis of Mathematical Creative Thinking Skills

N A D Wulandari[1] and Y L Sukestiyarno[2]

[1, 2]Department of Mathematics, FMIPA, Semarang State University, Indonesia.

[2]Corresponding author: sukestiyarno@mail.unnes.ac.id

Abstract. This research aims to develop an OQALE based reference module for school geometry subject that meetsthe criteria ofa validandpractical. OQALE approach is learning by of O = observation, Q = question, A = Analyze, L = Logic, E = Express. Geometry subject presented in the module are a triangle, the Pythagorean theorem, and rectangular. Mathematicalskills ofcreative thinkingshownfrom four aspects: fluency, flexibility, originality and elaboration. Research procedures in the development of reference module using a strategy of the investigation and development described by [2], which is limited to the sixth stage is leading field testing. The focus of this research is to develop a reference module that is valid, practical and able to increase the mathematicalcreative thinkingskillsof students. The testing is limited to three teachers, nine students and two mathematic readers using purposive sampling technique. The data validity, practicality, and creative thinkingskills upgradingcollected through questionnaires, observations, and interviews and analysed with a valid test, practical test, gain test and qualitative descriptive.The results were obtained (1) the validity of the module = 4.52, which is $4.20 \leq V_m < 5.00$ included in the category of very valid; (2) the results of the questionnaire responses of teachers = 4.53, which is $4.20 \leq R_g < 5.00$ included in the category of very good; (3) the results of the survey responses of students = 3.13, which is $2.80 \leq R_{pd} < 3.40$ included in the category of good with an average percentage of 78%; and (4) increasing skills of creative thinking mathematically nine students through the test of the gain included in the high and medium category. The conclusions of this research are the generatedOQALE basedreference module for school geometry subjectis valid and practical.

1. Introduction

In the sphere of education schooling, mathematics is one of the subjects were considered to have a significant role because math can improve the knowledge of students in logical thinking, rational, critical, thorough, effective and efficient. Besides, mathematics also has a significant role in the development of education. Mathematics showed as a basic science for the development of other disciplines. But unfortunately, mathematics is still regarded as a scourge for students. They assume that math is complicated and abstract. But in fact, math is very close to our daily lives.

The goals of national education is to develop the ability, character, and creative potential [17]. Therefore, the ability and creative thinking skills must also be owned by every student in the mathematics learning. Mathematics requires logical reasoning, efficient and has a vocabulary of ideas and creative patterns and fantastic. Therefore, mathematics is often called the art of creative thinking [4]. Creativity is the ability to think about things in a new way and unusual and produce a unique solution to a problem [15]. Creativity can be measured by four criteria, namely; (1) fluency; (2) flexibility; (3) originality; and (4) elaboration [8].

There are some learning subjects that students often difficult to understand or teachers are difficult to explain. The difficulties can happen because of the issues abstract, complicated, or unfamiliar. These problems can be resolved, and it is necessary to develop a book or appropriate teaching materials [3]. One of the subjects mathematics which is considered challenging and abstract by students is geometry. The results of the national exam (UN) in 2012 shows that the average absorption students junior high school in Central Java for the material geometry, especially the Pythagorean theorem, flat figure, and the triangle is still below the mean of the national absorption capacity. That is supported by observations and interviews of researchers in some schools in Semarang city. Based on observations and interview researchers, obtained by the fact that there are still many problems in the learning of mathematics.

Regarding students, most students still have a low level of maturity subjects and the level of seriousness learn less. Most of the students are also difficulties in working on different sample questions given by the teacher when in fact the same concept used. That indicates that the level of creative thinking of students is still small.

Regarding learning resources, existing textbooks are still felting dull, unattractive and difficult to be understood by students.

The fact, as mentioned above encourage authors to do research development reference module that can be used not only in the classroom but also can be used anytime and anywhere as a reference study. This reference module can also customise the speed of comprehension each student. The module is a book written with the purpose make the students can learn independently [3]. The reference module is a module that content and presentation could be used to obtain information about the science are deep and extensive. This module can also be used by anyone students who need. Learning to use module enables a student who has high speed in learning to be more quickly complete the module compared to other students [3].

Mathematics subject to be presented in a reference module that school geometry items are, (1) triangle, (2) the Pythagorean theorema, and (3) rectangle. The use of this reference module will facilitate the students more interested in learning geometry and increase the mathematical creative thinking skills of students.

Reference module that developed in this study is OQAME based reference module. The OQALE approach is learning by exploration, elaboration, and confirmation. Students are always taken on the atmosphere of O = observation, Q = question, A = analyze, L = logic, E = express. Observation means not only to see but also to read the issues to explore the whole concept to be learned. The question means asking questions about information that is not understood from what observed (exploration) or asking a question to increase the flexibility of information. Analyse means doing experiments, processing information from the results of observing and questioning activities. Logic ways to examine, to associate data with other data to make a conclusion of the analyse events. Express means ideas has compiled to get new ideas.

The development of this module is done to obtain OQALE based reference module of school geometry subject valid and practical. The components of educational products are said to be valid when based on state-of-the-art knowledge or strong theoretical rationale (content validity), and all elements must be associated consistently with each other (construct validity). While the elements of educational products said to be practical if the teacher can consider tools or subject that can be used and easier for teachers and students to use [11].

A module is one form of teaching materials packaged thoroughly and systematically, inside contained a set of learning experiences planned and designed to help students take control specific learning goals (Daryanto, 2013). A module is a complete unit that stands alone and consists of a series of learning activities which is structured to help students achieve some goals that are formulated specifically and explicitly [10].

To develop a reference module needs to pay attention to the characteristics of the modules include; (1) Self-instruction means enable someone to learn to be independent and does not depend on the other hand. (2) Self-contained means all the necessary learning materials contained in modules so that learners can learn thoroughly. (3) Stand alone means not dependent on teaching materials/other media or should not be used together with educational materials/other media. (4) Adaptive means if the module can adjust the development of science and technology, as well as flexible use in a wide range of hardware. (5) User-friendly means instruction and explanation of information in the module helpful and friendly to the wearer.

According to Regiero [6], thinks is all mental activities that help to formulate or solve problems, make decisions, fulfilling the desire to understand, search for answers, and achievement of meaning. According to Krulik and Rudnick [7], think categorised consists of basic thinking, critical thinking, and creative thinking. According to Saefudin [14], creative thinking is a single unit or a combination of logical thinking and divergent thinking to generate something new. Something new is one indication of creative thinking in mathematics, while the other indications related to logical thinking and divergent thinking. Munandar [8] suggests that creativity can be measured by four criteria, namely; (1) fluency; (2) flexibility; (3) originality; and (4) elaboration. Fluency means fluent in expressing his ideas. Flexibility ways to solve the problem with a variety of different ways. Creativity means creating an innovation with their thoughts. Elaboration means that details the steps to resolve the issue. Mathematical skills of creative thinking showed from psychomotor students through observation and interviews before and after using the reference module.

Learning geometry using the module is also associated with Brunner and Van Hiele learning theory. According to Brunner [12], there are three stages of learning progress are enactive, iconic and symbolic. Learning geometry using a reference module in agreement with Brunner's theory. Learning begins by linking the abstract geometry of subject with concrete objects that exist around the learner. Then the students are invited to visualise it in the form of images. Then the last, students can use mathematical symbols to solve the problems of geometry.

According to the theory of Van Hiele, someone will go through five stages of development thinking in learning geometry. The fifth stage of development think Van Hiele: stage 0 (visualization), stage 1 (analysis), stage 2 (deduction informal), stage 3 (deduction), and stage 4 (rigor).

Furthermore, according to the theory of constructivism [12], learning is an active process of learners in constructing meaning, discourse, dialogue, physical experience in the learning process to a process of assimilation and connecting experiences or information that has been learned. That is in line with the stage OQALE used in modules developed. Through this OQALE stage, learners will be more active in learning to use the module independently.

2. Methods

This type of research is a kind of the investigation and development, namely development of OQALE based reference module for school geometry subject. Development procedure used to research and development strategy described by [2], which includes ten steps. It will be taken to six level, namely; (1) research and information collecting; (2) planning; (3) develop a preliminary form of the product; (4) initial field testing; (5) elemental product revision; (6) main field testing.

Stage 1: Research and information collecting, analysing problems in learning mathematics, explaining the characteristics of students (creative thinking mathematically students), examining geometry subjects, investigating difficulty in learning geometry, literature reference module development, literature OQALE approach, and writing creative thinking mathematically.

Stage 2: Planning, conducted targeting reference module, determination subjects, a decision the purpose of learning using a reference module, the selection of the reference module format, and resolution components of the reference module.

Stage 3: Develop a preliminary form of product, done the preparation module (draft 1) and instruments of research.

Stage 4: Preliminary field testing, validation module (draft 1) and tools of research. Validation module is carried out by the three-person team of experts that are Supervisor 1, Supervisor 2, and a Lecturer in the Mathematics Department, Universitas Negeri Semarang while the instrument validation study carried out by supervisor 1 and supervisor 2.

Stage 5: Main product revision, revised module (draft 1) into a module (draft 2) and all the instruments that have been made based on advice from experts. This revision continues to obtain the module (draft 2) and valid instrument

Stage 6: Main field testing, conducted testing limited to 3 teachers, nine learners, and two mathematic readers to determine the practicality of the reference module. At this stage used instruments are questionnaire responses of teachers, questionnaire responses of students, and interview guide (for teachers, and mathematic readers). At this juncture also made observations of creative mathematical thinking skills of students before and after using the reference module.

To determine the validity of the reference module used instrument module validation. Besides, it can ascertain that the practicality of the reference modules used instruments questionnaire responses of teachers, questionnaire responses of students, and an interview guide. Meanwhile, it can determine the increase of mathematical creative thinking skills used guidelines observation skills.

The validation module reference calculated by the average score of each aspect indicators from the response of teachers and students [5].

Skills upgrading calculated using normalised gain by the following formula.

$$g = \frac{A \quad S \quad - B \quad S}{S \quad - B \quad S \; r}$$

g are a normalised gain. Criteria gain according to Hake [18] presented Table1.

Table1.CriteriaGainIndex

Gain Index	Criteria
$g \geq 0.7$	High
$0.3 \leq g < 0.7$	Medium
$g < 0.3$	Low

3. Results

3.1Research and Information Collecting

This stage obtained information and theories relating to the development of OQALE based reference module for school geometry subject (triangle, the Pythagorean theorem, rectangle), and mathematical creative thinking skills. These theories will help researchers to plan the development of a reference module.

3.2 Planning

OQALE based reference module consists of three sub-modules: module 1 triangle, module 2 Pythagorean theorem, and module 3 rectangle. The purpose of learning using a reference module that understands the material and improves the skills of creative thinking mathematically. Format reference module uses OQALE approach will help students learn from observing, questioning, analysing, logic, and expressing. Through a series of stages this OQALE, students will also be able to increase the mathematical skills of creative thinking with fluence thinking, flexible thinking, original thinking, and elaboration thinking. The validity of this reference module obtained through the stages of preliminary field testing and primary product revision. On stage initial field testing received validation expert and the reference module improvement suggestions from the experts.

Primary field testing was conducted to determine the practicality, usability, and legibility of module in the field. Data practicality reference module is obtained based on the response of teachers, students, and mathematic readers in the area. The design components of OQALE based reference module for School geometry subjects contain: 1. The initial part consists of a cover "ModulGeometriSekolahBerbasis OQALE", preface, guide the use of modules, table of contents, apperception, and a map of the module concept, and 2. The contents consist of the subjects presented in the module reference. Part of the substance of a reference sub-modules, comprising:

a. Map of sub-module concept contains the concepts of the subject to be learned from each sub-module and the relationship between the concepts presented.
b. InfoMath, in the form of information about the figures of mathematics, history of math and other information.
c. Description of the subject, which presented with a simple sentence.
d. Example question, in the form of questions.
e. You can do it, in the form of questions that test the ingenuity of students in solving a mathematical problem.
f. Summary, containing the main points of discussion in the submodules that completed studied.
g. Exercise matter includes questions to test the ability of students to understand the material they have learned.
h. Test formative contains questions to measure the subject understanding of specific sub-modules.
i. References.
j. Efforts to follow up is used to determine the level of mastery learners after taking the test formative.
k. Key answers developmental tests, such as formative test answers.

3.3 Develop Preliminary Form of Product

At this stage of the preparation of the reference module (draft 1) includes; the sheet of module validation; questionnaire responses of teachers; questionnaire responses of students; guidance teacher interviews; leadership student interviews; guidance mathematic readers interviews; the scoring guidelines mathematical creative thinking skills; and the sheet of instrument validation.

At this stage, the reference module has been through the validation and repeated revisions based on suggestions from the supervisor

3.4 Preliminary Field Testing

Atthis juncture,validationmodulereference(draft 1) andinstruments ofresearch. Results ofthe validationmoduleobtained an averageassessmentmodulebyan expertthirdvalidatorincluded inthe category of veryvalid, namely V_m=4.52(maximum score 5). So thatOQALE based referencemodulefor school geometry subjectis said to bevalid. Results ofthe validationmodule shown in Table2.

Table2.Results of Validation Module

The Validity ofModule	Validator Value		
	I	II	III
Validity	4.10	4.65	4.80
Criteria	Valid	Very Valid	Very Valid
V_m		4.52	
Criteria V_m		Very Valid	

It is suggested that each validator experts stated that the OQALE based reference module for school geometry subject is correct and may increase mathematical creative thinking skills. Depending on the results of expert validation, obtained recommendation that the module used with minor revisions. This review is done based on the comments and suggestions given by the experts.

Table 3. Results Validation Instrument

Instrument	averageResults ofValidation	Criteria
TeacherQuestionnaireResponse	4.50	VeryValid
Student QuestionnaireResponse	4.42	VeryValid
TeacherInterviewGuidelines	4.50	VeryValid
StudentInterviewGuidelines	4.50	VeryValid
Mathematic ReaderInterviewGuidelines	4.50	VeryValid
GuidelinesObservationSkills	4.17	Valid

Based on the validation results, obtained recommendation that the instrument used with minor revisions. This review is done based on the comments and suggestions given by the supervisor.

3.5 Main Product Revision

At this stage, revised the reference module (draft 1) into the module (draft 2) and all the instruments that have been made based on advice from experts. Revisions have been conducted by researchers to repair modules, namely (1) the addition of questions to think creatively and issues Olympics; (2) the improvement of grammar. especially preposition adjusted language; and (3) development of charts and Venn diagrams relations. After the module revised and declared valid, the module used in a practical test.

3.6 Main Field Testing

At this stage conducted testing limited to 3 teachers, nine learners, and two mathematic readers to determine the practicality of the module. Three teachers consisted of SD (G-1), SMP (G-2), and SMA (G-3). Nine students consist of 3 SD (PD-1, PD-2 and PD-3), 3 SMP (PD-4, PD-5, PD-6), and 3 SMA (PD-7, PD-8 and PD-9). Two mathematic readers (M-1 and M-2).

Based on the results of teachers' questionnaire responses, obtained average yield response of teachers = Rg = 4.53 (maximum score 5) and is included in the excellent category. Results of the teacher's response to the reference module shown in Table 4 below.

Table4. The teacher's response to the Module

Response	Validator Value		
	I	II	III
Teacher's Response	4.41	4.56	4.63
Criteria	Very Good	Very Good	Very Good
R_g		4.53	
CriteriaR_g		Very Good	

Based on the questionnaire responses of students, obtained average yield response of students = Rpd = 3.13 (maximum score 5) and including in right category. Results students response to the reference module shown in Table 5.

Table 5.Responseof Students

Students	Respons e	Criteria	PercentageRespon se	Average
PD-1	3.69	Very Good	92%	
PD-2	2.80	Good	70%	
PD-3	2.80	Good	70%	Rpd= 3,13
PD-4	3.44	Very Good	86%	(Good)
PD-5	3.36	Good	84%	
PD-6	2.82	Good	71%	Percentage Response
PD-7	3.09	Good	77%	= 78%
PD-8	3.00	Good	75%	
PD-9	3.16	Good	79%	

The third teacher gives suggestions: (1) language and writing system used in reference modules adapted to language; (2) the modules can be equipped with teaching VCD; (3) problems in the modules should be of a level is easy, rather easy, and challenging; (4) the language used in the module fixed to be simpler so easily understood by students; and (5) can be added high school subject so it used as a companion book high school students.

Ninth students advice that (1) exercises should also be added the answer key; (2) module is communicated to a wider student; (3) on the material triangle added about trigonometry subjects and speciality corners of the table; and(4) the words of motivation can be added.

Both mathematic readers advice that(1) it would be better if the module is separated from elementary schools, junior high schools and high school so that elementary school students are not confused, and(2) a mathematics problemmorevariedso that childrengrow accustomed toa lot of practiceandcreative thinking.

The average yield total score indicator mathematical creative thinking skills on the ninth students before and after using the OQALE based reference module for school geometry subject show an increase. Calculation test value gain of nine students presented in Table 6.

Table 6.Test Gain Mathematic Creative Thinking Skills Before and After Using Modules

Interviewees	The test resultsgain	Criteria
PD-1	0.76	High
PD-2	0.77	High
PD-3	0.72	High
PD-4	0.71	High
PD-5	0.53	Medium
PD-6	0.75	High
PD-7	0.77	High
PD-8	0.73	High
PD-9	0.50	Medium

4. Discussion

The research showed that through the stages Borg and Gall [2] obtained an OQALE based reference module for school geometry subject are valid, practical, and capable of improving the skills of creative thinking mathematically. Mathematical creative thinking skills improvement is due to the OQALE approach (Sukestiyarno, 2013) and the questions contained in the OQALE based reference module. The OQALE approach used in the module, in addition to constructing the way students learn from observing, questioning, analysing, associating, and expressing, also able to encourage students to improve mathematical skills of creative thinking. The questions contained in the reference module is also problem-solving questions and problems of creative thinking mathematically arranged to start from easy to difficult level. Problems solving this problem proved to increase mathematical creative thinking skills students. That is according with what stated Siswono [1] which mentions that the forms of matter and matter that can foster creative thinking are a form of mathematical problem solving, filing problems or a combination of both.

Increased creative thinking mathematically through the OQALE approach and problem solving is consistent with what raised Rothenberg [13] and the results Mustakim (2015). Rothenberg [13] argues that scientific discoveries demonstrate the significant role to enhance creative thinking in science by prioritising the investigation. Meanwhile, the research results Mustakim states that learning scientific approach to solving the problem with organised can improve the mathematical skills of creative thinking and learning outcomes of students [9].

Increased mathematical creative thinking skills students can be observed through the activities students when working on covering observations aspects of fluency, flexibility, originality, and elaboration [8].

5. Conclusion

To obtain a reference module that can be used independently and can enhance the creative thinking mathematically crucial stages learning which can bring students improve mathematical skills of creative thinking. The steps employed in this reference module is OQALE stages. OQALE stages are proven to bring students to observing, questioning, analysing, associating, and expressing and able to enhance creative thinking skills, mathematical students.

Based on the results of research and discussion, It concluded the OQALE based reference module for school geometry subject is valid. The average validity module and included in the category of very valid and OQALE based reference module for school geometry issue practical with a mean response of teachers, including in the excellent category and the average response of students included in both categories with a response percentage students.

References

[1] Azhari. 2013. Peningkatan Kemampuan Berpikir Kreatif Matematik SiswaMelalui Pendekatan Konstruktivisme di Kelas VII Sekolah Menengah Pertama (SMP) Negeri 2 Banyuasin III, *Jurnal Pendidikan Matematika,* 7 (2): 1–12.

[2] Borg R. W.& Gall M. D. 1983. Educational Research, An Introduction, Fourth Edition. New York: Longman.

[3] Depdiknas. 2008. Panduan Pengembangan Bahan Ajar. Dirjen Dikdasmen Direktorat Pembinaan SMA.

[4] Fathani, A. H. 2009. Matematika Hakikat dan Logika. Yogyakarta: Ar-Ruzz Media.

[5] Hobri. 2010. Metodologi Penelitian Pengembangan (Aplikasi pada Penelitian Pendidikan Matematika). Jember: Pena Salsabila.

[6] Johnson, E.B. 2002. Contextual Teaching and Learning. California: Corwin Press, Inc.

[7] Krulik, S, & Rudnick, J.A. 1999. The New Sourcebook for Teaching Reasoning and Problem Solving in Elementary School. Massachusetts: Allyn & Bacon A Simon & Schuster Company.

[8] Munandar, U.2009. Pengembangan Kreativitas Anak Berbakat. Jakarta: Rineka Cipta.

[9] Mustakim. 2015. Implementasi pembelajaran pemecahan masalah dengan pendekatan saintifik untuk meningkatkan kemampuan berpikir kreatif matematik dan prestasi belajar materi bangun datar segi empat bagi siswa kelas VII-A SMP Negeri 2 Patean semester ii tahun pelajaran 2013/2014, Jurnal Pendidikan, 16 (1): 19-33.

[10] Nasution, S. 2009. Berbagai Pendekatan dalam Proses Belajar dan Mengajar. Bumi Aksara. Jakarta.

[11] Nieveen. 1999. Prototyping to Reach Product Quality. Design Approaches and Tools in Education and Training (Van den Akker, Branch, Gustafson, Nieveen And Plomp). Netherlands: Kluwer Academic Publishers.
[12] Rifa'i, A. & C. Anni. 2011. Psikologi Pendidikan. Semarang: Universitas Negeri Semarang Press.
[13] Rothenberg, A. 1996. The Janusian Process in Scientific Creativity. Creativity Research Journal, 9, 207–209.
[14] Saefudin, A. A. 2012. Pengembangan Kemampuan Berpikir Kreatif Siswa dalam Pembelajaran Matematika dengan Pendekatan Pendidikan Matematika Realistik Indonesia (PMRI), Jurnal Al-Bidayah, 4(1): 37-48.
[15] Santrock, J. W. 2008. Psikologi Pendidikan (2^{th}ed.). Jakarta: Kencana Prenada Media Group.
[16] Sukestiyarno, Y. L. 2013. Pembelajaran Matematika dengan Pendekatan OQALE Berbasis Pendidikan Karakter Materi Operasi Bilangan Kelas IV dipresentasikan pada Seminar Nasional Pendampingan Implementasi Kurikulum 2013, Unnes Semarang, 26 Oktober.
[17] Undang-Undang Republik Indonesia Nomor 20 Tahun 2003 tentangSistemPendidikanNasional. 2003. Jakarta: Depdiknas.
[18] Fachrurazi. 2011. Pengembangan Pembelajaran Berbasis Masalah untuk Meningkatkan Kemampuan Berpikir Kritis dan Komunikasi Matematis Siswa Sekolah Dasar. Jurnal Edisi KhususNo.1:76-89 ISSN 1412-565X. Tersedia di http://jurnal.upi.edu/file/8-Fachrurazi.pdf [diaksestanggal 10 Februari 2015].

Mathematizing Process of Junior High School Students to Improve Mathematics Literacy Refers PISA on RCP Learning

Wardono, S Mariani, P Hendikawati, Ikayani
Universitas Negeri Semarang

Corresponding author: wardono@mail.unnes.ac.id

Abstract. Mathematizing process (MP) is the process of modeling a phenomenon mathematically or establish the concept of a phenomenon. There are two mathematizing that is Mathematizing Horizontal (MH) and Mathematizing Vertical (MV). MH as events changes contextual problems into mathematical problems, while MV is the process of formulation of the problem into a variety of settlement mathematics by using some appropriate rules. Mathematics Literacy (ML) is the ability to formulate, implement and interpret mathematics in various contexts, including the capacity to perform reasoning mathematically and using the concepts, procedures, and facts to describe, explain or predict phenomena incident. If junior high school students are conditioned continuously to conduct mathematizing activities on RCP (RME-Card Problem) learning, it will be able to improve ML that refers PISA. The purpose of this research is to know the capability of the MP grade VIII on ML content shape and space with the matter of the cube and beams with RCP learning better than the scientific learning, upgrade MP grade VIII in the issue of the cube and beams with RCP learning better than the scientific learning in terms of cognitive styles reflective and impulsive the MP grade VIII with the approach of the RCP learning in terms of cognitive styles reflective and impulsive This research is the mixed methods model concurrent embedded. The population in this study, i.e., class VIII SMPN 1 Batang with sample two class. Data were taken with the observation, interviews, and tests and analyzed with a different test average of one party the right qualitative and descriptive. The results of this study demonstrate the capability of the MP student with RCP learning better than the scientific learning, upgrade MP with RCP learning better compare with scientific learning in term cognitive style of reflective and impulsive. The subject of the reflective group top, middle, and bottom can meet all the process of MH indicators are then the subject of the reflective upper and intermediate group can meet all the MV indicators but to lower groups can only fulfill some MV indicators. The subject is impulsive upper and middle group can meet all the MH indicators but to lower groups can only meet some MH indicator, then the subject is impulsive group can meet all the MV indicators but for middle and the bottom group can only fulfill some MV indicators.

1. Introduction

Since PISA was first implemented in 2000 Indonesia has been involved from the beginning in the administration of PISA, the results achieved by students in the PISA Indonesia is far from satisfactory. In PISA 2000, Indonesia was ranked 39 out of 41 countries [1], in PISA 2003 was ranked 38 out of 40 countries [2], in PISA 2006 was ranked 50 out 57 countries [3], in PISA 2009 was ranked 61 out of 65 countries [4] and in PISA 2015 was ranked 62 out of 70 countries [5]. From the PISA results showed a low ability of Indonesian students MP means that students have not been able to interpret the mathematization ability in everyday life in various contexts.

MP is a process for mathematics phenomenon. Mathematics could be interpreted as a phenomenon mathematically model (in the sense of looking mathematics relevant to a phenomenon) or building a mathematical concept of a phenomenon [6]. [7] distinguishes mathematization into two kinds, namely MH and MV. [8] defines the MH is an activity change contextual problems into mathematical problems (symbol), while the vertical is to formulate mathematization problems in a variety of settlement mathematics by using some the corresponding mathematical rules.

According to Freudental in [6] a science will be meaningful to the learner if the learning process involves a realistic problem. One model of learning that emphasizes the significance of science is realistic mathematics education(RME). Judging from the use of MH and MV[7] distinguishes four approaches mathematics learning, namely mechanistic approach (mechanistic), structuralistic, empiristic, and a realistic approach. Mathematization mechanistic approach both horizontally and vertically is not used. In empiristic approach simply uses MH. The structural approach using only MV. While the realistic approach both horizontally and vertically MP used [7]. So as to improve the ability of students can use the MP realistic approach, a realistic approach to mathematics is the approach RME.

According to Gravemeijer (1994: 90) as cited [7] there are three key principles in designing learning realistic mathematics (1) (guided reinvention and progressive mathematizing) (2) didactical phenomenology) (3) (self-developed models). Three key principles of RME in implementation bore characteristics of mathematics learning realistic, namely: (1) the use of context, (2) the use of models, bridging by vertical instrument, (3) student contribution, (4) interactivity and (5) the intertwining [6].

Solving mathematical problems always occur MP, and students will use a variety of strategies. According to [10] mathematical problem solving strategy turned out to be heavily influenced by the student's cognitive style. According to Susan, as cited by [10], that "general problem solving strategies such as These are further influenced by cognitive style". When students have different cognitive styles, the way of solving mathematical problems are also different, so the difference it will also trigger different processes, mathematization students. According [11] cognitive styles are individual characteristics regarding feel, remember, organize, process, and problem solving. [12] cognitive styles are classified as follows: (1) differences in cognitive styles psychologically, including cognitive style field dependent and field independent, (2) differences in cognitive styles are conceptually tempo, include: cognitive style impulsive cognitive style reflexive, (3) differences in cognitive styles based way of thinking, including intuitive cognitive style-inductive and deductive logic.

Kagan as cited by [13], cognitive style grouping into two groups, are: (1) cognitive style are a reflective cognitive style that has the characteristics of slow in responding to the problem, but carefully or thoroughly, so the answer tends to be true. (2) the impulsive cognitive style is a child's cognitive style that has the characteristics of quick in answering the problem, but no or less carefully, so that the answers tend to be wrong.

2. Method

This type of research is the combination research (*mixed methods*) *embedded concurrent*. According [14], a combination of methods concurrent embedded or design is qualitative and quantitative research methods by mixing these two methods are not balanced. The research design used *pretest-posttest control group design*. The population in this study were all students of class VIII SMP 1 Batang in the academic year 2015/2016. The samples in this study using simple random sampling technique selected two samples is class VIII C as the experimental class was subjected with RCP learning and VIII F as a control class with scientific learning. Decision The subject of research in a qualitative study using purposive sampling technique of determining where the subject is done with certain considerations [15]. Taking the subject of the research is done with consideration of test results of cognitive style and a preliminary test value i.e. selecting students from the class of experiments each of two Subjects from the upper class, the middle and the bottom of the students with cognitive styles reflective and each of two Subjects from the upper class, the middle and the bottom of the students with cognitive style impulsiveness. The necessary data in this study were obtained through observation, documentation, tests, and interviews with the subject in depth. Analysis of initial data including normality test, homogeneity, and the equality test average. Final data analysis including normality test, homogeneity test, different test average right side.

3. Result And Discussion

Cognitive style measurement results as follows the number of students Reflective 38.24%, the number of students Impulsive 35.29%, the number of students slow-inaccurate 11.76%, while students fast-accurate 14.71%. This shows that the proportion of students who have the characteristics of a reflective-impulsive are 73.53% higher than the students who have the characteristics of fast and precise/accurate in answering or slower and less precise/less accurate in answering, are 26.47%. Then based on Reflective the number of students 35.29%, the number of students Impulsive 38.24%), the number of students slow-inaccurate 14.71%, while students fast-accurate 11.76%. This shows that the proportion of students who have the characteristics of a reflective-impulsive 73.53% higher than the students who have the characteristics of fast and precise/accurate in answering or slower and less precise/less accurate in answering 26.47%, These results are consistent with previous research, the study [13] the proportion of children reflective-impulsive 73.7% and the proportion of children who have the characteristics of fast and precise/accurate in answering or slower and less precise/less accurate in answering 26.3%. t test was conducted to determine the ability MP of the student with RCP learning better than the ability MP of student with a scientific learning. The results as in Table 1.

Table 1. Results Of The Test The Average Difference Mathematization Process Capability

t_{Count}	α	t_{table}	Conclusion	It Means
3,873	5%	1,669	$t_{Count} > t_{table}$	Ability to process mathematization math class VIII student on materials cubes and blocks with RCP learning the problem more than the ability of class VIII student mathematization process on the material cubes and beams with a scientific learning.

From Table 1, the research concluded that the ability MP student with RCP was learning more than the ability MP student with a scientific learning.The achievement was attributed to several factors one of the factors that influence it Problem Based Learning(PBL). According to [16] PBL as a student-centered learning that encourages students to construct their own knowledge, foster critical thinking skills, find and solve problems independently. PB requires students to be able to think critically about the information he got from the activity of thinking is emphasized in this study. Another factor influencing the approach is RME with the problems of daily life are used for learning where in line with the opinion of [8] who said that mathematics should be associated with reality and mathematics is a human activity. This is consistent with [17], [18]. [19] and [20] research that the RME category PBL approach is very good and positive effect of increasing the ML including MP.

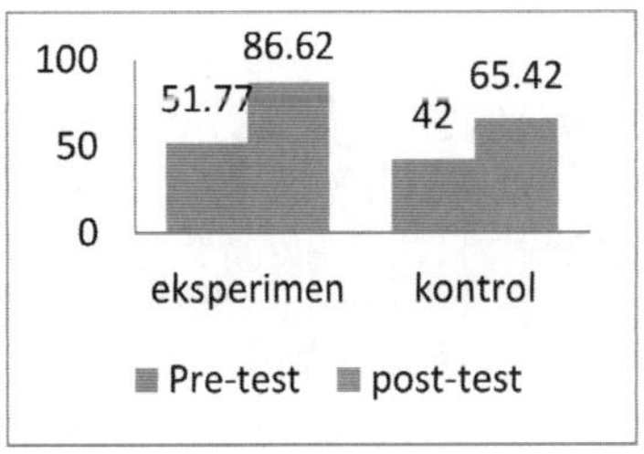

Figure1.Histogram Improving Ability MP of Students Reflective

Figure 1 describes bout the increase in MP. The increase in clas experiment(RCP learning) regarding cognitive style reflective reached 34.85 where the average pre-test 51.77 increased to 86.62 on an average post-test. Average index gain reflective students in the experimental class 0.72 are included in the high category. The increase in class control (Scientific Learning) regarding cognitive style reflective reached 23.42 where the average value of the pre-test 42 rose to 65.42 on an average post-test. The average index of reflective students gains the control class of 0.40 in the medium category. Improving the ability of MP represents the difference between the pre-test and post-test students' reflective MP further after the test is calculated by one side (right) using the t test result as in Table 2.

Table 2. The Result of The Difference In Average Increase Students' Reflective Ability MP

t_{Count}	α	t_{table}	Conclusion	It Means
3,205	5%	1,669	$t_{Count} > t_{table}$	The increase in the ability of class VIII student mathematization process on the material cube and beam-assisted learning with RME approach the problem better card than on improving the ability of class VIII student mathematization process on the material cubes and beams of learning the scientific approach regarding cognitive style reflective.

From Table 2, the research concluded that an increased the MP of the student with RCP learning better than the MP of the student with scientific learning in terms of cognitive style reflective.

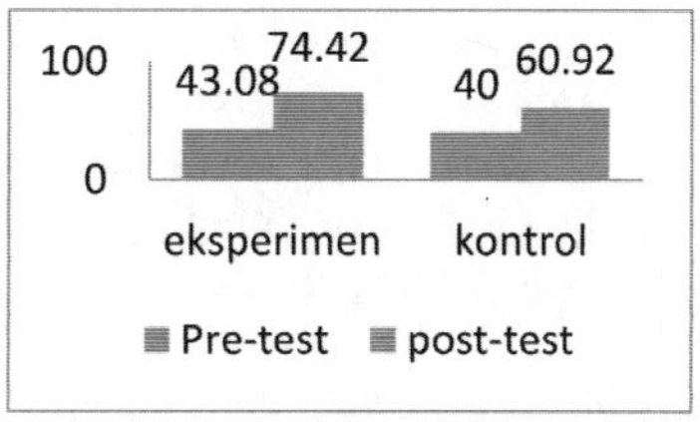

Figure 2.Histogram Improving Ability MP of Students Impulsive

Figure 2 describes the increase in MP. The increase in the class experiment regarding cognitive style impulsive reached 31.34 where the average pre-test 43.08 increased to 74.42 on an average post-test. Average index gains impulsive students in the experimental class 0.55 which is included in the medium category. The increase in class control regarding cognitive style Impulsiveness reached 20.92 where the average value of the pre-test 40 rose to 60.92 on an average post-test. Average index gain control class students impulsive at 0.35 in the medium category.

Table 3. The Result of The Difference In Average Increase Students' Impulsive Ability MP

t_{Count}	α	t_{table}	Conclusion	It Means
2,729	5%	1,71.	$t_{Count} > t_{table}$	The increase in the ability of class VIII student mathematization process on the material cube and beam-assisted learning with RME approach the problem better card than on improving the ability of class VIII student mathematization process on the material cubes and beams of learning the scientific approach regarding cognitive style impulsive.

From Table 3, the research concluded that an increased the MP of a student with RCP learning better than the MP of a student with scientific learning regarding impulsive cognitive style. Increasing the capability this occurs because the MP of learning processes that occur in class experiments under the theory of the Bruner points out that in the learning process of the child passes through stages of enaktif, iconic, and symbolic. In PBL learning with realistic approach aided problem cards, enaktif stage occurs when students manipulate objects through a matter of contextual and symbolic and iconic stage occurs when the students answer the questions on the student worksheet. Moreover, the Meaningfulness of this material can be obtained by directly connect him with daily life with the approach of the RME. Through the approach of the material can be accepted by RME students due to logical and relevant to a problem that is common in the environment students

3.1. Mathematizing Process Students Reflective

Based on an MP students reflective group of top, middle and bottom indicate that the indicator for identifying relevant math concepts with real world problems of subjects either group of top, middle, and bottom of the conduct of such activities. In the process of MH, indicator represents the problem with a variety of different ways, including organizing problem under the relevant mathematical concepts, and to formulate appropriate assumptions; either by making a drawing or restates the problem in their own words a good group of top, middle and bottom meet. On the indicators are looking for connections between "languages" problem with symbol and "formal mathematical language" so that the real problems can be understood mathematically, a group of top, middle or bottom meet. On the indicators look for regularity, relationships, and patterns that are associated with the problem; all groups either top, middle, or bottom of doing this activity.On the problem of translating the indicators into mathematical form, namely in the form of mathematical models, all groups either top, middle or bottom translate problem so that solutions are obtained from its perception is true.

In the process of MV, not all subjects fulfilling all the indicators. Using a variety of indicators of different mathematical representation, all the groups either top, middle and bottom using a representation. Indicators Using symbols, "language" and the process of formal mathematics; This relates to MH process, which, if in the process of linking the subject MH problems with it in the process of mathematization symbols vertical during the process of formal mathematics also use symbols in arithmetic operations. In the group of top, middle and bottom meet. In the group of top, middle and bottom meet Indicators make adjustments and development of mathematical models, combine and incorporate the various models on student reflective some difficulties in combining and incorporating a wide range of models tailored to the issues. In the group of top and middle of this indicator is met, but at the bottom of the group does not meet. Subject research not able to make any adjustments or development concept, but in combining and combining these three subjects experiencing errors. Indicators in making mathematical argumentation, the group of top and middle of this indicator is met, but at the bottom of the group does not meet. The group under its mathematical argumentation reveals the difficulty of the problems presented and its mathematical resolution so that the answer of the solution have not been correct. Indicators of generalities the indicator meets the top of the group while in the middle of the group and below do not meet as it does not make a generalization or not conclude appropriately.

3.1. Mathematizing Process Students Impulsive

Based on the MP impulsive student group, the top, middle and bottom indicate that the indicator for identifying relevant math concepts with real world problems of subjects either group of top, middle, and bottom of the conduct of such activities. On the process of MH indicators, represents a problem with a variety of different ways, including organizing the issue by the relevant mathematical concepts, and to formulate appropriate assumptions; either by making a drawing or restates the problem in their words, either group over the middle and bottom groups can meet. On the indicator for the relationship between the "language" problem with symbols and "language" of formal mathematics that the real problem can be understood mathematically, top and the middle group can meet these indicators, while the bottom group does not have an associate problem with the symbol. In indicator for regularities, relations, and patterns related to the problem, both groups of upper, middle, or bottom of this activity. In the indicator of translating the problem into mathematical form , namely a mathematical model, the groups either top, intermediate and lower translate problem to obtain the correct solution of perceptions.

In the process of MV, not all subjects fulfilling all the indicators. Indicators using a variety of different mathematical representation, a group of top, middle and bottom using a representation in finding a solution. Indicators Using symbols, "language" and the process of formal mathematics; This relates to MH process, if in the process of MH subject problem does not associate with it in the process of mathematization symbols vertical during the process of formal mathematical calculations directly on the operation. In the group of top and middle of the meet while in the bottom group did not meet because during the process of formal mathematical symbols and do not use directly on arithmetic operations. Indicators make adjustments and development of mathematical models, combine and incorporate the various models of most of the difficulties in combining and incorporating a wide range of models tailored to the problem. At the top of the group meet the indicator while the middle and bottom all the subjects did not meet either it is not appropriate when doing the adjustment, development, combine or merge them. Indicators in making mathematical argument in the top group meet these indicators, while the middle and bottom are not making the argument in solving the problem so that the obtained solution is not correct. Indicators on generalizing groups meet these indicators to make decisions with the right while in the middle, and bottom groups do not generalize or not concluded.

4. Conclusion

Based of research and discussion, the conclusions can be obtained. (1) the capability of the MP that one component of ML of student with RCP are better than the MP of student with scientific learning, (2) Students with cognitive style reflective the increase ability of the MP d with RCP learning are better than the MP of student with scientific learning, (3) Students with cognitive style impulsive the increased ability of the MP of student with RCP learning more good than the MP of student with scientific learning,(4) Subject reflective top and middle groups identified all indicators of MP meet both MH and MV; and reflective subject in a bottom group identified all MH process indicators are met and not all the indicators are met MV processes, (5) Subject impulsive group on MP identified all of the indicators are met both MH and MV process; Subject impulsive middle group identified all MH process indicators are met and not all the indicators are met MV process; Subject impulsive at bottom group identified that not all indicators of MH and MV processes are met.

References

[1] OECD.2003. Literacy Skills for the World of Tomorrow Further Results from PISA 2000 [Paris: OECD].
[2] OECD 2004 Learning for Tomorrow's World. First Result from PISA 2003 [Paris: OECD]
[3] OECD 2007 PISA 2006: Science Competencies for Tomorrow's World [Paris: OECD]
[4] OECD.2010. PISA 2009 Result: What Students Know and Can Do. Student Performance in Reading, Mathematics and Science Vol. 1 [Paris: OECD].
[5] Wijaya A 2012 Pendidikan Matematika Realistik Suatu Alternatif Pendekatan Pembelajaran Matematika (Yogyakarta: Graha Ilmu)
[6] Treffers A 1987 *Three Dimensions A Model Of Goal And Theory Description In Mathematics Education* (Dordrecht: Reidel The Wiscobas Project)
[7] Frudenthal H. 2002 *Revisiting Mathematics Education* China Lectures (Dordrecht: Kluwer)
[8] Murdani et all. 2013 Pengembangan Perangkat Pembelajaran Matematika dengan Pendekatan Realistik untuk Meningkatkan Penalaran Geometri Spasial Siswa di SMP Negeri Arun Lhokseumawe. *Jurnal Peluang* (Banda Aceh) Vol 1 No. 2
[9] Ningsih PR.2012. Profil Berpikir Kritis Siswa SMP Dalam Menyelesaikan Masalah Matematika Berdasarkan Gaya Kognitif.*Jurnal Gamatika* Vol 2 No2 year 212.
[10] Warli. 2008. Pentingnya Memahami Gaya Kognitif Impulsif-Reflektif bagi Guru. *Majalah Ilmiah Sain dan Edukasi*, Vol. 6, No. 2 Juli 2008. Lembaga Penelitian IKIP PGRI Jember
[11] Rahman A.2008. Analisis Hasil Belajar Matematika Berdasarkan Perbedaan Gaya Kognitif Secara Psikologis dan Konseptual Tempo pada Siswa Kelas X SMA Negeri 3 Makasar. *Jurnal Pendidikan Dan Kebudayaan* No. 072 Tahun ke-14, Mei p: 452-473.
[12] Warli 2010 Profil Krativitas Siswa yang Bergaya Kognitif Reflektif dan Siswa yang Bergaya Kognitif Impulsif dalam Memecahkan Masalah Matematika Disertation Not Published (PPs UNESA)
[13] Sugiyono 2013a *Metode Penelitian Kombinasi (Mixed Methods)* (Bandung: Alfabeta)
[14] Sugiyono 2010 *Metode Penelitian Pendidikan Pendekatan Kuantitatif, Kualitatif, dan R&D* (Bandung: Alfabeta)
[15] Huang KT & T Wang 2013 *Applying Problem-based Learning (PBL) in University English Translation Classes*
[16] Wardono and Scolastika Mariani. 2014. The Realistic Learning Model With Character Education and PISA Assessment to Improve Mathematics Literacy. *International Journal of Education and Research.* Vol 2 No 7 July 2014.
[17] Wardono et al. 2015. The Realistic Scientific Humanist Learning With Character Education to Improve Mathematics Literacy Based On PISA.*International Journal of Education and Research.* Vol 3 No 1 January 2015
[18] Wardono et al. 2016. Mathematics Literacy on Problem Based Learning with Indonesian Realistic Mathematics Education Approach Assisted E-Learning Edmodo. *Journal of Physics Conference Series.*
[19] Budiono and Wardono. 2014. PBM Berorientasi PISA Berpendekatan PMRI Bermedia LKPD Meningkatkan Literasi Matematika Siswa SMP. *Unnes Journal of Mathematics Education* Vol 3 No. 3.

IOP Conf. Series: Journal of Physics: Conf. Series **824** (2017) 012050 doi:10.1088/1742-6596/824/1/012050

Probabilistic Thinking Ability of Students Viewed from Their Field Independent and Field Dependent Cognitive Style

A Taram[1]

Mathematics Education Study Program of Ahmad Dahlan University, Yogyakarta

[1]Corresponding author: taromahmad@yahoo.com

Abstract. The aims of this research are to study: (1) probabilistic thinking ability of mathematics education students, (2) classification of the students' cognitive style, (3) levelling of the students' probabilistic thinking ability viewed from their cognitive styles. This research used thequalitative descriptive method and involved 74 subjects. The measured subjects were Group 1 with "fixed FD" classification consisted of 7 students, Group 2 with "mobile FD and mobile FI" classification consisted of 9 students, and Group 3 with "fixed FI" classification consisted of 5 students. The classification of cognitive styles into three groups revealed that there was suitability between cognitive style and probabilistic thinking ability from low to high level. These results could be analysed from theclassification of cognitive style and anaverage of their value of probabilistic thinking ability. The average of probabilistic thinking ability of Group 1 was 42.58; the average of probabilistic thinking ability of Group 2 was 54.44,and the average of probabilistic thinking ability of Group 3 was 68.6. Group 1 and 3 had small standard deviation for the value of probabilistic thinking ability, respectively are 11.36 and 12.30. Thus the data was relatively homogeneous. Meanwhile, Group 2 had a huge standard deviation for the value of probabilistic thinking ability, namely 19.36 which means that the data was relatively heterogeneous. Most of the probabilistic thinking ability level for Group 1 and 2 was Level 2, which is Transitional level, while the most of theprobabilistic thinking ability level for Group 3 was Level 4, which is Numeric level.

1. Introduction

Students' ability is influenced by several factors,namelythe external factors and the internal factors. Several studies have shown that the influence of the internal factor is predominant. One of the internal factors is the thinking ability, including the spatial thinking ability, mathematical rigourthinking ability [2] literacy thinking ability, probabilisticthinking ability, and statistical thinking skills [1, 3].

Studies related to probabilistic thinking is still progressing. Jones (1997, 1999, 2002), and Polaki (2002)suggest the levelling of probabilistic thinking and recommend further research to investigate the probabilistic thinking of students bytheir different backgrounds,languages, and cultures. Sharma (2006, 2012) indicates that many students use strategies based on the experience of culture (beliefs, everyday life, and school experiences) and intuitivestrategies [4-7]. Sujadi (2008) developed levels offered by Jones and add one level of probability thinking for junior high school students who have not been taught by probability material [14]. Maftuh (2014) suggests that probabilistic reasoning students in junior high school level in problem-solving associated with theprobability of occurrence is biased to respond to various situations within a context that include unsure- an element of uncertainty, while the reasoning in any problem-solving steps is useful decisionmaking [9].

The results of the studies are very interesting and need to be followed up at the college levelsince the results can be referred when the decision maker wants to determine the learning approach used, the lecture material, as well as the treatment in coaching and giving attention to the students.

The National Qualifications Framework of Indonesia (NQFI/KKNI) mentioned that the bachelor degree should possess Level 6. They have to master the theoretical area of knowledge in general and a special concept for the deeper theoretical part in science and be able to formulate settlement procedural problems. It leads to the idea that the students of bachelor degree need to possess the probabilistic thinking ability.

On the other hand, the trend of theindividual to receive, to process, and to organise information and then to present the information again, which is known as cognitive style, need the attention of the lecturer since the style will be associated with the cognitive process of students. Witkin (1976: 254) classifies cognitive style in several types, one of which is thecognitive style offieldindependent (FI) and field-dependent (FD) [15]. Classification of these cognitive styles regarding theindividual ability to distinguish relevant aspects in particular situation.This research is related to probabilistic thinking skills of students associated with cognitive styles of the students since it is important to improve the learning approach in mathematics education study program after learning our students' probabilistic thinking and their related cognitive style.

2. Study Theory

2.1. Probabilistic Thinking and Level

Probabilistic thinking processes of students is a higher order thinking processes, which is positionedon the stage of aformal operation process according to Piaget (Soeparno, 2001:25). Furthermore, Piaget explained that to understand the process of probability, a student needs to know two principal operations, namely system combination and proportion calculations. The materialsof probability theory which are fundamental to measure the probabilisticthinking includesample space, theprobability of an event, conditional probability, and case of independence (Jones,1999).The probabilistic thinking level proposed by Jones (1997, 1999) consists of four levels, namely: (1) Level 1 Subjective, (2) Level 2 Transitional, (3) Level 3 Informal Quantitative, and (4) Level 4 Numerical. The details are shownin Table 1 below.

Table 1 Probabilistic Thinking Level

Level	Characteristics	Indicators
1. Subjective	Students continuously think bonded on thesubjectivebackground.	• Signed an incomplete set of experimental results one level. • Predict events that are most likely or least likely, based on a subjective opinion • Recognising the unlikely occurrence and certainly • Comparing the odds of an event in two different sample chamber, usually based on a subjective opinion • Unable to distinguish situations probabilistic "fair" from "unfair."
2. Transition	A transition period between thinking objectively and thinkingquantitativelycharacterised by students' thinking naive and often changed in quantifying probability	• Apply a complete set of experimental results one level • Sometimes a complete register with experimental results using two-level strategies is limited and not systematic. • Predict events that are most likely or least likely, based on the opinions quantitatively but back on a subjective opinion • Making comparisons based on the chances of quantitative statement (probably not quantitative, and may have limitations in which the events of the adjacent engaged) • Start to distinguish situations probabilistic "fair" from "unfair."

Level	Characteristics	Indicators
3. Quantitative informal	Thinking at this level is indicated through the use of generative strategies in registering the second stage of experimental results, and can align and quantifying their thinking about the sample space and opportunities	• Apply the experimental results consistently using a two-level portion of the generative strategy • Predict events that are most likely or least likely, based on the opinions of quantitative includes situations which contain results that are not adjacent (noncontiguous outcomes). • Use the numbers on an informal basis for comparing the probability. • Distinguish the events certain, impossible and possible, and justify the choice quantitatively. • Making comparisons based on the chances of consistent quantitative opinion. • Give reasons to quantitative reasoning is valid but limited to events that are not adjacent (noncontiguous events) • Distinguish generator opportunities "fair" from "unfair", based on the numeric reasoning is valid.
4. Numeric	Students can make the proper relationship of the sample chamber and opportunities and can use appropriately sized numerically to describe the probability of occurrence	• Implementing and using generative strategy that allows registering the complete results of experiment two or three levels • Predict the most likely or least likely to experiment one level or two levels. • State the opportunity an event numerically (either real or opportunities that are uniquely shaped) • Determine the size of the opportunities in numeric and compare the incidence • Combining the results of adjacent (contiguous outcomes) and the results are not adjacent (noncontiguous outcomes) in determining opportunities • Determine equality of opportunity in numeric for the events that are likely the same.

Sharma (2012) suggeststhat students use strategies based on the experience of culture (beliefs, everyday and school experiences) and intuitive strategy. While the results of the study confirm the findings of other researchers, the findings beyond those discussed in the literature. Use of beliefs, everyday and school experience was far more common than is discussed in the literature.

2.2. Cognitive Style

2.2.1. Understanding Cognitive Style

Cognitive styles according to Koheznikov (2007) refer to the individual characteristics of the environment toorganise conceptually. Furthermore, Witkin (1976) defines cognitive style as an approach to receive,to process, and to organise information and present the information returnedThere are individuals who receive such information presented, while the other individual to reorganisethe information in his way [15].Witkin (1976) classify cognitive style in several types, one of which is cognitive style field-independent and field-dependent as shown in Table 2 [15].

Table 2 Different characteristics of the individualfield dependent and independent

Field Dependent	Field Independent
1. Oriented social.	1. Oriented impersonal
2. Stresses external motivation	2. Stresses internal motivation
3. More affected by external reinforcement	3. More affected by internal reinforcement
4. Object looked globally	4. Looking at objects composed of discrete parts and separate from the environment
5. Thinking globally	5. Thinking analytically
6. Tend to choose a profession that promotes social skills and humanities	6. Tend to choose a profession that promotes the ability to analyse.

3. Research Methodology

3.1. Type and Research Subjects
This research method is descriptive qualitative research which involved mathematics education students who take probability theory courses in the second semester of 2015/2016academic year.

3.2. Research Instruments
The instrument used was the test of probabilistic containing four problems.Question number 1 is to measure the ability of students to prove theorems of probability; Question number 2 is to measure the ability of students to determine sample space, probability of an event, random variables and their distribution; Question number 3 is to measure the ability of students to the conditional probability of an event; and the Question number 4 is to measure the ability of students to the problems associated with the probability of Bayes theorem.Identification of cognitive style subjects in this study carried out based on the test results cognitive style with instrument GEFT which consists of 25 items divided into three parts, of which 7 item in Part I of the exercise and 18 items in sections II and III are the core of GEFT. Each correct answer which means the subject can precisely shape thicken simple images which are hidden in the complex image, given a score of 1. In this study, subjects who score> 9 classed FI and subjects who score ≤9 classified FD.

4. Results and Discussion

4.1. Results
a. Classification of Cognitive Style
Group 1 is those who obtain the correct number of items in the test GEFT between "0" up to "3" with the classification "fixed FD" there are seven students.Group 2 are those who obtain the correct number of items in the test GEFT between "8" up to "11" with the classification "mobile mobile FD and FI" there are nine students. Group 3, namely those who obtain the correct number of items in the test GEFT between "16" up to "18" with the classification "fixed FI" there are five students. While the value of the measurement results of probabilistic thinking skills for each of the three groups of students is as follows. The average value of the Group 1is 42.86 with astandard deviation of 11.31. The average value of probabilistic tests of Group 2 is 54.44; with the standard deviation of 19.36.The average value of theprobabilistic test of Group 3 is 68.6; with astandard deviation of 12.30.

4.2. Discussion
From the results of research related to the classification of cognitive style, three groups of classification showed that there is a match between three groups with the probabili stic thinking ability. It can be seen from the classification of cognitive style,and the acquisition value of the probabilistic thinking ability.It is the one with the classification of fixed FD, the average value of probabilistic thinking ability is 42.58; group 2 with mobile classification FD and FI mobile has anaverage value of probabilistic thinking ability is 54.44, and group 3 with the classification of fixed FI has anaverage value of probabilistic thinking ability is 68.6. (1) Group 1, in this group there is an interesting fact that a student become an outlier among others, with the value of the probabilistic thinking ability is 67, this indicates that he does not belong to fixed FD group, but the probabilistic thinking ability is high enough, we concerned that in question 1 is at level 4, question 2 is at level 4, question 3 is at level 3, and question 4 is at level 2.Likewise, for group 1 is not located at level 1, it

draws attention to exploring further. The largest percentage of the level of probabilistic thinking skills are at level 2 by 71.43%, and the lowest percentage in the level 1 is 10.71%. These results indicate that the majority of this group at the level of probabilitythinking: transitional. The standard deviation of 11.31 These groups are the one most small compared with other groups, have demonstrated the ability to think of groups one is homogeneous. (2) Group 2 with a mobile classification FD and FI mobile, the number of subjects at most compared with other groups are 9 subject, which draws attention to the value range of probabilistic thinking ability is large enough that the smallest value and the greatest value 30 85, and the standard deviation 19.36; This demonstrates the ability to think of this group is very heterogeneous, this is in accordance with the mobile classification to their cognitive style. Probabilistic ability level for this group represented all for 4 existing level. The largest percentage of the level of probabilistic thinking skills are at level 2 is 44.44%, and the lowest percentage was at level 1 by 5.56%., These results indicate that the majority of this group at the level of probability thinking: transitional. (3) Group 3 with fixed FI classification, in this group the number of subjects at least compared to the other two groups that are 5. The largest percentage of the subject's level probabilistic thinking skills are at level 4 by 40%, while the lowest percentage at level 1 by 0%. It shows that the majority of these groups are at a numerical level, which is the highest level of thelevel that existed at the level of probabilistic thinking skills. The standard deviation is 12.30 these three groups are relatively small; this also shows the probabilistic thinking ability is arelatively homogeneous group. The findings are suitable with the finding of Fitriyani (2013) and Witkin (1976) that explained that the cognitive style influences how the students solve mathematics problems. The mathematics problem in this research is the probabilistic problem which is influenced by the FI and FD of the students' cognitive style [2,15].

5. Conclusion

In this section will be presented the conclusions of results and discussion in this study. (1) Classification of cognitive style into 3 groups showed match between level of compliance with probabilistic thinking ability of low level to high level, it can be seen from the classification of cognitive style and the average acquisition value of their probabilistic capabilities, namely the one with the classification of fixed FD average value of ability probabilistic is 42.58; group 2 with mobile classification FD and FI mobile average value of ability probabilistic is 54.44; and group 3 with the classification of fixed FI average value was 68.6 probabilistic capabilities. (2) Group 1 and Group 3 had a small standard deviation for the value of probabilistic thinking skills, respectively by 11, 36 and 12,30, suggesting that the ability to think probabilistic both groups are relatively homogeneous, while group 2 had a fairly large standard deviation for the value of probabilistic thinking skills in the amount of 19.36, this suggests that the probabilistic thinking ability is relatively heterogeneous group. (3) Level probabilistic thinking skills to groups 1 and 2 majority on level 2, with percentages respectively 71.89% and 44.44%, so both groups tended to be on level 2: transitional; while the majority of the group 3 level 4 with percentage by 40% so that these groups tend to be at level 4: Numerical.

References

[1] Bambang, A.P.M. 2014. Modifikasi MEAs dengan Menggunakan Didactical Design Research untuk Meningkatkan Kemampuan Berpikir Statistis Mahasiswa. *Cakrawala Pendidikan*, 33(2): 267-276.

[2] Fitriyani, H. 2013. Profil Berpikir Matematis Rigor Siswa SMP dalam Memecahkan Masalah Matematika Ditinjau dari Perbedaan Kemampuan Matematika.*AdMathEdu*, 3(1): 37-56.

[3] Helgason, C.M. & Jobe, T.H. 2010. Principled Versus Statistical Thinking in Diagnosis and Treatment of Stroke. *Current Treatment Options in Cardiovascular Medicine*, 12: 292–296. DOI 10.1007/s11936-010-0073-x.

[4] Jones, G.A., Langrall, C.W., Thornton, C.A. et al. 1997. A Framework For Assesing and Nuturing Young Children's Thinking in Probability.*Educational Studies in Mathematics*, 32(2): 101. doi:10.1023/A:1002981520728.

[5] Jones,G.A.1997. An Understanding of Student Probability Reasoning. Reston, Virginia: The NTCM.

[6] Jones, G.A.1999. Student's Probabilistic Thinking in Instruction. *Journal for Research in Mathematics Educational*, 30: 487-519.

[7] Jones, G.A. 2002. Elementary Student' Access to Powerfull Mathematical Ideas. *Handbook of Internasional Research in Mathematics Education*, London: The NTCM, 6: 113-141.
[8] Kozhevnikov, M. 2007. Cognitive Styles in the Context of Modern Psychology: Toward an Integrated Framework of Cognitive Style.*Psychological Bulletin*, 133(3): 464–481.
[9] Maftuh, S. 2014. Profil Penalaran Probabilistik Siswa SMP Laki-laki dalam Pemecahan Masalah Probabilitas.Paper presented on Seminar of Mathematics Education, Surabaya.
[10] Polaki, M.V. 2002. *Assessing and Tracing The Development of Basotho Elementary Students' Growth in Probabilistic Thinking*, Lesotho: National University of Lesotho.
[11] Sharma, S. 2006. Personal Experiences and Beliefs in Probabilistic Reasoning Implications for Research. *International Electronic Journal of Mathematics Education (IEJME)*, 1(1).
[12] Sharma, S. 2012. Cultural Influences in Probabilistic Thinking. *Journal of Mathematics Research*, 4(5).
[13] Soeparno,P. 2001. Teori Perkembangan Kognitif Jean Piaget. Yogyakarta: Penerbit Kanisius.
[14] Sujadi, I. 2008. Rekontruksi Tingkat-Tingkat Berpikir Probabilistik Siswa Sekolah Menengah Pertama.Paper presented in the Seminar on Mathematics and Mathemtics Education, FKIP UNS, Surakarta.
[15] Witkin, H. 1976. *Cognitive style in academic performance and in teacher-student relations*. In S. Messick and associates (Eds.), Individuality in learning. San Francisco:Jossey-Bass.

Elementary School Teacher Training Based on Needs and Interests of Teachers and The Effectiveness of The Improvement of Students Competence

A Suyitno[1], E Sugiharti and E Pujiastuti

Maths and Natural Sciences Faculty of Universitas Negeri Semarang
Gd. D-12 Kampus Unnes Sekaran Gunungpati, Semarang, Central Java - Indonesia

[1]Corresponding author: amin.mat@mail.unnes.ac.id

Abstract. Teachers need always to improve their competencebecause of the growth of science very rapidly as well as elementary school teachers. However, the provision of training will not produce an increase in the competence effectively if itis implemented without considering the needs and interests of teachers. The novelty factor which is highlighted through this research results, want to answer a problem, namely: how to conduct an effective training based on the needs and interests of teachers so that the effects of training can improve the competence of teachers? After going throughresearch for twoyears with aqualitative approach which was preceded by a visit to the school and a series of interviews, treatment of training model, FGD, and triangulation then have been produced a way to implement of training based on the needs and interests of teachers. The training model includes face to face training which structured and scheduled according to the needs of elementary school teachers, followed by workshops, simulations, then the coach holds guidance in the classroom, and ends with reflections.

Keywords: Teacher training, workshop, guidance.

1. Introduction

1.1. Background of Research

Teachers are agents of education. Science is also rapidly evolving. Therefore, teachers are always required to learn so that teachers can follow the pace of development of the science. If the teacher can keep pace with the development of science, the teacher will be able to bring their students to achieve the expected competencies, as well as elementary school teachers. Additionally, Regulation of the Minister of Education and Culture, No. 53/2015 of Republic of Indonesia, learning in elementary school should refer to the Integrated Thematic learning. Through integrated thematic, learners are invited and trained to associate one phenomenon with each other. Such occurrence in mathematics is known as mathematical literacy.

Teachers need to know about the mathematical literacy and integrated thematic. The teachers can gain this knowledge through training. However, the provision of training will not result in increased competence for teachers effectively if the training implemented without considering the needs and interests of teachers.

Vithal and Bishop (2006), and Doyran (2011) wrote that focus on teacher training was to raise issues that emerged at this time, for example, Mathematical Literacy. Vithal and Bishop also suggested that in solving problems related to mathematical literacy, teachers need to use an approach based on mathematical modelling. They said that to prepare the teachers who do not have adequate math skills was a challenge so that finally the teachers have the knowledge to become the teachers who master of mathematics literacy[1,2].

On the other hand, for teachers in Indonesia, teachers must teach the contents of a curriculum, known as Curriculum of 2013 Revised Edition. In the Curriculum of 2013 Revised Edition, all subjects including mathematics should be taught through a scientific approach. However, not a few of the primary school teachers who do not understand the significance of scientific approach. In fact, elementary school teachers must know how to teach mathematics through this scientific approach. As a result, there is a gap between the knowledge and skills required elementary teachers with the reality of the school.

Therefore, it is natural that elementary school teachers in Indonesia at this time is in need of knowledge, skills, and mastery of materials associated with mathematical literacy and applying of thescientific approach. Elementary school teachers need to be given an effective training based on searches of teacher competence to its ability in mathematics taught through scientific approach is characterised by mathematical literacy. An effective training and has an element of novelty needs to be discovered through good research results.

1.2. Research Problems

The novelty factors wanted to be highlighted by this research results, want to answer a problem, namely: How to implement an effective training based on the needs and interests of teachers so that the effects of training can improve the competence of teachers?

1.3. Research purposes

The purpose of this research conducted was to find a model of how to conduct effective training based on the needs and interests of teachers so that after effects of training can improve the competence of teachers. The requirement is to implement the scientific approach characterised by mathematical literacy.

2. Literature Review

2.1. The Importance of Effective Training for Elementary School Teachers

In any country, the government's efforts to improve the quality of learning for the teacher always executed. The final goal is that the students who are taught can reach the target competencies expected.

Harwell (2003), Kramarz (2008) and Burns (2011) who frequently wrote articles about the training of teachers said that teacher training was very necessary so that teachers had the ability to develop the profession as a teacher and had a high quality in the learning process [2,3,4].

2.2. Ability of Mathematical Literacy as a Need for Teachers and Students

The definition of mathematical literacy in PISA (OECD, 2014) was as follows. Mathematical literacy is an individual's capacity to formulate, work, and interpret mathematics in a variety of contexts. The capacity of this individual includes using mathematical reasoning and mathematical concepts, procedures, facts and tools to describe, explain, and predict of the phenomena.

In many countries, the term of mathematical literacy is different from each other. Among them, according to Brown (2003) in England mathematical literacy was called numeracy [5]. While in America, according to Steen (2001), mathematical literacy was called qualitative literacy[6]. In South Africa, according to the DOE (2003) and Vithal and Bishop (2006), mathematical literacy was defined by the mathematical activities that were able to provide students with the awareness and understanding of the role that mathematics was very useful in the modern world [6,7].

So, mathematical literacy is the subject of activity of math applications that related to life. This allows students to develop the ability and confidence to think numerically and spatially. Furthermore, students can interpret and analyse critically in everyday situations, and also solve the problem. Sahlberg (2010), Stacey (2011), UNESCO (2014), Tai and Lin (2015) also wrote about the need for mathematical literacy was mastered by the teachers and students[8-11].

Here is presented an example of a problem which is a form of mathematical literacy. The problemwas taken from the writings by Setiawan, Dafik, and Lestari (2014):

For a rock concert, a rectangular field measuring 100 meters long and 50 meters wide was prepared for visitors. Tickets were sold out even a lot of fans standing. What is the approximate number of visitors to the concert?

a. 2000 b. 5000 c. 20,000 d. 50,000 e. 100,000

Answer:

The problem above, require higher-level thinking for the level of the elementary students. Students need to read and understand the complex situation carefully. Students must be able to measure the area of the field and understand the situation that was because the tickets were sold out even a lot of fans standing [12].

Subsequently, in the next step, the student was required to evaluate options that may suit situation described in theproblem. The first step, looking for the area of thefield. Field area = (100 × 50) m2 = 5000 m2. After this stage, the students need to think a high level to continue the next process. The right step, the students need to evaluate multiple choices that possible. With the area of thefield was 5000 m2, students must be able to imagine, how many people might be at a site of each 1 m2?

Students must pay attention to the problem where many fans were standing. For alternative answers A, i.e., 2000 people unlikely to happen, because the problem was mentioned that the field was full and many fans standing. If it is only 2000 people, then each person occupies 2.5 m2, i.e., 5000: 2000 = 2.5.

This answer can not be happened because the problem was explained that the visitors are so many where the fans had to stand. For an alternative answer B, the 5000 is also unlikely, since 5000 means that each 1 m2 occupied by one person, namely 5000: 5000 = 1. For an alternative toanswer C, it was written that there were 20,000 people, so the meaning is that each 1 m2 were occupied by four people, which was obtained from 20,000: 5,000 = 4, and this answer was reasonable. For answers D and E, the student should see that the selection of D shows every ten people will occupy 1 m2, this was clearly not possible unless the audience was forced overlap, but the information was not the case. While alternative answer E is no longer possible because it means that there were 20 people in 1 m2, for 100,000: 5000 = 20. Thus, the correct answer was C.

So that students can work on the problems mentioned above with good, elementary school teachers and students need to practice with math problems containing elements of mathematical literacy. Thus, the ability of Mathematical Literacy is a necessity for elementary school teachers and students.

2.3. Scientific Approach in the Curriculum in Indonesia

The curriculum that conducted in Indonesia today is Curriculum of 2013 Revised Edition. Implementation was done with a scientific approach. It is based on the General Guidelines for Learning that stipulated in Regulation of the Minister of Education and Culture, No. 103/2015 of Republic of Indonesia. Scientific approach covering a range of activities to observe, ask, gathering of information, associate, and communicate. This approach must be controlled and implemented by teachers when teaching.

Description of the steps of the scientific approach is as follows. (1) Observing: activities such as reading, listening, viewing (without or with a tool) to identify the things that the students want to know. (2) Asking: students ask questions about things that the students were not understood from what that observed or making inquiries to obtain additional information about what was observed. (3) Gathering of information: that can do experiments, read other sources and textbooks, observing the objects/events/activities, or interviews with sources. (4) Associating: students process the information that had been collected was either limited results of collecting activities/experiments would also result from the activity observed and information gathering activities. This associating link the phenomena/information that was relevant to find a pattern and conclude. (5) Communicating: students communicate their product of observations or conclusions based on the analysis of oral, written, or other media. Students can present their report in the form of charts, diagrams, or graphs.

Thus, at this point, the knowledge and practice of implementing the scientific approach are the need for teachers, especially elementary school teachers in Indonesia.

3. Research Methods

3.1. Approach and Research Subjects

To answer the problem and achieve the goal of this research, the author had conducted research with a qualitative approach. Research subjects are selected from the elementary school teachers in the city of Semarang, Central Java, Indonesia.

According to Miles and Huberman (1994 and 2014), the research methods include (1) data reduction, (2) the display of data, (3) the interpretation of data, and (4) conclusion/verification [13].

The data obtained were analysed through a series of Focus Group Discussion (FGD), reinforced by the findings of the interviews and triangulation. After obtaining a draft of a training model based on searches of competence of the teachers related to mathematical literacy that taught through a scientific approach, it was then conducted the trial test of the draft. The result of the trial was analysed in FGD forum to get an effective training model, based on searches competence of the teachers related to mathematical literacy that taught through a scientific approach.

3.2. Trial against Prototypes

The trial, the prototype of training model, had been held at the beginning of the second year of the study. The training was based on searches of competence of the teachers related to mathematical literacy that taught through a scientific approach. The trial of training model was conducted in Semarang for elementary teachers. Figures 1 and 2 show the atmosphere of the trial implementation of the training model.

Figure 1: The executive committee of training, taking a picture together.

Figure 2: Atmosphere during the ongoing trialand a team of researchersof the implementation of its training models.

To deepen toanalyse the trial results, selected six of elementary school teachers as subjects of this research. As the follow-up, the trial results are re-analyzed in FGD forum to get an effective training model, based on searches competence of the teachers related to mathematical literacy that taught through a scientific approach.

Elements of FGD members are drawn from 3 of the elementary school teachers who become the subject of this research, two lecturers from the Department of Mathematics Education of Universitas Negeri Semarang (UNNES), and all members of the research team.

To strengthen the results of this study, then the result of research has also been presented in a seminar at the international level, namely in ICMSE at the Universitas Negeri Semarang - Indonesia, on 3 September 2016. Through seminars International level, researchers were expected to have an additional input from the audience.

4. The Results Achieved as A Form of Novelty

To test This research has been conducted for two years with a qualitative approach. The first year was preceded by a visit at the elementary school and a series of interviews, FGD, and triangulation. Furthermore, in the first year has produced a draft on how to implement the training based on the needs and interests of teachers.

In the second year, the draft of training model had been tested and passed with FGD back. The result, had produced a training model based on the needs and interests of teachers, which includes: (1) face to face training that structured and scheduled in accordance with the needs of elementary school teachers the training of mathematics characterized by mathematical literacy and a scientific approach,

(2) forwarded by workshop activity, (3) It is then forwarded to the simulation after the workshop is completed, (4) then trainer conducts mentoring in the classroom, and (5) training activities and then ends with a reflection.

Referred face to face training is structured and scheduled according to the needs of elementary school teachers are as follows. At first, conducted a survey to look at the needs of elementary school teachers. Also, the needs of elementary school teachers can also be obtained from various sources, for example through to the Head of Education Office, Supervisory of Elementary School, Principal of School, teachers, or parents.

The face to face training that structured and scheduled means that the implementation of such training should be designed carefully; therefore, it is not to harm the students in the learning, scheduled, and the teachers know what material will be received. The training is followed by a workshop, which means that teachers must be trained to be able to produce the products such as Lesson Plan or other product related to the improving of learning quality. For example, capable of generating problems that contain mathematical literacy, make a teaching aid as learning support, or a student activity sheet.

The product of workshop will look in vain if it is not implemented in the simulation of study. Therefore, it is then immediately conducted a simulation to practice one workshop products after the workshop was completed. Once teachers have adequate provisions for the implementation of the simulation, the trainer can provide guidance in the class, so the teacher is getting ready to implement their product of the training.

After the guidance is completed, then trainers and trainees can meet again for conducting reflection. With this reflection, teachers can discuss all of the weaknesses, strengths, and the benefits of their training.

5. Conclusions and Recommendations

5.1. Research Activities Performed

The conclusion is as follows. This research has produced a way to implement training based on the needs and interests of teachers. The training model includes: (1) Face to face training was structured and scheduled according to the needs and interests of elementary school teachers, (2) It is followed by a workshop, (3) after the workshop is completed then forwarded to the simulation, (4) after that the trainer accompanying a teacher in the classroom, and (5) then the training ends with reflections.

The recommendation, the training will not result in increased competence effectively if the training is done without considering the needs and interests of teachers. Therefore, before the training should be preceded by a study conducted to reveal the needs and interests of teachers.

References

[1] Vithal, Renuka and Bishop, Alan J. 2006 *Journal of Pythagoras* 64, December, 2006, pp. 2-5.
[2] Doyran, Feyza. 2012 ISBN: 978-960-9549-74-5.
[3] Harwell, Sandra H 2003 Published and distributed by CORD - P.O. Box 21689.
[4] Kramarz, Francis. 2008 *IZA Discussion Paper No. 3871*. December 2008.
[5] Brown, M. 2003 Proceedings of the 11th Annual Conference of the Southern African Association for Research in Mathematics, Science and Technology Education Research,11-15 Jan, Waterford Kamhlaba, Swaziland
[6] Steen, L.A. 2001 (Ed.) *Mathematics and Democracy: Case for Quantitative Literacy*. National Council on Education and the Disciplines, The Woodrow Wilson National Fellowship Foundation.
[7] Department of Education (DOE). 2003 National Curriculum Statement Grades 10 12 (General) Mathematical Literacy.Pretoria: Department of Education.
[8] Sahlberg, Pasi. 2010 *Research Brief.* September 2010. https://edpolicy.stanford.edu/sites/default/files/publications/secret-finland%E2%80%99s-success-educating-teachers.pdf
[9] Stacey, K. 2011 *IndoMS*. J.M.E. 2 (2), hlm. 95-126.

[10] UNESCO. 2014 *Literacy for All*. http://en.unesco.org/themes/literacy-all. Accessed on June 12, 2016.

[11] Tai, Wen Chun and Lin, Su Wei. 2015 *Academic Journals - Educational Research and Reviews*. Vol. 10(11), pp. 1480-1486, 10 June, 2015 DOI: 10.5897/ERR2015.2266. ISSN 1990-3839.

[12] Setiawan H, Dafik, dan Lestari NH. 2014. *Prosiding Seminar Nasional Matematika, Universitas Jember*, 19 November 2014.

[13] Miles, Matthew B & Huberman, A Michael. (2014). *Qualitative Data Analysis – A Methods Sourcebook*. Third Edition. London: SAGE Publication

Analysis of Mathematics Critical Thinking Students in Junior High School Based on Cognitive Style

A Agoestanto[1],YLSukestiyarno and Rochmad

Math Department of UniversitasNegeri Semarang
Gd. D-7KampusUnnesSekaranGunungpati, Semarang, Central Java - Indonesia

[1]Corresponding author: arief.mat@mail.unnes.ac.id

Abstract. The purpose of this research was to determine the critical thinking ability of mathematics from junior high school students based on FI and FD cognitive style. Data of this research were taken from students grade VIII at SMPN 2 Ambarawa. The research method used a descriptive qualitative approach. Data was taken with a testing method; the critical thinking was measuredwith WGCTA which is modified with mathematical problems, thecognitive style was measured with GEFT. The student's test result was analysed, then four students were selected, the two of them are FI cognitive style, and the others are FD cognitive style, forqualitative analysis. The result showed that the ability of mathematics critical thinking students with FI cognitive style is better than FD cognitive style on the ability of inference, assumption, deduction,and interpretation. While on the aspect of argument evaluation, mathematics critical thinking ability of students with FD cognitive style is a littlebetter than students with FI cognitive style.

1. Introduction

In mathematics learning, students not only learn mathematics material but also learn to deal with problems and challenges in social life.One of theprovisions which should be owned by the students when facing problems and challenges in social life is critical thinking. Critical thinking has become one of the important things in daily life to survive [1]. This skillsshould be owned by each student to be able to face the problems.

The main purpose of education today is to provide students with a key competencies knowledge, which will create the basis for further learning and actby the demands of social and professional [2]. Critical thinking is one of the things used in everyday life to solve some problems. This case proves the importance of equipping students to learn mathematics not only learn the theorybut also equip them with critical thinking skills.

When teaching mathematics in schools, we need to integrate and emphasise critical thinking in the curriculum.So the students could learn the skills and applied them to improve the performance and capabilities of reasoning [1]. The critical thinking should be developed.So the students can overcome their future and useful for companies where students work [3]. The students are not enoughto only hasknowledge or information.To be effective at work or in life, students should be able to think critically [4]. A man who can think critically bring up the vital questions and problems and formulate it clearly and precisely [5]. It means that the ability of think critically needs to be developed in learning.

There are several measuring instruments used to assess critical thinking such that Watson-Glaser Critical Thinking Appraisal (WGCTA) [6]. The WGCTA is a way to measure the ability of think critically. This instrument is a written test that mostly used in the education fields and professional occupations [7]. The WGCTA consists of the composition of issues, statements, arguments, and interpretation to assess critical thinking someone tested.

In developing the critical thinking, everyone has their ways are preferred in preparing what is seen, remembered, and thought. The differences in eachare how to prepare, and process information and experiences are known as a cognitive style. Cognitive style is a consistent way a person does in capturing stimulus or information, how to remember, think, and solve problems, respond to a task or various types of environmental situations. Cognitive style is an important variable that influences studentschoices in the academic field, the continuing development of academic, how students learn, and how students and teachers interact in the classroom.

There is some cognitive styles, which is field-dependent or field-independent developed by Witkin and Goodenough; global or analytic developed Dwyer and Moore; concrete or abstract developed Jonassen and Grabowski; sequential or random developed Summerville; risk-taking or cautions developed Jonassen [8]. Cognitive style field dependent and independent field has been extensively studied and has wide applications in education issues [9].Thus,this study used thecognitive style of field dependent and field independent.

Cognitive style is very influential on the choice of learning strategies learners [10]. People who have independent cognitive style field prefer to separate the parts of some patterns and analyse patterns based on its components and objectives can be achieved by strengthening its own [11]. Meanwhile, people who have field-dependent cognitive styles tend to look at a pattern as a whole, not separated into their parts and rely on information from the outside to reach the goal. Thus, by knowing the cognitive style of the students,appropriate learning strategies will be selected to get optimal student learning outcomes.

Based on the explanation above, theproblem in this research is how the critical thinking skills of mathematical students, especially students of SMP N 2 Ambarawaregarding cognitive style. Critical thinking skills include the ability to mathematically studied drawing inference, assumptions, deduction, interpretation and evaluation of arguments.

2. Research Methods

Research conducted a qualitative descriptive study. Descriptive qualitative meaning describes events that became the centre of attention that is characteristic of students 'critical thinking mathematically seen from cognitive style quantitative data were analysed descriptively only as supporting data that the average ratio of students' critical thinking skills in two cognitive styles. Sources of data in this study were students class VIII SMP Negeri 2 Ambarawa. Results of tests of Mathematic critical thinking analysed to determine the characteristics of students' critical thinking skills regarding cognitive style. Tests critical thinking based on the mathematical WGCTA modified with a mathematical question, especially algebra. To determine the cognitive styles of students used the Group Embedded Figure Test (GEFT) developed by Witkin.

In this research, atriangulation that is comparing the data written test results of students with data from interviews, as well as comparing and checking data from different subjects in the cognitive style. To meet the credibility of data is done by persistent observation. Researchers also held triangulation to validate the data.

3. Results and Discussion

3.1. Critical thinking mathematically junior high school students field-independent cognitive style (FI) and field-dependent cognitive style (FD)

From the results of tests, critical thinking skills of students in general mathematical field-independent cognitive style (FI) students obtained a score of 51.41,andfield-dependent cognitive style (FD) obtained a score of 45.35 with a maximum score of 100. It demonstrates the ability of critical thinking mathematically junior high school students still being. Critical thinking skills test results mathematically Students with first FI cognitive style (SFI-1) and the student with the second FI cognitive style (SFI-2) each scored 72 and 58, while students with a first FD cognitive style (SFD-1) and the student with the second FD cognitive style (SFI-2) each scored 45 and 39. These results are triangulated with interviews obtained results are still appropriate written test, so the critical thinking skills of students with cognitive style mathematical FI and FD in the medium category.

Seen from the cognitive style of critical thinking skills mathematically of junior high school students with FI cognitive style is higher than students with FD cognitive style. These results support the findings of [8] which showed that the students' mathematics achievement FI cognitive styles significantly higher than the FD cognitive style. It is also corroborated by the findings[12] which show that students with FI cognitive styles understand the problem better when compared with students with FD cognitive style, where as the ability to understand the problem is the initial capabilities required incritical thinking. Similarly with[13] which states FI students have better cognitive performance than students FD.

3.2. Critical thinking mathematically Aspects inference junior high school students field-independent cognitive style (FI) and field-dependent cognitive style (FD)

The ability to think critically about the inference FI students is higher than the FD. The inference is a conclusion that made the students to illustrate the observed fact. Measure the student's ability to draw conclusions given some questions. Each question begins with a statement of facts which are true. Furthermore, was given the proposed conclusion. Student task is to determine whether the conclusions are made true, probably true, additional information wasneeded, it may be wrong, and wrong.

From interviewing, students SFI-1 and SFI-2 still wrong in working on inference due fooled by the conclusions put forward by a statement about, so trying to validate the proposed conclusion, regardless of the statement. Also, the two subjects are still wrong in formulating the sentencing statement in mathematics, is not rigorous in calculating the final result although the previous steps are correct.

From the interview, students SFD-1 and SFD-2showed some difficulties in working on inference due to its origin in choosing an answer. They do not know how to solve problems, not thorough in scrutinising the matter of the request, do not understand the statements and conclusions presented in a matter that is not can formulate mathematical statements and conclusions in a sentence correctly.Also, less skilled in the conversion unit so that the results of the calculation to be wrong.

Differences in students' ability to draw inference SFI and other SFD, SFD which students have not been able to develop a comprehensive comparison of the rectangle beginning and end, while the SFI students have been able to construct a rectangular area ratio beginning and end correctly so that proper inferences were drawn.

3.3. Characteristics of critical thinking mathematical assumptions junior high school students field-independent cognitive style (FI) and field-dependent cognitive style (FD)

The ability to think critically about the capability assumptions FI students is higher than the FD. From interviewing, students SFI-1 and SFI-2 still wrong in working on the assumption because one of the two subtraction operations algebraic form, understand the meaning of the sign is equal to zero, do not try to resolve the matter of the proposed assumptions to prove the worth of the calculation results of the statement.From interview, students SFD-1 and SFD-2 still wrong in working on the assumption because one of the elaborations of the algebra multiplication, subtraction on two forms of algebra, understanding the meaning of the sign is equal to zero, do not try to resolve the matter of the proposed assumptions for proving the worth of the calculation results of the statement. Student SFD also was not thorough in number operation.

Differences in student's ability to make the assumption that students SFI and SFD has not been able to make assumptions in showing that the general formula proposed in the matter in accordance with the sequence of numbers, while the SFI students have been able to prove that the general formula proposed in the matter in accordance with the sequence of numbers, namely by entering a value of n in the general formula. Student SFD was wrong in the reduction operation two forms of algebra, and also not thorough in number operation.

3.4. Characteristics of critical thinking aspect mathematical deduction junior high school students field-independent cognitive style (FI) and field-dependent cognitive style (FD)

The ability to think critically on the ability aspect deductions FI higher than the FD. From interviewing, students SFI-1 and SFI-2 still wrong in working on deduction due after trying to prove the truth of the statement directly choose the answer that is not necessarily the conclusion of the proposed deduction is true. SFI also one error in the number operation due to forgetting the concept, if all numbers are divided 0 then the results are undefined. It resulted in SFI wrong in formulating conclusions.

From interviewing, students SFD-1 and SFD-2 still wrong in working on deduction due to not understanding the statements and conclusions presented in the matter so as to prove the statement, originally in selecting answers without validating the proposed conclusions. Also, one in operating numbers that $\frac{2}{0} = 0$, misunderstand 4a, lazy in the count because his number was not unanimous.

Differences in student's abilities of deduction, namely SFI and SFD students have not been able to understand the statements and conclusions presented in the matter, while the SFI students have been able to understand and try to prove the truth of the statements and conclusions presented on the matter. Student FD is less resilient when faced with problems involving non-integer number comparing with the others.

3.5. Critical thinking mathematically interpretation junior high school students field-independent cognitive style (FI) and field-dependent cognitive style (FD)

The ability to think critically on the ability to interpret aspects of interpretation FI students is higher than the FD. From interviewing, students SFI-1 and SFI-2 are still wrong in working on interpreting the interpretation because only able to find a relationship between the number of a sequence of numbers or geometrical patterns and can continue next parts in geometric patterns, but can not formulate common form tribestonne. Also, the origin of the answer as confused how to solve them. Another cause is the inaccuracy in the algebraic operation so that the results of the settlement system of linear equations into one.

From interviewing, students SFD-1 and SFD-2 are still wrong in working on interpreting the interpretation because only able to determine the relationship between the numbers of rows of numbers, yet can continue to the next rate in geometric patterns. In completing the system of linear equations, is still one of the algebraic operation so that the settlement be incorrect. In using algebra to solve the problem of the invention, SFD students only take their number by themselves and the variable was not included so wrong in formulating conclusions.

SFD studentshave not been able to prove the conclusion proposed in the matter, but can only pass on the next number of the sequence of numbers, while the SFI students have been able to prove the conclusion proposed in the matter. SFI students have been able to involve variable in the calculation, but students SFD has not been able to involve variable in the calculation.

3.6. Critical thinking mathematically argument evaluation junior high school students field-independent cognitive style (FI) and field-dependent cognitive style (FD)

The ability to think criticallyto analyse aspects of the argument evaluation FI students is lower than the FD students, although the difference is not great. From student, interviews provide results that students SFI-1 and SFI-2 are still wrong in working on argument evaluation because after proving that the statement is false matter, direct students chose the answer 'weak argument', despite the fact that the arguments put forward are not necessarily weak. In addition to completing the quadratic form of algebra, students still make mistakes in algebra and number operations.

From interviewing, students SFD-1 and SFD-2 are still wrong in working on argument evaluation because there is no accurate in number operations, originally in answering because confused how to solve them. Students SFD are wrong in solving aquadratic form of algebra, students still make mistakes in the operation of algebra and multiplication of numbers that describe two forms of algebra and operate tribes that are not similar.SFD students have not been able to understand the problem and one of the number operations, while the SFI students have been able to understand and able to complete but less thorough.

Seen from the critical thinking skills of all students have the ability to think critically mathematical being, it needs to get the attention of teachers to facilitate better students in learning to develop critical thinking mathematically because with the critical thinking students' understanding of mathematical concepts will also increase. According to [1] critical thinking skills is an effective way to improve students' understanding of mathematical concepts.

From interviews found most students with FD cognitive style more wrong than students with FI cognitive styles in the following points. Misunderstood the problem, one sentence in formulating statements in mathematics is not rigorous in calculating the final result, and one to operate the algebra, compared. It is consistent with findings [14] showed cognitive styles FI students with better understanding compared with the FD. In [15] also reported better FI students in the writing test.Good writing skillsrequire understandingandgood accuracy; it is abasic ability tothink criticallymathematically.

Results also show that both the students' cognitive style FI and FD highest ability to analyse arguments than the ability to draw conclusions, assumptions, deductions, and interpreting information. While the ability to draw conclusions showed the lowest results. It also needs attention because the teacher drew the conclusion is a very important thing and the end of the process of critical thinking or problem-solving.

From the aspect of student's critical thinking aspects with FI cognitive style is better than the FD on the viability inference, assumptions, deduction, interpretation, whereas argument evaluation of students with FD cognitive style is a little better than the FI. This is consistent with the nature of field-independent cognitive style (FI) that tend to be able to use the analogy approach in solving problems. Usually, they are more independent in organisingknowledge; it is easier to analyse a problem and rearrange the parts and more diligent in finding its solution. While students who have a field-dependent cognitive style is easy to remember information associated with social relationships, but it is difficult to process unstructured learning materials.

Critical thinking is a way of thinking that is tested, connect, and evaluate all aspects of a problem situation, including the ability to gather information, remember, analyse thesituation, to read and understand and identify the things that were necessary, so that critical thinking is analytical thinking. FI cognitive style characteristics that tend analytical thinking, therefore in mathematics students who have cognitive style independent field will show the level of achievement of the critical thinking skills that are higher than the FD.

Critical thinkingis notan innate abilitystudentsince birth. According toHemmingin[4],students are notborn with the abilityto think critically, andpreviouslearningexperienceoftendoes not require themtothink critically. Therefore,in the classroomteachershave to understandthe student's model of behaviourfirstly beforeintegratingthe thinking skills. Students withFDcognitivestyleprefer to workin groupsandrequireextrinsic motivation [11]. Thus tofoster FD student's mathematics critical thinking skills,teachers shouldgivea morestructuredstrengtheningandactivatingstudentwork group. It isin the opinion of[16]which statesthroughcollaborative learningFDstudentsgetgreat benefits.

4. Conclusions

Conclusions of this study are(1) the ability to think critically mathematical junior high school students are still; (2) in terms of cognitive styles critical thinking skills mathematical junior high school students with a higher FI cognitive styles of the students FD; (3) From the aspect of student's critical thinking aspect with FI cognitive style is better than the FD on the viability inference, assumptions, deduction, and argument evaluation

Therefore, (1) the ability to think critically mathematical see junior high school students are still being, the teacher needs to facilitate students in learning that develops critical thinking, (2) the need for special assistance for students with dependent cognitive style in developing the critical thinking aspect of inference, assumptions, deduction, interpretation.

References

[1] Nelson A 2013 *IOSRJournalofResearch&MethodinEducation***3** 18 (http://www.iosrjournals.org/iosr-jrme/papers/Vol-3%20Issue-5/D0351825.pdf?id=7370)

[2] LakomaE 2007 Learning Mathematical Modelling - From The Perspective Of Probability And Statistics Education. *Modelling and Applications in Mathematics Education.* The 14th ICMI Study. Springer.ISBN: 978-0-387-29820-7 (Print) 978-0-387-29822-1 (Online) DOI: 10.1007/978-0-387-29822-1_42

[3] Thomas T 2011*Asian Social Science*. 726DOI: http://dx.doi.org/10.5539/ass.v7n4p26

[4] Peter E E 2012*African Journal of Mathematics and Computer Science Research***5** 39 DOI: 10.5897/AJMCSR11.161

[5] Paul R and ElderL 2007. *The Miniature Guide to Critical Thinking Concepts and Tools.(*www.criticalthinking.org/files/Concepts_Tools.pdf)

[6] Bessick S C 2008*Improved Critical Thinking Skills as A Result of Direct Instruction and Their Relationship to Academic Achievement*. Dissertation Indiana University of Pennsylvania (https://books.google.co.id/books?hl=id&lr=&id=NCCfa07hFEwC&oi=fnd&pg=PA1&dq=Bessick+S+C+2008+Improved+Critical+Thinking+Skills+as+A+Result+of+Direct+Instruction+and+Their+Relationship+to+Academic+Achievement&ots=mtIRug_6uj&sig=k8rDSjn8Bl2ySGgjvdR90gTZj-g&redir_esc=y#v=onepage&q&f=false)

[7] Husband G 2006*An analysis of critical thinking skills in computer information technology using the california critical Thinking skills test*ThesisUniversity of Wisconsin-Stouthttp://www2.uwstout.edu/content/lib/thesis/2006/2006husbandg.pdf

[8] Umaru Y and Tukur H A 2013*Journal of Research in Education and Society***4** 50 (http://www.icidr.org/jresv4no2_content.php)

[9] Torres RM and Cano J 1995 *Journal of Agricultural Education***36** 55 (https://eric.ed.gov/?id=EJ514417)

[10] Shi C 2011 *Higher Education Studies***1** 20 DOI: http://dx.doi.org/10.5539/hes.v1n1p20

[11] Witkin H A, MooreC A, Goodenough D R and CoxP W 1977. *Review of Educational Research***47** pp 1-64 (http://www.jstor.org/stable/1169967)

[12] Ngilawajan DA 2013 *Pedagogia***2** 71 DOI: http://dx.doi.org/10.21070/pedagogia.v2i1.48

[13] Guisande MA, Páramo M F, Tinajero C and Almeida L S 2007*Psicothema***19**572 (https://www.ncbi.nlm.nih.gov/pubmed/17959109)

[14] Khodadady E andZeynali S 2012*International Journal of Linguistics***4** 622 DOI: http://dx.doi.org/10.5296/ijl.v4i3.2389

[15] Nosratinia M andAdibifar S 2014 *Procedia- Social and Behavioral Science***98** 1390http://dx.doi.org/10.1016/j.sbspro.2014.03.557

[16] Kuo F-R, Hwang G-J, Chen S-C andChen S Y 2012 *Educational Technology & Society***15**319 (https://www.learntechlib.org/p/75466)

Making a Math Teaching Aids of Junior High School Based on Scientific Approach Through an Integrated and Sustainable Training

E Pujiastuti[1] and Mashuri

Maths and Natural Sciences Faculty of Universitas Negeri Semarang
Gd. D-12 Kampus Unnes Sekaran Gunungpati, Semarang, Central Java - Indonesia

[1]Corresponding author: emi.mat@mail.unnes.ac.id

Abstract. Not all of teachers of Mathematics in Junior High School (JHS) can design and create teaching aids. Moreover, if teaching aids should be designed so that it can be used in learning through scientific approaches. The problem: How to conduct an integrated and sustainable training that the math teacher of JHS, especially in Semarang can design and create teaching aids that can be presented to the scientific approach?The purpose of this study to find a way of integrated and continuous training so that the math teacher of JHS can design and create teaching aids that can be presented to the scientific approach. This article was based on research with a qualitative approach. Through trials activities of resulting of training model, Focus Group Discussions (FGD), interviews, and triangulation of the results of theresearch were: (1) Produced a training model of integrated and sustainable that the mathematics teacher of JHS can design and create teaching aids that can be presented to the scientific approach. (2) In training, there was the provision of material and workshop (3) There was a mentoring in the classroom. (4) Sustainability of the consultation. Our advice: (1) the trainer should be clever, (2) the training can be held at the holidays, while the assistance during the holiday season was over.

Keywords:training, scientific approach, teaching aids.

1. Introduction

1.1. Background

The teacher teaches mathematics at JHS still need teaching aids. However, not all of mathematics teachers in JHS can design and make teaching aids that are needed. Moreover, the teaching aids should be designed so that it can be used in learning through scientific approaches.

Therefore, we need training for the mathematics teachers of JHS on how to design and to create teaching aids that can be presented through a scientific approach. The training is said to be very effective if the teacher can absorb and apply to training materials and after completion of the training. Thus, we need a model of training for teachers that effective, scalable, programmable, integrated, and sustainable after the training is completed. On the other hand, the current curriculum that applies is the Curriculum of 2013 Revised Edition. In this curriculum, learning is carried out through a scientific approach. Therefore, the teaching aids that is used by the math teacher is also highly recommended to be implemented through a scientific approach as well.

1.2. Problems of this Study

The problem as an element of novelty that resolved through this study was as follows. How to implement an integrated and sustainable training so that math teacher of JHS, especially in Semarang can design and create teaching aids that can be presented to the scientific approach?

1.3. Objectives of this Study

The purpose of this study to find a way or model of integrated and sustainable training so that the math teacher of JHS can design and create teaching aids that can be presented to the scientific approach.As an element of novelty that resolved through this study was as follows. How to implement an integrated and sustainable training so that math teacher of JHS, especially in Semarang can design and create teaching aids that can be presented to the scientific approach?

1.4. Benefits

The expected benefits to the findings of the training model were as follows. (1) For the math teachers of JHS, were expected to become more skilled and innovative in design, build, and utilize the manipulative mathematics presented with a scientific approach; (2) A greater insight will be owned by the training providers on a way the implementation of the training, the training associated with learning that utilizes the manipulative for the math teacher of JHS based on scientific approach which is innovative, integrated, applicable, and sustainable.

2. Literature Review

2.1. Definition and Utilisation of Teaching Aids

A teaching aids in mathematics learning is a form of the learning media. Learning media can be as textbooks, blackboards, teachers themselves, pictures, graphics, shapes geometry models, and so on. According to Suharjana (2009), instructional media were all objects that could be a mediator in the learning process. Based on the function of media can take the form the teaching aids and equipment. Therefore, the teaching aids in mathematics learning was a part of the learning media [11].

Government Regulation No. 19 of 2005 on Article 42 (1) stated that "Each educational unit has to have the means, which includes furniture, appliances education, media education, books and other sources, materials consumables, and other equipment needed to support the process regular and continuous learning ". It means the teaching aids are part of the means that must be owned by any educational institution.

The position of the teaching aids associated with the method component of teaching is one of the efforts to enhance the process of interaction between teachers and students in the learning environment. Objects in the mathematics learning of facts, concepts, principles, and skills is an object of the mind which are abstract and can not be observed by the senses. Therefore it is natural that some students do not easily understand mathematics.

Widyantini and Sigit (2009) wrote that to overcome the math easily understood by the students, then in studying an object in mathematics needs experiences through real objects (concrete objects) namely teaching aidsthat can be used as a bridge for students to think abstractly. Students will better understand abstract mathematical concepts that are presented in the form of concrete. The learning activities that use the teaching aids is a great significance for the success of student learning.

It is expected that by using a teaching aids students can see, feel, express by thinking directly to the object which is studied. Thus, the abstract concept that is being studied is absorbed in the minds of students. The use of teaching aids can be associated with the aspects of concepts planting, understanding of the concept, proof of a formula, as well as coaching skills. Also, the use of teaching aids also can increase student motivation.

In the book of Curriculum 2013 explicitly suggested that teachers use the media or alternative learning resources available in the school environment. Such facilities can be a person, material or events. Thus, the ability of a math teacher in thejunior high school to be able to make the teaching aids and use in learning is very necessary.

2.2. Requirements of A Good Teaching Aids

Teaching aids can be used properly if they meet several requirements. These requirements are as follows. (1) the teaching aids are madeby a concept or a mathematical formula that is being provided. (2) easy to present, so the abstract concepts become easily absorbed by students through the use of the teaching aids. (3) Students have become more active in learning and independent by using the teaching aids.

The teaching aids which made should also meet the following requirements. (1) Made from a good material tobelong durability. (2) Do not harm the wearer, for example; there is a sharp wire bonding or no spikes that can hurt a finger, and so on. (3) Made with a variety of shapes or interesting colours. (4) The size is fit so easily seen by the students who sit in the back. (5) can clarify the concepts of mathematics and not vice versa. (6) Easy to carry, used, and stored back.

2.3. Scientific Approach in Utilising the Teaching Aids

The current curriculum in Indonesia is Curriculum of 2013 Revised Edition. In Curriculum of 2013 Revised Edition, the learning is carried out through a scientific approach. Stages in the scientific approach, the learning is implemented through a series of learning activities of students to: (1) observe, (2) ask, (3) collect information, (4) associate, and (5) communicate. Because the math teachers of JHS need to use the teaching aids in their teaching, the presentation of these teaching aids is also conducted by ateacher with thescientific approach as well. Utilisation of the teaching aids in mathematics learning is done through a scientific approach, required the creativity of teachers so that the creativity of students is also growing and be able to absorb the material optimally.

In connection with the description above, if the junior high school math teacher will foster students' creativity in mathematics learning, it is necessary to look for the learning that can (1) make the students active in learning, (2) students brave to find or express ideas, (3) encourage students to develop their creativity, (4) math subject made the atmosphere become fun, and (5) can train students to help each other positively and politely.

Teachers can also obtain such learning during the previous lecture or through training. Figure 1 below is one of the training model followed by the teachers.

Figure 1. Teachers activities in training

2.4. Effective Training

Teachers need to follow a structured training in an integrated, scheduled, and sustainable. Jacob and Lefgren (2004), Siegle and McCoach (2007), also Cabrita et al. (2015) wrote that mathematics teachers followed the effective training can enhance students' ability in mathematics. Increased knowledge of the teachers, especially those related to the material of mathematics, mathematics learning, or training on the use of the teaching aids is needed. Thus, how to implement an effective training, integrated, sustainable, and meaningful need to be found. Figure 2 below, an illustration or description of the math teachers of Junior High School attended training on how to design and create off the teaching aids [1,3,10].

Figure 2. Teachers activities make a teaching aidsin a training

3. Research Methods

3.1. Approach and data analysis

This article is based on research with a qualitative approach. Analysis of the data, according to Miles and Huberman (1994 and 2014), and Moleong (2010), that the activity in qualitative data analysis performed interactively and lasts through to the end. Activities in the analysis of the data include data reduction, display data, interpretation of data, and conclusion drawing /verification [5,6].

3.2. Time and location

There are two stages in the research time. The first stage has been implemented in the Academic Year 2014/2015. Stage 2 has been completed in the Academic Year 2015/2016. Location of research: At the centre of MGMP activities of math teachers in Semarang and the centre of MGMP activities of math teachers of Salatiga district.

3.3. Activities

Based on the problems and objectives to be achieved, the activities of study as supporting this article includes the beginning assessment on the needs of the math teachers of JHS in Semarang in the teaching-learning process, the research team make a designing training activities, conduct of training on the teacher Training, Assessments test, analyzing the results of assessment, implement Focus Group Discussion (FGD), testing in the teacher testing, the selection of the research subjects, interviews, triangulation, and FGD order to obtain a training model or training method that allows the math teachers of JHS can design and create that teaching aids can be presented to the scientific approach.

FGD implementation involves the team of researchers, two lecturers/students who were interested, and three the math teacher of JHS who concerned, to found an integrated training model, programmed, sustainable, and applicable.

4. Results and Discussion

4.1. Result of this Study

Based on the results of the activities, the results of assessment tests, FGD, experiment, interview, and triangulation it has obtained the study results that conclusive were as follows. (1) The math teacher of JHS who became a trainee begin to understand the meaning of scientific approach and its application in learning. (2) The math teacher of JHS who became a trainee begin to understand the use the teaching aids mathematics used to teach with a scientific approach. Some teachers have been able to demonstrate the use of math learning a teaching aids used to teach with a scientific approach. Teaching aids prepared and taken by thetrainer as a model. (3) The math teacher of JHS who became a trainee has been able to create a lesson plan that includes the use of teaching aids of mathematics learning used to teach with a scientific approach. (4) Through the FGD, it has arranged a training model on how to make and to use teaching aids of mathematics learning that is used to teach with a scientific approach. This Training model then tested or implemented. These training activities include the provision of material, particularly on the application of scientific approach, design and making of lesson plans, learning models, workshops, learning simulation, and mentoring in the classroom through activities such as lesson study, namely the teachers teaching and lecturers trainer followed some teacher trainees participate viewed the learning activities in the classroom. After mentoring in the classroom, it was followed by reflection and evaluation. Training schedules to be created programmatically, integrated and scheduled well. (5) Consultations between the lecturer's trainer with math teacher trainees of JHS to do after the training period was completed (ongoing).

Also, through the FGD, successfully strengthened the trial results of training model in the creation and utilisation of scientific approach based on teaching aids for the math teacher of JHS. Thus, the problem of this study has been solved. The objective of study has also been achieved which has found a way or a training model that the math teacher of JHS can design and create teaching aids that can be presented to the scientific approach. According to Popa & Bucur (2014), teachers who want to improve themselves by thefollowing thetraining, for example of mathematics learning, will be able to improve it is competitive [9].

4.2. Discussions

According to Carbonneau, K.J. et al. (2013) and Moyer (2001), using teaching aids was very effective in learning mathematics. Thus the provision of training in the math teacher of JHS on how making and using teaching aids for mathematics learning that used to teach based on the scientific approach was very important [2,7].

Through training that programmed and scheduled properly, then the effectiveness of the training will be high. Training activities were carried out through the provision of material, especially on the approach scientific, preparation of lesson plans appropriate, the learning model who involving the trainer and teachers elected, forwarded the workshop of manufacture of a teaching aids, then the learning simulation, and mentoring in the classroom through activities such as lesson study, then this training model becomes effective and useful.

Also, the implementation of the FGD that is followed by the team of researchers, two lecturers/students who were interested, and three of the math teacher of JHS related, also the implementation of reflection and evaluation, making the training model that is resulting was becoming more applicable and useful.

5. Conclusions and Suggestions

5.1. Conclusions

Based on the result of this research and the discussion above, then the conclusions which were obtained that the research team has succeeded in constructing a model of training activities for the math teacher of JHS in Semarang about way of making and using a teaching aids of mathematics learning that is used to teach with a scientific approach.

Training models include the provision of training materials, teaching models by utilisingteaching aids, practice makes a lesson plans that appropriate, forwarded with aworkshop of manufacture of teaching aids, then the learning simulation, mentoring in the classroom, and reflection. Through the implementation of a training model that was programmed and scheduled then the results were as follows. (1) The teacher trainees have understood the meaning of scientific approach and its application in learning. (2) The teachers have understood the use of teaching aids of mathematics learning that used to teach mathematics with scientific approach. (3) Through the workshop, the teachers through work in thegrouphave been successfully in designing and creating teaching aids of the mathematics learning of JHS that can be presented through a scientific approach.

5.2. Suggestions

Suggestions that can be recommended are as follows. (1) When the training model was followed up by an institution, then the ability of trainers should be accountable. (2) The training can be held at the time of the holiday season, while mentoring can be carried out when the holiday period finishes. (3) In order,a teaching aid produced becomes good, the making of teaching aids given sufficient of breaks time. (4) Consultations between the trainer lecturers and teacher as trainees to do after the training period are completed (ongoing/ sustainable).

References

[1] Cabrita, I. et al. 2015 *Procedia - Social and Behavioral Sciences* 186. Availabed in https://www.researchgate.net/publication/279529293_Maths_in-service_Teacher_Training_and_the_Restructuring_of_Secondary_Education_in_East-Timor.

[2] Carbonneau, K.J. *et al.* 2013 *Journal of Educational Psychology* 5(2). Available inhttps://www.researchgate.net/profile/Scott_Marley/publication/248701204_A_Meta-Analysis_of_the_Efficacy_of_Teaching_Mathematics_With_Concrete_Manipulatives/links/00b7d51e026e53a3dd000000.pdf.

[3] Jacob, B.A. & Lefgren, L. 2004 *The Journal of Human Resources* 39(1). Availabed inhttp://www.jstor.org/stable/3559005?seq=1#page_scan_tab_contents .

[4] Miles, Matthew B & Huberman, A Michael. 1994. *An Expanded Sourcebook – Qualitative Data Analysis*. Second Edition. London: SAGE Publication.
[5] Miles, Matthew B & Huberman, A Michael 2014 London: SAGE Publication.
[6] Moleong, J. 2013 Bandung: PT Remaja Rosdakarya.
[7] Moyer, P.S. 2001 Mathematics. *Educational Studies in Mathematics* 47(2). Available inhttp://www.jstor.org/stable/3483327?seq=1#page_scan_tab_contents.
[8] Permendikbud No 103 Tahun 2 015. Tentang Pedoman Umum Pembelajaran Kurikulum 2013.
[9] Popa, O.R. & Bucur, N.F. 2014 *Procedia - Social and Behavioral Sciences 128*. Availabed inhttps://www.researchgate.net/profile/Norica_Felicia_Bucur/publication/262306441_Strengths_and_Weaknesses_of_the_Romanian_Preservice_Training_System_in_Prospective_Secondary-school_Teachers'_View/links/557eae5f08aeb61eae24a12c.pdf .
[10] Siegle, D. & McCoach, D.B. 2007 *Journal of Advanced Academics* 18(2). Availabed in http://files.eric.ed.gov/fulltext/EJ767452.pdf .
[11] Suharjana, Agus. 2009 Pusat Pengembangan dan Pemberdayaan Pendidik – Depdiknas. Yokyakarta: P4TK.
[12] Widyantini dan Sigit TG. 2009 Yogyakarta: P4TK Matematika. Kode Dok:F-PRO-016.

Faded-example as a Tool to Acquire and Automate Mathematics Knowledge

E Retnowati

Department of Mathematics Education, UniversitasNegeri Yogyakarta

Corresponding author: e.retno@uny.ac.id

Abstract. Students themselves accomplish Knowledge acquisition and automation. The teacher plays a role as the facilitator by creating mathematics tasks that assist students in building knowledge efficiently and effectively. Cognitive load caused by learning material presented by teachers should be considered as a critical factor. While theintrinsic cognitive load is related to the degree of complexity of the material learning ones can handle, theextraneous cognitive load is directly caused by how the material is presented. Strategies to present a learning material in computational learning domains like mathematics are anamely worked example (fully-guided task) or problem-solving (discovery task with no guidance). According to the empirical evidence, learning based on problem-solving may cause high-extraneous cognitive load for students who have limited prior knowledge, conversely learn based on worked example may cause high-extraneous cognitive load for students who have mastered the knowledge base. An alternative is a fadedexample consisting of thepartly-completed task. Learning from faded-example can facilitate students who already acquire some knowledge about the to-be-learnedmaterial but still need more practice to automate the knowledge further. This instructional strategy provides a smooth transition from a fully-guided into an independent problem solver. Designs of faded examples for learning trigonometry are discussed.

1. Introduction

Learning is a cognitive process to construct knowledge in working memory and store it permanently in long-term memory [1]. These two kinds of memory systems are the major components of our cognitive architecture. Having a well-constructed knowledge means acquiring and organising knowledge meaningfully thus enables students to understand deeply how the knowledge is applied. Moreover, learning will be enhanced when this knowledge can be automated. It is suggested that knowledge automation occurs by deliberate practice (i.e., planned, regular, aim into specificity) [2]. Knowledge automation benefits students to solve problems faster, perform less error and advance learning. Accordingly, it is important to note that facilitating learning is not only to assist students acquiring knowledge but also automating knowledge.

Mathematics is a computational learning domain that has a well-structured knowledge building. It consists of operations and algorithms on how to solve problems. Indeed, it has been recommended that mathematics should be learned by problem-solving[3]. Problem-solving is an activity to solve complex problems. Based on the given complex problem, students attempt to understand mathematical concepts underpinning the problem, and at the same time, apply the algorithms or procedures of solving it. However, solving complex problems are not always easy. Different students may treat complexity differently depending on many factors, such as prior knowledge or motivation. Hence they require different instruction [4]. Consequently, teachers should employ strategies of mathematics instructions by which students can master mathematics knowledge underpinning the problems solving.

Cognitive load theory is an instructional design theory that draws its principles based on empirical evidence [1]. One of the principles is that learning occurs effectively when cognitive load is at amanageable level. Cognitive load results when we are organising learning material in our working memory. Two types of cognitive load have been defined, intrinsic and extraneous cognitive load [5, 6]. While theintrinsic cognitive load is caused by the degree of interactivity of the elements of the learning material the student can organise, theextraneous cognitive load is triggered by the presentation of the learning material. Working memory resource has to deal with these cognitive loads simultaneously. The germane cognitive load is a cognitive load that is devotedto acquiring and constructing knowledge [6]. Its load depends on how much working memory manages intrinsic and extraneous cognitive load. Extraneous cognitive load tends to hinder learning, thus minimising extraneous cognitive load is necessary to increase the chance of germane cognitive load.

The complexity of a learning material is about student's prior knowledge. If the pre-requisite knowledge exists in student's long-term memory and can be retrieved automatically, intrinsic cognitive load in working memory may be lower. It occurs because the existing knowledge guides students to recognise, identify and organise the to-be-learned material. However, if the pre-requisite knowledge does not exist, then it is unlikely to recognise the presented learning material. Knowledge of what students have would assist teachers to present sufficient task for students and thus learning is facilitated.

About extraneous cognitive load, it is suggested that learning material is presented thoughtfully. For instance, it provides easy-to-follow guidance or worked-example if the material is presented to novice students, or it has sufficient problem-solving task if presented to master students. As previously noted, which instructional strategy to apply should refer to student's level of prior knowledge. Based on cognitive load theory, aninstruction that uses problem-solving as the major activity may cause high-extraneous cognitive load if presented to students who are alack of prior knowledge. On the other hand, an instruction based on worked example may cause high-extraneous cognitive load for students who have advanced prior knowledge [7]. In fact, there are students who have learned some knowledge but limited ability to transfer it into problem-solving. An alternative instruction for these students may be an instruction based on faded-examples. Faded-examples consist of partly-completed problem-solving. It may be applicable for students who have some of therequired knowledge bases with limited use. This paper discusses further this faded-example strategy for mathematics learning, as follows.

2. Worked-Example

A conventional teaching method, commonly a teacher starts with theintroduction of the topic, short explanation, then one or two worked examples followed by practice a set of problem-solving. Worked example is often used to provide adescription how to step-by-step solve a mathematics problem [8]. Atkinson, Derry [9] suggested that worked examples should demonstrate how a mathematician expert solves the problem, not simply a problem solution but a solution that assist students to understand and is powerful to imitate.

A worked example based instruction facilitates students to learn from a set of worked examples. This instruction may belong to a student-centered learning paradigm. In the classroom, teachers who apply this instruction begin thelesson with some introductory knowledge base for the to-be-learned material and explanation of what learn by illustrating or discussing a novel mathematics problem they are about to learn. It may challenge yet motivate students. The main learning activity is followed by giving a set of worked example instruction for students to learn by themselves; hence students have their phase to acquire new knowledge while completing the instructional task. After this learning phase, students are given theopportunity to confirm their newly constructed knowledge through thepresentation, correctness checking or further problem-solving elaboration with the other students or the teacher.

Worked example effect occurs when an instruction based on worked examples is proved to be more effective to facilitate learning compared to another minimally guided based instruction, such as problem-based learning. An explicit guidance decreases extraneous cognitive load by avoiding anirrelevant search of meaning during learning. Worked examples eliminate the use of means-ends analysis by asking students to learn a solution to a novel problem solving, rather than asking students to discover by themselves. In turn, knowledge acquired from worked examples will play a role as aknowledge base when given similar problem-solving. The existence of knowledge base is essential for learning and hence working memory is devoted to knowledge acquisition and automation.

Atkinson, Derry [9]summarised that research into the use of worked examples for learning new material had been carried out for more than six decades using various subject matters. A few evidence for the worked example effect in mathematicswas provided by Zhu and Simon [10] in their longitudinal study using a 3-year curriculum in algebra and geometry in a Chinese middle school, where they found that students studying the worked examples could complete the three-year course in only two years. Research by Tarmizi and Sweller [11] using geometry also found the worked example effect, once split-attention was avoided Further evidence of the worked example effect were provided by Chi, Bassok [12] using mathematics for physics, and Ward and Sweller [13] using geometric optics and kinematics.

2.1. Principles of creating worked examples

Effectively worked examples can be created based on principles developed by the cognitive load theory. These principles are usually explainedregarding whether the modified format of worked example based instruction decreases or increases either extraneous or intrinsic cognitive loads and thus manages germane cognitive load. These principles guide teachers how to create the format of worked examples, how many examples should be presented, its variation and how to ensure that students attend the task fully.

The first principle is to reduce theextraneous load. It canbe done by arranging the presented information in such a way connecting elements or relevant information can be simultaneously grasped. Evidence from the split attention or spatial–contiguity effects supports this principle [11, 14]. When learners are imposed to do anunnecessary search and connect elements of information to understand the meaning of presented information, this will cause high extraneous cognitive load. It happens when relevant information that must be considered simultaneously are spatially separated or successively delivered. Accordingly, providing a clear instruction will surely help students getting in the learning material. Moreover, redundant information must be omitted [15]. Redundancy effect may be caused by presenting unnecessary information, adding extra information to already self-contained information, using multiple formats for the same information or repeating similar information concurrently. Such information may distract students from understanding the learning material, and hence cause heavy extraneous cognitive load.To reduce extraneous cognitive load can also be done by considering the number of worked examples presented to students. First, it must be noted that this kind of instruction is suitable for anovel or complex learning material. In other words, it is given for students who have limited prior knowledge. Accordingly, the number of worked example has to be sufficient to students. The degree of complexity has to be considered, including variations of thecontext of the problems [16, 17]. For students who have sufficient prior knowledge, they may not benefit from worked examples. Instead, these may cause redundant information in their working memory and hamper learning [7].

The second principle is to manage intrinsic cognitive load, the complexity of the learning material. Its level of complexity may describe the characteristic of the learning material. The content of a learning material may have anunchanged level of complexity because it characterises this material[6]. However, from the view of prior knowledge possessed by the learner, level of complexity may change according to how much prior knowledge can be retrieved to organise the complexity of the learning material. When dealing with the same learning material, students with high prior knowledge may see it less complex, but those with low prior knowledge may see it very complex. Consequently, creating worked examples should base on how much prior knowledge possessed by the students. Usually, it is suggested to present material successfully from less to more complex. Particularly for students with thelow prior knowledge, the worked example should show correct solutions with clear steps of solution to help them understanding the problem and how to solve it.

The third principle is to increase thegermane cognitive load. When dealing with the interacting elements of learning material, students will experience germane cognitive load that directs students to organise, understand and construct knowledge. By using a worked example based instruction, thegermane cognitive load may be imposed by asking students to apply their metacognitive skills by monitoring their thinking process, doing self-explanation [12], or imagining the context [18]. Students may be asked to write their conclusion or summary of the key concept or procedures they learn from the worked example, or to refer to the key answer. These activities help students to clarify and reflect their knowledge construction and therefore understand more meaningfully.

Further, the format of worked examples should be paired with similar problems [19]. The problem-solving pairs facilitate students to practice and automate the knowledge acquired from the worked example. Technically, the instruction consists of pairs of worked example and a similar problem-solving. The number of pairs depends on the complexity level of each problem solving, and hence instruction designers should consider its intrinsic cognitive load. More importantly, the paired problem solving must have similarity regardingconceptual base or procedural solution steps to the provided worked example.

It has been discussed that teachers have to ensure that the learner fully attends to the worked example tasks during instruction to increase thegermane cognitive load. Chi, Bassok [12] found that most low prior knowledge students often simply look at a glance and did not attempt to study all parts of worked examples fully. Van Merriënboer [20] suggested the use of completion problems to engage students into the worked example. This completion problem requires them to complete some key solution steps in the worked examples by themselves. This strategy could be effective particularly when the problem requires long solution steps. The effectiveness of completion strategy is supported by Paas [21] who investigated the effect of completion problems for learning statistics. Moreover, Paas also found that the completion and worked example conditions required a significantly shorter study time than conventional condition, which is problem-solving instruction.

To explain why partially completing the example is effective, it is argued that students' attention is directed to the problem state and the provided key solution steps while completing the missing solution steps. It may also be said that completion problems are a combination of worked examples and problem-solving, which is an alternative format of worked examples based instruction. In turn, completion problems have been used to develop a strategy of presenting worked examples namely the fading guidance effect.

3. Faded-Example

The guidance fading strategy uses a combination of worked examples, completion problems, and problem-solving which are presented sequentially, and designed to facilitate a smooth transition from novice to more knowledgeable students [22-24]. This strategy is developed based on the expertise reversal effect. Novice students would be more advantaged by anexplicit instruction like the worked examples based instruction, but master students would be more advantaged by animplicit instruction like the problem solving based instruction.

Renkl, Atkinson [23] suggested two fading techniques; these are called backwards and forward fading examples. The *backwardsfading* strategy, the first worked example is fully completed, the second worked example has the solution to the final step removed, the third has the two last steps removed, and so forth, until the final example presents the whole problem-to-be-solved only. The task is to complete the removed steps, whose number increases as knowledge automation develops. For the *forward fading* strategy, the series of completion are provided in the opposite direction, that is the first step of the solution is incomplete, then the second step is incomplete, and so forth, in a forward direction until the full incomplete problem is presented.

Research evidence showed by Renkl, Atkinson [23] indicated that the backwards fading strategy is more favourable for low prior knowledge students because they are benefitted from studying a full worked example at the beginning of the learning phase. Reisslein, Sullivan [25] supports that slow fading strategy was more advantageous for low prior knowledge students transitioning from worked example stage to independent problem-solving. Slow fading strategy uses a backwards fading strategy

IOP Conf. Series: Journal of Physics: Conf. Series **824** (2017) 012054 doi:10.1088/1742-6596/824/1/012054

that provides students with a longer phase of knowledge acquisition. In contrast, a fast fading strategy was found to be more advantageous for high prior knowledge students.

3.1. Developing Faded-Example Instructions

In a problem-solving type that has at least, for instance, four steps to complete the answer, using *backwards fading* strategy means providing the beginning steps as the workedexamples, and the last steps as the task students need to accomplish. The following example shows the use of backwards fading examples when studying how to simplify a trigonometry formula which commonly has the same number of solution steps. Four problems are provided in which the first problem has the fourth step is faded, the second one has the third, and the fourth are faded, the third one has the first step only shown, and the last one has no steps shown (problem-solving). The solution is provided in a table to ensure it is easy to match between the explanation and the execution of the solution steps. Different colours are used to reduce search of information. Also, this series of faded examples have asimilar basic concept to solve the problem that is manipulating the trigonometry form into trigonometry identity or basic trigonometry. A Little modification in a series is suggested to minimise unnecessary cognitive load. Thus students are facilitated more on knowledge acquisition and automation.

Task: Study how to simplify trigonometric forms by completing the problem solution below.

1. Simplify: $\sin t + \cot t \cos t$

Answer:

Explanation	Execution
Step 1 Substitute $\cot t$ with $\left(\frac{\cos t}{\sin t}\right)$	$\sin t + \cot t \cos t = \sin t + \left(\frac{\cos t}{\sin t}\right)\cos t$
Step 2 Form into a fraction with denominator $\sin t$. By this way, the nominator makes the trigonometry identity $\sin^2 t + \cos^2 t$	⇔ $\sin t + \cot t \cos t = \frac{\sin^2 t + \cos^2 t}{\sin t}$
Step 3 Substitute $\sin^2 t + \cos^2 t$ by 1	⇔ $\sin t + \cot t \cos t = \frac{1}{\sin t}$
Step 4 Simplify $\frac{1}{\sin t}$ into $\operatorname{cosec} t$	⇔ $\sin t + \cot t \cos t = \ldots$

2. Simplify $\operatorname{cosec} t - \cot t \cos t$

Answer

Explanation	Execution
Step 1 Substitute $\operatorname{cosec} t$ with $\frac{1}{\sin t}$ and $\cot t$ with $\left(\frac{\cos t}{\sin t}\right)$	$\operatorname{cosec} t - \cot t \cos t = \left(\frac{1}{\sin t}\right) - \frac{\cos t}{\sin t}\cos t$
Step 2 Simplify the fraction on the right-handside, and hence the nominator makes trigonometry identity $1 - \cos^2 t$	⇔ $\operatorname{cosec} t - \cot t \cos t = \frac{1-\cos^2 t}{\sin t}$
Step 3 Substitute $1 - \cos^2 t$ with $\sin^2 t$	⇔ $\operatorname{cosec} t - \cot t \cos t = \ldots$
Step 4 ...	⇔ $\operatorname{cosec} t - \cot t \cos t = \ldots$

3. Simplify $\cos t + \tan t \sin t$

Answer

Explanation	Execution
Step 1 Substitute $\tan t$ with $\left(\frac{\sin t}{\cos t}\right)$	$\cos t + \tan t \sin t = \cos t + \left(\frac{\sin t}{\cos t}\right)\sin t$
Step 2 Simplify the fraction on the right-handside, and hence the nominator makes trigonometry identity $\cos^2 t + \sin^2 t$	⇔ $\cos t + \tan t \sin t = \ldots$
Step 3 ...	⇔ $\cos t + \tan t \sin t = \ldots$
Step 4 ...	⇔ $\cos t + \tan t \sin t = \ldots$

4. $\sec t - \tan t \sin t$

Answer

Explanation	**Execution**
Step 1 Substitute $tan\ t$ with $\left(\frac{\sin t}{\cos t}\right)$	$\sec t - \tan t \sin t \ = \ldots$
Step 2 ...	$\Leftrightarrow \ \sec t - \tan t \sin t = \ldots$
Step 3 ...	$\Leftrightarrow \ \sec t - \tan t \sin t = \ldots$
Step 4 ...	$\Leftrightarrow \ \sec t - \tan t \sin t = \ldots$

5. Conclusion: to simplify a trigonometry form, manipulate the given form by factorization, substitution, expansion or simplification based on basic trigonometry or trigonometry identity.
6. Create a trigonometry problem similar to the above problems, and then solve it. If it is still too hard, try to solve it again and give anexplanationof each step of thesolution.

The last two tasks above provide students to understand their learning more meaningfully, as discussed previously regarding the germane cognitive load. In a problem-solving type that has at least, for instance, four steps to complete the answer, using *forward fading* strategy means providing the last steps as the workedexamples and the first steps as the task students need to accomplish. In the backwards faded examples above, solution step 4 is faded, then step 3, 2 and 1. In the faded forward examples, the very first instruction contains solution step 1 is firstly faded while solution step 2, 3 and 4 are showed. Then, in the second instruction, step 1 and 2 are faded; in the third instruction, step 1, 2, and 3are faded. It has to be noted that students are instructed to complete the faded example from the first step while studying the pattern of the problem solution.

When aninstructionis focused on facilitating thetransition from fully-guided problem solver into independent problem solver, the faded example strategy can be implemented. It is critical that teachers develop this kind of instruction carefully. Thus, students can manage their cognitive load, either in terms of intrinsic cognitive load (i.e., the arrangement of the level of complexity), extraneous cognitive load (i.e., the arrangement of the presentation of the connected information), and germane cognitive load (i.e., how the task engage students in meaningful learning). It should be noted that creating this instruction requires adeep understanding of the content material as well as how to present problem solution examples that assist students in automating their knowledge more efficiently.

4. Conclusion

Learning is an activity to acquire and automate knowledge. When learning new or complex problem to solve, working memory has to load high demand of cognitive process because of lack of prior knowledge can be retrieved from long-term memory. For this category, students are better facilitated with an explicit instruction such as worked example based instruction. On the other hand, when students can organise learning material efficiently because the already well-possess knowledge base to be applied, students will improve their learning by an implicit instruction, such as discovery learning. When students have acquired some level of prior knowledge but limited in use, teachers could facilitate them with a faded example based instruction using either forward or backwards strategy. This instruction consists of a set of completion problems that require students to filling incomplete solution steps. The solution steps were faded gradually to ensure students attend the task fully and hence acquire and automate their knowledge. Eventually, theinstructional strategy must be developed by considering the cognitive load may be imposed whenstudents use it.

References

[1] Sweller, J 2004 *Instructional Science* **32**(1-2): p. 9-31.
[2] Ericsson, K.A., R.T. Krampe, and C. Tesch-Romer 1993. **100**(3): p. 363-406.
[3] NCTM, *Principles and standards for school mathematics*. 2000, Reston, VA: Author.
[4] Retnowati, E., 2016SEAMEO Qitep in mathematics: Yogyakarta.
[5] Kalyuga, S., 2011. **23**(1): p. 1-19.
[6] Sweller, J.,2010. **22**(2): p. 123-138.doi: 10.1007/s10648-010-9128-5
[7] Kalyuga, S., et al., 2003 Educational Psychologist. **38**(1): p. 23-31.
[8] Ayres, P. and J. Sweller,2013, Routledge: London, UK. p. 408-410.
[9] Atkinson, R.K., et al., 2000. **70**(2): p. 181-214.doi: 10.3102/00346543070002181

[10] Zhu, X. and H.A. Simon, 1987 *Cognition and Instruction*,. **4**(3): p. 137-166.doi: 10.1207/s1532690xci0403_1
[11] Tarmizi, R.A. and J. 1988. *Journal of Educational Psychology*. **80**(4): p. 424-436.
[12] Chi, M.T.H., et al., *Self explanations: How students study and use examples in learning to solve problems.* Cognitive Science, 1989(13): p. 145-182.doi: 10.1207/s15516709cog1302_1
[13] Ward, M. and J. Sweller, *Structuring effective worked examples*. Cognition and Instruction, 1990. 7(1): p. 1-39.doi: 10.1207/s1532690xci0701_1
[14] Moreno, R. and R. Mayer, 1999 *Journal of Educational Psychology*. **91**(2): p. 358-368.
[15] Chandler, P. and J. Sweller, 1991 Cognition and Instruction. **8**(4): p. 293-332.
[16] Paas, F. and J.J.G. van Merriënboer, 1994 *Journal of Educational Psychology*. **86**(1): p. 122-133.
[17] van Gog, T., F. Paas, and J.J.G. van Merriënboer, 2008. Learning and Instruction **18**(3): p. 211-222.doi: 10.1016/j.learninstruc.2007.03.003
[18] Leahy, W. and J. Sweller, 2005 *Journal of Experimental Psychology*, **11**(4): p. 266-276.doi: 10.1037/1076-898X.11.4.266
[19] Trafton, J.G. and B.J. Reiser. 1993. Hillsdale, NJ: Erlbaum.
[20] van Merriënboer, J.J.G 1990 *Journal of Educational Computing Research*,. **6**(3): p. 265-285.doi: 10.2190/4NK5-17L7-TWQV-1EHL
[21] Paas, F., 1992 *Journal of Educational Psychology*, (84): p. 429-434.
[22] Renkl, A. and R.K. Atkinson 2003 *Educational Psychologist*. **38**(1): p. 15-22.doi: 10.1207/S15326985EP3801_3
[23] Renkl, A., et al., 2002 *The Journal of Experimental Education*, 2002. **70**(4): p. 293-315.doi: 10.1080/00220970209599510
[24] Atkinson, R.K., A. Renkl, and M.M. Merrill, 2003 *Journal of Educational Psychology*, **95**(4): p. 774-783.
[25] Reisslein, J., H. Sullivan, and M. Reisslein, 2007 *Journal of Engineering Education*, 2007. **96**(1): p. 45-56.doi: 10.1002/j.2168-9830.2007.tb00914.x

IOP Conf. Series: Journal of Physics: Conf. Series **824** (2017) 012055 doi:10.1088/1742-6596/824/1/012055

The Anticipation: How to Solve Problem in Integral?

E Yudianto[1, 4], St Suwarsono[2] and D Juniati[3]

[1]Department of Mathematics Education – University of Jember
[2]Department of Mathematics Education - Universitas Sanata Dharma
[3]Department of Mathematics - Universitas Negeri Surabaya

[4]Corresponding author: erfanyudi@unej.ac.id

Abstract.This research is motivated by the interest of researchers to the phenomenon of the ability of thepreservice teacher to solve problems integral to how to anticipate. It isbecause a preservice teacher to solve the problem using the integral another way to illustrate a given problem then just finish it. Anticipation in this study consisted of (1) interpreting, (2) predicting a result, and (3) foreseeing an action. This research aims to know how the preservice teacher who have the cognitive style field independent in solving problems of integral. The method used is the method of test and interview. The test consists of an essay and interview used an unstructured interview. The results obtained by the preservice teacher to solve the problems of integral to analyse a given problem through an initial guess (predicting) then do the problem, not in detail (foreseeing), but the conclusions are true. Preservice teacher in this study can be categorised in explorative anticipation

1. Introduction

The mathematics still often considered difficult for students. Many researchers mathematics education focus to improving students' ability in solving mathematical problems well. However, theyhave not yet to find out about students' ability in solving mathematical problems. Piaget's notion of anticipation is about intelligence. The Intelligence is a particular instance of biological adaptation and drew parallels between intellectual adaptation, and organic (i.e. physiological) adaptation, an essential characteristic of both is that strive towards equilibrium[1]. Organic adaptation refers to the readjustment of the sensorimotor structures in response to pressures from the changing environment for survival. Intellectual adaptation refers to the reorganisation of the conceptual structures to eliminate cognitive conflicts. Notion anticipation by[2] in anticipation, sometimes students need also foresight/predict any mental act performed.

Every step a person in solving the problem has a shadow of the future (certain final solution) followed by way of thinking and way of understanding that interacting in the brain although the foreseeing/predicting is false. This certain final solution means that every step/certain stage of a problem that is solved, then it has a temporary solution before next stage proceeding.

Notion of anticipation by identifying three general types of anticipation: (a) implicit expectations that are present in our action, e.g., the preparation and control of our movements when we grope in the dark; (b) explicit expectation of an outcome based on certain cause-effect relationships (e.g., predicting that it will soon rain upon noticing that dark clouds are covering the sky); (c) anticipation of a desired event and the means for attaining it (e.g., a child's foresight of the means to get his parent to give in, say by throwing a temper tantrum in public)[1][3].

The three constructs mental act, away of understanding (WoU), and way of thinking (WoT) form a triad, as depicted in Figure 1[4][5].

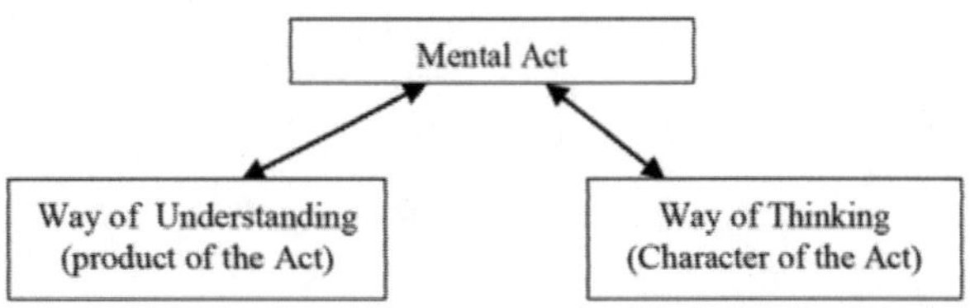

Figure 1. Harel's MA-WoU-WoT Triad

The researcher provide the triad of MA-WoU-WoT with a means analyzing based on student's statements and actions, what a student understands of a phenomenon or particular thing (i.e., what the product of the student's mental act is); and subsequently infer the manner (i.e., character of the act) in which the student engages a particular mental act to arrive at that way of understanding. Ways of understanding are neutral in the sense that it merely a certain product of a mental act and doesn't suggest what a student understands or doesn't understand. Nevertheless, a student's WoU may be desirable or undesirable about those that have been accepted by the mathematical community at large.

The students develop WoT only through the construction of WoU, and the WoU they produce are determined by the WoT they possess[6]. This principle importance of incorporating complementary WoU and WoT into cognitive objectives for instruction, which should help students reason independently and solve theproblem in mathematics. The mathematics teachers focus on imparting ways of understanding such as definitions, rules, algorithms, solutions, theorems, and proofs, without attempting to help students develop desirable WoT[6]:

> "We have observed that teachers often form, at least implicitly, cognitive objectives regarding ways of thinking, but their efforts to teach ways of thinking is often counterproductive because these efforts do not build on ways of understanding. Conversely, teachers often focus on ways of understanding but overlook the goal of helping student's abstract effective ways of thinking (WoT) from these ways of understanding (WoU)."

The implementation of WoU and WoT involves: (a) attending to students existing WoU and WoT; (b) identifying appropriate cognitive objectives, appropriate in the sense that they are aligned with student's current WoU and WoT, and that they preserve the mathematical integrity of the content; and (c) designing activities, with an understanding of the interplay among various WoU and WoT, to meet those objectives.

Teaching is the ways of thinking directly to students is unproductive[6]. According to the WoU and WoT [4], it is through the construction of WoU that students develop WoT; conversely, it is through the application of WoT that students develop their WoU. These WoU may be deficient initially but can be progressively refined towards those that are institutionalised (i.e., accepted as correct and useful by the mathematics community). Hence, the target WoT and WoU must complement each other, so that applying certain WoT will lead to the development of certain WoU, which may help to cultivate target WoT.Five anticipation that is: (1) impulsive anticipation is spontaneously proceeded with the idea that comes to mind without analyzing the problem situation and without considering the relevance of the idea to the problem situation, (2) interiorized anticipation is spontaneous proceeds with an idea without having to analyze the problem situation because one has interiorized the relevance of the anticipated action to the situation at hand, (3) analytic anticipation is analyzed the problem situation and establishes a goal or a criterion to guide one's actions, (4) explorative anticipation is explores an idea to gain a better understanding of the problem situation, and (5) tenacious anticipation is maintains and does not reevaluate one's way of understanding (prediction, problem solving approach, claim, or conclusion) of the problem situation in light of new information[2].

2. Method

The purpose of this research is to know how the preservice teacher who have the cognitive style field independent in solving problems of integral. The method used is the method of test and interview. The test consists of an essay and interview used an unstructured interview. During an investigation, the researcher acts as the main instrument that means where researchers cannotbe replaced by someone else or something else. In this research did not manipulate a variable too, but preferably things that do preservice teacher at the time of data collection activities.Thus, this research used a qualitative approach [7]. Subjects in this research were preservice teachers fourth semesters of mathematics education with field-independent cognitive style.

3. Result and Discussion

The problem was given to astudent is:

Which the larger between $\int_0^1 x\,dx$ or $\int_1^2 x\,dx$? Explain, please!

The predicting and foreseeing these research that is: (1) predicting (a result) is the mental act of conceiving an expectation for the result of an event without actually performing the operations associated with the event, and (2) foreseeing (an action) is the mental act of conceiving an expectation that the leads to the volition for an action, prior the performing the operations associated with the action[2].

The subject began reading the problem with mumbling and then explained the purpose of the repeat question (not detailed) given aproblem. When researchers re-confirm the aim of the question, the subject try to theunderstanding of problem for 2 seconds and then answer that "*di soaliniditanyakanhasilyangterbesarantara integral kiri dan kanan*". This is consider preservice teacher respons.

P : *Maksud soal apa mas? (menanyakan kedua kalinya)*
F : *Di soal ini ditanyakan hasil yang terbesar antara integral kiri dan kanan (menunjuk soal)*
P : *Integral kiri dan kananya?*
F : *Iya ... (berpikir 5 detik) hem ... maksudnya hasil integralnya pak*
P : *O hasilintegralnya ya?*
F : *Iya pak ... jadi jika ini diintegralkan .. dikerjakan yang kiri hasilnya sesuatu dan yang kanan juga hasilnya sesuatu nanti hasil keduanya dibandingkan. Yang paling besar itu jawabnya pak*

In the prediction phase, the subject answered that the answer of that question is right. This means the best results is $\int_1^2 x\,dx$. The consider preservice teacher respons below.

P : *Kira-kira menurut mas hasilnya apa ya?* ***Predicting***
F : *Hasilnya .... (berpikir 10 detik)yang kanan pak yaitu* $\int_1^2 x\,dx$
P : *Trus*
F : *Maksudnya pak?*
P : *Dari mana masbisamenjawabitu?*
F : *Ini pak kalau kita gambar (subjek tidak menggambar tapi menjelaskan dengan memainkan tangannya) yang kiri melalui nol dan sampai satu sedangkan yang kanan juga melalui nol tapi dari satu sampai 2. Jadi lebih luas kanan.*
P : *terus*
F : *Ya jelas jawabannya lebih besar yang kanan pak.*
P : *O...o...*

While foreseeing, the subject in finishing the problem for five seconds is donewhile thinking about the results obtained.It isevident from the answers of a subject that are integral to the right is larger. Then, when the subject asked to indicate the results which have been obtained, the subject began to explain the results that have been obtained (prediction). Consider the following interview excerpts.

P : *Menurut pendapat mas, kira-kira jawaban dari soal itu apa mas?* ***(Predicting)***
F : *(berpikir5detik) yang kananpak*
P : *Yakin?*
F : *Yakin pak*
P : *Coba tunjukkanmz!* ***(Foreseeing)***
F : *Ini yang integral kiri gambarnya seperti ini (gambar sebelah kiri dengan batas 0 sampai 1) sedangkan yang integral kanan gambarnya seperti ini (gambar*

sebelah kanan dengan batas 1 sampai 2)
P : *Yakin mas?*
F : *Iya yakin ...*

The subject explains the problem based on geometric shapes. Based on the excerpts of the interview above, subject verbally explains that "this integral picture left like this (*gambarsebelahkiridenganbatas 0 sampai 1*) while the integral right of the picture like this (picture to the right with a limit of 1 to 2)". The explanation of the subject can be seen in figure 2 below.

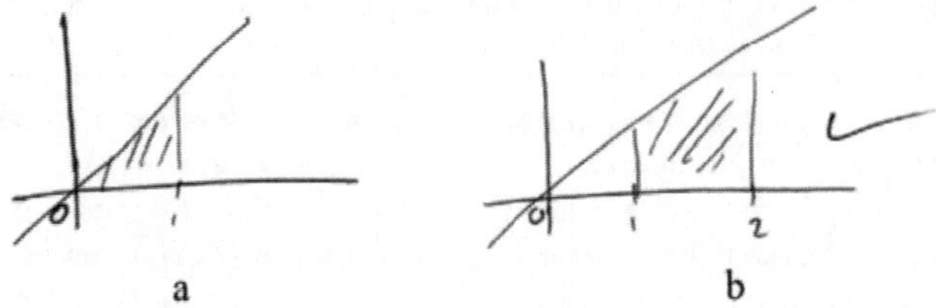

a b

Figure 2. Foreseeing of Subject (Geometry Shape)

In the Figure 2.a, subject describes the functions $\int_0^1 x\,dx$ and the figure 2.b; the subject describes the functions $\int_1^2 x\,dx$ based on geometric shapes. The subject compares both of these functions using a graphic illustration. He assumed that the area $\int_1^2 x\,dx$ wider than the area $\int_0^1 x\,dx$ So that the subject concluded that the function of the right side is larger than the function of the left side. Alsoby using the geometric form, the subject also calculations for convinced researchers on this problem. The result of the calculation subject can be seen in the following figure.

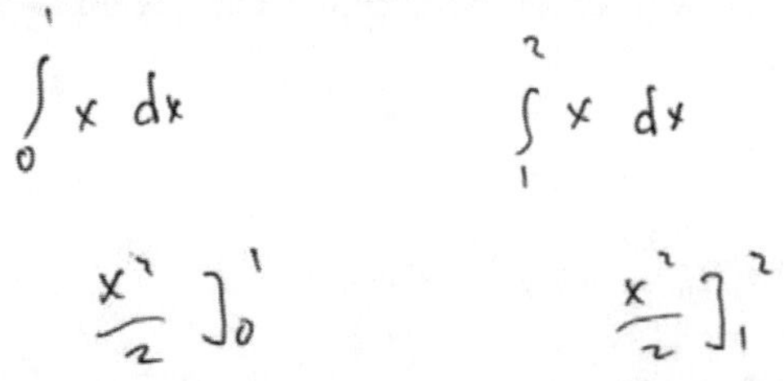

a b

Figure 3. Foreseeing of Subject

The subject explained that the calculation results together equal the results made of a sketch graph (Figure 2). When the subject is substituting the boundaries of function $\int_0^1 x\,dx$ and function $\int_1^2 x\,dx$, the subject goes on to explain the results to be obtained the larger is the right side. The importance of conceptual structures to solving the problem that "the basic and the most determine anticipation is a conceptual structure[8]. Building the confidence of students, [9] said that the importance of anticipating student responses in resolving the problem. This method is used to build a strategy on how students in solving mathematical problems. Furthermore, [10] said that current students need anticipation guide in solving a mathematics problem. However, students with high math ability are not necessarily anticipated analytically. These results were reported by [11] that student high mathematical ability internationalised anticipation. As a note, theunsuccessful anticipation of student thinking might reveal to teachers' need to rethink the ways in which learning occurred in the classroom. What is purely an error of students? It is also probably because of there are contributions from teachers. What is purely an error of students? No, it is also possible there are contributions from teachers. This case was reported by [12] that equally important is a teacher's ability of further accommodate students learning when there is a mismatch between the teacher and student perspectives.

Based on above that can be concluded the subject using explorative anticipation. It can be seen from the way the subject matter of counting and drawing the illustrations in front of the subject face. Based on research reports [13] regarding the anticipation of explorative (1) read the questions more than once, (2) the finding might ask and what is known, (3) outlining the problem in detail, (4) incorporate criteria known, and (5) solve the problem by considering an alternative solution. Further results of the study [14] if subject explorative anticipation, then the subject will try (trial and error) even though in the end the subject right in understanding it.

4. Conclusion

Based on the above results it can be concluded that preservice teacher to solve the problems of integral to analyse a given problem through an initial guess (predicting) then do the problem is not in detail (foreseeing) but the result is true. The results of this research subject can be categorised in explorative anticipation.

5. References

[1] J. Piaget, , 1985 IL: University of Chicago Press,.
[2] K. H. Lim, 2006 in *Proceedings of the 28th annual meeting of the North American Chapter of the International Group for the Psychology of Mathematics Education*, vol. 2, pp. 102–109.
[3] E. von Glasersfeld, 1998, pp. 1–10.
[4] G. Harel, 2006, pp. 1–26.
[5] G. Harel, 2014 *Intellectual Need*. Springer Science+Business Media New York.
[6] G. Harel and L. 2005 Sowder, *Math. Think. Learn.*, vol. 7, no. 1, pp. 27–50.
[7] L. J. Moleong, 2011Bandung: PT Remaja Rosdakarya.
[8] P. Cobb, 1985 *Educ. Stud. Math.*, vol. 16, pp. 111–126.
[9] S. Evans and M. Swan, 2014 *J. Int. Soc. Des. Dev. Educ.*, vol. 2, no. 7, pp. 1–31.
[10] A. E. Adams, J. Pegg, and M. Case, 2015. [Online]. Available: http://www.nctm.org/Publications/mathematics-teacher/2015/Vol108/Issue7/Anticipation-Guides_-Reading-for-Mathematics-Understanding/.
[11] E. Yudianto, *Kreano*, vol. 6, no. 1, pp. 21–25, 2015.
[12] D. Kotsopoulos and S. Lavigne 2008 *Int. Electron. J. Math. Educ.*, vol. 3, no. 1, pp. 1–23.
[13] E. Yudianto, "STUDI KASUS : KARAKTERISTIK ANTISIPASI EKSPLORATIF," vol. 6, no. 1, pp. 1–6, 2016.
[14] E. Yudianto, 2016 in *Seminar Nasional Matematika dan Pendidikan Matematika UNY*, pp. 327–334.

IOP Conf. Series: Journal of Physics: Conf. Series **824** (2017) 012056 doi:10.1088/1742-6596/824/1/012056

Representation Elements of Spatial Thinking

F R Fiantika[1]

Universitas Negeri Surabaya

[1]Corresponding author: fentfeny@gmail.com

Abstract.This paper aims to add a reference in revealing spatial thinking. There several definitions of spatial thinking but it is not easy to defining it. We can start to discuss theconcept, its basic a forming representation. Initially, the five sense catch the natural phenomenon and forward it to memory for processing. Abstraction plays a role in processing information into a concept. There are two types of representation, namely internal representation and external representation. The internal representation is also known as mental representation; this representation is in the human mind. The external representation may include images, auditory and kinesthetic which can be used to describe, explain and communicate the structure, operation, thefunction of the object as well as relationships. There are two main elements, representations properties and object relationships. These elements play a role in forming a representation.

1. Introduction

There are several definitions of spatial thinking, but it is not easy to define it. This study refers to the definition of spatial thinking by NRC [1] states "Spatial thinking is a collection of cognitive skills. The skills consist of declarative and perceptual forms of knowledge and some cognitive operations that can be used to transform, combine, or otherwise, operate on this knowledge. The key to spatial thinking is a constructive amalgam of three elements: concepts of space, tools of representation, and processes of reasoning". In the following descriptions will explain further and discuss the concept and it is the basic a forming representation.

Initially, the human five senses will detect and capture the natural phenomenon then transfer it into memory for processing. There is a coding process that involves abstraction information to carry out the relevant information and ignore the irrelevant information. Abstraction plays a role in processing information into a concept. The information that relevant gathers up into a concept and create a structure that calls a schema concept. "A schema (plural: schemata) is a mental model, a way of understanding the world, a set of Assumptions or understandings about reality" [2]. The concept and structure are placed in the internal memory. This concept and structure will have to mean if it is expressed by the good symbol in the form of visual, auditory and kinesthetic. In the process of internal concept into theexternal concept in the form of visual, auditory and kinesthetic will continue as the external of representation. There are three steps of spatial thinking is digging the spatial structure, spatial transformation and draw conclusions. Representation plays an important role in exploring the spatial structure. There are two types of representation items, namely internal representation and external representation. The internal representation is also known as mental representation; this representation is in the human mind. The external representation may include images, auditory and kinesthetic which can be used to describe, explain and communicate the structure, operation, thefunction of the object as well as the relationships. There are two main elements, representations properties and object relationships. These elements play an important role in forming a representation.

1.1. The Characteristics of Representation Object

According to NRC:41 [1] stated that Representations is whether in the mind or external, map elements and their relations in the world to elements and their relations in the represented world. It means that the representation happened started from our five senses that able to detect a phenomenon. The elements and the relationship of the detections phenomenon will form into code in human mind.

There is anabstraction in the process of coding which is the elements, and therelevant relationship will be used, but the elements and theirrelevant relationship will be ignored. Abstraction of the elements and the association will be named, and eventually formed a concept. Someone said to master a concept that can describe the properties or characteristics and relationships of the objects.

This research begins from digging the characteristics of student's representation object in spatial thinking in the topic material of "Cube". How the process of students identifies then catches the cube model and mentioned the elements how the formula of the cube formed and described the abstraction process up to forming in the cube concept in their mind. Further revealing about how students express a cube that exists in the minds of students into the external representation whether in the form of drawings, models or using words. Data obtained in the form of a direct description of the interview.

1.2. Relationships between Object

Relationships between objects need to be considered to expand the concept. There are two relationships between objects are studied in this paper is the relationship between static objects and dynamic relationships between objects. A static object is an object that does not have movement. Comparison between one or more static objects can be seen as the relationship between static objects. Determining the orientation, distance measuring, comparing the colour, texture, the location, the shape, direction, and other attributes as an indicator to determine the relationship of static objects [1].

While the relationship dynamic objects to compare two or more objects are dynamic or moving includes directions of movement, locomotion, speed, intersections and collisions [1]. The dynamic object will move in a way and a different direction, change the texture, shape and colour in a way that is orderly and predictable.

2. ResearchMethods

This research is a descriptive exploratory with a qualitative approach. The subjects of this research are three students, each of them has learning style visual, auditory and kinesthetic ability relative same. The subject is SDI AL-Irsyad Kediri, and they used Google SketchUp as geometry media teaching learning.

The spatial representation of the students explored the way students are assigned to draw a cube models, to determine the orientation of students in drawing the next cube models, students are assigned to compare with any other cube models drawn to the orientation of different viewpoints. Students compare shapes, elements owned by the two pieces of the cube models, and the viewpoints of the two models of the cube. Students evaluate, compare and then summed up the results of such comparisons.

To find out how these dynamic processes of object relationships, students are assigned to analyse two models of the cube rotated in a way and a different direction. Students identify whether by way of rotating the different elements of the two cubes remain and how these relate to the viewpoints of both the cube. The results obtained by these students were exhumed by using direct interviews. Information obtained by investigators then analysed and presented in the form of a description of the representation of students in fulfilling the terms of the two elements of the spatial representation.

3. Result and Discussion

The preliminary studies have shown that every studenthas different kinds of style to construct it and the representation of their understanding. However, it also depends on their habits and experiences. Preliminary pilot studied have been shownin Table 1.

Table 1. Visual, auditory and kinesthetic student representations

Visual Student	Auditory Student	Kinesthetic Student
• Students draw a cube model I using the menus from isometric	• Students draw a cube model I using the menus from the front	• Students draw a cube model I using the menus from the left

IOP Conf. Series: Journal of Physics: Conf. Series **824** (2017) 012056 doi:10.1088/1742-6596/824/1/012056

Visual Student	**Auditory Student**	**Kinesthetic Student**
looks.  (a) • Students draw a cube model II using the menus from the on top looks. 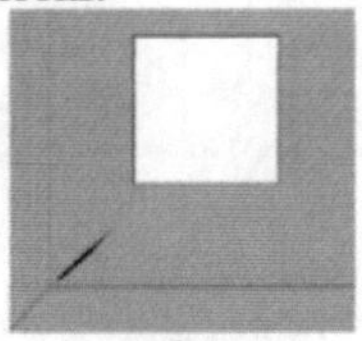 (b) **Figure 1** (a) cubes isometric looks, (b) cubes top looks	looks.  (c) • Students draw a cube model II using the menus from the back side looks.  (d) **Figure 2** (c) cubes front looks, (d) cubes back side looks	side looks.  (e) • Students draw a cube model II using the menus from theback side looks.  (f) **Figure 3** (e) cubes left side looks, (f) cubes back side looks
• Students identify the cube's elements using pictures, more than one picture's model.	• Students identify the cube's elements with images and mention the words.	• Students identify the elements of the cube with pictures and explain it with body movements.
• They mention the elements of thecube by using the pictures more than one picture and accompanied by the necessary words.	• They mention the elements of thecube in detail using words.	• They mention the elements of acube with movement, words and occasionally show a picture.
• In explaining the relationship between these elements, it does not mention the origin clearly due to work according to their intuition.	• They mention the relationship between these elements in detail and clearly	• They mention the relationship between these elements by showing a picture followed by body movements.
• They can rotate the model from different angles and use the toolbar menu to another by using two images for reference rotate the model image.	• They rotate the model image within 2-3 rounds only because sometimes they still confused by references used	• They rotate the model image within 2-3 rounds with theoccasional demonstration.
• They can mention the orientation and direction of the cube which rotated.	• They can mention the rotary directions but sometimes confused to determine its orientation.	• They canmentiongrabbing orientation. To demonstrate and determine the direction of therotary.
• Students conclude elements of cube consistently despite cube rotate objects. The worldview affects the shape of a cube when drawn.	• Students conclude elements of thecube that built from flat elements and perspective of someone that affect the image of acube.	• Students conclude the elements of the cube that did not see all in the picture but depending on someone's perspective on the picture.

4. Conclusion

Arguing from thebasic definition of spatial thinking, they have three basic elements. One of them is representations. Representation has two elements that students able to show it when they construct their concepts. The preliminary studies have shown that every studenthas different kinds of style to construct it and the representation of their understanding. It depends on their habits and experiences. The activities in this research were designed that students able to interact and to handle all of the problemsencounter with their capabilities by using google sketchup and their spatial experiences. They can handle all of themanipulatives such as how to connect their spatial experiences in the form of apicture (cubes) using google sketchup. The different result showed by visual, auditory and kinesthetic students can be seen on the differences of how they represent their capabilities. To clarify of this research to be useful, it has to be extended into subsequent representation research.

References

[1] Abidin, A. A, Rezaee, A. A. , Abdullah, H. N, Sigh, K. K. B. , 2011. *Learning Styles and Overall Academic Achievement in a Specific EducationalSystem.* Malaysia: International Journal of Humanities and Social Science Vol. 1 No. 10; August 2011.
[2] Ding, C. S. , Song, K. , Richardson, L. I. , 2007. Do Mathematical Gender Differences Continue? A Longitudinal Study of Gender Difference and Excellence in Mathematics Performance in the U.S. America: Eric Web site: http://www.LEAonline.com.
[3] Embree M C 2009 *PSY 202 - Introductory Psychology* University of Wisconsin (http://online.uwc.edu/academics/courses/psy-202)
[4] Ernest, Paul, 2014. *Gender Equity for Mathemativs and Science; Invite Faculty Presentation; Elizabeth Fennema*. UK: Philosophy of Mathematics Education Jurnal No 28 (October 2014). Online.
[5] Hegarty, M., &Tarampi, M. R. 2016. *Teaching Spatial Thinking: Perspectives from Cognitive Psychology*. In H. Burte, T. Kauppinen, & M. Hegarty (Eds.), Proceedings of the Workshop on Teaching Spatial Thinking from Interdisciplinary Perspectives *(TSTIP 2015) at COSIT 2015 in Santa Fe, NM* (pp. 36-44). CEUR-WS.org, online: ceur-ws.org/Vol-1557/.
[6] NRC 2010 *Learning to think Spatially*(Washington DC: The National Academies Press)
[7] Turgut M 2009 *Spatial Ability of a Mathematics Teacher*: The Case of Oya. Turkey: IBSU Scientific Journal .
[8] Turgut M , 2015. Individual Differences in the Mental Rotation Skills of Turkish Prospective Teachers. Turkey: Faculty of Education Eskisehir.
[9] Uttal H, Miller DI, and Newcombe NS 2015. *Exploring and Enhancing Spatial Thinking*: Link to Achievement in Science, Technology, Engineering, and, Mathematics (STEM)?. NorthWestern University: SAGE.
[10] Van der, Merwe F 2011 *Concepts of Space in Spatial Thinking*. South Africa: University of Pretoria.
[11] Vincent A & Ross D 2001 *Learning Style Awareness*: A Basis For Developing Teaching and Learning Strategies. University of Louisiana at Lafayette.

Creating Dynamic Learning Environment to Enhance Students' Engagement in Learning Geometry

Sariyasa

Department of Mathematics Education, Ganesha University of Education
JalanUdayana, Singaraja, Bali, Indonesia

Corresponding author: sariyasa64@yahoo.com

Abstract. Learning geometry gives many benefits to students. It strengthens the development of deductive thinking and reasoning; it also provides an opportunity to improve visualisation and spatial ability. Some studies, however, have pointed out the difficulties that students encountered when learning geometry. A preliminary study by the author in Bali revealed that one of the main problems was teachers' difficulties in delivering geometry instruction. It was partly due to the lack of appropriate instructional media. Coupling with dynamic geometry software, dynamic learning environments is a promising solution to this problem. Employing GeoGebra software supported by the well-designed instructional process may result in more meaningful learning, and consequently, students are motivated to engage in the learning process more deeply and actively. In this paper, we provide some examples of GeoGebra-aided learning activities that allow students to interactively explore and investigate geometry concepts and the properties of geometry objects. Thus, it is expected that such learning environment will enhance students' internalisation process of geometry concepts.

1. Introduction

Improving the effectiveness of geometry instruction, such as improving students' engagement, providing learning condition to stimulate cognitive activities, encouraging students to discuss and to share, is not always easy to be implemented. There are some obstacles that challenge teachers, one of them is the tendency to use the traditional approach. In a traditional geometry course, students are told definitions and theorems and assigned problems and proofs; they do not experience the discovery of geometric relationships, nor invent any mathematics [1]. In line with this, Reys et al.[2] suggest that geometry is best learned in a hands-on, active manner, one that should not rely on learning about geometry by reading from a textbook. A study by the author revealed that most teachers have difficulties in delivering some geometry concepts [3]. It was partly due to the lack of availability of media that can visualise geometry concepts. Teachers are aware of the importance of media, but they do not fully utilise media in their geometry instruction. If they use media, then their media is static ones, it cannot be manipulated dynamically.

One possible solution to overcome this problem is to create a technology-aided learning environment. The integration of computer in the instruction will assist students to imagine geometry concept and to make observation [4]. Some studies have shown that technology can be use in the instruction and give positive impacts on the quality of the instruction (see for instance [5]; [6]; [7]). Moreover, technology can be used to strengthen student learning and enhance pedagogy [8]. The use of technology in the instruction has some benefits such as providing students with greater chances of study [9].

There are two forms of technology use in mathematics instructional, i.e., virtual manipulatives and mathematical software tools [10]. Due to the limitation of virtual manipulative to do mathematical experiments, many teachers use educational software packages that enable both teachers and students to visualise and explore mathematical concepts in their creative ways (Barzel, 2007 in [10]). Examples of educational software are dynamic geometry software, computer algebra system, and spreadsheet [10].

Technology is not intended to replace mathematics. It is intended to support the learning and teaching of mathematics instead. Empirical results obtained from the use of technology give a better insight into the writing of a formal proof. Furthermore, the use of technology in mathematics teaching should support and facilitate conceptual development, exploration, reasoning, and problem-solving[5]. Also, NCTM has stressed that teachers should use technology to enhance their students learning opportunities by selecting or creating mathematical tasks that take advantage of what technology can do efficiently and well-graphing, visualising, and computing [11].

2. Dynamic Learning Environment

Integration of technology in geometry instruction necessarily needs a computer software to support the learning process effectively. We choose the second form of utilising technology where a dynamic geometry software (DGS) is employed. DGS has some tools that can be activated by clicking the tool buttons with the mouse. Afterwards, the tools can be applied to do construction, exploration, and investigation. DGS has three main features that usually cannot be found in computer algebra system or spreadsheet, i.e. drag mode, customizable tools, and trace or locus of objects (Graumann, et al., 1996 in [10]).

With the aid of DGS, a dynamic learning environment (DLE) can be established to facilitate geometry instruction. In DLE, students can quickly and accurately explore geometrical objects, discover patterns, and formulate a conjecture. Delivering geometry instruction in a way that encourages exploration and stimulates active participation in the learning are expected to improve the quality of students' learning. Hence, one goal of using DGS is to provide a learning environment which allows students to explore geometry objects and investigate their attributes and properties while preserving the rigorous nature of mathematics. In DLE, students are offered the opportunity to explore, to make a conjecture, to discover, and to experiment in a way that cannot be done with just pencil and paper [12]. DLE enables extensive and instant manipulation through the power of drag mode. This feature allows students to instantly experiment with as many examples as possible and get immediate feedbacks.

In this paper, we use GeoGebra as our DGS because it is free, is easy to operate, and has features to do dynamic exploration. GeoGebra allows teachers and students to conduct exploration easily and interactively, thus encourage students to involve actively in the learning process. In this way, understanding and internalisation of geometry concepts and their properties become easier. GeoGebra positively affects students' learning achievement and improves motivation [4]. The study has shown that students' understanding of mathematics improved after using DGS [13].

3. Examples of Exploration with Geogebra

In the following, we provide some examples of activities that utilise GeoGebra to carry out classroom investigation. These examples provide a dynamic learning environment so that the geometry instruction becomes meaningful and joyful for students.

3.1. Median of a Triangle

The following discussion is adapted from [14]. In GeoGebra graphic window, draw the triangle ABC. Mark a point on one of the sides say side AB. Call this point D. Draw a line through D and parallel to BC. This line intercepts side AC at point E. Draw segments BE and CD and name the intersection of these two segments M. Finally, draw the line AM and extend it to meet side BC at point N. See figure 1a. Ask students to measure and record the length of BN and CN. Students are then requested to move point D with the mouse along the side AB and observe the length of BN and CN. Now, change the shape of the triangle by dragging one of its vertices with the mouse. One possible result of this action is shown in figure 1b. Guide students to conclude that the point N is the midpoint of the side BC, regardless of both the position of D on the side AB and the shape of the triangle. Hence, the segment AN is always a median of the triangle. The next task is to prove that AN is always a median. Teachers can invite students to engage in a discussion to mathematically prove this result.

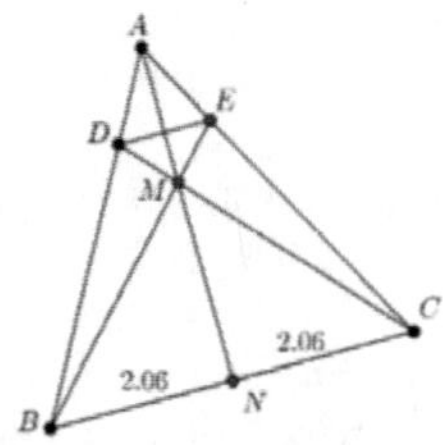

Figure 1a. Segment *AN* is a median

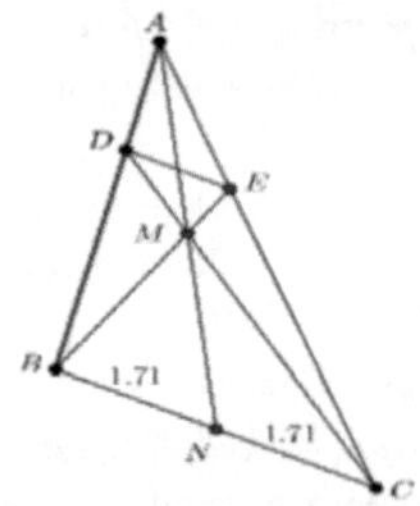

Figure 1b. Different shape of triangle, *AN* is still a median

The investigation can be continued further on. Teachers may ask students to explore the properties of three medians of the triangle (such as the three medians of any triangle are concurrent; the centroid of a triangle divides each median in the ratio 1:2). We notice that each median divides the triangle into two smaller triangles. Students easily verify using GeoGebra that the area of these two triangles is the same. The three medians divide the triangle into six smaller triangles. The area of these six triangles is always equal regardless of the type of the original triangle. Using GeoGebra, this property can be reinvented by students just by dragging the vertices of the triangle and observe the changes that happen. Students can investigate the type of the triangle for which the six smaller triangles are congruent to each other. An assignment can be given to students to prove the results obtained in this investigation formally.

3.2. Midpoint Quadrilateral

Consider an arbitrary convex quadrilateral. Connect the midpoints of the sides of the quadrilateral to form a new quadrilateral which is known to be a parallelogram. See figure 2.Teachers can provide learning activities supported by GeoGebra-based instructional media to investigate this fact. In GeoGebra, start with four points and construct a convex quadrilateral with these four points as vertices. Mark the midpoint of each side and then join them to form a new quadrilateral.

Figure 2. Midpoint quadrilateral is a parallelogram

Ask students to manipulate the original quadrilateral by dragging one of its vertices. Notice that as the shape of the original quadrilateral changes, the shape of the resulted quadrilateral also changes. Students are instructed to observe the resulted quadrilateral and record the measure of all angles, the length of each side, and the position of two opposite sides. In a group, students can discuss and decide that the resulted quadrilateral is a parallelogram. Ask students to replace the original quadrilateral with a square, a rectangle, a rhombus, a trapezium, or a kite. Then, observe and discuss what happen to midpoint quadrilateral. Furthermore, students can investigate the conditions for which the midpoint quadrilateral is a rectangle, a square, a rhombus, or a kite.

3.3. Midsegment of a Triangle

Following the discussion of this topic in [15], we shall use GeoGebra to assist students to formulate the midsegment theorem of a triangle. In the graphic window of GeoGebra, construct an arbitrary triangle with vertices A, B, and C. Put one point, say M, on any side of the triangle, in this case, the side AB. Construct a line segment through M and parallel to the side BC and cut the side AC at N. See figure 3.

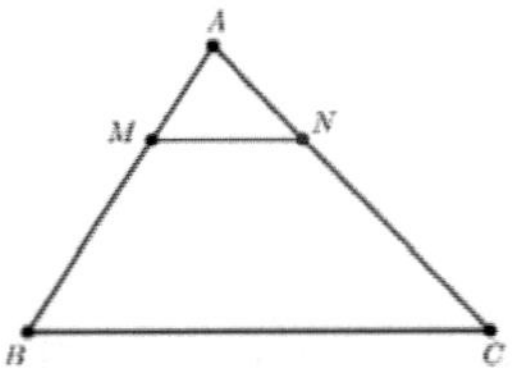

Figure 3. Segment *MN* is parallel to *BC* with *M* is an arbitrary point on *AB*

Determine the ratio AM/BM and AN/CN. Record this ratio on a piece of paper or the worksheet in GeoGebra window. Also, determine the ratio MN/BC. Now, here is the dynamic part of the activities. Teachers instruct students to use the mouse to drag the point M to a new position and note the three ratios. Repeat this process until AM/BM = AN/CN = 1. When this happened, teachers ask students to find the ratio MN/BC. Students can repeat these activities for a different type of triangle by dragging one of the triangle vertices to change the shape of the triangle. It is expected that students, after group discussion under teacher's guidance, can state the midsegment theorem.

4. Conclusion

Geometry is a good start to learn more about axiomatic systems and the deductive thinking process. One way to improve the effectiveness of geometry instruction is the integration of technology. Coupling with dynamic geometry software, a dynamic learning environment can be established to provide explorative and multi-representation learning, thus make learning more meaningful. With the aid of GeoGebra-based instructional media supported by well-designed lesson plan, it is possible for students to visualise and investigate geometry concepts and their properties dynamically. It is thus expected that students engage in the learning process more actively and deeply. Therefore, such learning environment will enhance students' internalisation and understanding of geometry concepts.

References

[1] King J R and Schattschneider D 1997 *Geometry turned on!* (Washington DC: MAA)
[2] Furner J M and Marinas C A 2007 *Eurasia Journal of Mathematics, Science & Technology Education* **3** 83-91
[3] Sariyasa 2015 *Proc. 1st Int. Conf. on Innovative Research Across Disciplines (Denpasar)* pp 32-35
[4] Doğan M and Içel R 2011 *International Journal of Human Sciences* **8** 1441- 1458
[5] Garofalo J, Drier H S, Harper S and Timmerman M A 2000 *Contemporary Issues in Technology and Teacher Education* **1** 66-88
[6] McPherson R F and Tyson V 2006 *Contemporary Issues in Technology and Teacher Education* **6** 293-301
[7] Keong C C, Horani S and Daniel J 2005 *Malaysian Online Journal of Instructional Technology (MOJIT)* **2** 43-51
[8] Dede C 2000 *Journal of Curriculum Studies* **32** 281-303
[9] Roberts G R 2012 Technology and learning expectations of the net generation *Educating the Net Generation*, eds D G Oblinger and J L Oblinger
[10] Preiner J 2008 *Introducing dynamic mathematics software to mathematics teachers: the case of GeoGebra* (Dissertation, University of Salzburg)
[11] NCTM 2000 *Principles and Standards for School Mathematics* (Reston, VA: National Council of Teachers of Mathematics)
[12] Jones K 2000 *Educational Studies in Mathematics* **44** 55–85
[13] Haciomeroglu E S and Andreasen J B 2013 *Mathematics and Computer Education* **47** 6-18
[14] Johnston-Wilder S and Mason J 2005 *Developing thinking in geometry* (London: Paul Chapman Publishing) pp 39-43

[15] Billstein R, Libeskind S and Lott J W 2016 *A Problem solving approach to mathematics for elementary school teachers 12th edition* (Boston: Pearson Education International) p 798

IOP Conf. Series: Journal of Physics: Conf. Series **824** (2017) 012058 doi:10.1088/1742-6596/824/1/012058

Density, Viability Conidia And Symptoms of *Metarhizium anisopliae* infection on *Oryctes rhinoceros* larvae

D R Indriyanti[1,*], R I P Putri[1], P Widiyaningrum[1], and L Herlina[1]

Biology Department, Faculty of Mathematics and Natural Sciences, Universitas Negeri Semarang, Gedung D6 1st floor, Jalan Raya Sekaran-Gunungpati-Semarang-Indonesia 50299

*Corresponding author: dyahrini36@gmail.com

Abstract. *M. anisopliae* is parasitic fungus on insect pests; itis used as a biocontrol agent. *M. anisopliae* can be propagated on maize or rice substrate. *M. anisopliae*is currently sold in the form of kaolin powder formulations. Before it is used to check the density, viability and pathogenicityof *M. anisopliae*.However the problem is the kaolin powder very soft, so it difficult to distinguish between kaolin and conidia.This article gives information on how to calculate conidia density, viability and symptoms of *M.anisopliae*infection on *Oryctes rhinoceros* larvae.The study was conducted in the laboratory to determine the density and viability.The pathogenicity testing was done using pots. The Pot is containing soil substrate mixed with *M. Anispoliae* and ten tails *O. Rhinoceros* larvae per pot. The results showed that the density of *M. anisopliae* conidia was 1.81 x 10^8 conidia mL^{-1} and the viability was 94% within 24 hours. The larval mortality began to emerge in the 1stweek, and all larvae died at thesixth week. The symptom of *M.anisopliae*infection on *Oryctes rhinoceros* larvae, there was ablack spot on the larval integument. The larvae movements become slow and poor appetite; it will die within 3-7 days. The larvae diehard,and the white hyphae grow on the body surface that turns green.

1. Introduction

Oryctes rhinoceros beetle is one of the major pests attacking coconut crops. The *O. rhinoceros*attack by 2015 in Central Java reaches 10,240.86 Ha [1]. The beetle attacks base of the stem top of the coconut palm which makes the damaged on growing point. The beetle cuts unopened fronds causing the "V" shape cut on open fronds. It is the characteristic of *O. rhinoceros'* attack. The beetles lay eggs on the ground, while the larvae eat the organic matter such as the dirt animal, organic waste, and the trunks of coconuts rotted and the remnants of sugarcane [2].

One of the biological agents to control the larvae of *O.rhinoceros*isis *Metarhiziumanisopliae,*parasitic fungi. *M. anisopliae*can infect many insects. The best host for breeding *M. anisopliae*is the *O.rhinoceros*larvae [3-5].

Laboratory-scale research suggests that *M. anisopliae*is effective in controlling the populations of some insects such as*Spodopteralitura* i.e. [6,7], *Coptotermescurvignathus*[8], *Stibaropusmolginus* [9], *Lepidiota stigma* [10]. The fungi *M. anispoliae* can be grown on corn or rice medium [11]. Currently,*M. anispoliae* sold in the form of kaolin powder formulation. It means that the conidia *M. anispoliae* mixed with kaolin. The density test and viability conidia are needed to find out if the fungus conidia quality is good or not. The problem is kaolin's powder is very soft; it is difficult to distinguish between kaolin's grains and conidia. If the solvent is water, then it is difficult to distinguish. Therefore, this article gives information on how to calculate the conidia density, viability of *M.anispoliae* and *M. anisopliae*infection symptoms on larvae of *O. rhinoceros*.

2. Methods

The density and viability conidia of fungi *M. anisopliae*testingwas done at BalaiProteksiTanaman Perkebunan (BPT-BUN) in Central Java, Salatiga. The infection symptoms of larvae *O.rhinoceros*observed in larvae fed on *M. anisopliae*, the experiment was done at Desa Jeruk Wangi's garden in Bangsri, Jepara, on March – May 2015.

2.1. Preparation of Fungi M. anisopliae

The fungi *M. anisopliae*was obtained from BalaiProteksiTanaman Perkebunan (BPT-BUN) Salatiga with the commercial name ZIUM OR WP. This product was floured, or kaolin media are containing *M. anisopliae*fungiconidia. Treatment doses used were 2 grams. The test of density and viability conidia of *M. anisopliae*fungi were donebefore used and applied to control the larvae of *O. rhinoceros*. There was a simple way to differentiate between conidia and kaolin; the flour was dissolved in a solvent special kaolin called *alkilarilpoliglycolether* (one brand is Agristick).

Conidia density calculation of *M. anisopliae*was done by 1 gram of *M. anisopliae*thatwas dissolved in 100 ml of aquadest and 1 ml of*alkilarilpoliglycol ether* (Agristick). The addition of *alkilarilpoliglycol ether* (agristick) was used as smoothing agent because of the difficulty to differentiate the kaolin and conidia. The suspension of the fungi *M. anisopliae*stirred with a glass stirrer for 2 minutes. Conidia suspension was taken as much as 1 ml using the eyedropper while stirring. Conidia suspension melted on the haemocytometer and then covered with glass cover and wait up to 1 minute until the location of the conidia stable. Conidia density calculated using a microscope with a magnification of 200 x on the haemocytometer (a + b + c + d + e). The treatment was repeated four times. Density was calculated by using the conidia formula:

$$S = \frac{t \, x \, d}{n \, x \, 0{,}25} x \, 10^6$$

Description:
S = number of conidia per gram of media culture
t = the number conidia which is calculated on the ' count media " (a, b, c, d, e)
d = degree of dilution
n = the number of small boxes observed (i.e. 5 x 16 = 80 small box)

The calculation conidia viability of *M. anisopliae* calculated with growing 0.05 ml conidia solution on *Potato Dextrose Agar* (PDA) media that has melted on glass objects using the eyedropper tool, wait until solidified (each one drop) and incubation for 8 hours, 12 hours, 24 hours in room temperature. The treatment was repeated three times. Conidia viability were observed and calculated using a microscope at a magnification of 200 x conidia between the life and the dead conidia using the formula:

$$V = \frac{a}{a + b} x \, 100\%$$

Description:
V: conidia viability percentage
a : the number of live conidia
b: number of dead conidia

2.2. The Larvae of O. rhinoceros

The *O.rhinoceros*larvae retrieved from the field at DesaJeruk Wangi, Sub-District Bangsri, Jepara. Three instar larvae used as many as 240, long size 7-10 cm long and weighs 9-11 grams per larvae.

2.3. *Media preparation*

The *O.rhinoceros*larvae needed organic soil media for maintenance. The media inserted into the black pot with 34 cm diameter, 22 cm height. Every pot in media content as much as 5 kg of ground mixture, coconut palm' sawdust, and manure fertiliser in the comparison of 1:1: 1.

2.4. Test Execution

Each pot contains 5 kg of mixed media with 2 grams of *M. anisopliae*was giventen *O.rhinoceros*larvae and the treatment was repeated six times. The observation was done every day to see the infection symptoms of *M. anisopliae*on larvae. Data collected: conidia density, conidia viability and infection symptoms of the *O.rhinoceros*larvae due to *M. anisopliae*. Data infection symptoms of *M. anisopliae*on larvae *O.rhinoceros*analysed descriptively.

3. Results and discussion

3.1. Density and Viability of Conidia M. anisopliae

Test results of density conidia *M. anisopliae*on kaolin media will be presented in Table 1.

Table 1.Density conidia *M. anisopliae*in kaolin

Count Field Box	Time			
Haemocytometer	1	2	3	4
a	10	9	15	8
b	9	2	9	5
c	8	6	8	5
d	11	2	4	7
e	9	10	6	2
Total	47	29	42	27
Density conidia	2.35×10^8	1.45×10^8	2.10×10^8	1.35×10^8
Average	1.81×10^8 conidiamL^{-1}			

Table 1 depicts that density conidia *M. anisopliae*1.81×10^8mL^{-1}conidia belongs to the good, about the standard quality of the fungi *M. anisopliae*as a biological agency must be $\geq 10^6$conidia/ml [12].Conidia viability of *M. anisopliae*kaolin media for 8 hours, 12 hours, 24 hours of incubation, presented in Table 2.

Table 2. Conidia viability *M. anisopliae*in the medium of kaolin

Time	Repetition	Total of observed conidia			Viability (%)	Average (%)
		Not Growing	Grow	Total		
8 hours	1	11	39	50	78	
	2	14	36	50	72	72.67
	3	16	34	50	68	
16 hours	1	7	43	50	86	
	2	8	42	50	84	84.67
	3	8	42	50	84	
24 hours	1	2	48	50	96	
	2	4	46	50	92	94.00
	3	3	47	50	94	

Table 2 shows that the viability of conidia *M. anisopliae*after 24 hours incubation was 94%. The results categorised as a good result, referring to the standard quality of mushrooms *M. anisopliae*as a biological agency (BPTBUN). Conidia viability is affected by temperature, humidity, pH, and the sunshine [13].

3.2. Symptoms of O.rhinoceroslarvae infected with M. anisopliae

The infection mechanism of *M. anisopliae*on*O.rhinoceros*larvae occurred in four stages: 1) the stage of inoculation, 2) snapping and germination, 3) penetration, 4) invasion and destruction at the point of penetration [14]. In inoculation stage, the contact occurred between fungi propagules (conidia) and the body of the larva. The second stage was the process of snapping and propagule germination of fungus on insects integumentary. The third stage was penetrating and invasion; it penetrated the integumentary and formed a sprouts tube (appressorium). The fourth stage was the destruction at the point of penetration, and the formation of blastospore spread out into hemolymph that formed secondary Hypha to invade other tissues. After insects die, mould will continue to cycle in a phase of continuing depression saprophytic. Mildew will form colonies around the body of the host. After the body of an insect host full of the colony of mould, then conidia will be produced to infect another host [15].

Early symptoms, contact and snapping occurred when fungi *M. anisopliae*mixed in media containing larvae.If humidity and moisture content high on a substrate, then conidia will germinate. Conidia stick to the integumentary germinates,then hypha continued to spread throughout larval' body. Dextrusin toxins released by the fungi *M. anisopliae*will spread throughout the body of the larva. Hypha of *M. anisopliae*fungiutilised the nutrients in the body of the larva [16].

Penetration symptoms and the invasion of *M. anisopliae*on*O.rhinoceros*larvae were indicated by ablackish spot on the abdomen of the larvae. The larvae with blackish spot will be dead within 3-7 days. The blackish spot was the sign that showed larvae had been infected by *M. anisopliae* [11].

Other symptoms of infected larvae could be seen through slow movement of the larvae and appetite reduced. Mould toxic affected on theparalysis of larva's body [17]. The paralysis led larva to lose its motion systems coordination became achaotic movement, gradually weakening the total paralysis occurs later (dead).

The symptoms of infected larvae were characterised with the dead larvae became harder. The surface of the body of the larva grows white hypha become green by the age conidia an increasingly mature. The *O.rhinoceros*larval body became hard, because of the entire system and the larval body fluids have been exhausted utilised by the fungus *M. anisopliae*. As a result, the dead insect's body became hard like Mummy [18].

Temperature Conditions, humidity, light intensity and pH of soil were also greatly affected by the success of the fungi *M. anisopliae*in infecting larvae. The field temperature while the study held was in the range 26 ° C - 28 ° C and 76-87 % humidity. The range of temperature and humidity supported the growth of the fungi *M. anisopliae*, as evidenced by the existence of the mortality of the larvae of *O.rhinoceros*due to the growth of fungi *M. anisopliae*.The optimum temperature for the growth of the fungi *M. anisopliae*occurred at a temperature of 22-27 °C, whereas the formation of sprouts on conidia occurred at humidity 80-92% [13]. It wasby the current condition of research.

4. Conclusion

The level of conidia density of fungi *M. anisopliae*in kaolin media was 1.81 ×**1** ^{8}Conidia/ml with 94% conidia viability in 24 hours. Symptoms of early infection were the blackish spots on the larval integument, larvae dead, and the larvae became hardened and embossed white hypha and gradually turn green.

References

[1] HidayantiE 2015 Surabaya: BBPPTP Surabaya.
[2] Mulyono. 2007 Tesis. Surakarta: Program Pascasarjana Universitas Sebelas Maret.
[3] Sambiran, W.J. & Hosang, M.L.A. 2007 Buletin Palma, (33): 9-17.
[4] Heriyanto &Suharno. 2008 *J. ilmu-ilmu Pertanian*. 4(1): 47-54.
[5] Manurung, E.M., Tobing, M.C., Lubis, L. & Priwiratama H. 2012 *Jurnal Online Agroekoteknologi*, 1(1): 47-63.
[6] Trizelia, My Syahrawati, & Aina Mardiah. 2011 *J. Entomol. Indon,* 8(1):45-54.
[7] Tampubolon, D.Y., Pangestiningsih, Y., Zahara, F. & Manik, F. 2013 *Jurnal Online Agroekoteknologi*, 1(3): 783-793.
[8] Khairunnisa, Atria Martina &Titrawani. 2014 *JOM FMIPA* ,1(2) : 430-438.
[9] Rosmayuningsih, A., Rahardjo, B.T. & Rachmawati, R. 2014 *Jurnal HPT,* 2(2): 28-37.
[10] Harjaka T, Arif Wibowo, F.X. Wagiman & Muhammad W. Hidayat. 2011 *Prosiding Semnas Pesnab IV*, Jakarta, 15 Oktober 2011.
[11] Gusmara BH. *(Skripsi)*. Depok: Universitas Indonesia.
[12] Balai Proteksi Tanaman Perkebunan. 2012 Salatiga.

[13] Prayogo, Y, W. Tengkono & Marwoto. 2005. *Jurnal Penelitian dan Pengembangan Pertanian* 94(1): 19-26.
[14] Freimoser FM, S Screen, S Bagga, GHu& St Leger RJ. 2003 *Journal of Microbiology* 149: 239-247.
[15] Marheni, Hasanuddin, Pinde&Wirda Suziani. 2010 Sumatra Utara: Agroekoteknologi Fakultas Pertanian USU.
[16] Yasmin,Y& Fitri L. 2010 *Jurnal Natural,* 10(1): 31-35.
[17] Wahyudi, P. 2008 *Jurnal Ilmu kefarmasian Indonesia*, 6(2): 51-56.
[18] Moslim, R., N. Kamaruddin, Ang, B. Na, S. R. A. Ali& M. B. Wahid. 2007 *J. Oil Palm Res*. 19 (1): 319-331.

IOP Conf. Series: Journal of Physics: Conf. Series **824** (2017) 012059 doi:10.1088/1742-6596/824/1/012059

Cordia dichotoma G. Forst. : Bioecology and population density

E S Rahayu[1,*], P Martin[1], N K Dewi[1], and F H Kurniawan[1]

[1]Department of Biology, Faculty of Mathematics and Natural Sciences, Universitas Negeri Semarang, Sekaran Street, Gunungpati Semarang Indonesia, 50229

*Corresponding author: enni_sr@mail.unnes.ac.id

Abstract. Kendal tree is declared as an identity flora of Kendal Regency, Central Java, Indonesia. It is predicted as a rare species; most of the local people do not know the existence of this tree. The study aimed to describe some aspects of bioecology and a population density of Kendal tree. An explorative study was conducted from March to July 2016, through interview, observation, and literature review. The respondents were determined by purposive and snowball sampling methods. The data were analysed descriptively. Results showed that there were very limited numbers (only five trees) of Kendal tree throughout Kendal Regency. The species was identified as *Cordia dichotoma* G. Forst. The tree is 3–15 m in height. The leaves are simple and arranged spirally, inflorescence dichotomous, bisexual, with five gamosepalous and gamopetalous. Generative reproduction occurred at a low rate. The tree grew optimally in a dusty, sandy loam soil, pH 6.0-6.5 with a temperature ranged from 27-34 °C, a light intensity of 450-1580 lux, and an altitude of about 10 meters above sea level. In conclusion, *C. dichotoma* is a rare plant in Kendal Regency. This plant needs an alternative method of propagation, regeneration and conservation using *in vitro* technique.

1. Introduction

Kendal Regency is one of 35 regencies/cities in Central Java Province, Indonesia. As other cities and regencies, Kendal has a plant species declared as an identity flora, namely Kendal tree. This declaration is based on the historical value of Kendal tree. The data compiled by Environment Agency of Central Java in 2015 show that the existence of the identity plant has not fully and well documented. In fact, most of the local people or people had lived for a long period in Kendal Regency also do not know this plant. As a flora of identity, the data of Kendal tree should be available completely.

The population density in an area is determined by the capacity of plants growth and development. The growth and development of plants are influenced by internal factors and external factors. External factors include biotic factors (competitors, pollinators, symbionts, and others) and abiotic (edaphic and climatic) factors. The width of the optimum range of abiotic factors determines the spaciousness of plant habitats. Edaphic factors can be divided into physical and chemical properties of the soil, whereas the climatic factors include temperature, rainfall pattern per year, and others. Also to the edaphic and climatic factors, people density also influences plant population through the intensity of cultivation. The more intensive cultivation of a species, the higher the plant population will grow.

The total area of Kendal Regency is 1,002.23 ha, which can be divided into lowland area with an altitude range from 0-10 m above sea level, and the Highland on altitude between 10-2.579 m above sea level. The lowland area is located in the northern as well as the coastal with temperature ranges around 27 °C, while the highland or mountainous area is located in the southern area has a temperature around 25 °C [1]. Kendal Regency is suitable for the growth and the development of Kendal tree. In the other countries, Kendal tree is abundant at low altitudes from sea level upwards to about 500 m altitudes, on the dry side of mangroves, coastal hills below 100 m, inland forests along rivers, and on limestone [2].

This study aimed to describe the bioecology aspects of Kendal tree and the population density at Kendal Regency. The results can be used as a recommendation of the plant germplasm preservation to get more beneficial to the society and environment. Also, the present study provides a baseline data for future pharmacological studies.

2. Materials and methods

The study was conducted in Kendal Regency during March-August 2016. The population studied was all (20) of districts of Kendal Regency, which can be divided into two groups, i.e. the lowland and highland districts. The lowland districts located along the northern coast consisted of 12 districts, namely Weleri, Rowosari, Kangkung, Cepiring, Gemuh, Ringinarum, Pegandon, Ngampel, Patebon, Kendal City, Brangsong, and Kaliwungu. Whereas, the highland district located in the southern part which includes eight districts, i.e. Plantungan, Pegeruyung, Sukorejo, Patean, Boja, Limbangan, Singorojo, and South Kaliwungu (Fig. 1).

Proportional random sampling took the area samples (30%). Area sample for lowland taken by random sampling was four districts, namely District of Kendal City, Patebon, Cepiring, and Kaliwungu. Area sample of the Highland taken by random was three districts, namely Plantungan, Boja and Limbangan. From each district selected as the sample, the observation areas were selected purposively based on the result of the interview before. The variables observed in this study were the biology, ecology and population density of Kendal trees. The biological aspects include morphological characters, reproduction techniques, and utility for human lives. The observed ecological aspects were altitude, soil texture, soil pH, temperature, and light intensity. The population density was determined by the number of the plant.

An early interview was carried out through face to face interviews with 30 informants in the age of 17 to 62 years; these included males and females to get information about the number of Kendal tree. They are the student, lecturer, and other person living in Semarang City who was born and had been lived for several years in Kendal Regency.

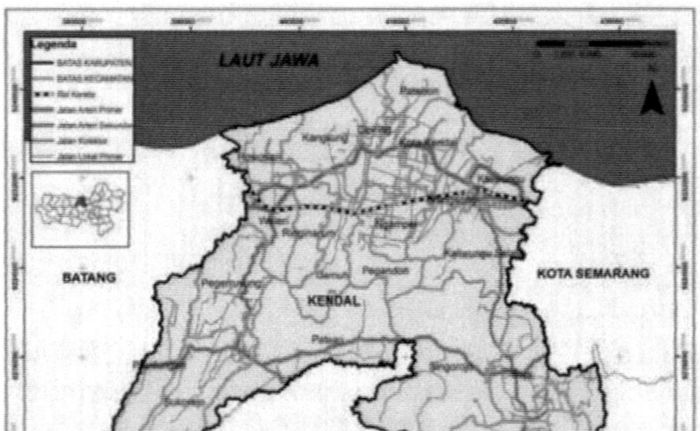

Figure 1. Map of Kendal Regency bordered by Batang Regency, Java Sea, Semarang City, and Temanggung Regency

The interview at an early step of the research was followed by some observations using observation sheet and depth interview with local people in the area selected as the sample. The observation methods were done to obtain some information including the morphological characters, reproduction techniques, the level of knowledge of the existence and utility of Kendal tree. The respondents of depth interview were selected from local people through purposive and snowball sampling methods; by considering the status and role in society and their adequacy of information about Kendal tree. They were employees of Regency office, village heads, and community leaders over 50 years old.

Snowball method was implemented by using recommendation of initial informants to other informants who considered know about Kendal tree.

The morphological characteristics of Kendal tree were observed carefully, taken photos in place or partially transported to the laboratory to be photographed using stereomicroscope. Environmental factors measured at a point of plants grow to a radius of 10 m from the outer of the plant canopy. After collecting the specimens, we represented these specimens to different people to confirm the accuracy of the results. Subsequently, specimens were identified at Bogor Botanical Garden, West Java.

Data collection tool were GPS (Global Positioning System), altimeter, thermohydrometer, soil tester, and digital camera. Finally, the data were analysed using descriptive statistics.

3. Results

The early interview showed that only one person of 30 respondents knew about Kendal tree. She said that there is only one Kendal tree in a sacred tomb located in the town of Kendal. Based on this information the observation was conducted directly to the location. The observation was also carried out at the seven districts which selected as area samples. We found five Kendal trees only, i.e., two trees in the area of the tomb of Wali Gembyang, Sub-district of Patukangan; one tree in the West Square, and two trees in the yard the Office of Kendal Regency. Observations conducted in the District of Patebon and Cepiring did not find any Kendal tree. It seems that Kendal tree name was a homonym which was the local name (i.e. Kendal tree) was used for some species. People in the District of Patebon and Cepiring called Kendal tree for *Ficus sp*. In Kaliwungu, Plantungan, Boja and Limbangan we also did not find any Kendal trees. Some respondents questioned it, including the people of old age did not know anything about the tree.

The five Kendal trees were found well grow and develop in a dusty, sandy loam soil, pH 6.0-6.5 with a temperature of 27-34 °C, a light intensity of 450-1580 lux at an altitude of about 10 meters above sea level.

Kendal tree has many synonym scientific names, i.e. *Cordia obliqua* auct. non. Willd., *C. dichotoma*, *C. blancoi* Vid., *C. griffithii* C.B. Clarke, *C. myxa* auct. non. L., *C. suaveolens* Bl., *C. suaveolens* Vidal, *C. subdentata* Miq. [4]. The taxonomic position of this species is as follows.

Kingdom	: Plantae
Subkingdom	: Tracheobionta
Superdivision	: Spermatophyta
Division	: Magnoliophyta
Classis	: Magnoliopsida
Subclassis	: Asteridae
Ordo	: Lamiales
Family	: Boraginaceae
Subfamily	: Cordioideae
Genus	: Cordia
Species	: *Cordia dichotoma* G. Forst
Local names	: cordia tree (English), nona burung, petekat, sekendal (Malaysia), kanonang, kendal, knadate, lantolo, mampapu, manonang, toteo (Indonesia)

IOP Conf. Series: Journal of Physics: Conf. Series **824** (2017) 012059 doi:10.1088/1742-6596/824/1/012059

Figure 2. The habitus, stem and leaf morphology of *C. dichotoma* which grown at Kendal Regency. Tree grew in the Kyai Gembyang tomb (a) and the yard of Kendal Regency Office (b), stem bark of old tree (c), lenticellate branchlet (d), leaf buds (e), and the leaf shape (f)

Kendal tree forms a rounded canopy. The stem erect, the old tree could reach about 15 m as found in the yard of Wali Gembyang tomb (Fig. 2a), but the relatively younger just only 3 m (Fig. 2b). Its branching is sloping upwards (Fig. 2 a, b). The bark of old stem shallowly fissured or wrinkled longitudinally, brown or blackish (Fig. 2c). The branchlets are slender; its lenticellate is grey, greyish brown to brownish (Fig. 2d). The leaf buds are longitudinally folded, flat, and brown to orange (Fig. 2e). The leaves are simple, arranged spirally, generally lanceolate, dentate leaf edge, acute at apex, and obtuse at base. The venation pinnate, lateral nerves 3–5 pairs (Fig. 2f). Petioles are 1.5-2.5 cm long, and leaf blade ranged at 6–10 x 3–4 cm.

The inflorescence located terminally (Fig. 3a) or axillary (Fig. 3b) of branches. It is a cyme, forking repeatedly or dichotomous, 4-11 cm in length, with 4 - many flowers (Fig. 3c). Flowers are bisexual, complete, short-stalked, and actinomorphic. A fully open flower is 3-5 mm in average diameter. The calyx is cup-shaped. Sepals are about 3-4 mm in length, light green and gamosepalous. The corolla has five yellow whitish colour petals which are 4-5 mm in length and gamopetalous. Androecium contains five stamens, each having a long filament and epipetalous. The ovary is located inferior (Fig. 3d), one locus (Fig. 3e), 4 mm in length and having a globular shaped ovary. In recent years, the Kendal trees formed many flowers, but only a few of flowers develop into fruits (Fig. 3f).

(a) (b) (c)

(d) (e) (f)

Figure 3. The Flower and Fruit of Kendal tree. A terminal inflorescence (a), axillary inflorescence (b), dichotomous flowering (c), longitudinal section of the ovary (d), the cross-section of one locus ovary (e), and among some flowers just one fruit is formed (f).

The breeding effort of *C. dichotoma* has been initiated by the Government of Kendal Regency by planting a few trees in the yard of regency office. Based on the interview with employees of the office, the obstacles encountered were the difficulty of breeding this plant. Seeds are difficult to obtain. The propagation by cuttings was also difficult. The people around the tomb of Wali Gembyang had tried to propagate through cuttings, but the success rate was only 3%.

4. Discussion

Based on the survey results, Kendal tree is very rarely used by local people. Therefore, the people almost have never been cultivated this plant. Some respondents stated that many people from outside Kendal Regency often seek the leaf, bark and fruit of Kendal tree for medicinal usage, especially for the diseases because of degeneration of human body function.

Kendal tree organs have proved to produce many secondary metabolites which useful in various fields. The leaves of *C. dichotoma* contains arabinoglucan and linolenic acid [5], apigenin, hentricontanol, hesperidin, octacosanol, lupeol, amyrin, robonine, rutin, betulin, dihydrorobenetin [6, 7, 8]. From the seed extract 2 compounds alpha-amyrin and 5-dirhamnoside have been isolated [2].

There are some studies evaluating many activities of organs of *C. dichotoma*, such as anti-microbial, anti-inflammatory, anti-fertility, anti-snake bite, hypolipidemic, insecticidal and antioxidant

[5, 9]. Its bark is often used for treating headaches and fevers [4]. The anti-inflammatory activity of the seeds extract at a dose of 500 mg/kg, aqueous and ethanol extracts show maximal inhibition of oedema [10]. It indicates the efficacy of the seed extract as an anti-inflammatory agent. Extract of the bark inhibits the growth of three common pathogenic fungi (*Aspergillus niger, Aspergillus clavatus,* and *Candida albicans*) [11]. The acetone extract of seed also shows significant antibacterial activity against *S. aureus* [12]. Meanwhile, the extract obtained from leaves has anti-fertility activity. A potential anti-implantation activity has observed. At a dose of 800 mg/kg, pregnancy was blocked 100%, and there were no alterations in the behaviour of test animal [6].

Besides being used for medicinal purpose, organs of *C. dichotoma* are also useful for various purposes. Fruit contains a sticky gum and can be used as paper glue. The wood is used for house construction or basic material for making agricultural tools (e.g. ploughs). The leaves yield good feed. The seed kernel contains a high proportion of fatty oils and proteins (46 and 31%, respectively) which have potential as cattle feed. Fruit extract has evaluated to suppress larval hatching of *Meloidogyne incognita* [2].

Considering the many uses of Kendal tree, it is feasible to breed this species in the area of Kendal Regency. Moreover, this tree has been declared as the flora of identity. Kendal tree should be grown in all areas of the regency. The ecological conditions of Kendal region are suitable for the growth and development of Kendal tree. *C. dichotoma* is a quick-growing tree, performing well under semi-arid conditions and suitable for planting along boundary and farm roads [2].

When a tree produces fruit, it is suggested to collect ripe fruits and rubbed to remove the flesh. The healthy seed is dried in the shade and kept in tin containers. The seed can be stored for one year in airtight containers kept in a dry place to avoid insect attack. Furthermore, seed germination requires special conditions. Young seedlings are suffered by exposure to direct sunlight. However, the healthy seedlings usually grow fast become young trees. On optimal sites, the trees reach a height of 4 m in 4 years and a diameter of over 20 cm in 8-9 years [2].

Because the seeds are not easily formed, it is recommended to perform vegetative reproduction by cutting or grafting. For this matter, it is necessary to study the optimisation techniques of vegetative reproduction. The aspects need to be examined are the cutting or grafting medium, the addition of growth regulators, the modification of physical environment (temperature, pH, moisture), and others. On the other hand, it also needs to consider alternative techniques to breed Kendal tree with an innovative method, using *in vitro* technique. The advantages of this technique are producing pathogen-free plantlets and efficient in terms of time and place. The efficiency of this technique will be achieved when the protocol which has been found suitable for certain species.

5. Conclusion

In conclusion, *C. dichotoma* was a rare plant at Kendal Regency. The plant was not well known and commonly found. Its distribution is very limited because it is underutilised by local people. As a result, there are no significant cultivation activities. There are many types of research exploring the utility of *C. dichotoma* organs. Because *C. dichotoma* proved as a very beneficial and have a role as identity flora of Kendal Regency, it is necessary to planting it on a large scale. This activity requires the optimum breeding techniques. The generative reproduction is difficult to implement. Therefore, it is recommended to develop a vegetative reproduction technique through cuttings, grafting and tissue culture.

References

[1] BPS-Statistics of Kendal Region, Kabupaten Kendal dalam Angka (Kendal Regency in Figures) 2015. 291 p. Kendal, BPS-Statistics Agency of Kendal Region (2015).

[2] C. Orwa, A. Mutua, R. Kindt, R. Jamnadass, S. Anthony, Agroforestree Database: a tree reference and selection guide version 4.0 (http://www.worldagroforestry.org/sites/treedbs/treedatabases.asp) (2009).

[3] E.F.F. Matias, E.F. Alves, M.K.N. Silva, V.R.A. Carvalho, H.D.M. Coutinho, J.G.M. da Costa, Revista Brasileira de Farmacognosia **25,** 542–552 (2015).

[4] N. Balachandran and K. Rajendiran, *Cordia ramanujamii* (Cordiaceae): new species from Tamil Nadu, India. Taiwania **61**(2), 74–77 (2016).

IOP Conf. Series: Journal of Physics: Conf. Series **824** (2017) 012059 doi:10.1088/1742-6596/824/1/012059

[5] J. Guil-Guerrero, F. Garcıa-Maroto, M. Vilches-Ferrón, D. López-Alonso, Gamma-linolenic acid from fourteen Boraginaceae species. Ind. Crops Prod. **18**, 85–89 (2003).
[6] P. Bhattacharya, A. Saha, Evaluation of reversible contraceptive potential of *Cordia dichotoma* leaves extract. Rev. Bras. Farmacogn. **23**, 342–350 (2013).
[7] N. Hussain and B. Kakoti, Review on ethnobotany and phytopharmacology of *Cordia dichotoma*. J. Drug. Deli. Ther. **3**, 110–113 (2013).
[8] P.G. Jamkhande, S.R. Barde, S.L. Patwekar, P.S. Tidke, Plant profile, phyto-chemistry and pharmacology of *Cordia dichotoma* (Indian cherry): a review. Asian Pac. J. Trop. Biomed. **3,** 1009–1012 (2013).
[9] K. Thirupathi, S.S. Kumar, V.S. Raju, B. Ravikumar, D.R. Krishna, G.K. Mohan J Nat Remedies. **8**:1–10 (2008).
[10] V.K. Sharma and D.P. Asati, Pediatric contact dermatitis. Indian J. Dermatol. Venereol. Leprol. **76**, 514–520 (2010).
[11] P.B. Nariya, N.R. Bhalodia, V. Shukla, R. Acharya, Antimicrobial and antifungal activities of *Cordia dichotoma* (Forster F.) bark extracts. Ayu **32**, 585 (2011).
[12] M. Panghal, V. Kaushal, J.P Ann. Clin. Microbiol. Antimicrob. **10**, 1–11 (2011).

IOP Conf. Series: Journal of Physics: Conf. Series **824** (2017) 012060 doi:10.1088/1742-6596/824/1/012060

Effects of Rambutan Peel Extract to The Number of Erythrocytes and Haemoglobin in Rats Exposed to Cigarette Smoke

Lisdiana[1,*] **and F K Dewi**[1]

[1]Department of Biology, Faculty of Mathematics and Natural Sciences, Universitas Negeri Semarang, Sekaran Street, Gunungpati Semarang Indonesia, 50229

*Corresponding author: lisdiana@mail.unnes.ac.id

Abstract. Cigarette smoke is one of the exogenous free radicals sources. When it is inhaled, its activity may damage the structure of erythrocyte membrane function. The impacts of free radicals can be reduced through the provision of antioxidants. Rambutan fruit peel contains the phenolic compound in the form of polyphenols that are antioxidants. The purpose of this study is to determine the effect of rambutan fruit peel extracts to the number of erythrocytes and haemoglobin in rats exposed to cigarette smoke. This design used Post Test Control Group Design. A sample of 25 rats was divided into five groups, each group consisting of 5 rats. The positive control group (K+) were given a standard food and drinking water. The negative control group (K) by three cigarettes, the treatment group (KP1, KP2, KP3) by three cigarettes and skin extract of rambutan each treatment group with a dose 15 mg/kg, 30 mg/kg and 45 mg/kg for 30 days. Data on the number of erythrocytes and haemoglobin in rat blood was analysed with LSD and to determine the optimum dosage was analysed by using regression test. Research results shown that the content of rambutan fruit peel extract may increase the number of erythrocytes and haemoglobin of blood. Conclusions from this research are the rambutan fruit peel extract at a dose of 45 mg/kg body weight can increase and maintain the number of erythrocytes and haemoglobin in the blood of rat exposed to cigarette smoke.

1. Introduction

In some people, smoking is a habit that is hard to remove and harmful to health. Smoking is not only harmful to smokers but also for people who are exposed to cigarette smoke surrounding. Combustion will generate smoke cigarettes are divided into cigarette smoke (mainstream smoke) containing 25 % levels of hazardous materials and smoke the side (sidestream smoke) containing 75 % levels of hazardous substances [1].

The Cigarette smoke, both primary and side smoke when inhaled into the respiratory system can get into the blood circulation system, causing Reactive Oxygen Species (ROS) which cause oxidative stress [2], especially in erythrocytes. In the erythrocytes containing haemoglobin (Hb). If the content of cigarette smoke, especially carbon monoxide reacts with haemoglobin, it forms carbon mono oxyhemoglobin (HbCO). The affinity of haemoglobin for oxygen is much lower than its affinity for carbon monoxide so that the carbon monoxide replaces oxygen in haemoglobin and decreases the oxygen-carrying capacity of the blood as. HbCO relatively slow decomposition causes the cells work is hampered in its function of carrying oxygen throughout the body. Conditions such as these can result in serious, even fatal, as it can lead to poisoning [3].

Erythrocyte membranes composed of carbohydrates, proteins, oligosaccharides, and lipids. Phospholipids are one of the lipids that were the most numerous and is susceptible to oxidative stress [4]. If free radicals are not stopped, can damage all types of cellular macromolecules, including carbohydrates, proteins, lipids and nucleic acids [5]. Free radical attack on membrane lipids can lead to the formation of lipid peroxidation.). Lipid peroxides of erythrocyte membranes may result in loss of membrane fluidity and increase the erythrocyte membrane fragility which in turn resulted in erythrocytes will be easily broken or hemolysis. Lysis of erythrocyte membranes causing haemoglobin into plasma, thus dwindling. It resulted in levels of haemoglobin contained in erythrocytes low. As a result, the cells will be starved of oxygen [6].

When the erythrocyte membrane damage continues, it is likely to cause anaemia. Effects of free radicals from tobacco smoke against erythrocytes and haemoglobin can be reduced through the provision of antioxidants. Antioxidants play a role in preventing the oxidative stress caused by exposure to free radicals. Antioxidants can be obtained from the results of the chemical compounds of various plant secondary metabolites. One type of skin of the fruit is efficacious as a drug and has antioxidant activity that is rambutan (Nephelium lappaceum L).

Rambutan is a fruit commonly eaten in Indonesia, easily obtained, as well as widely known. Rambutan is one of the fruits that all the parts (bark, leaves, seeds and roots) can be useful as a drug [7]. Currently, rambutan peel is still a waste. His peel was red still not fully utilised. Rambutan peel is known to have antioxidant activity containing phenolic compounds, alkaloids, steroids and terpenoids [8], flavonoids [9] and anthocyanins with the highest content of phenolic compounds [10]. Rambutan fruit peel contains phenolic compounds in the form of polyphenols with the main component of acid, ellagic, geraniin and origin [8]. These polyphenols act to protect body cells from damage by free radicals by binding with free radicals [11].

Based on the research results test the antioxidant activity quantitatively showed that the methanol extract of the peel of rambutan IC50 value of 0.412 mg / mL; whereas the IC50 value of vitamin C of 1.776603 mg / mL. The results of research ethanol extract of rambutan showing the IC50 value of 4.29 mg/mL and vitamin E IC50 value of 8.48 mg / mL. Therefore, the extract of rambutan fruit peel has a stronger antioxidant activity than the compound of vitamin C and vitamin E. Based on the above research efforts are needed to determine whether the rambutan peel extract can have an effect on the number of erythrocytes and blood haemoglobin levels as a result of being exposed to cigarette smoke.

2. Materials and methods

This study is an experimental research design with The Post Test Only Control Group Design. The population of rats (Rattus norvegicus) Wistar, samples of 25 male rats were aged 2-3 months with a body weight of 180-200 grams. The samples were divided into five groups. K+ (positive control), K- (negative control), KP1, KP2 and KP3 (treatment group). Rats were taken at random and divided into five groups, the cages for five rats. Rats are adapted for one week, given the standard feed and drink ad libitum. KP, KP2 and KP3 treated rambutan fruit peel extract at a dose of 15 mg /kg, 30 mg /kg, and 45 mg /kg orally at 10.00 hrs for 30 days, 1x per day. Then the negative control group, KP1, KP2 and KP3 treated three cigarettes at 08.00, 12.00 and 14.00 pm for 30 days. Blood sampling performed on day 6, 12, 18, 24 and 30 through orbital sinus eye with a hematocrit of 1 ml and collected in Eppendorf tubes containing anticoagulant that has ethylenediaminetetraacetic acid (EDTA), then measure the blood sample with Hematology Analyzer BC 2600. Measurement data on day 30 were statistically analysed using One-Way ANOVA if the results are significantly different then tested further with LSD. The optimum dose of rambutan fruit peel extracts was analysed using regression test.

3. Results

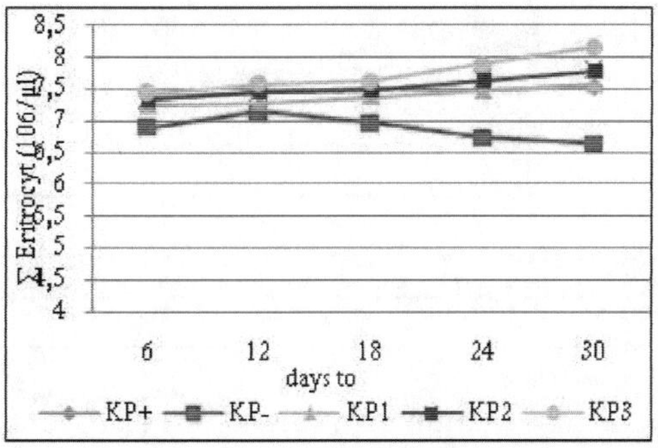

Figure 1. Number of Red Blood Cells (Erythrocytes)

Based on the graph in Figure 1 shows that the negative control showed the lowest average result than the number of erythrocytes in normal conditions ranging 6.64-7.15×106/ml of blood, when compared to the positive control group, KP1, KP2 and KP3. The results showed the highest number in the KP3, ranging 7.42-8.16×106/ml of blood.

The number of erythrocytes in normal rats ranged 7.2-9.6×106/ml. In the treatment group KP1, KP2 and KP3 by rambutan fruit peel extract increased when compared to the negative control group and positive control group.

Table 1. LSD test results on the number of erythrocytes (106/ml). The letters different in the same column shows the difference in each group with a level of accuracy of $p < 0.05$

Group	Treatment	Σ Erythrocyte (Mean ± SD)
K+	Control I	7.53 ± 0.200^{b}
K-	3 cigarettes	6.64 ± 0.339^{a}
KP 1	Rambutan fruit peel extracts 15 mg/kgBB + 3 cigarettes	7.58 ± 0.516^{b}
KP 2	Rambutan fruit peel extracts 30 mg/kgBB + 3 cigarettes	7.79 ± 0.347^{b}
KP 3	Rambutan fruit peel extracts 45 mg/kgBB + 3 cigarettes	8.16 ± 0.082^{c}

However, from the graph, the average number of erythrocytes in Figure 1, that group KP1, KP2 and KP3 has increased the number of erythrocytes up to 30 days. It suggests that the increase in the number of erythrocytes in line with the increase in the concentration of rambutan fruit skin extracts in the treatment group. Results of linear regression data on the number of erythrocytes showing the relationship between the dose of rambutan fruit peel extracts and the number of erythrocytes with the linear regression model are $Y = 7.26 + (0.29) X$, with R2 values of 0.320 = 32 %.

Based on the graph on figure 2. showed that the average haemoglobin levels from day 6 to day 30 in the positive control group, KP1, KP2 and KP3 have elevated levels of haemoglobin. When compared to the negative control group decreased haemoglobin levels. On day 30, the average haemoglobin levels higher KP3 group compared with positive control group, the negative control group, KP1 and KP2. The negative control group had an average of the lowest haemoglobin levels below normal levels of haemoglobin in conditions ranging from 10.88 g/dL of blood. Haemoglobin levels when compared with that measured in the positive control group and the group KP1, KP2, KP3 shows the results tend to be constant and within the normal range. Normal haemoglobin level in rat between 11.1 to 18 g/dl.

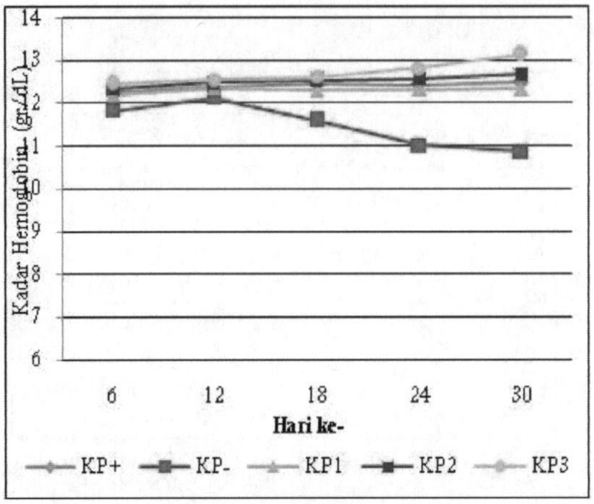

Figure 2. Average Levels of Haemoglobin (g/dl)

The statistical test by using One Way ANOVA shows that there are significant differences in the negative control group with a significance level of 0.000 less than 0.05 ($p < 5\%$) of the positive control group, KP1, KP2 and KP3. It is because in each group KP1, KP2, KP3 given rambutan fruit peel extracts with different doses, while the negative control without rambutan peel extracts only exposure to cigarette smoke. In the group KP1, KP2 and KP3 no significant difference where it is shown that the concentration of the rambutan fruit peel extracts has a similar effect in increasing haemoglobin levels within the normal range (Table 2). However, from the mean haemoglobin level diagram in Figure 2 an increase of up to 30 days. It suggests that the increase in haemoglobin levels with increased concentrations of rambutan fruit peel extracts in the treatment group. Accordance with the results of the linear regression test that the greater the dose of rambutan fruit peel extract is given, then it can increase haemoglobin levels. Results of linear regression test haemoglobin levels of data showing the relationship between dose and rambutan fruit peel extract with haemoglobin levels. The linear regression model are $Y = 11.92 + (0.40) X$, with R2 values of 0.297 = 29.7%.

Table 2. LSD test results on the Levels of haemoglobin (gr/dL). The letters different in the same column shows the difference in each group with a level of accuracy of $p < 0.05$.

Group	Treatment	Hemoglobin Levels (Mean ± SD)
K+	Control	12.48 ± 0.506^{b}
K-	3 cigarettes	10.88 ± 0.725^{a}
KP 1	Rambutan fruit peel extracts 15 mg/kgBB + 3 cigarettes	12.34 ± 0.114^{b}
KP 2	Rambutan fruit peel extracts 30 mg/kgBB + 3 cigarettes	12.68 ± 0.496^{b}
KP 3	Rambutan fruit peel extracts 45 mg/kgBB + 3 cigarettes	13.14 ± 0.826^{b}

4. Discussion

Based on the chart pattern and the overall value obtained from this study looks in line with the decrease and increase in the number of erythrocytes and haemoglobin concentration of blood. The decline in the number of erythrocytes and haemoglobin levels in the negative control group occurred because one of the contents of the dominant cigarette smoke is carbon monoxide (CO) Carbon monoxide is a toxic gas that is colourless and odourless. Carbon monoxide can cause a reduction in the delivery and utilisation of oxygen in body tissue [1].

The content of cigarette smoke, especially carbon monoxide reacts with haemoglobin to form carbonmonoxyhemoglobin (HbCO). The affinity of haemoglobin for oxygen is much lower than its affinity for carbon monoxide, so the carbon monoxide replaces oxygen in haemoglobin and decreases as the oxygen-carrying capacity of the blood. HbCO relatively slow decomposition causes the function of erythrocytes to carry oxygen throughout the body is inhibited. Conditions such as these can result in serious, even fatal, as it can lead to poisoning [3].

Erythrocyte membranes composed of carbohydrates, proteins, oligosaccharides, and lipids. Phospholipids are one of the lipids that was the most numerous and is one of the lipid membranes are susceptible to oxidative stress. With the oxidative stress will cause lipid peroxidation [6]. Peroksida lipids in erythrocyte membranes are resulting in loss of membrane fluidity and increase the erythrocyte membrane fragility or friability which further lead to easily broken or erythrocyte hemolysis. Lysis of erythrocyte membranes causes free haemoglobin into plasma, thus dwindling. It resulted in levels of haemoglobin contained in erythrocytes low. As a result, the body's cells will be starved of oxygen [6].

Based on the results of the linear regression test showed that the magnitude of the dose of rambutan fruit peel extracts, it can increase the number of erythrocytes and haemoglobin levels. The increase in the number of erythrocytes and haemoglobin levels occurred in the group KP1, KP2 and KP3. Flavonoids are one of the active compounds from the group of polyphenolic compounds that can be found in the peel of rambutan as an antioxidant [12]. Flavonoids are the active compounds polyphenols that act as antioxidants, which can increase erythropoiesis (the process of formation of erythrocytes) in the bone marrow [13]. It is consistent with the statement that the flavonoid on rambutan fruit peel extract can increase the number of erythrocytes and haemoglobin levels. By way of flavonoids stimulate renal plasma globulin cells to secrete a hormone called erythropoietin. Erythropoietin is a glycoprotein hormone that is found in the blood, then the hormone erythropoietin circulating in the blood vessels that stimulates the bone marrow to increase red blood cell formation or erythropoiesis.

Polyphenol compounds are known to have various biological effects such as antioxidant activity through a mechanism as reducing agents, catcher of free radicals, chelating the metal and the formation of singlet oxygen absorbers as well as the electron donor [8]. Besides flavonoids also may prevent oxidative stress. Flavonoids are found in the peel extract of rambutan is thought to inhibit the occurrence of lipid peroxidation, so that free radicals can not develop into a free radical new way of donating one electron hydrogen atoms in unpaired electrons in free radicals so many free radicals become less and less reactive [14].

The relationship between flavonoids and reduction activity of free radicals (free radical scavenging) shows that there is a difference between flavonoid compound activity. The difference depends on the structure and substituents on the heterocyclic ring C and B. There are two major functional groups in flavonoids that determine the potential reduction of free radicals, namely: (a). Hydroxyl group 3', 4' (ortho-dihydroxy) in the Bring flavonoids, which has the properties as an electron donor and radical targets; (B). 2.3 conjugated double bonds with 4-oxo clusters (groups 1,4-Piron) on rings C and (c). the hydroxyl group in position 3 and five on the heterocyclic ring which acts on the electron delocalization.

5. Conclusion

Based on the result it was concluded that *C. dichotama* was a rare plant at Kendal Regency. The plant was not well known and commonly found. Its distribution is very limited because it is underutilized by local people. As a result, there are no significant cultivation activities. There are many types of research exploring the utility of *C. dichotoma* organs. Because *C. dichotoma* proved as a very beneficial and have a role as identity flora of Kendal Regency, it is necessary to planting it on a large scale. This activity requires the optimum breeding techniques. The generative reproduction is difficult to implement. Therefore it is recommended to develop a vegetative reproduction technique through cuttings, grafting and tissue culture.

References

[1] A Inayatillah IR. 2014 *Jurnal Respirasi Indonesia,* Vol. 34 No. 4.
[2] Haliwell B & Whiteman M. 2004 J Pharmacol 142:231-255
[3] Yanbaeva DG, Dentener MA, Creutzberg EC, Wesseling G & Wouters EF. 2007. Systemic effect of smoking. Chest 131(5): 1557-1566

[4] Lima VR, Morfim MP, Teixeria A, Crecszynski TB. 2004 *Chemistry and physics of lipids* 132: 197-20

[5] Abdollahi M, Ranjbar A, Shadnia S, Nikfar S & Rezale A. 2004 *Med Sci.monit,* 10(6): 141-147.

[6] Ahumibe AA & Braide VB. 2009 *Nigerian Journal of Physiological Sciences,* 24 (1): 47-52.

[7] Syamsidi A. 2014 *Journal of Natural Science,* Vol. 3(2): 1-9.

[8] Thitilertdecha N, Teerawatgulrag A, Rakariyatham N & Kilburn JD. 2010 *Journal Molecules,* 15: 1453-1465.

[9] Fidrianny I, Sari PI & Wirasutisna KR. 2015 *International Journal of Pharmacognosy and Phytochemical Research*, 7(2); 280-285.

[10] Fila WO, Johnson JT, Edem PN, Odey MO, Ekam VS, Ujong UP & Eteng OE. 2012 *Annals of Biological Research*, 3 (11): 5151-5156.

[11] Guo C, Yang J, Wei J, Li Y, Xu J & Jiang Y. 2003 *Nutrition Research*, 23: 1719-1726.

[12] Dewi AK, Umie L & Sri RL. 2013. The Effect of Kulit Buah Rambutan *(Nephelium lappaceum L.)* Peel Extract on Lipid Peroxidation in Liver of Obese Rats. Makalah disajikan dalam International Conference Biologi Sciences, di UGM tanggal 20-21 September.

[13] Palanisamy U, Cheng HM, Masilamani T, Subramaniam T, Ling LT & Radhakrishnan AK. 2008 *J. Food Chemistry* 109(1): 54-63

[14] Suparmi, Hady A & Niche D. 2012 *Jurnal Ilmiah Farmasi*, Vol. 9 No. 1.

Nest Records of Wreathed Hornbill (*Rhyticeros undulates*) in Gunung Gentong Station, Mount Ungaran Central Java

M Rahayuningsih[1,*], NE Kartijomo[1] , A Retnaningsih[1], M Munir[1], and J Dahlan[1]

Biology Department, Mathematics and Natural Sciences Faculty, Universitas Negeri Semarang, Indonesia

*Corresponding author: etak_sigid@mail.unnes.ac.id

Abstract. The remaining forest of Mount Ungaran, Central Javais the suitable habitat of Wreathed Hornbill (*Rhyticeros undulatus*), especially for a nesting site.The objective of the study was to analyse thenest record and characteristics of habitat around the nest, especially in Gunung Gentong station. The research was conducted from 2010-2016 using exploration method. The methodhabitat profile of the vertical structure tree canopy was taken by plot size 60 x 20 m. Measurements were taken to the standing of vegetation, canopy closure, the direction of the canopy, height canopy, a former branch of the vegetation height, and stem diameter.The Result of the study showed that Gunung Gentong is one of the research station that we have been recorded for nesting site on 2010-2015. Atotal of the nest record on Gunung Genting station was 10 nests. Estimate the elevation of nest location between 939-1240 AMSL. The tree species that used for nesting was *Syzygium glabatrum, Syzygium antisepticum, Ceratoxylon formosum,* and *Ficus* sp

1. Introduction

Three of Hornbills species (Fam. Bucerotidae) existing in Java, one of them, Wreathed Hornbill (*Rhyticeros undulates*) was found in Mount Ungaran [1]. Mount Ungaran area has the last remaining forest area especially in Central Java and established for Important Bird Area (IBA) [2] and alsoThe Alliance for Zero Extinction (AZE). The main factors that Hornbills' species in the world are under the threat of extinction because they are specialised about habitat, food and nesting sites. Hornbills have a very important role in the forest ecosystem as seed dispersers [4]-[6]. The population of Wreathed Hornbill especially in Java was tendency threatened by habitat loss because of illegal logging, fragmentation, hunting pressure, and trading of the bird. Based on the last research of the Wreathed Hornbill on Mount Ungaran [1][7] showed that Mount Ungaran is a suitable habitat for the availability of food, water, shelter, perching and nesting site forWreathed Hornbill species. But, Based on the landcover data on 1991-2009 the research also showed that the natural forest on Mount Ungaran was decreased almost 53.31%, while the secondary forest, plantation and the settlement became increase[8]. Hornbills nesting in cavities, usually in large trees, they cannot excavate their nest holes.

Hornbill usually used nest cavities by created capable excavators (e.g. woodpeckers and barbets), or natural cavities that form after branches breaks or after other injuries are inflicted upon forest tree [9]. When the breeding season, the female enters the nest and while sitting within the nest cavity. The femalehas prepared the cavity and plastering of the cavity entrance. The female keeps her bill sliding over the cavity entrance with the mud lump being smeared into layers of semi-solid soil. Only a narrow opening hole for malewillallow delivery feed for female and the chicks [10]. Hornbills rely entirely on the availability of natural tree cavities for nesting and also dependon their food resource, any loss of habitat means loss not only food resources but also breeding sites [11] There for, the study of hornbill ecology in Mount Ungaran must be intensive and sustainability, including for the nest

IOP Conf. Series: Journal of Physics: Conf. Series **824** (2017) 012061 doi:10.1088/1742-6596/824/1/012061

record and their habitat.The objective of the study was to analyse the nest record and characteristics of habitat around the nest on Gunung Gentong station.

2. Material and Methods

The study was carried out in Gunung Gentong station,Mount Ungaran 07009'26.8" S 110020'11.8" E (Figure 1) and was conducted from 2010-2016. The materials and equipment for this research are GPS Garmyn e-Trex 12 channel, camera, tally sheet, label, climbing equipment, glove, and stationeries. Data Nest characteristics especially vegetation structure and profile of the habitat. Nest record method by exploration survey especially in Gunung Gentong and information from the citizen which live around Gunung Gentong village. The method of vegetationstructure was taken nested plot sampling, and habitat profile by the vertical structure of tree canopy closure used plot size with measuring 60 x 20 m. Measurements were taken to the standing of plant canopy closure, the direction of the canopy, height canopy, a former branch of the plant height, and stem diameter.

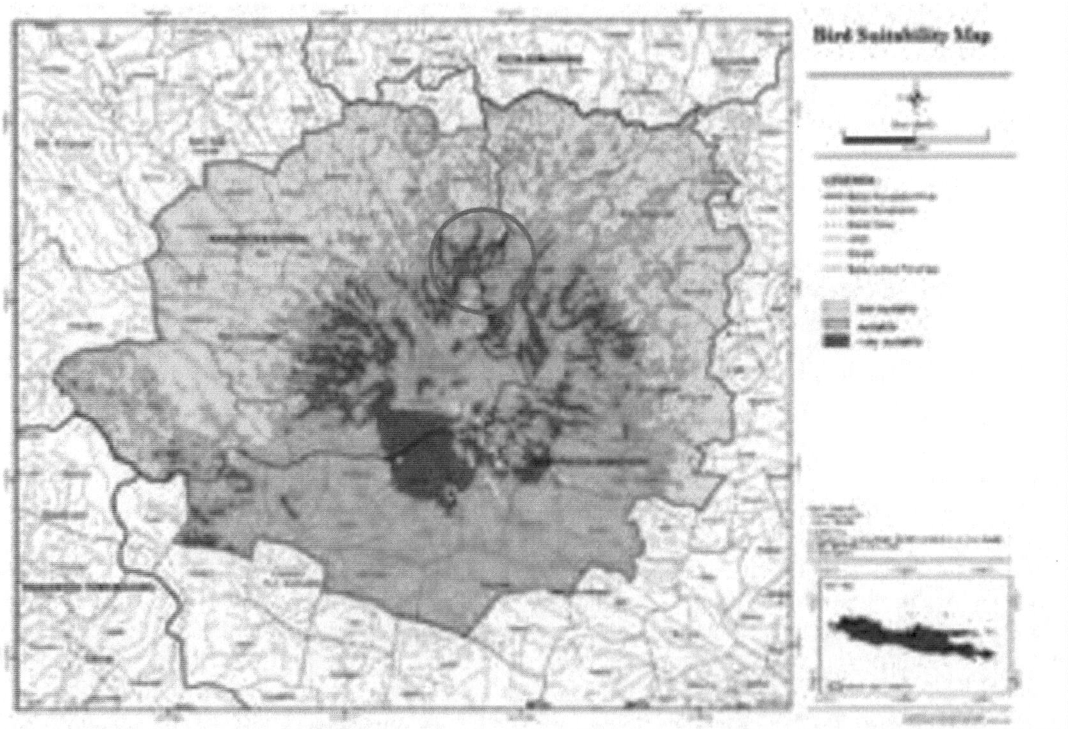

Figure 1. The location in Gunung Gentong, Mount Ungaran

3. Result and Discussion

The breeding period of Wreathed Hornbill in Mount Ungaran were reported on Agust-December. The Result of the study showed that total10nestshadbeen recorded in Gunung Gentong since 2010-2015(Table 1), six nests have been recordedin active and four nests not active. Estimate elevation of the nest was located between 939-1240 AMSL. According to Mackinnon et al. [12], the Wreathed Hornbill usually found at elevation 0-2000 meters AMSL. Tree hight of the nest between 16-27 m, diameter 0.83-1.75 m, and the high of the nest 9-27 m. Meanwhile, some researchers explained that nesting trees for Wreathed Hornbill have height ranged more than 20 meters and diameter about0.4-2,67 m [13]. The high tree was chosen because when there was a storm, the tree tend to be damaged including broken branches and in some cases would naturally form a hole. Also, the strongest trees can survive longer than other trees, so the possibility of using it as a nest was higher than other trees [14].The research also showed, the tree species that used for nesting was *Syzygium glabatrum, Syzygium antisepticum, Ceratoxylon formosum,* and *Ficus sp*. Based on Poonswad recorded [14], higher and larger trees will attract hornbills including Wreathed Hornbill to use it as a nesting place and the selected tree was Dipterocarpus and *Eugenia*. While in Mount Ungaran, Wreathed Hornbill usually used the nests at *Syzygium glabratum* and *Syzygium antisepticum*. Both species are still the same family with *Eugeniasp* (Fam. Myrtaceae).

Table 1. Nest record in Gunung Gentong station

Nest no	Elevation	Tree species	Tree hight (m)	Tree diameter (m)	Nest hight	Year	Status of nest
1	939	*Syzygium glabatrum*	27	1.75	18	2010-2013	active
2	1056	*Syzygium antisepticum*	24	1.15	9	2010 - 2013	active
3	1241	*Syzygium antisepticum*	32	1.37	14	2013-2015	active
4	1185	*Syzygium antisepticum*	30	0.83	16		active
5	1105	*Syzygium antisepticum*	26	0.89	16		active
6	1112	*Cratoxylon formosum*	34	1.62	17	2015	active
7	1039	*Syzygium antisepticum*	30	1.15	27	2015	non active
8	972	*Syzygium antisepticum*	16	1.10	12	2015	non active
9	982	*Ficus* sp	17	1.05	9	2015	non-active
10	950	*Ficus* sp	16	1.00	8	2015	non-active

In this study, we try to analysed the profile of vegetation around the tree of nest existing. Analysis of vegetation profiles in Gunung Gentong station showed the dominant tree vegetation is*Syzygium* , *Litsea* sp, and *Ficus* sp (Figure 2). Based on the study also showed that of the tree species was the main feed source for Wreathed Hornbill in Gunung Gentong.Therefore, the existence of tree species is important to the survival of the Wreathed Hornbill. Such as, *Ficus* spis a primary species the source of food for Wreathed Hornbill. As seen during the study the *Ficus*sphave been fruiting and Wreathed Hornbilloften perching, restingon the tree and also looking for a feeding. The *Ficus sp* holds many important roles for many species of fruit eater especially for the Wreathed Hornbill. Wreathed hornbill in GunungGenting always perched on the top canopy (A) and the second canopy (B), while in the third canopy (C) was used as a nesting site. Most fruit consumed by Wreathed Hornbill in Mount Ungaran is from family Lauraceae, Moraceae, Myristicaceae, Rutaceae, Myrtaceae, and Euphorbiaceae [15].*Litsea* sp include the family of Lauraceae and *Ficus* sp include in the family of Moraceae. Hornbill from all studies combine especially in Asia ate 497 fruit species from 135 genera in 46 families [7]. The top five families ranked of species consumed are Lauraceae, Moraceae, Meliaceae, Myristicaceae, dan Annonaceae.

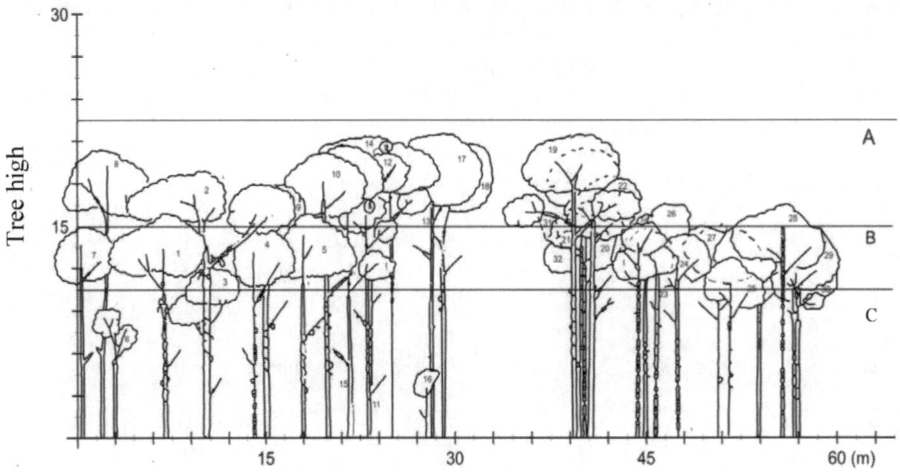

Figure 2. The vertical structure of tree habitat profile: (A) Top Canopy, (B) Second Canopy, (C) Top Third Canopy

The problem such as deforestation and habitat fragmentation poses the greatest threat to hornbills in Moun Ungaran. Unlimited demand, poor planning, and unsustainable use of this valuable and natural capital resource have resulted in a high depletion of forest resources and hence of hornbill habitat.

4. Conclusion

Gunung Gentong Station in Moun Ungaran Central Java was the suitable habitat for perching, roosting, food source from the some of fruit tree species, and nesting site. The tree species for nesting site are *Syzygium glabatrum, Syzygium antisepticum, Ceratoxylon formosum,* and *Ficus* sp.Most fruit consumed by Wreathed Hornbill in Mount Ungaran is from family Lauraceae, Moraceae, Myristicaceae, Rutaceae, Myrtaceae, and Euphorbiaceae.

References

[1] Rombang, M, Rudyanto Bogor: PKA/Birdlife International-Indonesia , 77, 1999
[2] Rahayuningsih M, Nugroho Edi K IJESD, 5 (1): 492-495, 2013.
[3] Krishna cm, Kuladip S, Awadhesh K, *India*, JoTT, 4(14): 3342–3348,2012.
[4] Kinnaird, M. F., O'Brien, T. G *Conserv. Biol*, 12: 954–956, 2007.
[5] Kimura K. A. *Global Environmental Research7*: 113-122, 2003.
[6] Balasubramanian P, R. Aruna 2, C *India*, JoTT 3(5):1775–1777, 2011.
[7] Rahayuningsih M, Nugroho Edi K, Erni S. IJESD, 6(6) :474-477, 2015
[8] Rahayuningsih, Nugroho Edi K, Siti Alimah. IJEE, 2 (8), 2016.
[9] Kinnaird M.F, O'Brien, T.G The University of Chicago Press, 2007.
[10] Charde P, Raju K., Jeevan L. Tarar *The Raffles Bulletin of Zoology*. 24: 59-64. 2011
[11] Poonswad P, Vijak C, Narong M and Sittichai M. *Conservation Hornbill in Thailand. Conservation Biology: Voices from the Tropics*, 2013
[12] MacKinnon J, Philips, Karen, S. van Balen Puslitbang, Jakarta, 2010.
[13] Marthy W, *A review: Breeding ecology and nest site characteristics of Hornbills in-situ and ex-situ. Harapan Rainforest,* 2008.
[14] Poonswad P., *Thailand.IBIS.* 137:183-191.1995
[15] Dahlan J, Rahayuningsih M. *Unnes J Life Sci* 3 (1), 2015

IOP Conf. Series: Journal of Physics: Conf. Series **824** (2017) 012062 doi:10.1088/1742-6596/824/1/012062

Optimising Sterilisation Techniques and Callus Induction of Nodes Durio Zibethinus Murr in Vitro Method with Various Media

N Hermayani[1,*], A Retnoningsih[1], E S Rahayu[1]

[1]Department of Biology, Faculty of Mathematics and Natural Sciences, Universitas Negeri Semarang, Sekaran Street Gunungpati Semarang 50229-Indonesia

*Corresponding author: novitahermayani@gmail.com

Abstract. The application of in vitro propagation method needs an aseptic or sterile condition. The objective of the study was to get an optimal sterilisation techniques and medium of callus induction of *Durio zibethinus*. Sterilisation treatments studied were NaClO and $Ca(ClO)_2$. The three kind of callus induction medium studied were Gamborg (B5), Woody Plant Medium (WPM), and Murashige and Skoog (MS) with the addition of auxin and cytokinin. The experiment unit was three bottles with nodes as explant. Cultures were kept in the culture room for 16 hours daily by LED light intensity of 1000 lux. Parameters investigated of sterilisation technique development was the percentage of contamination, the percentage of explant browning, and the percentage of life explants; whereas the callus induction measured were the rate of callus formation, percentage of callus covered, callus texture, colour and diameter of callus. The results showed that the code of C3 is soaking in calcium hypochlorite ($Ca(ClO)_2$) 50% for 3 minutes, calcium hypochlorite 40% for 2 minutes, and 70% alcohol for 30 seconds separately were suitable for sterilisation of nodes explant. The development of callus on B5 medium with the addition of auxin (2,4-D) 2 ppm and cytokinin (Thidiazuron) 1 ppm was best compared to the other.

1. Introduction

Durian (*Durio zibethinus* Murr.) or King of Fruit, is the seasonal exotic tropical fruit plant that has unique scent and flavour, it is popular in the community. This fruit became one of the commodities that have high economic value. Constraints experienced in cultivating the durian is thevery limited availability of quality seeds. Therefore, it is necessary to repair the characters, among others through micrografting. Micrografting requires in vitro shoots which can be developed from the callus. Also, callus is also required for repair techniques besides micrografting, namely the induction of somaclonal variation, mutation and recombination of genes. Thus callus is needed in breeding efforts durian. Callus induction is done by growing the explants in the form organs or tissues. The success of callus induction was determined by the sterility of the explants, the type of media and plant growth regulators.

Plants that grow on the field is a lot contaminated with microorganisms. Microbial contamination on explants has animpact on the success of the culture because the microbes can be a competitor in the absorption of nutrients, increasing the death of plant tissue, tissue undergoes necrosis, reduce cell proliferation, buds and roots[1]. Contamination is not always visible in the early phase of the culture;sometimes contamination can also be caused by contaminants from endogenous plants where it is difficult to do the sterilisation process.The internal contamination can be controlled to a certain extent by frequent transfer to a fresh medium or by the use of a low concentration of antibiotics in the medium. Overexposing tissues to decontaminating chemicals can also kill tissues, so there is a balancing act between sterilising explants and killing the explants themselves [2],[3]. External contamination is easily removed with running water for 30-90 minutes and soaked in disinfectant which functions as tween added to the sterilising agents such as fungicides, alcohol, HgCl2 and NaOCl. Some the sterilising agents found to be highly effective as a disinfectant explants taken from the field. To obtain a proper sterilisation procedures required special research. A wide variety of sterilising materials that are used for sterilisation, including an ingredient that is toxic to the plant tissue so that need to find concentration, sterilisation of time, and use the proper resistance standards explant it can survive [4].

Sterile proper material identification has been made on some plant, but the results cannot be applied to theothers.The effectiveness of disinfectant in controlling contamination on the growth of a zygotic culture of *Toona sinensis* showed that disinfectant HgCl2 0.05% and Clorox 20%-30% produce germination value (63.3%) highlywithout contamination [5].[6] reported sterilisation by using plant preservative mixture (PPM) with a concentration of 4.6% for 4 hours by periodic shaking are more effective as sterilisation, endurance and the ability of nodes in shaping the development of *Cestrum nocturnum* Lbuds.

The objective of the study was to get optimal sterilisation techniques and to identify a suitable media to induce callus from explants nodes durian. Sterilisation technique applications and in vitro technique on durian plant is expected to help the propagation of superior plants in a short time.

2. Material and Methods

Plant materials

Research has been conducted in tissue culture laboratory, Department of Biology, Faculty of Mathematics and Natural Science (FMIPA), Universitas Negeri Semarang in March until August 2016. The study was divided into two phases: 1) the optimisation of sterilisation and 2) optimisation of callus induction media. The nodes explants were collected sprouts *exvitro* durian seeds of local varieties grown at Biology fields.

Sterilization Preparation

Optimisation of sterilisation through 2 method that is 1) the influence of the type and concentration of the sterilising single and 2) the influence of the type and concentration of the sterilising terraced. Explants surface sterilisation freshly harvested emerging young nodal, cut about 5 cm to obtain ± 3-5 nodes segments were cleaned, and dead parts separated that were washed with the running tap water for 5 minutes to remove the dust or sand particles. Surface sterilisationwas stand soaking (capillarity of crops) that uses bactericide (2000 mg/L)) and fungicide (2000 mg/L) for 60 minutes, then used detergent (3 ml) for 15 minutes. Nodes are inserted into culture bottles, then added disinfectant (3 ml + 100 ml sterile of water). Shaking with a speed of 170 rpm for 30 minutes, then washed with sterile water as manythree times, continued with bactericide (1000 mg/L) and fungicide (3000 mg/L) shaking with a speed of 170 rpm for 60 minutes.

The first method of research was carried out experimentally, completely randomised design with three factors. The tree factors are the type sterilising agent (NaClO and $Ca(ClO)_2$), the concentration of the sterilising (20%, 30%, 40%, 50%) and the exposure of time (10 and 15 minutes). After explant surface sterilisation, then inside the laminar air flow treat with sodium hypochlorite (NaClO) and calcium hypochlorite ($Ca(ClO)_2$) with different concentration of 20, 30, 40, 50% with add tween 20® (2 drops) at adifferent time of 15 and 10 minutes. The research variables measured were the percentage of sterile explants and speed of contamination. Data were analysed descriptively.

The second method of research conducted with acompletelyrandomised design with two factors that is the type of sterilising agent (NaClO and $Ca(ClO)_2$) and the concentration of the sterilising terraced (code S1, S2, S3, C1, C2, C3). In The two types of optimisation studies of this sterilisation technique, the research units are one bottle contained one explant and repeated three times. The code of S1 showed sodium hypochlorite 50% with theaddition two drops tween 20® for 3 minutes, then sodium hypochlorite 40% with theaddition two drops tween 20® for 2 minutes and the last alcohol 70% for 1 minutes. Every turn issterilising, washed with sterile water three times. The code for S2 and S3 has thesame method with thecode of S1, but the concentration is different that is, the code of S2 showed sodium hypochlorite 40% and 30% whereas code of S3 showed sodium hypochlorite 30%

and 20%. While for the code of C1 showed thesame method with S1, only the sterilising agent replaced with calcium hypochlorite and the code of C2, C3 had same. The research variables measured were apercentage of sterile explant, thepercentage of living explants, and percentage of browning explants. Data were analysed descriptively.

Media preparation

Then explants were inoculated aseptically on B5 [7], MS [8] and WPM woody plant media [9]. Media supplemented with either various combination of auxin and cytokinin were analysed for callus induction. The three media were supplemented with 3% sucrose (w/v) and solidified with 0.8% agar. The pH of the medium was adjusted to 5.6-5.8 before autoclaving at 121°C and 1.2–1.3 kg/cm2 pressure for 20 min. Media was dispensed into small culture bottles; every bottle contained one explant. The cultures were incubated under growth room conditions (20 ± 3°C, 16 hours photoperiod and light intensity of 1000 lux provided by LED light (Phillips 18 Watt). After eight weeks of plant culture, the data were recorded on callus induction. The research variables measured were variable speed callus formation (days after planting), the extent of callus (cm^2), the percentage covered of callus, callus texture, and colourcallus.

Statistical Analyzed

The two method of optimisationsterilisation was analysed descriptively, whereas the optimisation of callus induction media wasanalysed used ANOVA by significance level of 5 %. Data were analysed used ANOVA by significance level of 5% was continued LSD.

3. Result and Discussion

Observation sterility of the explants on the type and concentration single treatment was carried out on days 3. The observation results showed that the use the sodium hypochlorite for all concentration has result nothing (0%) explants sterile (Figure C1. The treatment on days of 3th until days of ninth had contaminant for all explants. The used of calcium hypochlorite 40% and 50% all of theexplant sterile (100%), but in theconcentration of 20% and 30% has result nothing (0%) explants sterile. On day 9th until 14th all of theexplants had acontaminant in concentration 20% and 30%, whereas the explant had sterile until 30 days in concentration 40% and 50% (Table 1).

The sterile of explants from the best result treatment in thefirst method maintained into three media that addition by auxin and cytokinin. The code of C3 used for thesecond method to produce callus.

Table 1. Percentage of sterile explants, thelife of explants, and browning explants on agent sterilising terraced consist.

Sterilising Agent of code	Percentage of sterile explant (%)	Percentage of living explants (%)	Percentage of browning explants (%)
S1	40	0	40
S2	60	0	60
S3	100	60	40
C1	100	0	100
C2	100	40	60
C3	100	100	0

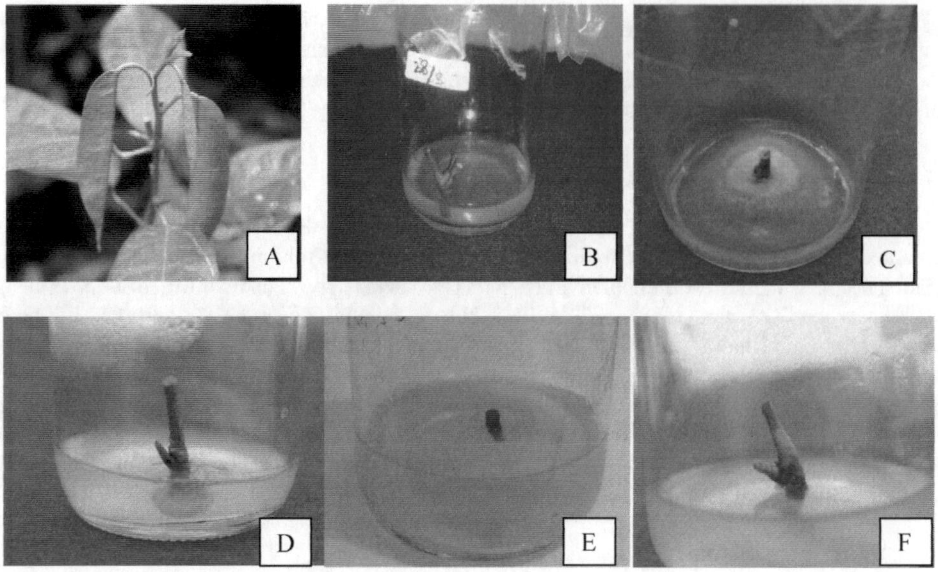

Figure 1. A. nodal from *exvitro* sprout of seed durian. B. The nodal has sterile of theexplant. C. the contaminant explant by fungi. D. The nodal forming callus inside of media. E. the nodal has sterile but undergo browning explant. F. the one of explant has expansion phase

Table 2. Time of callus appear, percentage covering callus, thearea of callus, thetexture of callus, and colour of callus *Durio zibethinus* Murr in various media.

Treatment	Time of callus appear (Mean ± SD)	Percentage covering callus (%) (Mean ± SD)	Area of callus (mm^2) (Mean ± SD)	Texture of callus (Mean ± SD)	Colour of callus (Mean ± SD)
WPM	32.00± 7.106	41.00±30.083^a	254.00±235.296^b	compact	white
B5	32.80±6.496	41.00±29.665^a	350.20±253.791^a	crumb	white
MS	33.80±8.526	25.00±26.926^b	187.60±207.921^c	crumb	white

*The means followed by the same letters along the columns are not significantly different from each other (α=0.05) LSD test.

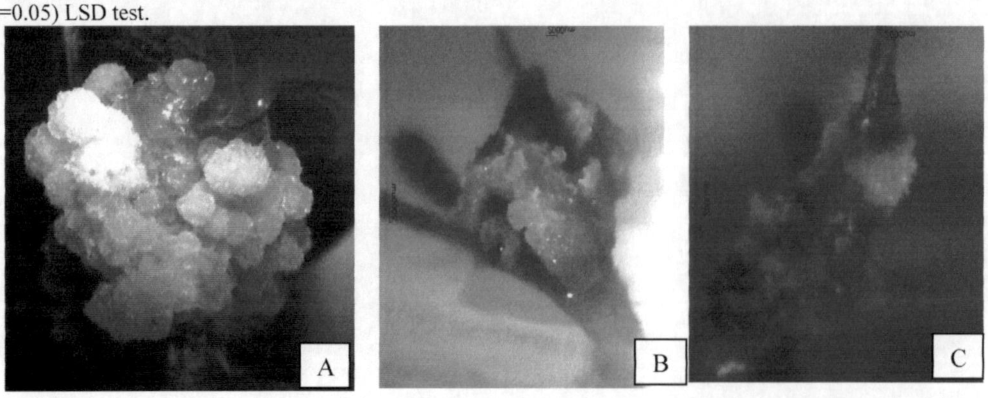

Figure 2. A callus formed on media B-C. The callus has been browning on B5 and MS media

The result analysed showed the type of media affect to the percentage of covered and area of callus formed, however, does not affect the appearedtime of callus. The WPM and B5 produced callus 41% from anexplant of area surface, however the B5 media to induction of callus wider than WPM media. The WPM media has to produce compact texture whereas B5 and MS media produce crumb texture (**Table 2**).

The first of callus formed, all of themedia produce thewhitecolour of callus (**Figure 2A**). However, after twodays, the colour of callus on B5 and MS media turn be brown colour (**Figure 2B and 2C**). The crumb of callus has a wavy structure to grows apart into small fragments, can be continued to formed embryogenic callus due to theaddition of growth regulators 2,4-D in media [10]. The media used in tissue culture is thedevelopment of a nutrient solution for used crops such as rooting media. The media containing some or all of components macro, micro, vitamin, amino acid/nitrogenous supplement, carbon source, organic supplement, growth regulator and gelling agent [11]

Cytokines are added to the three kinds of the medium in research is TDZ 1 mg/L. Cytokinins are adenine derivative which acts as a hormone to regulate various processes in the plant, but it was found as a factor in the tobacco tissue culture. It regulates the development of other parts of the plant [12]. Calli is a collection of undifferentiated plant cells and is formed under the control of plant growth regulator that is available in the medium. Growth regulator substances with thespecific concentration required for callus induction, but it depends on the plant species ([13].

Auxin used in this study was 2,4-D 2 mg/l. Auxin is used to stimulate callus formation and growth of cells, initiation of shoots and roots, induction of somatic embryogenesis, stimulation of apical buds and axillary buds. [14]Reportedan MS medium supplemented with 2,4-D could induce callus the epidermis and the transport tissue *Elaeis quineensis* Jacq.

Optimisation of the type of medium for callus induction was performed on several species. *Cinnamomum Tamala is*grown on WPM medium with the addition of BA and IBA to form a callus [15]. Medium and growth regulators alike also stimulates somatic embryogenesis in *Quercus semecarpifolia* [16]. [17] Reported that best callus induction *Ruscus hypoglossum* occurred on MS media plus TDZ.

4. Conclusion

Research shows that the optimum sterilising agent is calcium hypochlorite (the code of C3) with terraced exposure sterilisation of a concentration of 50% plus two drops of Tween 20® for 3 minutes and 40% Tween 20® plus two drops for 2 minutes followed by 70% alcohol for 1 minute. The three media had same affect the formation of callus (Table 3), callus produced has crumb texture and compact and has a white colour.

Acknowledgements

Project supported by a project grant from The Ministry of Research, Technology and Higher Education, Indonesia (HIKOM 2014).

References

[1] Oyebanji, O., et al., 2009 *African Journal of Biotechnology*
[2] Qin H, Gu Q, Zhang J, Sun L, Kuppu S, Zhang Y, Burow M, Payton P, Blumwald E, Zhang H 2012 *Plant and Cell Physiology* **52** 1904-1914
[3] Waswala-Olewe B. M, Andanje S. A, Nyagito M.M, Njoka J.T, Mutahi S, Abae R, Amin R 2014 *Tanzania* ***J****of Forestry and Nature Conservation* **83** 30-43
[4] Dhingra, M., et al., 1992 *Tech. Bull.* 39
[5] Hidayat Y 2008 ***J*** *Wanamukti forestry***16** 36-44
[6] Sameer NM, Nabeel K, Al-Ani 2016 *IntJ of Research Studies in Biosciences* ***4****4-9*
[7] Gamborg OL, Miller RA, Ojima K 1968 *Exp Cell Res***50**15158
[8] Murashige T, Skoog F 1962 *Physiology Plant***15**473-97
[9] Lloyd G, McCown B 1980 *Plant propragation***30** 421-27
[10] Lizawati, NeliyatidanDesfira R 2012 *J Sains Dasar FMIPA UNY* 2 **1**19-25

IOP Conf. Series: Journal of Physics: Conf. Series **824** (2017) 012062 doi:10.1088/1742-6596/824/1/012062

[11] De Fossard R., Myint A., Lee E 1974 *Plant Physiology* **31** 125–130
[12] Barciszewski J, Massino F, Clark B. F. C 2006 *J of Biological Macromolecules* **40** 182-192
[13] Yadav, K., & Singh, N 2011 *J Analele Universității din Oradea- Fasicula Biologie* **18** 151-156
[14] Thawaro S & Te-chato S 2009 *J of Agricultural Technology* **5** 167-177
[15] Sharma G, Nautiyal BP, Nautiyal AR 2009 *J. Natural and Science* **9** 1-7
[16] TamtaS, Palni LMS, Vyas P, Bisht MS 2009 *J. of American* **1** 70-76
[17] Ivanova, T., D. Dimitrova, G. Angelov, Ch. Gussev, Y. Bosseva & T. Stoeva 2013 Callus Cultures

IOP Conf. Series: Journal of Physics: Conf. Series **824** (2017) 012063 doi:10.1088/1742-6596/824/1/012063

Extractive Fermentation of Sugarcane Juice to Produce High Yield and Productivity of Bioethanol

U Rofiqah[1,*], T Widjaja[2,3], A Altway[2] and A Bramantyo[2]

[1]Department of Chemical Engineering, Universitas Islam Indonesia Jl. Kaliurang KM 14,5, Yogyakarta, Kode Pos 55501 Indonesia.
[2]Department of Chemical Engineering, Sepuluh Nopember Institute of Technology Jl. Raya ITS, KampusSukolilo Surabaya, Kode Pos 60111 Indonesia.

*Corresponding author: papatri2003@yahoo.com

Abstract. Ethanol production by batch fermentation requires a simple process and it is widely used. Batch fermentation produces ethanol with low yield and productivity due to the accumulation of ethanol in which poisons microorganisms in thefermenter. Extractive fermentation technique is appliedto solve the microorganism inhibition problem by ethanol. Extractive fermentation technique can produce ethanol with high yield and productivity. In this process raffinatestill, contains much sugar because conversion in the fermentation process is not perfect. Thus, to enhance ethanol yield and productivity, recycle system is applied by returning the raffinate from the extraction process to the fermentation process. This raffinate also contains ethanol which would inhibit the performance of microorganisms in producing ethanol during thefermentation process. Therefore, this study aims to find the optimum condition for the amount of solvent to broth ratio (S: B) and recycle to fresh feed ratio (R: F) which enter the fermenter to produce high yield and productivity. This research was carried out by experiment. In the experiment, sugarcane juice was fermented using *Zymomonasmobilis* mutant. The fermentation broth was extracted using amyl alcohol. The process was integrated with the recycle system by varying the recycle ratio. The highest yield and productivity is 22.3901% and 103.115 g / L.h respectively, obtained in a process that uses recycle to fresh feed ratio (R: F) of 50:50 and solvents to brothratio of 1.

1. Introduction

Production of bioethanol is always interesting to be developed and researched. Needs and benefits of bioethanol as an alternative fuel or for food grade encourage further studies. Indonesia has the potential natural resources, one of which is sugar cane, which has been widely cultivated. Sugarcane juice is a source of glucose and sucrose. Sugarcane juice fermented to ethanol by *Zymomonasmobilis* mutant is known to be able to produce ethanol with higher yield within a shorter time than by *Saccharomyces cerevisiae* [1]. The use of sugarcane juice as araw material is better than molasses because molasses contains high salt concentration that would inhibit the growth of *Zymomonasmobilis* mutant [2][3].

Continuous fermentation using cell immobilisation technique in a packed bed bioreactor is thought to be able to solve the problem of inhibition in fermentation batch, which resulted in lower productivity, and lower concentration of ethanol during the production process. In batch fermentation, the microorganisms are poisoned by the accumulation of ethanol in the fermenter. The main inhibitor to *Zymomonasmobilis* mutant in the fermentation process was the accumulation of ethanol itself [4][5]. Extractive fermentation is a continuous fermentation to produce bioethanol with high yield and productivity. In this process, the compound inhibitor in the fermenter was separated. The specific growth of microorganisms decreased when the ethanol concentration in the fermentation broth was at 12% (v / v) [6]. Thus, with this integrated separation, microorganisms will grow well and can produce ethanol in large quantities. Additionally, this process provides advantages that the process of recovery and purification of ethanol would be easier [7].

During the extractive fermentation process, there was still unconverted sugar in the fermenter; therefore, a portion of the raffinatewas returned into the fermenter. Extractive fermentation process with recycle system was done to increase the amount of ethanol produced. In an organic solvent, amyl alcohol has the best ability in extracting ethanol compared dodecanol, and octanol [8]. However, there are problems with their systems that recycle solvent, ethanol diffuses into the bead, and it will poison the microorganisms. Productivity and ethanol yield increased with the increase in recycling ratio. When raffinate is recycled or returned to the fermenter, ethanol dissolved in the raffinate will add an inhibitory effect on the solvent itself.

This research aims to find the influence of solvent to the broth in the fermenter, and the ratio between the amount of raffinate returned to the fermenter to the incoming fresh feed fermenter (R: F) to the yield and productivity of ethanol from sugarcane juice.

2. Material and Methods

2.1.Materials

This study used sugar cane juice raw material with a sugar concentration of 176.3 g/*Zymomonasmobilis* mutant and k-carrageenan as a supporting matrix fermented L. Sugarcane juice. $(NH_4)_2SO_4$, $MgSO_4$, KH_2PO_4, H_2SO_4, and Yeast Extract were added as microorganism nutrition.

2.2. Method

Sugarcane juice preparation. Sugarcane juice was sterilised from other microorganisms in an autoclave at a temperature of 121 °C for 15 minutes. The sugar cane juice was adjustedto a pH of 4-5 by adding H_2SO_4.

Cell immobilisation.Zymomonasmobilis mutantwas put into 100 mL of nutrient medium (sugar cane juice that has been added 1 gram of $(NH_4)_2SO_4$, 1 gram of KH_2PO_4 1 gram of $MgSO_4$, 0.5 grams of $7H_2O$, 10 grams of yeast extract). The media was cultivated in an incubator shaker at a temperature of 30°C for 15 hours. Ten grams of κ-carrageenan in 450 mL was dissolved in distilled water, then heated to a temperature of 70° C to form a gel. Carrageenan solution was cooled to a temperature 50°C. Then 50 ml of nutrient medium was added to 450 ml of theκ-carrageenan solution so that the mixture solution concentration reached 2%. 500 mL of this mixture solution was formed in 1000 mL of 3.5% KCl solution to form the desired bead. The bead was then washed 0.85% of NaCl solution. Furthermore, it was incubated in a shaker incubator for 24 hours. Finally, it was stored at 4 °C until the cells were used.

Extractive fermentation. Bead immobilisation cell was inserted into the fermenter. Sterile cane juice was pumped with a peristaltic pump into the packed bed bioreactor. Fermented (broth) was taken as a sample every 6 hours to know the concentration of residual sugar and ethanol by using a DNS test and Gas Chromatography (GC). This step was conducted until the steady state was reached. It would occur when decrementing of sugar concentration and ethanol formation was stable over time. After fermentation had reached the steady state, it was integrated with the extraction. The brothwas overflown from thefermenter into extractor packed column. Packing used in the extraction process was Raschig rings. Amyl-alcohol as a solvent was passed into a/ the packed column in counter-current accommodating extract coming out of a packed column extractor and draining portion of the raffinate to thefermenter with a recycle ratio of 30%, 50%, 60% and 80%.

3. Result and Discussion

This study uses sugarcane juice with aninitial sugar concentration of 176.3 g/L with a microorganisms *Zymomonasmobilis* mutant and k-carrageenan as a supporting matrix. Integrated fermentation process was carried out using an extraction process using recycle system in which a fermentation process was operated at room temperature 30° C, pH 4-5, feed flow rate of 8 ml/min and a dilution rate of 1.95918 / hour. Solvents used were amyl alcohol in the ratio S: B is 1: 1.

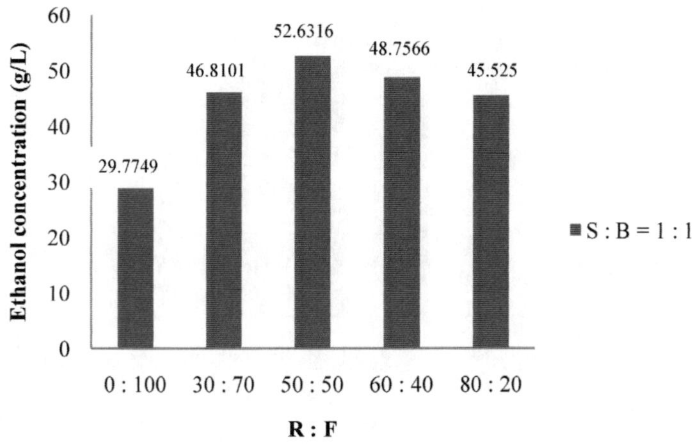

Figure 1. The ratio of R: F to concentration of ethanol

Figure1 illustrates that the ratio of R: F(recycle: fresh feed) had a significant effect on the concentration of ethanol. There was an increase of ethanol concentration with increasing R:Fratio of 0: 100 to 50: 50. At higher ratio, the ethanol concentration decrease with increasing the ratio. It happens because the amount of raffinate returned to the fermenter is greater than the amount of fresh feed fermented which results in a fewer fermented fresh feed (initial sugar). It is possible that the raffinate contains ethanol and a small amount of residual sugar;hence, this affects the amount of ethanol from theextractive fermentation process. The ethanol carried in the flow of raffinate would poison the microorganisms in the fermenter. The ethanol concentration obtained at R:F ratio of 50: 50was 52.6316 g / L.

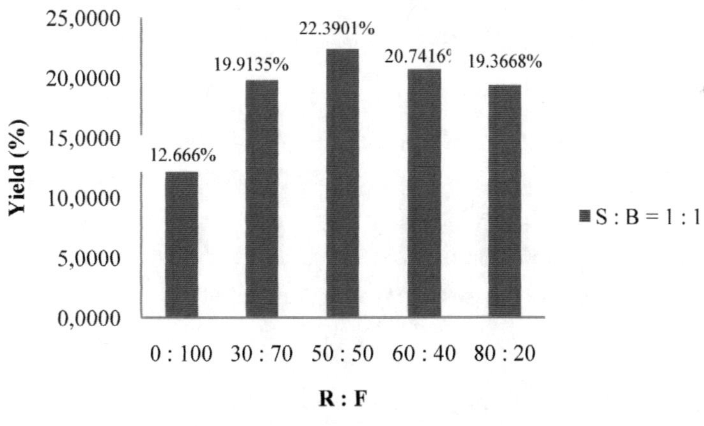

Figure 2. The ratio of R: F to the yield

Figure 2 depicts that the greater the ratio of R: F was, the greater the yield would be. The biggest yield was obtained from the process with a ratio of R: F = 50: 50, of 22.3901%. Yield decreased in R: F = 80: 20 when the amount of raffinate returned to the fermenter was greater; thus, the residual sugar in the raffinate, which was not maximally converted, would be re-fermented which resulted in greater yield. However, the greater the ethanol dissolved in the raffinate returned to the fermenter, it would be toxic to microorganisms that will diffuse into the bead.Widjaja et al. [3] by using immobilisation*Zymomonas mobilis* mutant on batch system reach ayield of 40,92%. This show that continuous system has less yield productivity than batch.

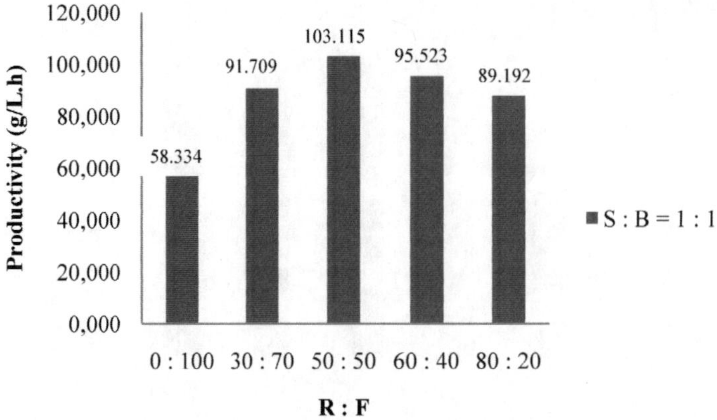

Figure 3. The ratio of R: F to productivity

Figure 3 explains the influence of theratio of R: F on productivity. The greatest productivity was gained from the process with a ratio of R: F = 50: 50, which amounted to 103.115 g / L.hour. Productivity declined in R: F = 60: 40 when the amount of raffinate returned to the fermenter was greater; thus, the residual sugar in the raffinate which was not maximally converted would be re-fermented; as a result, affecting the ethanol produced per time. However, the greater the ethanol dissolved in the raffinate returned to the fermenter; it would be toxic to microorganisms that will diffuse into the bead.Dimple et al.[10] using sweet sorghum on 120 hours of batch fermentation can reach maximum ethanol 79g/ L or productivity of 0.658 g/ L h. Lakkana et al. [11] also conducted anexperiment under fed-batch fermentation can reach productivity of 1.11 g/ L h. This show that the continuous process has higher productivity than abatch process.

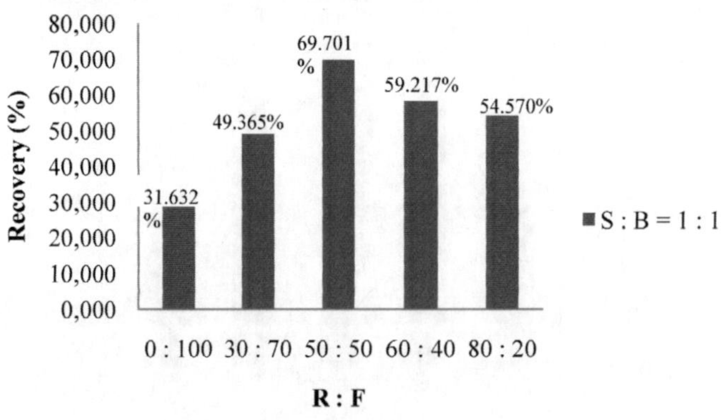

Figure 4. The ratio of R: F to the% recovery

Figure 4 illustrates that the % recovery was best obtained from the process with a recycle ratio of R: F = 50: 50 of 69.701%, while the lowest % recovery was obtained from the process without the use of recyclingof 31.632%.

4. Conclusion

The extractive fermentation process can produce high yields and productivity. The biggest yield and productivity were obtained from processes using R: F = 50: 50 of 22.3901% and 103.115 g / L.h with % recovery of 69.701%.

IOP Conf. Series: Journal of Physics: Conf. Series **824** (2017) 012063 doi:10.1088/1742-6596/824/1/012063

Acknowledgments

The authors would like to thanks to LPPM of InstitutTeknologiSepuluhNopember for the financial support in project number 016449.6/IT2.7/PN.01.00/2015 – Kemenristek DIKTI as 'PenelitianPendukungUnggulan' project from Indonesia Government, the lecturers and all of Biochemical Laboratory crews for the support.

References

[1] Bai JW, Anderson WA, and Moo-Young M. 2008 *Biotechnology Advances* **26** 89–105
[2] Van Vuuren HJJ and Meyer L 1982 *Biotechnol. Letters* **4** 253-256
[3] Widjaja T, Altway A, Ni'mah H, Tedji N, Rofiqah U 2015 *AIP Conference Proceed* 1699.030010
[4] Burrill H., Doelle HW and Greenfield PL 1983 *Biotechnol. Letters* **5** 423-428
[5] Ingram LO, Carey VC, Dombek KM, Holt AS, Holt WA, Osman YA, Walia SK 1984 Biochemical and genetic improvement of *Zymomonas mobilis Biomass* **6** 131-143
[6] Minier M and Goma G 1982 *Biotechnology and Bioengineering* **25** 1565-1579
[7] Kim CH, Kim SW, and Hong SI 1999 *Process Biochemistry* **35(5)** 485-490
[8] Widjaja T, Altway A, Permanasari A, and Gunawan S 2014 *Applied Mechanics and Materials* **493** 300-305.
[9] Dimple K. Kundiyana, Danielle D. Bellmer, Raymond L. Huhnke, Mark R. Wilkins, P.L. Claypool. 2010 *Biomass and Bioenergy* **34** 1481-1486.
[10] Lakkana L, Pornthap T, Prasit J, Pattana L2007 *World Journal of Microbiology and Biotechnology* **23** 1497

Effect of Inoculum Dosage *Aspergillus niger* and *Rhizopus* oryzae mixture with Fermentation Time of Oil Seed Cake (*Jatropha curcas* L) to the content of Protein and Crude Fiber

T Kurniati, L Nurlaila and Iim

Department of Biology Education, Mathematics Education, Faculty of Tarbiyah UIN Sunan Gunung Jati Bandung. No. 105, Street of AH. Nasution Bandung Phone./Fax. 022-7803936.

*Corresponding author: tutikurniati1959@gmail.com

Abstract. *Jatropha curcas* L already widely cultivated for its seeds pressed oil used as an alternative fuel. This plant productivity per hectare obtained 2.5-5 tonnes of oil/ha / year and jatropha seed cake from 5.5 to 9.5 tonnes/ha/year, nutrient content of *Jatropha curcas* seed L potential to be used as feed material, However, the constraints faced was the low crude protein and high crude protein. The purpose of the research was to determine the dosage of inoculum and fermentation time of Jatropha seed cake by a mixture of *Aspergillus niger* and *Rhizopus oryzae* on crude protein and crude fibre. The study was conducted by an experimental method using a Completely Randomised Design (CRD) factorial design (3x3). The treatment consisted of a mixture of three dosage levels of *Aspergillus niger* and *Rhizopus oryzae* (= 0.2% d1, d2 and d3 = 0.3% = 0.4%) and three levels of fermentation time (w1 = 72 hours, 96 hours and w2 = w3 = 120 hours) each repeated three times. The parameters measured were crude protein and crude fibre. The results showed that dosages of 0.3% (*Aspergillus niger Rhizopus oryzae* 0.15% and 0.15%) and 72 hours (d2w1) is the dosage and the optimal time to generate the highest crude protein content of 21.11% and crude fibre amounted to 21.36%.

1. Introduction

One of the byproducts of agriculture which used as feed material is seed cake of Jatropha (*Jatropha curcas L)*. Along with the processing and development of Jatropha curcas L oil, as an alternative energy producer of biodiesel. The oil extraction process results *from Jatropha* curcas L., potentially leading *Jatropha curcas L.* seed cake, amounting to 1 tonne/ha of production of 5 tonnes/ha of Jatropha curcas L. fruit, with oil production *Jatropha curcas L.*, amounting to 2 tonnes/ha [1]. *Jatropha* seed cake is an agro-industrial waste from the pressing jatropha producing mainly oil used as an alternative fuel. The potential of *Jatropha* seed cake was excellent because it does not compete with human needs if it used as feed material.

Agro-industry waste used as feed material has a low quality, including Jatropha seed cake containing only 17% and crude fibre content is high at 24.66%. In order Jatropha seed cake can be used further as feed material, the need for treatment to be done before. One other alternative in the treatment of organic waste is through fermentation by using bioprocess technology of microorganisms that can improve the quality of feed materials [2], as well as fermentation by moulds. Through the process of fermentation, moulds will produce certain enzymes that are expected to break down complex organic materials to be digested to be a simple organic ingredient that easily absorbed. Mould used for fermentation includes *Aspergillus niger* and *Rhizopus oryzae*. *Aspergillus niger* produces the enzyme amylase, protease and cellulose, while *Rhizopus oryzae* produces the enzyme amylase, protease, and lipase. Both types of these fungi have the same ability to break down protein and crude fibre.

The fermentation process is affected by the dosage and time. Dosage-related to the number of microbial populations and time related to the number of microbial growth on the substrate formed on the substrate. Fermentation using a mixture of both the fungus expected to improve the nutritional quality of Jatropha seed cake, especially crude protein and crude fibre, so it eventually used as feed material alternative sources of protein. But so far it has not been a lot of information about the use of Jatropha seed cake as feed and the use of a mixture of two types of mould as an inoculum in the fermentation process.
Based on the above, the author is interested in conducting research on "*Effect of Inoculum Dosage Aspergillus niger and Rhizopus oryzae mixture with Fermentation Time of Oil Seed Cake (Jatropha curcas L) to the content of Protein and Crude Fiber*".
".

Jatropha curcas L

Jatropha curcas L is a plant that is already familiar to the people of Indonesia. This plant is now widely cultivated for its seeds pressed oil used as a lubricating oil and alternative fuels. The productivity of these plants ranges between 3.5-4.5 kg of seeds/plant / year, with the rate of the population between 2500-3300 trees/ha. If the oil yield by 35%, then each hectare of land can be obtained 2.5-5 tonnes of oil/ha / year and Jatropha seed cake from 5.5 to 9.5 tonnes / ha / year.

Fermentation

Fermentation is an oxidation-reduction reaction in biological systems that produces energy in which the donor and acceptor of electrons in organic compounds, so that the resulting product is typical [3]. The moisture level of the solid substrate fermentation process varied between 30% -85%, however, the value of the optimum moisture is required for solid substrate fermentation depending on the organism and the substrates used. Application solid substrate fermentation has grown for the production of antibiotics, alcohol, hormones, biofuel, enzymes, organic acids, aroma compounds, also for the bioremediation of hazardous substances, waste detoxification Agroindustry, bio-pesticides, and maintenance of nutrition [4].

Aspergillus niger

Aspergillus niger has a characteristic that is a collection of single threads called hyphae or a collection of solid threads called mycelium white initially and then develops into yellow and heterotrophic life. The character of aerobic resulting in growing needs oxygen in sufficient quantities and multiple vegetative and generative through cell division and spores, the spores formed within an ascus (spore box). Meanwhile, another example is for a *α-amylase* activity of the highest in the koji fermented by Aspergillus oryzae requires a humidity of about 35% [5].

Rhizopus oryzae.

Rhizopus is a member of Zygomycetes. Rhizopus members often used in the fermentation process is Rhizopus oligosporus and Rhizopus oryzae. Rhizopus oryzae hyphae are not septate and colourless (clear), specialised into three forms, rhizoid, sporangiophore, and sporangium.

Whitish colonies and become gray-brown with age raised, and reached a height of approximately 10 mm, length 150-2000 μm and a diameter of 4-6 lm. Sporangia is round or semi-round, spiked walls, dark brown and a diameter of 50-200 lm. Columella is ovoid-shaped or spherical, 30-120 μm diameter and smooth walled or slightly rough. Sporangiophores round, ovoid, often shaped polygonal, striped on the surface and had a length of about 4-10 μm chlamydospore round shape, diameter 10-35μm. This species has an optimum growth temperature of 35 0C, minimum and maximum 50-70C 350-440 (-490)C. Ikasari and Mitchell (1994) states that as an example of the optimum moisture for protease production by *Rhizopus oligosporus* through solid substrate fermentation is 47%. According to Sarette et al. (1992), other studies show, during fermentation by *Rhizopus oligosporus* entire activity of the enzyme produced by this fungus is strongly affected water activity (a_w), such as enzymes polysaccharides.

Crude protein

Proteins are complex organic compounds that have a high molecular weight. As well as carbohydrates and lipids, proteins contain the elements carbon, hydrogen and oxygen, but also all proteins contain nitrogen.

Crude fibre

Crude fibre contains cellulose and lignin-hemicellulose. Hemicellulose cellulose and is a component of the cell walls of plants and can not be digested by monogastric animals (single bellied). Meanwhile, Medium ruminant animals have substances microorganisms, so the cattle have a greater ability to digest cellulose and hemicellulose, which is enzymatical. Lignin did not include in carbohydrates, but it is in the plant and part or unity in carbohydrates. These substances together form the cellulose component called lignocellulose, which has digestibility coefficient is tiny. In young plants, lignin content will be low but will increase with increasing age of the plant and reached the highest level when the plants are grown [6].

2. Methods

The first stage process, Jatropha curcas L. seed cake sterilised by autoclaving at a temperature of 121^{0}C at 1 atm pressure for 15 minutes. After sterile and then drained until it reaches a temperature of 30-350C, and then inoculated with the inoculum mix. Then incubated in a fermentation chamber at a temperature of 300C for a long 72 hours, 96 hours and 120 hours, and each treatment repeated three times.

The second stage is an experimental method, namely by applying experimental design Completely Randomised Design (CRD). The treatment consisted of a mixture of three dosage levels of *Aspergillus niger* and *Rhizopus oryzae* (= 0.2% d1, d2 and d3 = 0.3% = 0.4%) and three levels of fermentation time (w1 = 72 hours, 96 hours and w2 = w3 = 120 hours) each repeated three times. The parameters observed are crude protein and crude fibre. To see the effect of the treatment of the observed variables were statistically analysed by the method of variance followed by Duncan's multiple range tests.

3. Result

3.1 Effect of Treatment of Crude Protein Content

Protein is an organic compound of containing the elements carbon, hydrogen, oxygen and nitrogen as a constituent. The use of crude protein known in the analysis of animal food ingredients, all the nitrogen present in a feed material calculated as crude protein contained in the material. The higher the nitrogen content of material, the greater the amount of crude protein will be.

The results of the analysis of the crude protein content of Jatropha seed cake fermented in each treatment presented in Table 1.

Table 1. Crude protein content of Jatropha curcas Fermentation results.

Dosage	Time	Repeating 1	Repeating 2	Repeating 3	average
		..........................%....................................			
d1	w1	19,42	19,68	19,46	19,52
d1	w2	19,31	19,63	19,56	19,50
d1	w3	20,08	19,72	19,96	19,92
d2	w1	20,92	21,22	21,19	21,11
d2	w2	19,11	19,23	19,17	19,17
d2	w3	17,98	18,19	17,92	18,03
d3	w1	18,66	18,78	18,75	18,73
d3	w2	18,84	18,75	18,69	18,76
d3	w3	19,97	20,05	19,89	19,97

Note : d1= inoculum dosage 0, 2%
d2= inoculum dosage 0, 3%
d3= inoculum dosage 0, 4%
w1= fermentation time 72 hours
w2= fermentation time 96 hours
w3= fermentation time 120 hours

Table 1 shows that the 0.3% dosage of inoculum fermentation time of 72 hours (d2w1) produces the average value the highest crude protein content of 21.11%, while 0.3% inoculum dosage fermentation time of 120 hours (d2w3) produces the protein content rude lowest at 18.03%. Based on the analysis of variance is known that the dosage of inoculum, fermentation time and interaction significant ($P < 0.05$) crude protein substrate. Furthermore, to determine the differences among the treatments, Duncan's multiple range tests are done as in Table 2.

Table 2. Duncan's Multiple Range Test Effect of Treatment on Crude Protein of Substrates Content

Treatment	mean	significance α 0,05
	.........%............	
d2w3	18,03	a
d3w1	18,73	b
d3w2	18,76	b
d2w2	19,17	c
d1w2	19,50	d
d1w1	19,52	d
d1w3	19,92	e
d3w3	19,97	e

d2w1	21,11	f

Note: a different letter on the significance column indicates significantly different (P <0.05).

Duncan's multiple range test results (Table 2) shows that the crude protein content of the substrate d2w1 significantly (P <0.05) higher than other treatments. The crude protein content of the substrate and d1w3 d3w3 treatment show no significant differences (P> 0.05), but both are significantly (P <0.05) higher than treatment d1w1, d1w2, d2w2, d3w2, d3w1 and d2w3. The crude protein content of the substrate and d1w2 d1w1 treatment show no significant differences (P> 0.05), but both are significantly (P <0.05) higher than treatment d2w2, d3w2, d3w1 and d2w3. Crude protein substrate at d2w2 treatment is significantly (P <0.05) greater than treatment d3w2, d3w1 and d2w3. The crude protein content of the substrate and d3w1 d3w2 treatment show no significant differences (P> 0.05), but both are significantly (P <0.05) higher than d2w3 treatment. Crude protein substrate at d2w3 treatment is significantly (P <0.05) lower than other treatments.

3.2 Effect of Treatment of Crude Fibre Content

Crude fibre is a structural carbohydrate composed of cellulose, hemicellulose and lignin which are difficult to be digested by the digestive system of poultry. Microbes producing cellulolytic enzymes can degrade crude fibre, so the crude fibre content of a substrate decreased.

The results of the analysis of crude fibre content of Jatropha seed cake fermented in each treatment shown in Table 3.

Table 3. Crude Fibre Content of *Jatropha curcas* of Fermentation.

Dosage	time	Repeating 1	Repeating 2	Repeating 3	average
		........................%		....	
d1	w1	20,49	20,19	19,68	20,12
d1	w2	20,01	19,21	19,31	19,51
d1	w3	21,11	20,02	19,50	20,21
d2	w1	20,63	21,64	21,81	21,36
d2	w2	22,29	21,63	21,36	21,76
d2	w3	21,18	20,42	20,71	20,77
d3	w1	21,92	21,84	21,97	21,91
d3	w2	21,44	22,86	21,19	21,83
d3	w3	21,24	20,55	20,49	20,76

Table 3 shows that the 0.2% dosage of inoculum fermentation time 96 hours (d1w2) produces the average value of crude fibre content of the lowest of 19.51%, while 0.4% inoculum dosage fermentation time of 72 hours (d3w1) produces fibre highest rough that is equal to 21.91%. Based on the analysis of variance is known that inoculum dosage significantly (P <0.05) to the crude fibre content of the fermented substrate, but the fermentation time is not significant (P> 0.05) to the crude fibre content of the substrate and no interaction.

Furthermore, to determine the differences among the treatments, then Duncan's multiple range tests as shown in Table 4.

Table 4. Duncan's Multiple Range Test Effect of Treatment of Crude Fibre Content of Substrates

Treatment	Mean	significance α 0,05
	.........%...........	
d1	19,95	A
d2	21,30	B
d3	21,50	b

Note: a different letter on the significance column indicates significantly different ($P < 0.05$).

Duncan's multiple range test results (Table 4) shows that the content of crude fibre d1 dosage significantly ($P < 0.05$) lower than the dosage d2 and d3, respectively. The crude fibre content of the substrate on d2 and d3 is not shown significant differences ($P > 0.05$).

3. Discussion

The optimal dosage of the fermentation of Jatropha seed cake that is a dosage of 0.3% (*Aspergillus niger* 0.15% and *Rhizopus oryzae* 0.15%) with a time of ferment for 72 hours (d2w1). If a dosage is lower or smaller, thus it is lower crude protein content. Similarly, if the fermentation time increased to 96 and 120 hours (d2w2 and d2w3) the crude protein content is decline. It is caused by a mixture of two microbes that Aspergillus niger and *Rhizopus oryzae* are on the dynamics of the different growth rate. Rhizopus oryzae mycelia were growing faster, followed by mycelial growth of *Aspergillus niger*. Aspergillus niger uses proteins which are formed by *Rhizopus oryzae* for growth. So the addition fermentation time affects the decline in crude protein content of the fermented substrate.

The increased crude protein which occurs during the process of fermentation is caused by the work of the mould and the addition of protein donated from cell growth of mould [7]. The increase in crude protein content of fermented also caused due to falling moulds use food ingredients such as carbohydrates and fat for growth and breeding. The fungus can synthesise a protein by taking carbon source of carbohydrate (glucose, sucrose and maltose), a nitrogen source of inorganic or organic material and minerals from its substrate [8]. Also, because during the incubation period, foods derived from the substrate used for growth of mould forming body protein derived from *Rhizopus oryzae*. The more the growth of mould, then the protein substrate will increase [9]. The protein content increases proportionally substrate caused by the reduced carbohydrate content [10] Mechanical fermentation done through the utilisation of biochemical properties of a particular microbe or a mixture of some bacteria [11]. The longer of the fermentation time caused an increasing chance of mould colonies to growth and fermentation so that the number of colonies has increased and will increase the protein content [12].

The increase in crude protein content in the substrate research results come from the ability of *Aspergillus niger* and *Rhizopus oryzae* in overhauling protein complex into simple amino acids and nitrogen. Microbial fermentation by mould and yeasts can improve the nutrition of agricultural waste fermentation products and agriculture industry [13].

Inoculum dosage of *Aspergillus niger* and *Rhizopus oryzae* fermentation of Jatropha seed cake significantly affect crude fibre content of the substrate fermented. The higher dosage of inoculum, the higher crude fibre content of the substrate will be. It is caused by *Rhizopus oryzae* mycelial growth which is faster than the mycelia *Aspergillus niger*. *Rhizopus oryzae* mycelia wall contains a lot of cellulose and chitin substances (compounds that have the same function with the plant cell wall), so that is an increase in crude fibre content of the substrate fermented. The production of cellulolytic enzymes produced by *Aspergillus niger* is not optimal because of the slow growth of the fungus, so the addition of fermentation time does not affect the crude fibre content of the substrate. Mycelia is a collection of hyphae, hyphae surrounded by a cell wall composed of polysaccharides, and the highest content of cell wall in most of the mold are cellulose [8]. The high content of crude fibre on the substrate fermented due to the crude fibre of mould mycelia. Along the slow growth of mould mycelia result can not be opened and a coarse substrate is not accompanied by the opening of other nutrients (fat and starch) led to increasing content of crude fibre fermentation results [14].

4. Conclusion

Based on the results of this study concluded that a dosage of 0.3% (Aspergillus niger 0.15% and *Rhizopus oryzae* 0,15%) and 72 hours (d2w1) is the dosage and the optimal time to produce the highest crude protein content of 21.11 % and amounted to 21.36% crude fibre.

References

[1] Becker, K., dan H.P.S Makkar, 2000. Makalah pada Daimler Chrysler/The world Bank Environment Forum Magdeburg 1999.
[2] Djunaidi, I.H. Natsir, M.H Hardini, D. Tastra, I.K Fatah, G.S.A. 2002. Jurnal Ilmu-Ilmu Hayati 14 (2)
[3] Winarno, F.G. 1980. Kimia Pangan dan Gizi. PT Gramedia, Jakarta.
[4] Perez-Guerra, N., Tarrado-Agrasar, A., Lopez-Macias, C., Pastrana, 2003. Electronic Journal of Environmental
[5] Jung-Yin, L. (2005). J.Fish
[6] Tillman, A. D. H. Hartadi., S. reksohadoprojo., S. Prawirokusumo dan S. Lebdosukojo. 1991. *Ilmu Makanan Ternak Dasar*. Cetakan ke enam. Gadjah Mada University Press, Yogyakarta.
[7] Tangendjaya, B. 1993. Jurnal Ilmu Peternakan Vol.6 No.2; 34-38
[8] Fardiaz, S. 1992. Mikrobiologi Pangan 1. PT Gramedia Pustaka Utama, Jakarta.
[9] Setiyatwan, H. 2007. Jurnal Ilmu Ternak. 7 (2): 113-116
[10] Kurniati, Tuti. 2011. ISSN 1970- 891 Volume V No. 1 -2
[11] Salmah. 2004. Analisa Pertumbuhan Mikroba pada Fermentasi. USU.
[12] Tampoebolon, B.I.M.2009. *Kajian Perbedaan Aras dan Lama Pemeranan Fermentasi Ampas Sagu dengan Aspergillus niger Terhadap Kandungan Protein Kasar dan Serat Kasar*. Seminar Nasional Kebangkitan Peternakan, Semarang Hal 235-243.
[13] Ringpfeili, M.negeli, B.Kreuteri, T.H. Moo Young, M.Rolz , C.1987. *Recommend Method for Caracterization of Agricultural Residues and Feeds Products Derived Through Biocenversation*. Pure and Appl. Chem.59 (5):723-730.

IOP Conf. Series: Journal of Physics: Conf. Series **824** (2017) 012064 doi:10.1088/1742-6596/824/1/012064

[14] Nurhayati, O.Sjofjan dan Koentjoko.2006. J. Indon. Trop. Anim. Agric. 31.

Developing Multimedia-assisted Inquiry Learning Instruments for Basic Biology Intended to Foster Students' Scientific Inquiry

R. Cahyani[1]

[1]Universitas Islam Nusantara, Jl. SoekarnoHatta No.530, Bandung.

[1] Corresponding author: cahyanirianti@yahoo.com

Abstract. Seasonal Influenza is one of disease that outbreaks periodically at least once every year. This disease caused many people hospitalized. Many hospitalized people as employers would infect production quantities, distribution time, and some economic aspects. It will infect economic growth. Infected people need treatments to reduce infection period and cure the infection. In this paper, we discussed a mathematical model of seasonal influenza with treatment. Factually, the disease was held in short period, less than one year. Hence, we can assume that the population is constant at the disease outbreak time. In this paper, we analyzed the existence of the equilibrium points of the model and their stability. We also give some simulation to give a geometric image about the results of the analysis process.

1. Introduction

In 2009 avian influenza was become epidemic and followed by swine flu epidemic [1]. Those epidemics made some death cases and many people hospitalization. The strain of H5N1 virus was identified as a cause of avian influenza epidemic, and Strain of H1N1 virus was identified as a cause of swine influenza while the cause of seasonal influenza is H3N2 virus. The symptoms of all influenza disease are similar. Influenza viruses were responsible for many death cases and many people hospitalization. Seasonal influenza usually doesn't cause death, and it has some effects economically i.e. productivity of employee will decrease, some production process needs longer time, some distribution of goods was suspended [2].

Factually, the infected persons need some treatments to cure infection like drugs assimilation and hospitalization. In [3], it was stated extending oseltamivir to infected children (1 – 3 years old) which don't exceed 24 hours after the symptoms give the significant result. Oseltamivir is known as the only one exact drug for the children under 15 years old. Committee on Infectious Diseases ([4]) also stated treatment should be offered as early as possible. In [5], it was stated that early treatment with antiviral medications may reduce the severity and duration of symptoms, hospitalizations, and complications (otitis media, bronchitis, pneumonia), and may reduce the use of outpatient services and antibiotics, extent, and quantity of viral shedding, and possibly mortality in certain populations. In [6], it was stated that giving the drug in earlier time will be reduced the infection period. In [6], it was also stated diagnosis and treatment would be faster the result of oseltamivir therapy. Urashima [7] suggested that one of the actions to reduce the incidence of influenza A, especially in specific subgroups of schoolchildrenwas vitamin D3 supplementation. In [8], it was stated that using oseltamivir prophylaxis can shorten the durationof influenza outbreaks. In [9,10], there were mutated viruses that have immunities of oseltamivir.

2. Methods

This study employed Research & Development (R&D). The R&D stages undertaken in this study included: (1) preparation; (2) development; and (3) field test [9]. In this research validity test was conducted through analysis of validation evaluation conducted by the expert validator. Meanwhile, a classroom test was conducted to measure the validity, reliability, difficulty level, and discriminating power of items intended to measure scientific work skills. Practicality was measured through evaluation of observation sheets over lecturer activities, and effectiveness was measured through observation sheets of students' scientific inquiry.

2.1 The Subjects

The subjects of this study were 27 preservice teachers at Science department in Nusantara Islamic University, Bandung, Indonesia. An expert validation test was conducted by two experts who held a doctoral degree and Associate professor level.

2.2 Data Analysis

An expert validation sheet consists of five choices of answers as follows: 1, 2, 3, 4, and 5. It refers to guidelines Liker's scale [9]. 1 = very poor, 2=poor, 3= fair, 4= good and 5= very good. The validation sheets were used to evaluate the relevance of depth, completeness, details, and readability of the materials being developed. The expert validation questionnaires consist of 54 items. The data was then analyzed and calculated a percentage of the total score. Practicality and effectiveness of the learning activities were analyzed based on observation sheets on students and lecturer activities. Meanwhile, the effectiveness criterion for the learning process was using Normal Curve Distribution [10,11,15], where highly effective learning process is between 85%-100%, effective learning process is valued between 70%-84%, fairly effective is between 55%-69%, less effective is between 50%-54% and not effective is between 0%-49%.

3. Results and Discussion

3.1Description of Learning Instruments Development

This study has been through several steps from planning to development. On the development stage, validation and revision I was conducted at one class. The expert validator did an evaluation on the reliability and validity of the learning instruments and materials. The validation process covers all of the instruments and materials such as syllabus, lesson plan, learning materials, scientific inquiry tests, students' worksheets, evaluation rubric, observation sheets, observation sheets on attitude, multimedia materials, and students' responses observation sheets. The study developed materials and instruments for three chapters: two chapters about the environment and one chapter about reproduction. The multimedia materials were developed about students' need for tools to collect data and information for the inquiry learning. The scientific work indicators that are developed include data collection, hypothesis formulation, problem formulation, variable definition, prediction, counting, graphics, communication, and make conclusions [2].

Multimedia assisted inquiry learning, to begin with, students asking a question in their worksheets. Students were then given opportunities to read the problem presented in the worksheets to help them more focus on the given problem. Lecturers play the multimedia materials while students pay attention closely to it and take some notes. Discussion and information exchange activities are critical in students' scientific works. Students discuss the problems and make some scientific inquiry using the worksheets while lecturer acts as a facilitator who helps students understand the topic at hand.

3.2 Analysis of Validity Test Results on Learning Materials conducted by Expert Validator

The analysis of validity test results on learning materials conducted by expert validators is shown in the following Table 1:

TABLE 1. Recapitulation of Expert Validator Analysis on Learning Materials

Types of Materials/Learning Instruments	Scores Validator I	Scores Validator II	Average Scores	%	Description
Syllabus (7 item)	30	26	28	80	Valid
Lesson Plan (6 item)	25	27	26	86.7	Valid
Learning Material Module (7 item)	27	28	27.5	78.6	Valid
Scientific inquiry Test (6 item)	22	24	23	76.7	Valid
Student Worksheets (4 item)	16	16	16	80	Valid
Evaluation Rubrics (5 item)	18	20	19	76	Valid
Observation Rubrics (5 item)	21	22	21.5	86	Valid
Students' Attitude Observation Sheets (4 item)	15	16	15.5	77.5	Valid
Multimedia Learning (6 item)	24	23	23.5	78.3	Valid
Students' Responses Sheets (4 item)	15	16	15,5	77,5	Valid
Average				79,8	Valid

Table 1 exhibits that all learning instruments are valid with the scores in 79.8%. Therefore, the instruments are deemed to be valid but need minor revisions. The revision was conducted based on notes and suggestions from both expert validators. These are 1) to pay attention more on the appropriateness of competencies sought by the course and the content in syllabus; 2) to manage the gradual improvement of the difficulties from easy to hard; 3) to not be concerned with the conceptual questions due to it being created by human being; 4) to include objectives and competence that will be pursued in the learning materials; 5) to revise items that have low validity; 6) to minimize the observation on students' attitude in order to make the observation more focused; and, 7) to provide a high degree of thoroughness in the evaluation of a rubric.

3.3 Analysis of Validity and Reliability Test Results on Scientific Work Skill

The analysis of validity test results on scientific work skills is shown in Table 2 below.

TABLE 2. Validity, Reliability, IK and KD Test Results on Scientific Inquiry

Number of Items	Validity	IK	DP	Reliability	Correlation Test	Description
43	Valid	Medium	Medium to high	0.96 (High)	0.78 (High)	No revision
4	Low	Difficult	Low			Revised

Analysis of validity test results as shown above in Table 2 indicates that the 43 items of the test have been classified as valid with high-reliability level (0.96) and high correlation scores (0.78). Therefore, the items are deemed to be reliable and valid to use as instruments for data collection. However, there were 4 items that have low validity (<0.25) with low discriminating power, and high difficulty level. Therefore, the 4test items were then recommended to be revised. The revision was conduct because there was a lack of necessary information to relate two variables being asked. Poor learners were not able to retrieve necessary information to compare both variables. Thus, by revising the chart into a clearer table, the information can be more retrievable by the students. It is in line with Zimmaro's suggestionthat students need to be equipped with a good knowledge about what being questioned [12]. Success at increasing students' ability to ask questions is made more difficult if students do not already possess personal experience or prior knowledge of the topic to be studied. For these students, the purpose of inquiring may simply be to gather more information [13].

3.4 The Analysis of Lecturer and Students' Activities

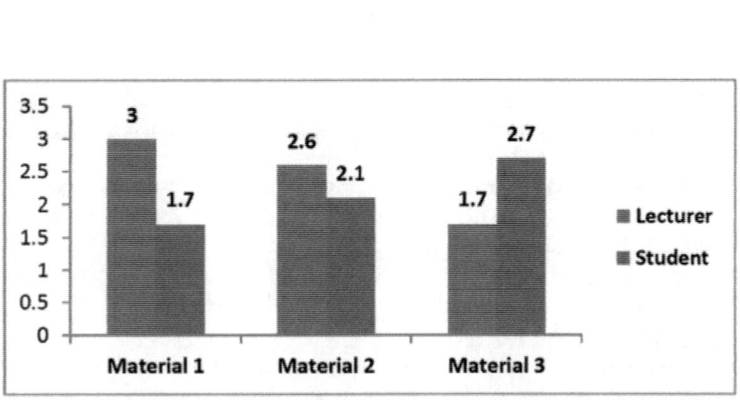

PICTURE 1. The Analysis of Lecturer and Students' Activities

In Picture 1, the average scores of lecturer's activities in Material I was higher than the students. In Material I, the lecturer acted as a facilitator that explain and provide examples of scientific inquiry the students need to do. In the early stage, since the students had to make themselves adapt to scientific inquiry, they made several errors in working on worksheets. They made errors most frequently in the formulizing problem, developing a hypothesis, defining variables, developing graphics, and making conclusions. In following sessions, the lecturer has become less active while the students were more engaged in classroom inquiry activities. In Material II session, the students have developed understanding about the learning process through scientific activities. It indicated that the lecturer has successfully solved the problems in the class session where the students now have become highly motivated to conduct inquiry Lecturer acted as a facilitator has a positive impact on science teacher skills to organize student scientific inquiry [14].In other words, the learning session has become practically implemented with the practicality level in 81.8%. The effectiveness of learning session can be seen from the observation results where they can work on the worksheet effectively. The observation on the students' worksheet found that the learning activities have 72.73% level of effectiveness. Based on the analysis, there are 3 (three) inhibiting factors:1) time management has become critical issue in developing and implementing the materials; 2) lecturer needs to motivate students to learn harder to avoid time extension; and, 3) learning activities highly depend on audio equipment, LCD, and computer.

4. Conclusion

This study concludes that there are some improvements made in the development of learning materials, including syllabus, teaching materials of scientific based basic biology, lesson plan, multimedia learning, students' worksheet, evaluation rubrics, scientific work test, observation sheets over students and lecturer's activities, and evaluation sheets on students' attitude, and students' responses sheets. The materials covered three chapters about awareness of environmental issues and one chapter about reproductions. The multimedia teaching aids were developed about students' needs to search for information and data for their inquiry. The indicators of scientific work this study developed to include the ability to collect data, formulize problem, formulize hypothesis, define variables, make predictions, counting, develop graphics, communicate findings, and draw conclusions.Validation test indicates that the learning materials are valid, reliable, practical, and effective.The study also found that time management, audio equipment, and computer were some of the factors that inhibit the process of the material development.

5. Acknowledgments

The author would like to thank Dr. Mimin Nurjhani and Dr. Okke Rosmaladewi for their help in the study. This work was supported by Indonesian Ministry of Research, Technology, and Higher Education under grant no. 0299/E3/2016.

References

[1] National Research Council, *Inquiry and theNational Science Education Standart: A Guide for*

Teaching and Learning, Washington, DC: National Academic Press, 2000.

[2] National Research Council, *Inquiry and the National Science Education Standard*, Washington DC: National Academies Press, 1996.

[3] Thoron, A.C.and Myers, B.E.,''Effects of Inquiry –based Agriscience Instruction on Student Scientific Reasoning,'' Journal of Agricultural Education, vol.53, No.4, pp.156-177,2012, DOI:10.5032/jae.2012.04156.

[4] Zeidan, A.H. & Jayosi, M.R, " Science Process Skills and Attitudes toward Science among Palestinian Secondary School Students World Journal of Education Vol. 5, No. 1, pp.13-14, 2015, DOI:10.5430/wje.v5n1.

[5] Suwondo and Wulandari, S, " Inquiry-Based Active Learning: The Enhancement of Attitude and Understanding of the Concept of Experimental Design in Biostatics Course,'' Journal Asian Social Science; Vol. 9, No. 12; 2013, doi:10.5539/ass.v9n12p212

[6] Chen, J and Cowie, B,'' Use of Digital Videos in New Zealand Science Classrooms: Opportunities for Teachers and Students,'' Curriculum and TeachingVol. 31, No. 2, pp. 71-86, 2016,

[7] Y. Rotbain, R. Stavy, and Marbach, "Using a computer animation to teach high school molecular Biology," *Journal of Science Education and Technology*, vol. 17, pp. 49-58, 2007.

[8] Gilakjani, A.P, Ismail, H.N. and Ahmad, S.M, ''The Effect of Multimodal Learning Models onLanguage Teaching and Learning,'' Theory and Practice in Language Studies,ACADEMY PUBLISHER Manufactured in Finland. Vol. 1, No. 10, , pp. 1321-1327 doi:10.4304/tpls.1.10.1321-1327

[9] Gelbart and A. Yarden, "Learning genetics through an authentic research simulation in bioinformatics," *Journal of Biological Education*, vol. 40, no. 3, pp. 107 112, 2006.

[10] Creswell, ''Educational Research Planning, Conducting, and Evaluating Quantitative and Qualitative Research''. University of Nebraska–Lincoln, 2012

[11] Gordon, S "The normal distribution"Mathematics Learning Centre, University of Sydney, pp. 14-15. NSW 2006.

[12] D. M. Zimmaro, "Writingmultiple-choice exams," MeasurementandEvaluation Center. The university of Texasat Austin. The website, vol. 16, 2004.

[13] Cuevas, P., et al.'' Improving Science Inquiry with Elementary Students of Diverse Backgrounds,'' : JOURNAL OF RESEARCH IN SCIENCE TEACHING, Vol. 42, No. 3,pp. 337–357, 2005. DOI 10.1002/tea.20053

[14] Volkinstein, J. and Namsone, D,'' Latvian science teacher experience in learning team for Improvement of Inquiry Teaching Practice,'' *Proceedings of ICERI2016 Conference 14th-16th November 2016, Seville, Spain pp.3911* doi:10.4304/tpls.1.10.1321-1327.

[15] Brock, S.E, '' Descriptive Statistics and Psychological Testing'', California State University, Sacramento. Pp.1-7[Online]

Meta-Analysis of Inquiry-Based Instruction Research

N Hasanah[1], A P B Prasetyo[1,*], and E Rudyatmi[1]

[1]Biology Deparment, FMIPA Universitas Negeri Semarang Bld. D6 1st floor,Jalan Raya Sekaran-Gunungpati-Semarang-Indonesia 50299

*Corresponding author: apbudiprasetyo@gmail.com

Abstract. Inquiry-based instruction in biology has been the focus of educational research conducted by Unnes biology department students in collaboration with their university supervisors. This study aimed to describe the methodological aspects, inquiry teaching methods critically, and to analysethe results claims, of the selected four student research reports, grounded in inquiry, based onthe database of Unnes biology department 2014. Four experimental quantitativeresearchof 16were selected as research objects by purposive sampling technique. Data collected through documentation studywas qualitatively analysed regarding methods used, quality of inquiry syntax, and finding claims. Findings showed that thestudent research was stillthe lack of relevant aspects of research methodology,namely in appropriate sampling procedures, limited validity tests of all research instruments, and the limited parametric statistic (t-test) not supported previously by data normality tests. Their consistent inquiry syntax supported the four mini-thesis claimsthat inquiry-based teaching influenced their dependent variablessignificantly. In other words, the findings indicated that positive claims of the research results were not fully supported by good research methods, and well-defined inquiry procedures implementation.

1. Introduction

Inquiry-based instructions have been recommended by science education experts to be one of science teaching approaches [1,2]. Inquiry-based instruction has been the focus of many science education types of research. On the other hand, some science educators argued that it not be easy to prove the effect of the nature of inquiry teaching under different school conditions [3-6]. Limited training on inquiry teaching was also taken into serious consideration of why it is not easy to be implemented.

Since 2011 there have been manymini-thesis on inquiry-based teaching, written by biology education students under the supervision of university teachers, at Biology Department, State University of Semarang. Many students believed that inquiry-based teaching brought significant effects on the quality of science education, however, in reality, there is limited implementation under which the claims were questioned. It was not easy to conduct inquiry-based teaching because teachers did not follow the recommended syntax of teaching such as problem formulation, experiments, and hypothesis formulation [7]. Parametric statistic methods could test the effect of inquiry-based instruction upon student learning achievement without previous normality test of data. In other words, these research were not supported by best methods and inquiry syntax [8,9]. As a result, the findings could not prove the effect of inquiry methods of teaching upon student learning.

Similar research was also undertaken by some researchers [10,11]. It was indicated that some aspects of research report were not well presented. The reasoning of why the problem was conducted, the theoretical framework was developed, and methods were chosen, was not well presented. This quality of research provided a background to do similar research on the related topics. Students' mini-thesis on inquiry-based teaching were systematically analysed again.

Meta-analysis of research is a quantitative research method chosen to providea more balanced and comprehensive research findings [12]. It uses statistical tools to analyse the result of several similar types of research critically. (19; 20;21) Science researchers did research on quantitative meta-analysis(19; 20;21) to assess the impact of instructional methods on student learning critically. On the other hand, some researchers use the similar terms (meta-analysis) to evaluate research findings quantitatively.This method is also used as the qualitative research method. A qualitative method is used to assess research findings in both education and market economy [13; 14]. Qualitative meta-analysis follows such procedures as quantitative ones, but it provides more interpretive and aggregate data. Qualitative methods are more appropriately chosen to assess the aspects of research methods, and claims of findings since all data determined are textual. Based on the explanation, this study was aimed at identifying and describing the quality of methodological aspects, inquiry-syntax and research finding claims of student mini thesis at Biology Department of Universitas Negeri Semarang 2014.

2. Methods

This study used qualitative methods to analyse the quality of inquiry-based teaching research reports undertaken by university students, Department Of Biology, Universitas Negeri Semarang, inthe Year 2015- 2016. The population of this study was all mini thesis reports (document) focused on inquiry teaching strategies (16 reports). Data was collected purposively from the four mini-thesis 2014 reports focused on both inquiry-based teaching and experiment designs. Instruments were developed to assess some aspects of research reports such as title, background, problem formulation, constitutive and operational definitions, research purposes, literature reviews, hypothesis, methods used, and research finding claims, as well as inquiry syntax used. Data were analysed qualitatively.

3. Research finding and discussion

Regarding a title, the mini-thesis reports were more likely to use a wordy title, more than 14 words used. According to the university scientific report guideline, this title writing was unappropriated. A title should be presented 'concise,' using key terms, and less than 14 words. Students seem unfamiliar with the principle of scientific writing that should be accurate, brief, and clear. Accuracy and clarity should be the focus of any scientific writing (18).

Regarding variables measured, all these scientific reports had already mentioned them in their titles. This result indicated that all reports underlined the best formula of writing mini thesis title under quantitative research paradigm, " the Effect of....on." The formula of writing a quantitativeresearch titleexpress the relationship of variables, between and or among independent and dependent variables. It is good that student researchers are aware of the formulation of the variable. Otherwise, they are not able to measure them. Understanding of variables is an important aspect of doing quantitative research (17, 19).

Concerning background presentation, all research reports already presented a gap between the expected and real ones. They then addressed the gap by developing a solution. However, it was not easy to identify the relationship between the fact and the expected one. These structures were supported by Setyosari underlining the importance of gap and solutions [14]. She suggested that research reports should be based on the real difference, what is to be done ideally and the problems existed. Biology education research report could be started by the gap and the suggested solutions. For example, many biology teachers are teacher-centered oriented, but they should be more student oriented. Researchers should address the gap. In reality, all research reports could not provide the clear gap and solution. As mentioned by Creswell [15], it was not always easy to present the structures; many research reports did not provide the clear reasoning. Readers were pushed to imagine the gap. If research reports did not provide the real problem and solution, then the significant implication of the research was limited. Regarding research purposes, the findings showed that all mini-thesis presented their objectives in line with their problem statements. This writing was supported by Nasution [16] who argued that research questions write research goals.

All research variables under the four mini thesis were already defined, and they were ready to be measured. Inquiry approaches as the dependent variable were defined as teaching approaches, or strategy or models or methods. However, they were not measured. Biology learning achievement as dependent variables was strictly defined and measured. Nasution [16] argued that problems should be clearly defined and the focus of research would be clear.

The findings showed that the four mini-theses presented the constitutive definition of concepts studied, but all of them did not provide their operational definition. Inquiry-based teaching was constitutively defined, but not yet operationally defined. It was their limitation of the research. Ary et al. suggested that any concept is both constitutively and operationally defined [17]. Otherwise, they were not properly measured. Constitutive definitions link a theory and facts. They help researchers to identify what is to be studied. This definition is important to provide a focus of the survey. Operational definition help researchers easily measure their variables. Without the proper definition, it is hard to measure concepts and or variables. The findings also showed there were the different definition of inquiry-based teaching by researchers and other references.

Regarding literature reviews, the results showed that all student research presented context of research, both dependent and independent variables, and the biology topic discussed. They did not provide theoretical reasoning of why variables were related. Critical reviews on related references were not presented. These findings were different from what itwas meant by the nature of literature reviews. Literature reviews could be an instrument for predicting or creating the hypothesis.

Regarding hypothesis, under the four mini-thesis, the hypothesis was not clearly presented. Hypothesis did not describe the relationship between variables, but only concepts. Cresswell [15] suggested there were two kinds of hypothesis, directional and non-directional, both of whom describe the relationship between two variables, not concepts.

Regarding research design, the mini-thesis 1, 3, and 4 implemented the true experimental design, but the mini-thesis 2 used quasi-experimental design. These design used could be questioned for several reasons. First, in the empiric research in education, all necessary requirements for experiments are never met. The educational variable is not easily controlled. Quasi-experiments were better designed, compared to true experiments. What about location? All mini thesis provide the precise description of the research sites, and readers will be easily well informed. Research area was to be clearly mentioned. In contrast to his ideas, many quantitative researchers believe that the investigation sites were not to be clearly presented since the quantitative study was objective, impersonal, and out of context. In conclusion, there are the different paradigm of writing location. It depends on the paradigm of research used.

How is sampling techniques implemented? The mini thesis 1 used random sampling technique, but the mini-thesis 2, 3 and 4 implemented the cluster random sampling. All these random samplings used could be wrong regarding the nature of subjects involved. There is no similar student in class, because of this random sampling could be in jeopardy. Simple random sampling could be implemented if every research subject could be involved. There is a variety of student's background, and consequently, purposive random sampling is better than the random sampling one.

The unit of study should be a student, but not a class. Under the mini-thesis 1, sampling methods were implemented on a class rather than a student. Under the mini-thesis 2, 3, and the unit of study was a class (a group of students). If the number of replication is limited, then it is hard to generalise the findings for population [17].

Regarding measurement of variables, all mini thesis developed instruments to measure the variables. Test of validity and reliability of the instruments were also conducted. However, they were not implemented for dependent variables. These are critical incidents of why they were not validated. Many experts suggested that all instruments be validated [17].

Regarding statistics tests used, all mini thesis used parametric statistics, namely t- tests. However, they did not undertake the test of normality, as the most basic requirement of using t-test. Normality test should be first conducted to make sure whether data was distributed normally or not. If the data was not normally, the use of t-test was questionable.

Regarding inquiry teaching syntax, there were much different understanding of it. Because of the limited constitutive definitions, inquiry teaching models were not operationally translated well. It may happen due to limited English of young researchers to read English science textbooks. Limited access to best textbooks can be one of the factors responsible for the weak definition of inquiry-based instructions.

Regarding research claims, all mini thesis reported that there was a significant effect of the use of inquiry teaching models upon the quality of biology learning achievement. These claims were now questionable when they were not supported by better methodologies and query syntax.

What does it mean? To some extend all scientific reports addressed the interesting topic in biology education (Inquiry-based instruction). However, they are still the lack of precise quantitative methodology, starting from writing a title, background of research, concept measurement, experimental design used, and finding presentation. It is critical incident for biology department teachers to provide their students as young researchers with better resources and guidances.

4. Conclusion

By this analysis, some conclusions could be withdrawn. First, there was a tendency that the writing of methodological aspects was still limited. Titles were written in wordy phrases. Background of research was not supported by the real gap in both literature and empiricism. Literature reviews were presented in the context of critical reviews and theoretical framework. The method of research was not supported by the proper design and statistic tools. Second, the claims of findingswere questioned because of limited methodological designs. All instruments used were not tested for their validity and reliability. Parametric statistics (t- tests) was not supported by the assumption tests (normality test). Finally, the claim of the significant effect of the use inquiry-based instruction upon the quality of biology learning achievement was also questionable, because of the limited definition of inquiry syntax of teaching and weak quantitative research designs. It is now suggested that students who conduct research for scientific writing be accompanied by better resources and guidances.

References

[1] National Research Council. 1996. *National Science Education Standards.* Washington: National Academy Press.

[2] Trowbridge, L.W. & R.W. Bybee. 1981. *Becoming a Secondary School Science Teacher.*Ohio: Merril Publishing Company.

[3] Becker, K. P., K. (2011). *Jurnal of STEM Education, 12*(5&6).

[4] Balaney, C. R., CE. (2013). *International Journal of Biology Education, 3*(1).

[5] Srisawasdi & Panjaburee, P. (2015) *Jurnal of Computer Education, 2*(3), 323 325

[6] Soprano, K. L.-l. Y. (2013) *International Journal of Science and Mathematics Education, 11.*

[7] Rizkiyana, A.Z. 2014 Skripsi. Semarang: FMIPA Unnes.

[8] Kurniawati, A. 2014 Skripsi. Semarang: FMIPA Unnes.

[9] Mukminin, A.A.2014 Skripsi. Semarang: FMIPA Unnes.

[10] Ardhana, W.S.F. &Wagiran. 2007 *SeminarHasilPenelitian Fakultas*. Yogyakarta: UNY.

[11] Indarto, W. 2012 *Educhild*, 1(1): 59-66.

[12] Siswanto. 2010 Surabaya: Pusat Penelitian dan Pengembangan Sistem dan Kebijakan Kesehatan.

[13] Wijayanti, E. H. Susilo & H. Suwono. 2013 *Seminar Nasional X Pendidikan Biologi FKIP UNS.* Surakarta: UNS.

GRETZEL, Y. A. P. A. U. 2007 *Journal of Travel Researc*h, 46(August), 46-43.

[14] Setyosari, P. 2010 Jakarta: Kencana.

[15] Creswell, J.W. 2009 Los Angles: SAGE.

[16] Nasution, S. 2001 Jakarta: Bumi Aksara.

[17] Ary, D., L. C. Jacobs, &A. Razavieh. 2004 Yogyakarta: Pustaka Pelajar.

[18] Ben, D. 2012 Writting the Empirical Journal Articles: Cornel University.

[19] Barbara J. Guzzetti, T. E. S., Gene V Glass. 1993 Reading Research Quarterly, 28(2), 116 -159

[20] Chauhan, S. 2017 Computers & Education105(Feb), 14-30.

Zahira Merchanta, Ernest T. Goetzb, Lauren Cifuentesc, Wendy Keeney-Kennicuttd, & Davisa, T. J 2014. Computers & Education 70(January), 29-24.

The Correlation Between Metacognition Level with Self-Efficacy of Biology Education College Students

S Ridlo[1] and F Lutfiya[1]

[1]Biology Department, Faculty of Mathematics and Natural Science, Semarang State University, Indonesia, Sekaran Gunungpati Semarang Indonesia 50229

*Corresponding author: saiful_ridlo@mail.unnes.ac.id, lutfiyafina@yahoo.com

Abstract. Self-efficacy is a strong predictor of academic achievement. Self-efficacy refers to the ability of college students to achieve the desired results. The metacognition level can influence college student's self-efficacy. This study aims to identify college student's metacognition level and self-efficacy, as well as determine the relationship between self-efficacy and metacognition level for college students of Biology Education 2013, Semarang State University. The ex-post facto quantitative research was conducted on 99 students Academic Year 2015/2016. Saturation sampling technique determined samples. E-D scale collected data for self-efficacy identification. Data for assess the metacognition level collected by Metacognitive Awareness Inventory. Data were analysed quantitatively by Pearson correlation and linear regression. Most college students have the high level of metacognition and average self-efficacy. Pearson correlation coefficient result was 0.367. This result showed that metacognition level and self-efficacy has a weak relationship. Based on linear regression test, self-efficacy influenced by metacognition level up to 13.5%.The results of the study showed that positive and significant relationships exist between metacognition level and self-efficacy. Therefore, if the metacognition level is high, then self-efficacy will also be high (appropriate).

1. Introduction

College students according to the laws of 2012 No. 12 defined as learners at higher education level (college). Higher education is an education after secondary education. It includes diploma, undergraduate, master program, doctoral and professional programs, as well as specialist courses, organised by the college based on the culture of Indonesia. College students, as members of civitas academic, positioned as an adult who being developing their self-awareness potential in college to become intellectuals, scientists, practitioners, and professionals. As college students, every individual has an academic purpose, which reaches the best educational achievement appropriate with the expectations. College students of Biology Education Studies Program, Faculty of Mathematics and Natural Science, Semarang State University (UNNES) also have the same purpose. Academic achievement can be predicted by student's confidence about their capacity to achieve the desired results in all subjects [1,2]. This determination is called self-efficacy.

Self-efficacy is an individual belief regarding the ability to organise and complete a task even determinate of future behaviour, which can influence the selection of activity, effort, persistence, and achievement [3,4]. Preliminary research indicated that 56% of the college students of Biology Education 2012 have self-efficacy in the average category. Thus, most of the college students have enough confidence in their ability to achieve the desired academic expectations. The confidence still easily influenced by various factors and will be reflected in student results. Self-efficacy of students in the future need to be considered because it is predicting academic achievement.

Research by [5] demonstrated that their metacognition could influence the self-efficacy of a person. [6] defined metacognition as awareness of the cognitive processes and the ability to monitor and control the cognitive process. Awareness of their thought processes assist the student in designing, monitoring, and evaluating learning strategies used. The learning process will do better. Successful experience influence college student's self-confidence.

IOP Conf. Series: Journal of Physics: Conf. Series **824** (2017) 012067 doi:10.1088/1742-6596/824/1/012067

College students of Biology Education are individuals in the phase of late teens to early adulthood who is developing their metacognition through education. It can be observed in experiment activity. Along with experiments during the first semester to the fifth semester, college students have more experiences in planning, monitoring, and evaluating. The success or failure as a result of the metacognitive activity is a direct experience that reinforced or weaken the student's confidence in their competence. In other words, student's metacognition related to self-efficacy.

Results of [5] was a person's self-efficacy influenced by his metacognition. A preliminary study by involving 50 college students of Biology Education 2012 reinforced Adicondro & Purnamasari's research [5]. Preliminary studies showed that 46% of student's metacognition level appropriate with self-efficacy. This study aims to determine the relationship between metacognition level with self-efficacy of Biology Education students 2013.

2. Research Methods

The research took place in the Biology Department, Faculty of Mathematics and Natural Science, UNNES at Even Semester Academic Year 2015/2016. The population of this study was all college students of Biology Education 2013. The sampling technique used is saturated sampling. The research sample consisted of all members of the population. The total samples were 99 (N=99). As a quantitative study, a self-report questionnaire was used to gather data. The questionnaire was based on relevant studies done previously, consists of two instruments including the measurement of metacognition level and self-efficacy. The respondents answered the questionnaire manually, and we asked them to return it upon completion.

2.1 Metacognitive Awareness Inventory (MAI).

The MAI [7] that was used to measure metacognitive awareness in several previous studies about metacognition. In this research, the MAI was modified into Indonesian version. The MAI consists of 52 items which provide alternative responses as being false and true. Within the scale, there are two metacognitive components, the cognitive knowledge, as well as the cognitive regulation, were assessed. The range of validity coefficient from 0.58-0.65. The reliability of this instrument is 0.95.

2.2 Self-Efficacy Scale (E-D Scale).

E-D scale was an instrument that intended to assess the level of college student efficacy. Consists of 26 items this scale was developed based on self-efficacy dimensions/components from Bandura [8], which reflect the Magnitude, Strength, and Generality of a person as a college student. The responses alternatives were presented of five grades scale that extended from strongly agree into strongly disagree. The range of validity coefficient from 0.286-0.652. The reliability of this instrument is 0.832.

Based on the assumption test consisting of normality test and linearity test, can be evidenced that data were normally distributed (0.087> 0.05 and 0.122> 0.05) and linear (0.765>0.05). Thus, the relationship between the variable X (metacognition level) and variable Y (self-efficacy) as can be tested by the Pearson correlation test and linear regression. Then, analysis of the results of hypothesis testing explained descriptively.

3. Result and Discussion

Results of this research are the identification of metacognition level and self-efficacy, also the relationship between metacognition level with self-efficacy of Biology Education 2013college students. The number of respondents in this study was 99 college students from three classes. Identify college students metacognition level, and self-efficacy are presented in Table 1.

Table1 Identification of metacognition level and self-efficacy

Attribute	High		Average		Low	
	Frequency	Percentage	Frequency	Percentage	Frequency	Percentage
Metacognition level	84	85%	15	15%	0	0%
Self-efficacy	39	39%	59	60%	1	1%

Based on Table 1, 85% of college students have high metacognition, 15% of college students have the average level of metacognition, and no college student has the low level of metacognition. College students already have good awareness and ability to monitor and control their cognitive process. All college students age is in the range 19 to 21 years. Thus, this research sample at the phase of late adolescence to early adulthood. At this time, there is increased activity to monitor and control their cognition effectively as a consequence of tasks complexity in the learning [9].

Metacognition had two dimensions/components; those are cognitive knowledge and cognitive regulation. Subcomponents of cognitive knowledge consist of declarative knowledge, procedural knowledge, and conditional knowledge. Cognitive regulation has planning, information management strategies, comprehension monitoring, debugging strategies, and evaluation subcomponents. In general, the percentage of cognitive learning dimension greater than the percentage of cognitive regulation, while the proportion of each metacognition subcomponents shown in Figure 1.

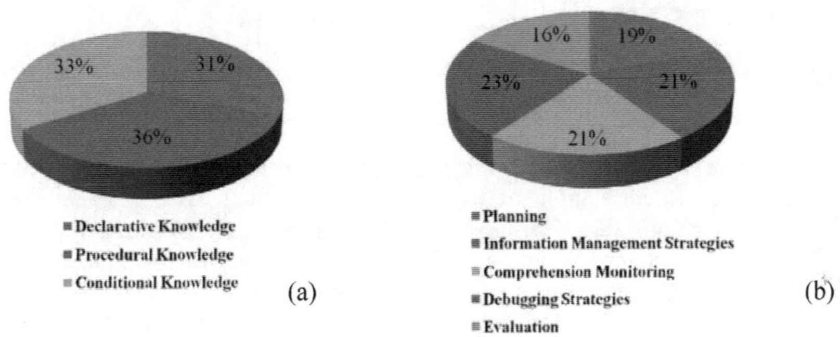

Figure 1 Percentage of (a) cognitive knowledge and (b) cognitive regulation

Based on Figure 1, declarative knowledge, planning, and evaluation subcomponents have the lowest percentage. Most college students had difficulty in assessing their learning ability. Capacity to plan, set goals, and analyse the results of learning, as well as evaluate the effectiveness of learning strategies used is still not optimal. It is a reason that students rarely monitor their progress in learning [10]. High metacognition for Biology Education 2013 college students as future teachers is critical because teachers need to monitor their process of teaching. This monitoring includes how to develop the education process, the efficiency of teaching strategies, and modify teaching strategies if needed [11].

Based on Table 1, self-efficacy of most Biology Education 2013 College students are in common category (60%). The college students felt confident enough to do the work, regulating the activity of their learning, and established academic expectations with their capabilities and competencies.The percentage of college students with high self-efficacy is 39%. College students have a strong belief that they can achieve their academic expectations.High self-efficacy will lead one to a better performance in various fields because of self-efficacy will activate the physiological changes that reduce stress. One college student has low self-efficacy. It can be interpreted that the student does not believe that he can do a challenging work with his competence at the certain moment.

Self-efficacy had three dimensions/components; those are magnitude, generality, and strength. Based on the analysis of each aspect, obtained a description that most of college student's magnitude and power were in the high category, while the generality dimension was in the average category. College students had great confidence in their capacity to resolve any difficult task given. Also, college students have the steadiness against his conviction to accomplish the variety of tasks [12]. Generality component at average category indicates that college students do not have enough confidence to take control of some tasks all at once, especially if the student follows the student organisations. In other words, their self-confidence definite to one particular situation.

The results of correlation between metacognition level with self-efficacy shown in Table 2. There is a positive and significant correlation between metacognition level and self-efficacy of Biology Education 2013 college students. The college student's confidence to achieve desired academic expectations related to their ability to organise and manage the cognitive process. This correlation enhanced college student's confidence about their competence, encourage students to learn, and consider the difficulties or obstacles as a challenge, not as a threat to be avoided [13].

Table 2 Result of Pearson correlation test

Variable X	Variable Y	*r*	Sig.
Metacognition level	Self-efficacy	0.367^{**}	0.000
Cognitive knowledge		0.324^{**}	0.001
Cognitive regulation		0.344^{**}	0.000
Declarative knowledge		0.287^{**}	0.004
Procedural knowledge		0.187	0.064
Conditional knowledge		0.216^{**}	0.032
Planning		0.417^{**}	0.000
Information management strategies		0.126	0.214
Comprehension monitoring		0.224^{**}	0.026
Debugging strategics		0.212^{**}	0.035
Evaluation		0.228^{**}	0.023

After further analysis with simple linear regression test, metacognition level affects the self-efficacy significantly. Pearson correlation coefficient of 0.367 indicates that metacognition level and self-efficacy has a weak relationship. This weak relationship reflected from the recapitulation of metacognition level and self-efficacy identification. High metacognition college students also have high self-efficacy, but this is not always happening. Summary of data recapitulation showed that 50 of 99 college students (50.5%) had a conformity between their metacognition level and self-efficacy, while 49 other college students do not have a conformity between metacognition level and their self-efficacy. Simple linear regression test results are summarised in Table 3.

Table3 Result linear regression test.

Model	Sum of square	F	Sig.	*r*	*r square*
Regresi	741.07	15.14	0.00	0.37	0.14
Residual	4749.11				
Total	5490.18				

[14] assert that a weak correlation between metacognition level and self-efficacy can be affected by other self-efficacy factors. Four factors affected self-efficacy, i.e., mastery experience, vicarious experience, verbal/social persuasion, and physiology-emotional arousal [12]. Metacognition activities (monitoring and controlling cognitive processes) assist students to learn effectively so that the academic expectations can be achieved. The successful experience can be applied to other situations. Metacognition experience that produces the favourable result is a form of mastery experience. The mean score for each self-efficacy factors is 3.56 and 4.12 for mastery and vicarious experience, 3.92 for social persuasion, and 3.67 for emotional arousal.

In this study, the largest source of college student confidence is the vicarious experience. Students prefer to observe model's behaviour to strengthen his conviction to be able to do the same tasks as well. Although metacognition as mastery experience had the low mean score, that is not to say metacognition factors can be ignored. Metacognition with all their activities is very beneficial to facilitate the learning process. Also, college students cannot rely on others everytime he has not been able to master skills.

Cognitive regulation dimension has a stronger correlation to the self-efficacy than cognitive knowledge. Furthermore, subcomponents of cognitive control that had the highest correlation to the weakest with self-efficacy are the planning, evaluation, comprehension monitoring, debugging strategies, and information management strategies. Based on these data, the ability of plan was a major component in metacognition that reinforces college student's self-efficacy.

Results of the analysis showed that the planning subcomponents of Biology Education college students need to be optimised because although thus subcomponents correlated most strongly with self-efficacy, the percentage and meant score is the second lowest from the other cognitive regulation subcomponents. In other words, most students planning ability is not as good as the other metacognitive skills. Planning is a major step in the learning process because the activities in the plan include planning, goal setting, and collection of reference sources to learn [7]. The right plan enhances the success of learning and college student's self-efficacy influenced.

Planning ability of college students could be optimised through metacognitive learning. Skills to design, monitor, and evaluate the learning process and results are expected to improve students' confidence, especially in their ability to achieve positive results. Metacognitive learning can be applied to specific teaching models. An example was done by [15,16]. They incorporate metacognitive activity in the jigsaw, problem-based and inquiry learning. The learning model that can be applied to optimise the biology learning outcome and self-directed learning.

4. Conclusion

In conclusion, Biology Education 2013 College students have high awareness and ability to monitor and control their cognitive processes. Also, most students felt confident enough to do the work, regulating their learning activity, and achieve academic expectations. The relationship between metacognition level with self-efficacy is straight and positive.

References

[1] Coutinho S A 2007 *Educate Journal* 7(1) 39-47
[2] Hermita M and W P Thamrin 2015 *Procedia Social and Behavioral Science* **171** 1075–1080.
[3] Myers D G 2010 *Psikologi Sosial* Translated byTusyani A 2012 10th ed (Jakarta: Salemba Humanika)
[4] Zakeri A, R Rahmany and E Labone 2016 *Journal of Language Teaching and Research* 7(1) 158-167
[5] Adicondro N and A Purnamasari 2011 *Humanitas* **3**(1) 17-27.
[6] Wade C and C Tavris 2007 *Psikologi* Translated by Widyasinta B 2008 9th ed (Jakarta: Erlangga)
[7] Schraw G and R S Dennison 1994 Assessing metacognitive awareness *Contemporary Educational Psychology* **19** 460-475
[8] Bandura A 2012 *Journal of Management* **38**(1) 9-44
[9] Santrock J W 2011 *Educational Psychology* 5th ed (New York: The McGraw-Hill)
[10] Reynolds W M and G E Miller 2003 *Handbook of Psychology* (Hoboken: John Wiley & Sons Inc)
[11] Wen Y H 2012 *The Journal of Human Resource and Adult Learning* **8**(1) 90-91
[12] deNoyelles A, S R Hornik, and R D Johnson 2014 *Journal of Online Learning and Teaching* **10**(2) 255-271
[13] Cera R, M Mancini, and A Antonietti 2013 *Journal* **7** 115-141
[14] Hasan M Z, T B Hossain, and A Islam 2014 *Advances in Environmental Biology* **8**(9): 695-705
[15] Putri W A, A P B Prasetyo, and Supriyanto 2012 *Unnes Journal of Biology Education* **1**(3): 266-271
[16] Aisyah S and S Ridlo 2015 *Unnes Journal of Biology Education* **4**(1): 22-28

IOP Conf. Series: Journal of Physics: Conf. Series **824** (2017) 012068 doi:10.1088/1742-6596/824/1/012068

Guided Inquiry Facilitated Blended Learning to Improve Metacognitive and Learning Outcome of High School Students

H Suwono[1], S Susanti[2], and U Lestari[1]

[1]Biology Department, Faculty of Mathematics and Natural Science, Universitas Negeri Malang, Jalan Semarang 5 Malang, Indonesia
[2]Undergraduate Student of Biology Education, Faculty of Mathematics and Natural Science, Universitas Negeri Malang, Jalan Semarang 5 Malang, Indonesia

*Corresponding author: hadi.suwono.fmipa@um.ac.id

Abstract. The learning activities that involve the students to learn actively is one of the characteristics of a qualified education. The learning strategy that involves students' active learning is guided inquiry. Learning problems today are growing metacognitive skills and cognitive learning outcomes. It is the research and development of learning module by using 4D models of Thiagarajan. The first phase is Define, which analyses the problems and needs required by the prior preparation of the module. The second phase is Design, which formulates learning design and devices to obtain the initial draft of learning modules. The third stage is Develop, which is developing and writing module, module validation, product testing, revision, and the resulting an end-product results module development. The fourth stage is Disseminate, which is disseminating of the valid products. Modules were validated by education experts, practitioners, subject matter experts, and expert of online media. The results of the validation module indicated that the module was valid and could be used in teaching and learning. In the validation phase of testing methods, we used experiments to know the difference of metacognitive skills and learning outcomes between the control group and experimental group. The experimental design was a one group pretest-posttest design. The results of the data analysis showed that the modules could enhance metacognitive skills and learning outcomes. The advantages of this module is as follows, 1) module is accompanied by a video link on a website that contains practical activities that are appropriate to Curriculum 2013, 2) module is accompanied by a video link on a website that contains about manual laboratory activities that will be used in the classroom face-to-face, so that students are ready when doing laboratory activities, 3) this module can be online through chat to increase students' understanding. The disadvantages of this module are the material presented in the modules is limited. It is suggested that for the better utilisation of the online activities, students should be present at every meeting of the activities, so as to make all the students participate actively. It is also suggested that school set up facilities to support blended learning.

1. Introduction

Learning activities should involve students to think actively and act. The activeness of the students in learning is one of the factors affecting the success of learning. The Partnership for 21st Century Skills has identified metacognitive as one of the life and career skills necessary to prepare students for post-secondary education and the workforce [1]. Metacognitive skill is one dimension that is emphasised in the school curriculum [2].

Metacognition is someone's awareness about his thinking process. In other words, metacognition is typically characterised by monitoring and regulating their thinking processes to produce an effective process for himself [3]. Metacognition can simply be defined as thinking critically about what the people think, or thinking about thinking [4].

Metacognition involves tasks, strategies, and cognition of oneself [2]. The strategy refers to the process of thinking, completing tasks using their cognition. Metacognition is the awareness of thinking to understand, monitor, about what is known and what is done [5]. Students must have the metacognitive knowledge to have the metacognitive ability. Thus, the metacognitive strategy is a very critical component. In metacognition are known the terms metacognitive awareness, that is what one believes and how one knows, and metacognitive strategies which are useful for information processing [6].

Tanner [7] states metacognition is knowledge about one's learning or knowing how to learn and monitoring one's learning behaviours to determine the degree of progress and strategies needed for accomplishing instructional goals. Based on the statement, it can be interpreted that metacognition is a person's knowledge about learning, and monitor the behaviour of their learning to determine the development and learning needed to achieve the learning objectives [8]. Metacognition is a reflective cognition, a higher order thinking process which involves active control over cognitive processes [9]. The statement means that metacognition is a form of cognition or high-level thinking processes or involving the control of the cognitive activity. Therefore, metacognitive can be regarded as a person's thinking about his thinking, or a person's cognition about his cognition.

Metacognition has three components for problem-solving in learning, namely: (a) metacognitive knowledge, (b) metacognitive skills, and (c) metacognitive beliefs [10]. However, the most common differences in metacognition are separating metacognitive knowledge from metacognitive skills. Metacognitive knowledge refers to declarative knowledge, procedural knowledge and conditional knowledge of a person on problem-solving. While metacognitive skills refer to the prediction skills, planning skills, monitoring skills, and evaluation skills.

Metacognitive skills of each are different, depending on the knowledge and experience [10]. The efforts to improve self-control in learning are as follows: a) to encourage students to select and monitor several learning objectives they expect, (b) to provide the opportunity for students to learn and perform optimally without any pressure from outside, (c) to provide the opportunity for students to do activities that involve their learning goals, management strategies and learning time, d) to help complete the learning equipment needed, e) to provide the opportunity for the students to think critically and to provide feedback on their efforts, f) to ask the students to evaluate the results that have been obtained, whether or not the results are suitable for the objectives and the standards that have been set [1].

The research showed that the Indonesian students' metacognitive ability was still low, both of the elementary school students [11] and the high school students [12]. The research results showed that low levels of metacognitive skills had an effect on the low learning results. The results of the observation on the level of the activeness of the students in Public Senior High School 1 Malang were quite low. It could be seen when the teacher gave the students questions. Learning problems that the students had been such as, 70.97% of 35 students found it difficult to learn biology because of the confusing scientific terms in the textbooks, and 83.87% of the students said that biology lessons were tedious because a lot of biology material required memorization learning. It was one of the causes that the students had not reached the KKM (the minimum passing grade). Also, 48.39% of the students did not realise the importance of learning biology for themselves, so that the students felt lazy to learn biology deeper. The results of the interview showed that the students had not been able to connect the biological concepts with their real life so that the students had not realised the importance of learning biology. The low awareness of the students showed that the students' metacognitive skills needed to be improved.

Metacognitive skills determine the learning results. Metacognitive skills are related to thinking how to think, knowing what is known and the unknown, so that it is useful for problem-solving [13]. These metacognitive skills are required to understand a concept so that the concept gained by students is better because it is gained through a process which involves metacognitive skills.

Metacognitive skills and metacognitive strategies can be learned [14]. Therefore, it is necessary to find learning strategies that improve metacognition. In recent years, research has been conducted to investigate the inquiry-based learning in supporting metacognitive skill development [15].

Inquiry and experimentation learning fosters metacognition and scientific reasoning [16]. The inquiry with questions will direct the learning in an inquiry climate that fosters self-dependency in mastering new things [17]. The inquiry is an important component in science education [18]. Biology learning in Indonesia has not yet fostered the process of inquiry in students. Guided inquiry-based learning makes students involved in learning so that learning becomes meaningful for students. The main characteristic of guided inquiry is that students are directed to search and find the concepts independently, critically, and logically. Guided inquiry learning by changing the passive learning patterns into active learning and critical learning patterns [18].

Inquiry learning has complex learning situations so that scaffolding is needed to engage students to inquire, manage investigations and solve problems; help students make sense of their thinking processes [19]. High school students are interested in using the web-based scientific resources), so that the sources on the web can be used as a scaffolding of inquiry [20]. Also, the worksheet can also be an effective scaffold in problem-solving learning [21]. Learning by utilising online computer can enhance the ability of self-explanation so that it can help learning by doing [9].

Blended learning is a learning method that combines the face-to-face session and online learning session [22]. Blended learning is also called a hybrid online [23] that combines the online digital content, multimedia training compact disks, supplemental learning content online discussions, and live broadcasting to improve the learning processes and learning outcomes [23]. In general terms, blended learning combines the online delivery of educational content with the best features of classroom interaction and live instruction to personalise learning, allow thoughtful reflection, and differentiate instruction from student to student across a diverse group of learners [24]. The virtual meetings between the teachers and learners were carried out online because they might be in two different worlds or places, but they could give feedback to each other, asked or answered questions. Long distance also calls it instructed learning or virtual instructor led training which is guided by an instructor virtually since the instructor and the student are in a different place. Regardless of the name, this learning utilises IT technology through the medium of video conferencing, phone conferencing, or chat online [25].

Metacognitive competence is needed for students. Biology learning in high schools uses scientific investigation to help students develop depth understanding of biological knowledge and problem-solving through metacognitive strategy. This research aimed at investigating the effect of guided inquiry facilitated blended learning on metacognitive skills and learning outcomes of high school students. This research was designed to test the hypothesis that the inquiry learning facilitated blended learning improves metacognition compared with inquiry learning. This research was conducted on senior high school students in the eleventh grade to prove the hypothesis.

2. Research Methodology

This research used the Research and Development approach using the 4D model [26]. The steps are as follows.

1. *Define* analysing problems, and the other required needs before the drafting module.
2. *Design:* formulating a lesson plan as an outline/draft of learning.
3. *Develop:* developing and writing modules, expert appraisals, product testing, revision, and producing the final product of the *module* development.
4. *Disseminate:* spreading the products that have been developed and have been validated.

Dissemination was done using quasi-experimental approach and a nonequivalent control group pretest-posttest design [27]. This research aimed at determining the effect of guided inquiry facilitated blended learning on the empowerment of students' metacognitive skill in biology learning in senior high schools in Malang, Indonesia. The independent variables were teaching strategies consisting of guided inquiry facilitated blended learning (GIqBL) and guided inquiry (GIG). The dependent variable was students' metacognitive skill. This research was conducted in the odd semester of academic year 2013/2014.

The research was conducted at public senior high school 1 Malang, Indonesia. The research population consisted of five classes of the second year students in academic year 2013/2014 as many as 165 students. The total samples of this research were 70 students. The number of the students in the experimental class (GiqBL) was 35 students, and the total number of the students in the control class was 35 students.

Before the research was conducted, the researcher developed a lesson plan, modules, and worksheets. The module about circulatory system consisted of two topics, namely the components of blood and blood circulation organs. Each module contains the explanation of the materials, blood circulation physiology processes and blood abnormality. The module was equipped with 1) Introduction, 2) Table of Contents, List of Tables, 3) List of Figures, 4) How to Use Module, 5) Objectives, Concept Map, 6) Topic I components of blood, 7) problems, 8) What is your hypothesis, 9) is your hypothesis correct? Find Out!, 10) Virtual Laboratory, 11) Conclude Results of the activity!, 12) Topic II blood circulation organs, 13) Material Description, 14) Feedback, 15) Summary, and 16) Competency Test.

GIqBL learning and GIQ learning were implemented for 6 weeks in different classes. The learning topics in both these strategies are blood system, lymph, and the circulation of blood and the circulation of lymph. Scientific inquiry learning used the stages/phases of learning referring to Llewellyn [28], namely exploring phenomena, formulating investigative / experimental questions, planning investigation/experiment, conducting investigation/experiment and collecting data, analysing the data and evidence, building new knowledge and communicating new knowledge. In GiqBL models, the learning materials were uploaded on the website which can be downloaded by the students. The online discussion was done online through Facebook.

All the learning materials have been reviewed by animal physiology lecturer of State University of Malang, Dr Abdul Ghofur and biology teacher of Public senior high school 1 Malang, Endah Sari Dewi, M. Pd (Table 1).

Table 1. Results of the expert review of the learning materials

Learning Media	Average Score of Validation	Qualification
Completeness of lesson plan components	2.5	Good
Indicator	2.6	Good
Learning objectives	2.9	Good
Learning materials	2.8	Good
Learning methods	3	Good
Learning Activities	2.9	Good
Assessment of learning results	3	Good
Time Allocation	3	Good
Learning Resources	3	Good
Online Forum Design	2.7	Good

Score: 1 = Bad; 2 = Enough; 3 = Good

The data collected in this research were metacognitive skills, learning results and students' response toward the learning process. The data of metacognitive skills were collected using a pretest and posttest by using ten essay questions. The data of the learning results were obtained from pretest and the posttest in the cognitive domain. Metacognitive and Cognitive test items have met the content validity and have good reliability.

The data were then analysed by using Analysis of Covariance (ANCOVA) and Least Significance Difference (LSD) test, the which were previously performed the prerequisite tests namely, the normality test using one-sample Kolmogorov-Smirnov test and homogeneity test using Levene's Test of Equality of Error Variances.

3. RESEARCH RESULT

3.1. Metacognitive

Based on the results of the pretest and posttest, the data of metacognitive skills both in the experimental class and in the control class were obtained. The mean value of metacognitive skills in the experimental class in the pretest was 59.14, and the mean value of metacognitive skills in the posttest was 83.40. In the control group, the mean value of the metacognitive skills in the pretest was 63.97, and the mean value of the metacognitive skills in the posttest was 68.93 (Table 2).

Table 2. Summary of data description of metacognitive skills based on the results of the pretest and posttest

Treatment	Mean		Std. deviation		N	
	Pretest	Posttest	Pretest	Posttest	Pretest	Posttest
Control	63.97	68.9 3	8.9294	8.9436	35	35
Experiment	59.14	83.40	8.7897	7.4338	35	35

The results of the ANCOVA test show that the probability value was 0.00 less than 0.05 so that the null hypothesis was rejected and the research hypothesis was accepted. Thus it can be said that the learning using the module of blood circulation system with guided inquiry blended learning has an effect on metacognitive skills. The Summary of ANCOVA of the results of the data calculation of metacognitive skills both on the pretests and posttests are shown in Table 3.

Table 3. Summary of ANCOVA on the data calculation results of metacognitive skills on both pretest and posttest

Source	Type III sum of squares	df	Mean Square	F	Significance
Corrected Learning	4834.088[a]	2	2417.044	47.223	.000
Intercept	3096.477	1	3096.477	60.498	.000
Treatment	4551.010	1	4551.010	88.916	.000*)
Pretest	1169.198	1	1169.198	22.843	.012
Error	3429.273	67	51.183		
Total	414333.250	70			
Corrected Total	8263.361	69			

*) significance at $p<0.05$

The comparison of the corrected mean score of the experimental class and the control class shows that the experimental class with the learning using blood circulatory system with guided inquiry-based facilitated blended learning gives the corrected mean score of metacognitive skills bigger than that of the control class (Table 4).

Table 4. The corrected mean score of metacognitive skills

Treatment	Pretest	Posttest	Difference	Corrected scores
Control	63.97	68.93	4.96	67.80
Experiment	59.14	83.40	24.26	84.53

3.2. Learning Outcome: Cognitive

The mean scores the cognitive pretest and cognitive posttest of the experimental class were 63.17 and 85.63 respectively. The increase in the cognitive mean scores achieved by the experimental was 35.21%. In the control class, the mean scores of the pretest and posttest were 65.71 and 73.26. Thus, it can be seen that the increase of the mean score of the students' learning outcome was 11.49% (Table 5).

Table 5. Summary of thc Data Dcscription of thc prctcst and posttcst of cognitivc lcarning outcomc

Treatment	Mean		Standard deviation		N	
	Pretest	Posttest	Pretest	Posttest	Pretest	Posttest
Control	65.71	73.26	6,479	6532	35	35
Experiment	63.17	85.63	7213	6890	35	35

IOP Conf. Series: Journal of Physics: Conf. Series **824** (2017) 012068 doi:10.1088/1742-6596/824/1/012068

The results of ANCOVA showed that the probability value was 0.00 less than 0.05 so that the null hypothesis was rejected, and the research hypothesis was accepted. Thus, it can be concluded that the learning through guided inquiry blended learning improve the learning outcome. Summary of the results of ANCOVA on the cognitive learning outcome both in the pretest and posttest can be seen in Table 6.

Table 6. Summary of ANCOVA on learning outcome

Source	Type III sum of squares	df	mean Square	F	Sig.
Corrected Learning	4044.844a	2	2022.422	79.781	.000
Intercept	1059.479	1	1059.479	41.795	.000
Treatment	3328.868	1	3328.868	131.318	.000*)
Pretests	1366.429	1	1366.429	53.903	.000
Error	1698.428	67	25.350		
Total	447525.000	70			
Corrected Total	5743.271	69			

*) significance at $p<0.05$

The comparison of the mean corrected of the experimental class, and the control class shows that the experimental class with the learning using inquiry learning blended learning module gives bigger cognitive learning results than the control class (Table 7). In addition to quantitative data, the researcher also obtained qualitative data in the form of students' feedback and suggestions on the blood circulatory system module with guided inquiry-based blended learning during the tryout. Feedback and suggestions can be seen in Table 8.

Table 7. The corrected mean score of cognitive learning outcome

Treatment	Pretest	Posttest	Difference	Corrected scores
Control	65.71	73.26	7.55	72.43
Experiment	63.17	85.63	22.46	86.46

Table 8. Students' feedback and suggestions on module with guided inquiry-based blended learning

Code number of students	Suggestions and feedback
1	It is better to have an online discussion, but it would be best if the schedule is determined in advance Link to the video is interesting to learn
2	Links to the videos are very helpful to my understanding during the lab practice It will be better if the text is colourful The language in the module is so simple that I can understand it more easily The module is difficult to be open via handphone; it was probably due to the Handphone Online discussion was fun, but it is with a friend, it takes a long time to answer
3	The text will be better if colourful Do not use PDF format; it cannot be copied When accessed in Explorer, it is difficult to be opened Overall the module has been good The Lab activities were not in a good order The Module is difficult to be opened in HP I feel more comfortable to have an online discussion because I do not feel shy to ask
4	Link to youtube is often difficult to access Difficult to access unless using Google Link to videos are lacking; it will be a lot better if there are more lick to videos Online discussion is more fun Difficult to access using handphone
5	The Online module helps me learn Chatting is good, but it depends on the internet connection at home
6	The text is less artistic Modules are difficult to access in schools Link to youtube is difficult to access at school. Students should be told to download the video at home so that students can just open it at school I love Online discussions and class discussion
7	The link is very good for learning, but unfortunately, it depends on the internet connection If it is possible, do not use PDF format. Tasks are easier if the tasks are given in class The online class should be scheduled in advance
8	The colour of the text and background is not contrasting The correcting of the learning task is better to be done during the classroom session The module is faster to be opened using a laptop than using an handphone The discussion with the teacher is better than that with friends because friends are slow response Learning using a module is simpler

4. Discussion

The research results show that the guided inquiry facilitated blended learning improves metacognitive skills. The research results show that the students who were taught by using blended learning had better academic performance and learning outcomes than those taught by using traditional method [29]. The research [30] showed that the blended learning models contributed more to students' attitudes toward geography course when compared to the traditional learning models; Blended learning models contributed more to students' critical dispositions and levels when compared to the traditional learning models. The research [31] indicates that blended learning has the potential to provide a better learning outcome. The research [32] showed that the blended learning support metacognitive processes, authentic learning activities, collaboration, motivation, individualised learning, and access to information on the Internet.

Students will get the best learning outcomes if they become reflective learners [33]. Metacognition refers to "thinking about thinking" [4]. Metacognition is an important objective of learning because students who do metacognition demonstrate the ability to reflect on what is known, what is not known, and how to manage and control the thinking processes, so that they can make a decision about what they want to learn more and evaluate whether they have been on the right process [34].

Currently, the students do not only compete with the students around them, but they compete with all the students from around the world [35]. Thus, today's technology needs to be utilised in learning activities. The technology utilised in education is to help the learning process and to improve the learning outcomes, not to learn the technology [36]. A similar opinion is also expressed by [34] that the application of technology in education improves learning opportunities for students, learning quality, the education quality of workers, and provides the opportunities to decline the injustice in education.

ICT (information and communication technology) creates motivation, has a positive effect on interaction and collaboration, and supports students' ability to reflect on their learning [37]. Web-based inquiry learning helps students with cognitive tools and procedural guidance in the learning process [38]. Blended learning does not completely leave face to face session. The design is there is face to face activity and online activity. In general, the proposed proportion of face to face and online activities is half-and-half [29]. Through blended learning, the difficulties encountered in online learning activities can be solved by face to face discussion. Conversely, the difficulties encountered in face to face activity can be solved through online discussion.

Blended learning has the potential to improve learning, but there is also a challenge, especially in response to the complexity of the two environments, namely the online and face to face activities, as well as the acknowledgement of the online learning as a legitimate learning environment [39]. To date, the blended learning has been implemented in universities, but only a few high schools implement the blended learning. The next challenge is to strengthen the teacher as the blended learning teacher who should master the learning materials and the mastery of ICT as the medium of online learning [28]. Teachers need to gain access to information and technology [40]).

In the process of blended learning, teachers have to design the activities carried out at the online and face to face session, so that both media can make the students learning process effective [41]. Teachers who facilitate the blended learning effectively use information, communication, and modern learning materials; promote in-depth dialogue about the learning experience through online media; encourage participation and communication among students through online media; even help students who have difficulties in speaking with technology [40].

5. CONCLUSION

Biology learning with guided inquiry facilitated blended learning improves metacognitive skills and learning outcomes of high school students. In connection with the inquiry processes, metacognition is a reflective process to regulate and control the thinking processes to be in the right process of inquiry in solving problems. Inquiry through blended learning is made easier with the help of offline and online modules that have been developed by the teacher. Teacher-student communication barriers in the traditional inquiry process can be overcome with the online activity. The ease of the students to have a discussion with the teachers that is not bound to places and opportunities helps students solve the difficulties they have.

References

[1] Lai E R 2011 *Metacognition: A Literature Review*□ Pearson Research Report pp 40 http://images.pearsonassessments.com/images/tmrs/metacognition_literature_review_final.

pdf

[2] Ellis A K , Denton D W and Bond J B An 2014 Analysis of Research on Metacognitive Teaching Strategies *Procedia - Soc. Behav. Sci.* vol. 116 p. 4015–4024.

[3] Schwartz D L, Chase C, Chin D, Oppezzo M, Kwong H, Okita S, Roscoe R, Jeong H, Wagster J, and Biswas G 2009 To appear in D.J. Hacker, J. Dunlosky, & A.C. Graesser (Eds.), *Handbook of Metacognition in Education* pp. 340–358.

[4] Mahdavi M 2014 An Overview : Metacognition in Education *Int. J. Multidiscip. Curr. Res*. P. 529–535.

[5] Lin X 2001 Designing metacognitive activities *Educ. Technol. Res. Dev*. pp 23–40.

[6] Kuhn D 2000 Metacognitive Development *Current Directions in Psychological Science*, Vol. 9 no. 5 p. 178-181

[7] Tanner K D 2011 Promoting Student Metacognition *CBE-Life Sciences Education* Vol.11 p. 113-120.

[8] Aleven V A W M M and Koedinger K R 2002 An effective metacognitive strategy: Learning by doing and explaining with a computer-based Cognitive Tutor *Cogn. Sci*. vol. 26 no. 2, pp 147–179.

[9] An Y J and Cao L 2014 Examining the Effects of Metacognitive Scaffolding on Students' Design Problem Solving and Metacognitive Skills in an Online Environment *Journal of Online Learning & Teaching* vol. 10 no. 4 p. 552–568

[10] Efklides A 2006 Metacognition and affect: What can metacognitive experiences tell us about the learning process *Educ. Res. Rev*. vol. 1 no. 1 p. 3–14

[11] Danoebroto S W 2008 Meningkatkan Kemampuan Pemecahan Masalah Melalui Pendekatan PMRI dan Pelatihan Metakognitif *Jurnal Penelitian dan Evaluasi Pendidikan* No.1 p. 73-87.

[12] Suratno 2011 Kemampuan Metakognisi dengan Metacognitive Awareness Inventory (MAI) pada Pembelajaran Biologi SMA dengan Strategi Jigsaw, Reciprocal Teaching (RT), dan Gabungan Jigsaw-RT *Jurnal Pendidikan dan Pembelajaran*, Vol 18 p. 11-18.

[13] Şahİn S M and Kendir F 2013 The Effect of Using Metacognitive Strategies for Solving Geometry Problems on Students ' Achievement and Attitude *Educ. Res. Rev*. vol. 8 no. 19 p. 1777–1792.

[14] Schraw G Crippen K J and Hartley K 2006 Promoting self-regulation in science education: Metacognition as part of a broader perspective on learning *Res. Sci. Educ.*vol. 36 no. 1–2 p. 111–139.

[15] Eilam E 2015 Measuring the level of complexity of scientific inquiries: The LCSI index *Int. J. Environ. Sci. Educ*. vol. 10 no. 1 p. 1–20.

[16] Zimmerman C 2007 The development of scientific thinking skills in elementary and middle school *Dev. Rev*. vol. 27, no. 2 p. 172–223.

[17] Cavagnetto A R Hand B and Norton-Meier L 2010 Negotiating the Inquiry Question: A Comparison of Whole Class and Small Group Strategies in Grade Five Science Classrooms *Res. Sci. Educ*. vol. 41, no. 2 p. 193–209.

[18] Odegaard M Haug B Mork S and Sorvik G O 2015 Budding Science and Literacy. A Classroom Video Study of the Challenges and Support in an Integrated Inquiry and Literacy Teaching Model *Procedia - Soc. Behav. Sci*. vol. 167 no. 1877 p. 274–278.

[19] Hmelo-Silver C E Duncan R G and Chinn C A 2007 Scaffolding and achievement in problem-based and inquiry learning: A response to Kirschner, Sweller, and Clark (2006) *Educ. Psychol*. vol. 42 no. 2 p. 99–107.

[20] Zhang M 2013Supporting middle school students' online reading of scientific resources: Moving beyond cursory, fragmented, and opportunistic reading *J. Comput. Assist. Learn* vol. 29 no. 2 p. 138–152.

[21] Choo S S Y Rotgans J I Yew E H J and Schmidt H G 2011 Effect of worksheet scaffolds on student learning in problem-based learning *Adv. Heal. Sci. Educ*. vol. 16 no. 4 p. 517–528.

[22] Ginns P and Ellis R 2007 Quality in blended learning: Exploring the relationships between on-line and face-to-face teaching and learning *Internet High. Educ*. vol. 10 no. 1 p. 53–64.

[23] Tsai A 2011 A Hybrid E-Learning Model Incorporating Some of the Principal Learning

Theories *Soc. Behav. Personal. an Int. J.* vol. 39 no. 2 p. 145–152,.
[24] Watson J 2008 Blended Learning : The Convergence of Online and Face-to-Face Education *North Am. Counc. Online Learn*. vol. 572 p. 16.
[25] Cheung W S and Hew K F 2011 Design and evaluation of two blended learning approaches : Lessons learned *Australasian Journal of Educational Technology* vol. 27 no. 8 pp. 1319–1337.
[26] Thiagarajan S Semmel D S Semmel M I 1974 *Instructional Development for Training Teachers of Exceptional Children* (University of Minnesota: the Center for Innovation in Teaching the Handycapped)
[27] Creswell J W 2012 *Educational Research: Planning, Conducting, and Evaluating Quantitative and Qualitative Research* (Boston: Pearson)
[28] Llewellyn D 2013 *Teaching High School Science Through Inquiry and Argumentation* (Thousand Oaks, California: Corwin)
[29] Greenberg B Medlock L and Stephens D 2011 Lessons Learned from a Blended Learning Pilot Blend My Learning," p. 25,.
[30] Ozgen K and Ufuk K 2009 The impact of blended learning model on student attitudes towards geography course and their critical thinking dispositions and levels *Turkish Online J. Distance Educ.*vol. 8 no. 4 p. 51–63.
[31] Means B Toyama Y Murphy R Baki M 2013 The effectiveness of online and blended learning: a meta-analysis of the empirical literature *Teachers College Record* vol. 115 p. 1-47
[32] Delialioglu O and Yildirim Z 2007 Students' perception of effective dimensions of interactive learning in a blended learning environment *Educational Technology & Society* vol. 10 no. 2 p. 133-146
[33] Bryce J, Frigo T, McKenzie P, and G. Withers 2001 The Era of Lifelong Learning: Implications for Secondary Schools http://research.acer.edu.au/lifelong_learning/1
[34] Sternberg R J 2009 Foreword to Handbook of Metacognition in Education *Handbook of Metacognition in Education* eds D J Hacker, J D Dunlosky, A C Graesser (Abingdon, Oxon: Roudlege) pp viii-ix
[35] Wang Y Han X and Yang J 2015 Revisiting the Blended Learning Literature: Using a Complex Adaptive Systems Framework *Educational Technology & Society* vol. 18 no. 2 pp 380–393
[36] Bach S Haynes P and Smith J L 2006 *Online learning and teaching in higher education* (Buckingham, UK: Open University Press).
[37] Svendsen L P and Mondahl M S 2013 How social-media enhanced learning platforms support students in taking responsibility for their own learning *J. Appl. Res. High. Educ.* vol. 5 no. 2 pp 261–272.
[38] Lim B R 2004 Challenges and issues in designing inquiry on the Web *Br. J. Educ. Technol.* vol. 35, no. 5 pp 627–643.
[39] Jeffrey L M Milne J Suddaby G and Higgins A 2014 Blended learning: How teachers balance the blend of online and classroom components *Journal of Information Technology Education: Research* vol. 13 p. 121-140. Retrieved from http://www.jite.org/documents/Vol13/JITEv13ResearchP121-140Jeffrey0460.pdf
[40] National Research Council 2007 *Enhancing professional development for teachers: Potential uses of information technology* (Washington, DC: The National Academies Press)
[41] Marsh D 2012 *Blended Learning Creating Learning Opportunities for Language Learners* (Cambridge: University Press)

Contour Detection of Leukocyte Cell Nucleus Using Morphological Image

R Supriyanti[1], G P Satrio[1], Y Ramadhani[1], and W Siswandari

[1]Electrical Engineering Department, Engineering Faculty, Jenderal Soedirman University Kampus Blater, Jl. Mayjend Sungkono KM 5, Blater, Purbalingga

[2]Department of Medicine, Jenderal Soedirman University Kampus Kedokteran, Jl. Gumbreg, Mersi, Purwokerto

*Corresponding author: retno_supriyanti@unsoed.ac.id

Abstract. Leukocytes are blood cells that do not contain color pigments. Leukocyte function to the tool body's defenses. Abnormal forms of leukocytes can be a sign of serious diseases such example is leukemia. Most laboratories still use cell morphology examination to assist the diagnosis of illness associated with white blood cells such example is leukemia because of limited resources, both infrastructure, and human resources as happens in developing nations, such as Indonesia. This examination is less expensive and quicker process. However, morphological review requires the expertise of a specialist clinical pathology were limited. This process is sometimes less valid cause in some cases trying to differentiate morphology blast cells into the type of myoblasts, lymphoblast, monoblast, or erythroblast thus potentially misdiagnosis. The goal of this research is to develop a detection device types of blood cells automatically as lower-priced, easy to use and accurate so that the tool can be distributed across all units in existing health services throughout Indonesia and in particular for remote areas. However, because the variables used in the identification of abnormal leukocytes are very complex, in this paper, we emphasize on the contour detection of leukocyte cell nucleus using the morphological image. The results show that this method is promising for further development.

1. Introduction

Leukocytes are white blood cells that work with the immune system to defend against infectious disease. Damage to the kidneys, ureters, urethra or bladder can cause leukocytes to appear. They can also be triggered when foreign materials are found in the body. There are five types of leukocytes. If a high level of a particular type of leukocyte is found in elevated levels, this can be a sign of disease. The types of leucocytes are (i) Monocytes: They have a longer lifespan than many white blood cells and help to break down bacteria. (ii) Lymphocytes: They make antibodies to defend against viruses, bacteria, and other potentially harmful invaders. (iii) Neutrophils: They kill and digest bacteria and fungi. They are the most numerous type of white blood cell and the first line of defense when infection strikes. (iv) Simple tools the blood. They secrete chemicals such as histamine, a marker of allergic disease, which help control the body's immune response. (v) to [1]. In order to determine an abnormality of leukocytes, there are some kinds of blood test, such as complete blood count, morphological of peripheral blood, morphological of bone marrow, Cytogenetic, double nucleic acid (DNA), and detection of cluster of differentiation (CD) using flow cytometry [2] [3] [4].

Most laboratories still use cell morphology examination to assist the diagnosis of diseases associated with white blood cells such example is leukemia because of limited resources, both infrastructure, and human resources as the Indonesia. This examination is less expensive and quicker process. However, morphological examination requires the expertise of a specialist clinical pathology were limited. This examination is sometimes less valid because in some cases difficult to differentiate morphology blast cells into the type of myoblasts, lymphoblast, monoblast, or erythroblast thus potentially misdiagnosis. According to the condition as mentioned above, it is necessary to develop a detection device types of blood cells automatically as lower-priced, easy to use and accurate so that the tool can be distributed across all units in existing health services throughout Indonesia and in particular for remote areas.

According to the research about leucocytes using image processing, there is some research that discussed this topic. Ko [5] applied seam carving to cell images, an SSM is initially generated using a visual attention model, and the structural properties of white blood cells are then used to create an energy map for seam carving. As a result, the energy map maximizes the energies of the white blood cells, while minimizing the energies of the red blood cells and background. Thus, the use of an SSM allows the proposed method to reduce the image size efficiently while preserving the important white blood cells. Madhloom [6] presented a new method that integrates color features with the morphological reconstruction to localize and isolate lymphoblast cells from a microscope image that contains many cells. The localization and segmentation are conducted using a proposed method that consists of an integration of several digital image processing techniques. 180 microscopic blood images were tested, and the proposed framework managed to obtain 100% accuracy for the localization of the lymphoblast cells and separate it from the image scene. Li [7] implemented a series of digital image processing methods, such as gray stretch, median filter, threshold segmentation, edge extraction and detection, the variations of red blood cells and the goal of identifying the shapes of variable red blood cells. Eric [8] presented an application that integrates visualization and quantitative analysis of 5-D (*x*, y, z, *t*, *c h a n n e l*) and large montage confocal fluorescence microscopy images. The image sequences show stem cells together with blood vessels, enabling quantification of the dynamic behaviors of stem cells about their vascular niche, with applications in developmental and cancer biology. Evans [9] developed an automated method to measure and compare the dynamic movement of cell membranes. Using the red blood cell as a common example the method locates the edge of the cell, with sub-pixel precision at multiple points on the periphery. Koltsov [10] considered the problem of the segmentation of microscopy images of blood cells. Krasheninniko [11] proposed some algorithms for processing images of blood serum facies for automated medical diagnosis, i.e., determination of the boundaries of facies, identification of their morphological type, and detection of certain markers indicating pathology. The mathematical modeling of such images is considered. Kinahan [12] presented retrieval of Peripheral Blood Mononuclear Cells by density-gradient medium based centrifugation for subsequent analysis of the leukocytes on an integrated microfluidic "Lab-on-a-Disc" cartridge. Isolation of white blood cells constitutes a critical sample preparation step for many bioassays. Centrifuge-pneumatic siphon valves are particularly suited for blood processing as they function without the need of surface treatment and are 'low-pass,' i.e., holding at high centrifugation speeds and opening upon reduction of the spin rate. Wojewodzka [13] used biodosimetric methods to measure the effects of radiation are critical for estimating the health risks to irradiated individuals or populations. The direct measurement of radiation-induced γ-H2AX foci in peripheral blood lymphocytes is one approach that provides a useful end point for triage. Panchbhai [14] developed an automated for the identification of malaria parasites. The given scheme based on used of RGB color space, G layer processing, and segmentation of Red Blood Cells (RBC) as well as cell parasites by auto-thresholding with the offset value and use of morphological processing. According to all research mentioned in this paragraph and also based on our previous research [15] [16] [17] [18] [19] [20] [21], we conclude that the image processing technique is widely used in applications of microscopic images of blood cells. However, the application of previous research was not emphasized on the identification of abnormal leucocytes. The goal of this research is developing simple tools for identifying abnormal leukocytes to support the diagnosis of leukemia quickly and easily in the rural areas are widely available in developing countries such as in Indonesia. However, because the variables used in the identification of abnormal leukocytes are very involved, in this paper, we emphasize on the contour detection of leukocyte cell nucleus using the morphological image.

2. Methods

2.1. Designing Contour Detection System of Leukocyte Cell Nucleus
In this experiment, we used the digital image of leukocytes coming from Margono Soekardjo Hospital, Banyumas, Central Java, Indonesia. The images have *jpg format with size 640x480 pixels. Images data have different characteristics. It is to be used to test the extent to which this system can work. To facilitate the use of algorithms that we developed, we developed a Graphical User Interface (GUI) based on the programming language Matlab as shown in Figure 1. In the image, acquisition phase begins with selecting the "Select Image" in the original image panel. Figure 2 shows the display GUI application when opening the blood image file.

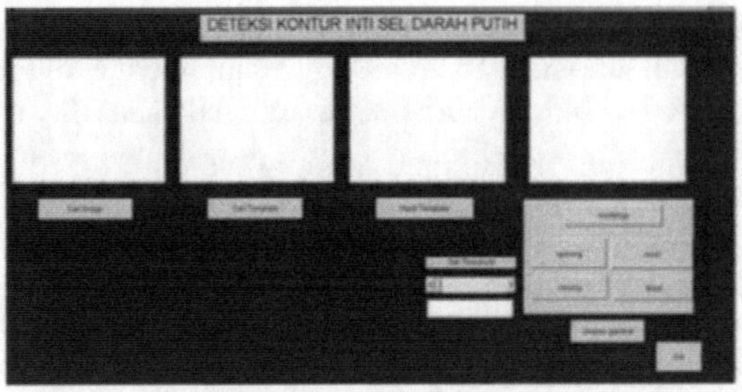

Figure 1. Initial display of our GUI

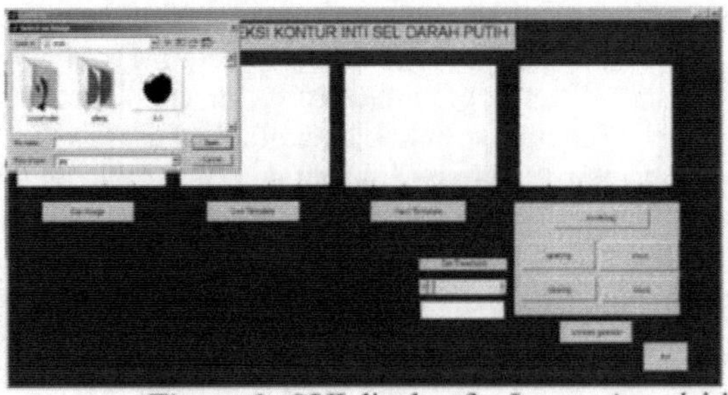

Figure 2. GUI display for Image Acquisition

2.2. Template Matching
Template matching is a technique in digital image processing to find small parts of the picture that matches the template image. Template matching is one idea that is used to explain how our brains recognize return forms or patterns. Templates in the context of pattern recognition refer to the internal constructs which if matched with sensing stimuli lead to the recognition of an object or pattern recognition occurs if there is a match between sensory stimuli with internal mental shape [22]. In this research approach towards partition division of white blood cell image by using a template. In the classification stage, the template is matched with the image of the white blood cells. Template matching with the image of the white blood cells produces a percentage match between templates with white blood cell nucleus.

2.3. Morphological Operations
Morphological operations are image processing techniques that are based on the shape of a segment or region in the image. Because it is focused on the shape of the object, then surgery is usually applied to the binary image. Usually, the segment was based on the object of concern. Segmentation is done by distinguishing between the object and the background, such as by utilizing a floating operation that changed the image of the color and gray-scale binary image. The binary value of the image represents the result of two circumstances: the object and not the object (background). Although it is more widely used in the binary image, morphological operations are often used in the image of gray scale and color. The results of morphological operations can be used for decision-making by further analysis. These operations include dilation, erosion, closing and the opening [22].

Erosion is a way to minimize the object. In this operation, the size of the object is reduced by eroding the surrounding objects. The first way is to change all the border points into the background and the second-way point by setting all points around the point of the background becomes foreground point. Dilation is done to enlarge the size of the object segment by adding a layer around the object. There are two ways to do this operation that is by changing the background all the points neighboring the point boundary a point of the object or simply set each point that his neighbor is a point object into a point object. The second way is to change all points around the boundary points a point of the object, or simply set all the points neighbors a point object into a point object. Closing is a combination of dilation and erosion operations are performed sequentially. Closing is a combination of dilation and erosion operations are performed sequentially. This operation used to close or eliminate the small holes that exist in the object segments, combine objects are adjacent and soften the boundaries of large objects without changing the object significantly. Opening a combination of erosion and dilation operations are carried out sequentially. This procedure used to sever portions of the object which is only connected to one or two dots, or eliminate small objects and softens the boundaries of large objects without changing the object area significantly [22].

3. Results and Discussions

In research we've done, there are five image templates that will test the performance of its success is ultimately just use one template only. Table 1 and Figure 3 described this results.

Table 1. Template's name

Template's name	Template's image
Template1.jpg	
Tcmplatc2.jpg	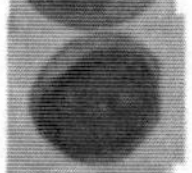
Template3.jpg	
Template4.jpg	
Template5.jpg	

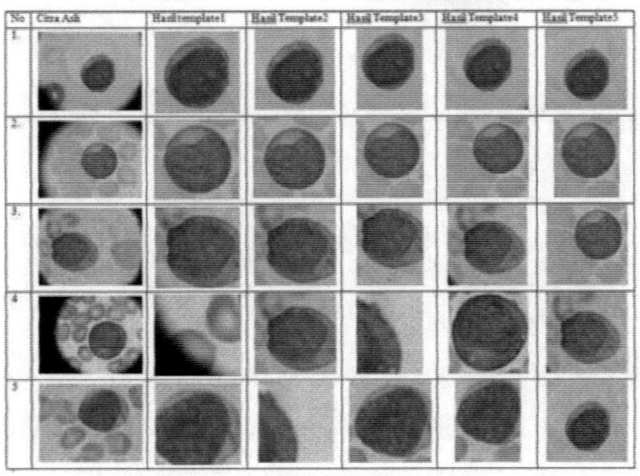

Figure 3. Template's result

Based on the above data, selected template five as a reference for template matching with the performance calculation as described in Table 2.

Table 2. Calculation of performance success template

IOP Conf. Series: Journal of Physics: Conf. Series **824** (2017) 012069 doi:10.1088/1742-6596/824/1/012069

Template	Performasi perhitungan
Template1. jpg	$\frac{2}{5}x100\% = 40\%$
Template2. jpg	$\frac{2}{5}x100\% = 40\%$
Template3. jpg	$\frac{2}{5}x100\% = 40\%$
Template4. jpg	$\frac{3}{5}x100\% = 60\%$
Template5. jpg	$\frac{5}{5}x100\% = 100\%$

Each image takes time that is different at the time of this template matching process. It was in because the size of the pixels of an image and also the intensity of the colors in the image, the larger the size of an image will be longer in the process of template matching. Table 3 shows the length of time the process template matching on the image of the white blood cells.

Morphology operation involves two components: the first component in the form of images that will be subjected to the morphology operation in this experiment is the image of blood cells, while the second component called a kernel or structuring element. We chose the disk-shaped structural element because its form is most appropriately used to detect the contours of the core blood. Because of its shape like a disc that conforms to the shape of blood cells. Table 4 shows the result of structure element comparison.

Each image has a different outcome after converted to binary. Thus requiring different morphological operations for each image. To get a perfect contour core, it is necessary to carefully search the Threshold value to get maximum results. Table 5 shows the results of this case.

IOP Conf. Series: Journal of Physics: Conf. Series **824** (2017) 012069 doi:10.1088/1742-6596/824/1/012069

Table 3. The time required for template calculation

No	Nama Citra	Ukuran citra	Citra darah	Waktu template matching
1	Citra1	640x480 pixel	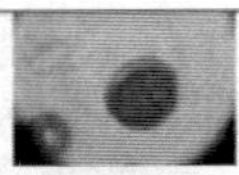	2 menit
2	Citra2	640x480 pixel	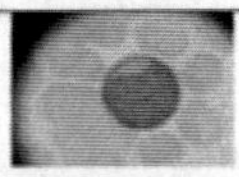	3 menit
3	Citra3	640x480 pixel	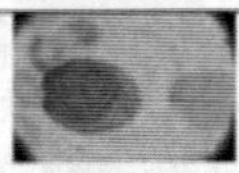	3 menit
4	Citra4	920x720 pixel	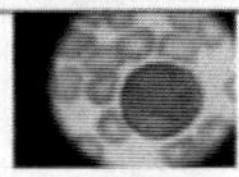	20 menit
5	Citra5	640x480 pixel	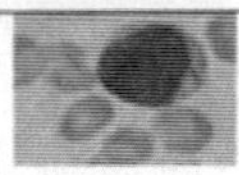	3 menit

Table 4. Comparison of structure element

No	Citra grayscale	Citra biner	Jenis streel
1	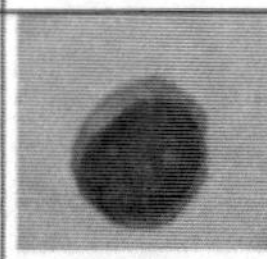		disk
2	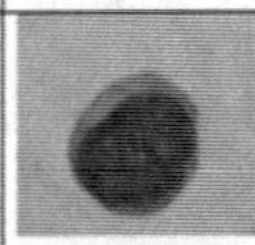		line
3	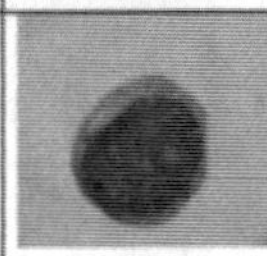		Rectangle

Table 5. Template's result

No	Nama File	Citra Asli	Citra Hasil Deteksi	Nilai Threshold	Citra Biner
1.	Citra1.jpg	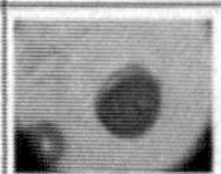	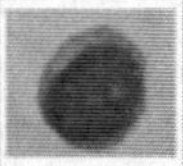	0.37	
2.	Citra2.jpg	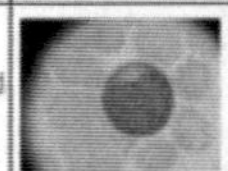	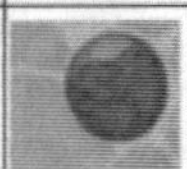	0.49	
3.	Citra3.jpg	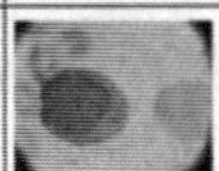	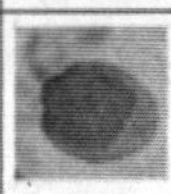	0.4	
4.	Citra4.jpg	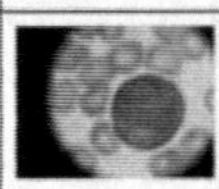		0.44	
5.	Citra5.jpg	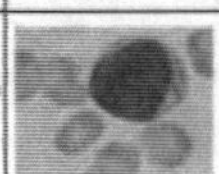		0.41	

Table 6. Morphological operation

No	Nama File	Citra Biner	Operasi Morfologi				Citra Hasil
			Erosi	Dilasi	Opening	Closing	
1.	Citra1.jpg		3x	3x	1	1	
2.	Citra2.jpg		3x	3x	1	-	
3.	Citra3.jpg		14x	14x	1	1	
4.	Citra4.jpg		6x	22x	1	1	
5.	Citra5.jpg		3x	3x	1	-	

In image1.jpg, using a threshold value of 0:37. This value is not necessarily applicable to another image. With the value of the acquired binary image of white blood cell nucleus that needs to be perfected again because there is still noise around the image. Therefore, it is still necessary morphological operations,order to obtain better. image2.jpg, using a threshold value of 0:49. With these values, they obtained a binary image that there are patches or perceived as noise. Therefore still needs to be done morphological operations to obtain better results. Image3.jpg using a threshold value of 0.4. With these values, they obtained a binary image that there are patches or perceived as noise. Therefore still needs to be done morphological operations. Image 4. .jpg using a threshold value of 0:44. With these values, they obtained a binary image that there are patches or perceived as noise. Therefore still needs to be done morphological operations. Image5.jpg using a threshold value of 0:41. With these values, they obtained a binary image that there are patches or perceived as noise. Therefore still needs to be done morphological operations. The use of the threshold value and the amount of the morphological operations are in wear on the image can be different between the images of one another. It is in use by the needs for each image, if there are noise or rudimentary, then search threshold value and also morphological operations will continue to be done. To find the perfect contours.

4. Conclusions
According to the results as discussed above, we can conclude that: (i) Based on the test results of some samples of the image of the leucocytes are used, the system can identify the morphology of white blood cell nucleus. Tests were performed using the operation dilation, erosion, opening, and closing with the structure element 'disk.' (ii) The method of adaptive threshold, i.e., setting the threshold value depends on the value of each image pixel for each image has a threshold value is different. (iii) The use of less appropriate threshold value will affect the morphology. Therefore, it is necessary threshold value with the best performance, which is approaching the contours of leukocyte cell nucleus

Acknowledgment
This work supported by BLU Universitas Jenderal Soedirman through Penelitian Unggulan Scheme

References
[1] Leukocytes in Urine *MD Health*, 2016 [Online] Available: http://wwwmd-healthcom/Leukocyteshtml [Accessed: 09-Jun-2016]
[2] Pokharel M 2012 Leukemia : A Review Article *Int J Adv Res Pharm Bio Sci* **2(3)** pp 397–407
[3] Roquiz W, Gandhi P, and Kini A R 2016 Acute Leukemias, in *Rodak's Hematology: Clinical Principles And Applications*, 5th Ed, Missouri: Elsevier Saunders
[4] Anastasi J and Hoffman R 2013 Progress In The Classification Of Myeloid Neoplasms: Clinical Implications, in *Hoffman, R eds Hematology: Basic Principles And Practice*, 6th Ed, Philadelphia: Elsevier Saunders
[5] Ko B, Kim S, and Nam J 2010 Image resizing using saliency strength map and seam carving for white blood cell analysis, *Biomed Eng Online* **54(9)**
[6] Madhloom H T, Kareem S A, and Ariffin H 2012 An image processing application for the localization and segmentation of lymphoblast cell using peripheral blood images *J Med Syst* **36(4)** pp 2149–58
[7] Li J, Mu H, and Xu W 2013 A method of using digital image processing for edge detection of red blood cells *Sensors & Transducers* **159(11)** pp 1–6
[8] Wait E, Winter M, Bjornsson C, Kokovay E, Wang Y, Goderie S, and Cohen A R 2014 Visualization and correction of automated segmentation, tracking and lineaging from 5-D stem cell image sequences *BMC Bioinformatics* **15(328)**
[9] Evans J J 2008 An image-processing technique for measuring the dynamic movement of cell membranes *Comput Biol Med* **38(5)** pp 545–554
[10] Koltsov P P, Kotovich N V, Kravchenko A A, Kutsaev A S, Kuznetsov A B, Osipov A S, Sukhenko E P, and Zakharov A V 2015 A segmentation method for the microscopy of images of blood cells *Pattern Recognit Image Anal* **25(2)** pp 167–173
[11] Krasheninnikov V R and Kopylova A S 2012 Algorithms for automated processing images of blood serum facies Pattern Recognition and Image Analysis *Pattern Recognit Image Anal* **22(4)** pp 583–592
[12] Kinahan D J, Kearney S M, Kilcawley N A, Early P L, Glynn M T, and Ducrée J 2016 Density-gradient mediated band extraction of leukocytes from whole blood using centrifugo-pneumatic siphon valving on centrifugal microfluidic discs *PLoS One* **11(5)**
[13] Wojewodzka M, Sommer S, Kruszewski M, Sikorska K, Lewicki M, Lisowska H, Wegierek A - Ciuk, Kowalska M, and Lankoff A 2015 Defining Blood Processing Parameters for Optimal Detection of c-H2AX Foci: A Small Blood Volume Method *Radiat Res Radiat Res* **184(184)** pp 95–104
[14] Panchbhai V V,Damahe L B, Nagpure A V, and Chopkar P N 2012 RBCs and Parasites Segmentation from Thin Smear Blood Cell Images *Int J Image, Graph Signal Process* **4(10)** pp 54–60
[15] Supriyanti R, Fariz A, Septiana T, Murdyantoro E, Ramadhani Y, and Widodo H B 2015 A simple screening for high-risk pregnancies in rural areas based expert system *Telkomnika* **13(2)** pp 661–669

IOP Conf. Series: Journal of Physics: Conf. Series **824** (2017) 012069 doi:10.1088/1742-6596/824/1/012069

[16] Supriyanti R, Habe H, Kidode M, and Nagata S 2009 Extracting appearance information inside the pupil for cataract screening in *11 th IAPR Conference on Machine Vision Applications* pp 342–345

[17] Supriyanti R, Habe H, Kidode M, and Nagata S 2009 Compact cataract screening system : Design and practical data acquisition, in *Proceeding of International Conference on Instrumentation, Communications, Information Technology, and Biomedical Engineering (ICICI-BME)* pp 1–6

[18] Supriyanti R, Habe H, Kidode M, and Nagata S 2008 A simple and robust method to screen cataracts using specular reflection appearance in *Proc SPIE 6915 Medical Imaging*

[19] Supriyanti R, Pranata E, Ramadhani Y, and Rosanti T I 2013 Separability filter for localizing abnormal pupil:identification of input image *Telkomnika* **11(4)** pp 783–790

[20] Supriyanti R, Putri D, Murdyantoro E, and Widodo H B 2013 Comparing edge detection methods to localize uterus area on ultrasound image in *International conference of Instrumentation, Communications, Information Technology, and Biomedical Engineering (ICICI- BME)* pp 152–155

[21] Widodo H B, Soelaiman A, Ramadhani Y, and Supriyanti R 2015 Calculating contrast stretching variables in order to improve dental radiology image quality *Int Conf Eng Technol Sustain Dev* **012002**

[22] Gonzales R C, Woods R E, and Eddins S L 2004 *Digital Image Processing using Matlab* United States: Pearson Prentice Hall

The Effects of Conceptual Understanding Procedures (CUPs) Towards Critical Thinking Skills of Senior High School Students

S Sukaesih and Sutrisno

Biology Deparment, FMIPA UniversitasNegeri Semarang
Bld. D6 1st floor,Jalan Raya Sekaran-Gunungpati-Semarang-Indonesia 50299

*Corresponding author: sri_sukaesih@mail.unnes.ac.id

Abstract. *The aim of the study was to analyse the effect of the application of Conceptual Understanding Procedures (CUPs) learning to the students' critical thinking skills in the matter of categorisaed in SMA Negeri 1 Larangan. This study was quasi-experimental design using nonequivalent control group design. The population in this study was entire class X. The samples that were taken by convenience sampling were class X MIA 1 and X MIA 2. Primary data in the study was the student's critical thinking skills, which was supported by student activity, the level of adherence to the CUPs learning model, student opinion and teacher opinion. N-gain test results showed that the students' critical thinking skills of experimental class increased by 89.32%, while the control group increased by 57.14%. Activity grade of experimental class with an average value of 72.37 was better than that of the control class with an average of only 22.69 student and teacher opinions to the learning were excellegoodnt. Based on this study concluded that the model of Conceptual UnderstandingProcedures (CUPs) had an effect on the student's critical thinking skills in the matter of protest in SMA Negeri 1 Larangan.*

1. Introduction

The quality of learning process which is implemented by a teacher is one of the important factors that determine the success of students' learning process. Teacher's mastery of teaching students represented about 30% of all the factors that influence the activity, the results the of learning process, and students' creativity [1].

Biology teaching process in SMA N 1 Larangan is using lecturing, discussing, and task-based learning methods. The material that employed in this teaching-learning process is not contextual, and teacher uses the textbook as a source to deliver the materials. The discussing technique that occurs in the class is just simply transferring knowledge from teacher to students. Lecturing method, in this case, puts students in a passive role in the cognitive. It makes students pay less attention in learning activities and teacher hard to check students' perception and understanding of the material.

From the observation conducted in SMA N 1 Larangan, we got the result that the material about Protista was the most difficult material at Biology subject of tenth-grade students in the first semester. It is seen from the percentage of the students that have passed the test with the minimum score of 75 for Biology subject in the first semester of tenth-grade students in the academic year of 2012/2013. The percentage of classical learning completeness in daily test in the chapter of "Biology as the Science" was 31%, the proportion of the learning completeness of the material about the virus was 24%, the percentage of the material about bacteria was 52%, while on the Protista was only 17%.

Students thought that the material about Protista was the most difficult one because in their opinion some of the concepts were abstract. The discussion process was still weak since only a few of the students were actively involved in learning process. Students' reading interest was also still low because the material presented in a sort of lengthy discourse. Students had difficulty in finding the meaning of foreign languages or scientific names. They tend to memorise a kind of abstract concepts without letting them figured out their concepts, and thus, it is resulting in a lack of ability to think in a critical way. The skills of thinking include critical thinking, creative thinking, and reflective thinking

[2]. Critical thinking is the basis of the ability to think creatively and reflectively. It means that critical thinking is necessary if a person needs to think creatively and reflectively.

The ability to think critically is essential for students as the next generation to face the challenges, problems, and competitions in free trade and global era since a significant amount of information, advertisement, conscious distortion, and even propaganda are continually sorted. Asean Free Trade Area (AFTA) which is planned to be held in 2015 is an actual effect of globalisation. AFTA is trading activity between countries in the ASEAN region without artificial obstacles (barriers that are created by the government), like the export-import tax, etc. [3]. Indonesian citizens are supposed to compete with the world community independently in various fields.

The development of critical thinking skills for students in the learning material, the use of language, the use of the logic structure of logical thinking, examining the accuracy of science, and the experience of the various aspects will influence the students to be independent [4]. Humans do not have any natural tendencies to think critically, so the critical thinking skills need to be improved [5]. To deal with these problems is the primary objective in teaching activities that encourages critical thinking to students. Learning material that is used in learning activities is supposed to cause the improvement of students' critical thinking skills.

The efforts to develop the essential thinking skills of students in the classroom require a contextual learning experience, encourage high levels of interaction and participation, and supports that will help students to understand these experiences [5]. Conceptual Understanding Procedures (CUPs) learning model consists of some methods that engage students' interaction and motivation [6]. Mills' research showed that basic tasks and cooperative approach, which used in CUPs learning model, provide many opportunities for students to improve their understanding of basic concepts. CUPs learning model is one of the learning models that uses a constructivist approach. Constructivist approach originated from a theory which developed by Jean Piaget, namely the theory of constructivism. This theory believes that every individual has the ability to construct their knowledge by encouraging interaction continuously with their environment. Constructivism approach is the importance of social interaction in the construction of knowledge. Social interaction among students is necessary for their development of knowledge [6].

The knowledge formed from individual's thought actively. People cannot simply get knowledge by passively receiving various information. Glasersfeld described the importance of social interaction in the formation of knowledge [7]. Social interaction among students is the main idea in building personal knowledge. In constructivist learning approach, students do not directly receive material from a teacher, but they proceed material in their way, so they are supposed to build their concepts. Every concept formed by each student may be different from one to another or even mistaken. It is teacher's duty to provide assistance and supports. Students' participation, which is directed by the facilitator, will ease students to gain a better understanding of the complex concepts [8]. Students' mistake is a part of a learning process that has to be appreciated to engage their interests in a learning activity. In constructivist class, the teacher encourages students to solve the problems in their way. The teacher does not only state a correct answer, but also she engages students in discussing, so the social interaction is built here. The discussion is going to lead them to the right concepts. Piaget says about constructivist theory requires students to actively develop their understanding concept by combining the experiences that they have had a new understanding [9].

CUPs learning model will support cooperative learning [6]. Collaborative learning aims to make students work together to achieve common goals [10]. Cooperative learning models are widely used to make teaching and learning activities centred on students, especially to make students have any eager to cooperate with others, and also to overcome the students who tend to be aggressive and do not care about others. Cooperative learning is one of the teaching models that leads the systems to learn and work collaboratively in small groups, and it will motivate them to learn [11]. Team working is a method that requires the teacher to put students in groups to learn a new material or to do any assignments. They interact with others in small groups to solve a problem. We can say that a team is efficient if every member of that group is responsible for their duty, they participate and cooperate effectively with other members, and they could be better in their behaviour.

Cooperative learning makes students more effectively in memorising materials that have been given rather than any traditional way. It can be seen from the posttest score of the experimental group, which increased after the treatment, compared to control group. Cooperative learning is student-centered learning, while the traditional one is teacher-centered learning. Teachers' conclusions and responses showed the importance of the transition phase from teacher-centered learning to student-centered learning [12]. The transformation of this learning model should be done gradually to make students easy to understand about the new learning model. Cooperative learning has its characteristics compared to the traditional one, as has been mentioned before, it has five characteristic elements.

These are the steps of CUPs learning models. That is students work individually, and then teacher divides students into several groups to discuss their individual works, each team consists of three persons, and the last step, all students in the classroom have to consider about their teamwork to determine the most appropriate and accurate one. The tasks are given in Conceptual Understanding Procedures (CUPs) are taken from our daily problems, which is also known as an authentic work, that is learning activities wich is resulting in understanding used in the daily activity outside the classroom [5].

The observation showed that Conceptual Understanding Procedures (CUPs) learning model applied in teaching Biology in SMA N 1 Larangan. Hopefully, the use of CUPs learning model can give positive contribution in students' critical thinking skills. Therefore, the research which is entitled "The Effects of Conceptual Understanding Procedures (CUPs) towards Critical Thinking Skills of Senior High School Students' should be done. The aim of this research is to analyse the effect of the application of CUPs learning model towards students' critical thinking skills in the chapter of Protista in SMA N 1 Larangan.

2. Methods

This research is a quasi-experimental design which used nonequivalent control group design [13]. The research is conducted in the first semester in SMA N 1 Larangan in the academic year of 2014/2015, from October through November 2014. The population is all of the tenth graders of SMA N 1 Larangan. The researcher used convenience sampling to select the research sample. X MIA 1 and X MIA 2 are the samples of this study. Independent variable in this research is Conceptual Understanding Procedures (CUPs) learning model, and the dependent variable is students' critical thinking skills.

In addition to the data of students' critical thinking skills, there are also supporting data like students' learning activities, the success rate of CUPs learning model, and student and teacher responses. Students' critical thinking abilities are presented in pretest and posttest result, it is analyzed by using homogeneity test, normality test, Wilcoxon's test, and n-gain test. N-gain test is aimed to determine the increase in critical thinking skills, while Wilcoxon's test is aimed to determine the difference between experimental and control class. The data of students' learning activities, the success rate, and students' responses toward CUPs learning models are analysed by using descriptive percentage. The data of teacher response toward CUPs learning model are analysed descriptively.

3. Result and discussion

3.1 Results of Students' Critical Thinking Skills

Based on the research results, we know that before the treatment, the average percentage of students' critical thinking skills from experimental and control class in the chapter of Protista are relatively similar, that is 43.47% of the experimental class and 45.09% of the control class. The average percentage of both classes are regarded low. The results of homogeneity test toward critical thinking skills before treatment (pre-test) showed that there is no difference variety in both class, so we can conclude that student's critical thinking skills of both experimental and control class are similar. This fulfils one of the characteristics of experimental that equivalency subject in a different group is required so that if there are different results, it is not caused by the panel equivalent, but the treatment [14]. During the learning process, both of the samples given different treatments. The experimental class was given CUPs learning model, while the control class was given the traditional one.

Table 1. Pre-test and post-test score classical students

Source	Score

Varian	Experimental		Control	
	Pretest	*Posttest*	*Pretest*	*Posttest*
mber of students	38	38	35	35
Highest score	72	82	64	84
Lowest score	24	63	15	39
age	43,74	77	45,09	63,91

Based on the results above, we can see that the average score of experimental class before the treatment was 43.74, which categorised as level, and after the treatment, the average score increased to 77. The improvement of critical thinking skills also happened in control class. The mean rating before they were given the treatment was 45.09, and after the treatment, the results increased to 63.91. The increase of students' critical thinking abilities of experimental class after the treatment is better than the control class. It is proved by the results of the tests showing that students' critical thinking skills of experimental class is different and better than students' critical thinking skills of the control class.

Based on N-gain test on Picture 1 below, we can see that 89.32% students in an experimental class are improved, 15.79% students are highly educated, while 73.53% are frequently developed. In control class, only 57.14% students developed, 8.57% are highly educated, while 48.57% are just average.

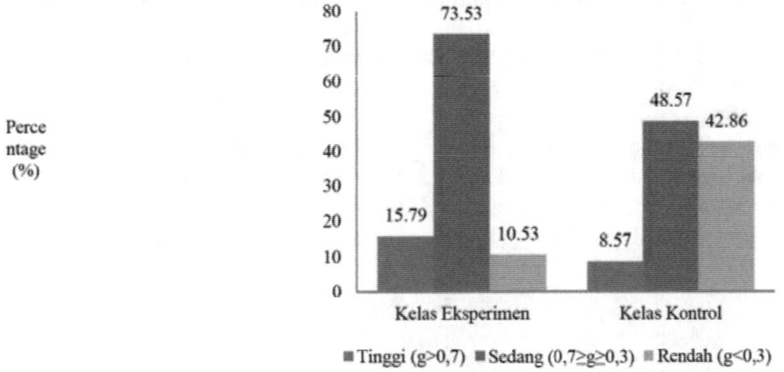

Picture 1. Results of n-gain test

Some factors that cause significant differences and improvement in critical thinking skills in the experimental class better than the control class are as follow:

a. CUPs learning model has several characteristics needed to develop critical thinking skills. CUPs learning model which applied in the preparatory course is more efficient compared to the traditional one which was used to teach in control class. This learning method works that way because it has the characteristics needed to develop critical thinking skills, while the traditional one does not have these features. Critical thinking requires a contextual learning experience, encouraging high levels of interaction and participation, and supports that help students to understand the materials [5]. Those statements appropriate with the characteristics of CUPs learning models. The application of CUPs learning models combining with cooperative learning and fundamental tasks can encourage active interaction and participation for all students, and also give them further understanding [6]. Direct learning model that applied in control class used lecturing method. That lecturing method is involving students passively and automatically, students tend just to memorise the theory that is given by teacher without knowing its function in their daily life [5]. Based on this statement, we can conclude that linear learning model using lecturing method does not have any characteristics needed to improve students' critical thinking skills since it does not make students participate actively in a learning process and it does not lead them to further understanding of the material.

b. Constructivist approach and cooperative value from CUPs learning model are supporting the improvement of students' critical thinking skills. In control class, students got information directly from the teacher by using lecturing method, while in experimental class, they were taught by CUPs learning method which encourages them to build their understanding of the materials. Information-transferring from teacher to students directly in control class makes them involve passively in the learning process so that it is impossible for them to develop their critical thinking skills. Students of an experimental class are more likely to learn new things. They gain real experience of the material given. It leads students of experimental class to be more active in the learning process and develop their critical thinking skills. CUPs learning models, which is supporting cooperative learning, also makes experimental class increase their critical thinking skills compared to control class used direct learning method.

c. CUPs learning model leads all students to be more active in the learning process. By using this learning model, the learning process in experimental class can be more active because every student has to involve in the discussion to solve the problem. Every step of CPUs learning method insisted students to engage in the learning process in the classroom actively. In the direct learning method, teacher divided students into big groups and gave the groups tasks. Some students did not participate in their group work and only being "free-rider" in the groups, while the smart ones did the tasks and being dominant in the groups. It leads students of experimental class to engage a better critical thinking skills than students of control class.
d. CUPs learning model gives an opportunity to students to develop their critical thinking skills compared to direct learning method. By using CUPs learning model, students have bigger chance to think critically, because this learning model consists of three steps where each step engage students in improving their critical thinking skills. In the direct learning method applied in control class, the discussion only occurred in the groups, and the groups in control class were also bigger than the experimental one. The amount of group member who was too large and the discussion which did not start with the building knowledge of the field caused students did not involve in doing group work. The smart students tended to be dominant in the group, and the rests were just being a free rider. These conditions made students lazy to develop their critical thinking skills.

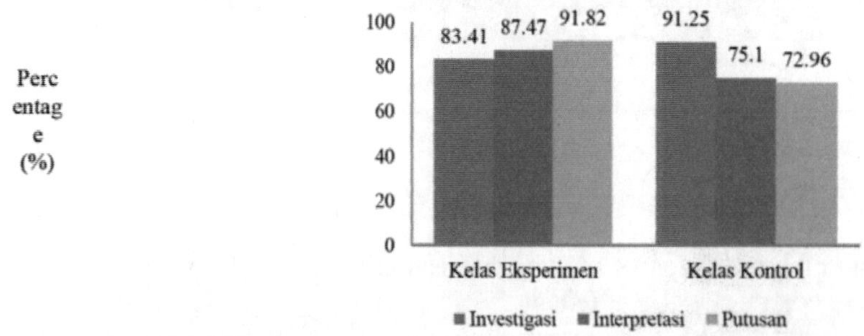

Picture 2. The score of each indicator of students' critical thinking skills test

This research used three aspects or indicators of critical thinking, that is investigation, interpretation, and decision [15]. Based on the test of each indicator, we can conclude that the result of the experimental class is better than the control class. CUPs learning model should be done step by step, and it automatically leads students to participate actively in the learning process. The first step is to make students work individually, so they require stating an opinion before they work in a group. Every task given in every step is similar, and students have to come out with an answer. Students insisted on thinking critically over and over again in one phase by using CUPs learning model so that they will get used to think critically and improve their critical thinking skills.

The results of three aspects of students' critical thinking test showed that on experimental class, decision point has the highest percentage compared to the two aspects, that is 91.82%, the percentage of interpretation issue is 87.47%, and the percentage of investigation aspect is 83.41%. On control class, the highest percentage occurred in research issues, that is 91.25%, while the interpretation issue is 75.1%, and the decision aspect is 72.96%. Students of the experimental class have bigger chance to state their opinion and also choose the most appropriate answer. Choice point occurred in the last session, where students have to choose the most correct answer. They do not just easily accept a reply, but they have also to provide the evidence and reasons of why they want that answer. Investigation and interpretation aspects mostly occurred in individual work. It made decision aspect on experimental class has the higher percentage than the two aspects.

On control group, students only had opportunities to think critically in group discussion. Each group consisted of more than three persons, without any building knowledge of the field, which made the group discussion less efficient since only smart students actively did the task. Students are more likely to get information from the teacher. Lecturing method made students tend to listen and to note any information, and then to memorise it [16]. This method is categorized in investigation aspect. These factors led students of the experimental class to have the critical thinking skills that dominant in the decision aspects, while students of control class are more likely dominant in investigation aspect. Therefore, a learning process which is experienced by students determines the level of their critical thinking skills. The experimental class, which is applied CUPs learning model, got better results compared to control class.

On CUPs learning model, students have to solve the problems themselves. They need to state their opinion and discuss it with their friends. Students do not just listen to the teacher's explanation like in direct learning method. This learning model is appropriate to Piaget learning theory, which stated that student-centered learning would gain better result [9].

3.2 Students' Learning Activity

In this research, students learning activity refers to three aspects of the primary activities of critical thinking that is investigation, interpretation, and decision [2]. Based on the test results, the percentage of students' learning activity on experimental class is 72.37%, and it categorised active, while on control class, the percentage is 22.69% categorized as less active one. The results showed that CUPs learning model improves students' learning activity compared to the direct learning method.

Differences between the experimental and control class learning activities occur because the students in the experimental class like and are interested in group discussion and task-based learning with a gradual learning step. If students are interested in the learning process, they will be more motivated in a participant teaching-learning process.

Table 2. Score of each aspect of students' learning activity

No	Aspects	Experimental class		Control class	
		Score	Percentage (%)	Score	Percentage (%)
1	Investigation	71	93,42	42	36,42
2	Interpretation	53,3	70,17	58	22,38
3	Decision	48,2	63,49	51	16,07

Based on the result in Table 2, we can conclude that investigation point has the highest percentage, that is 93.42% in the experimental class, while the percentage on interpretation issue is 70.17% and on the decision, aspect is 63.49%. The same thing occurred in control class, where the highest proportion is on investigation point, that is 36.43%, while the percentage on interpretation aspect is 22.38% and on decision issue is 16.07%.

According to research results, it can be seen that in both experimental and control class, the highest percentage is on investigation point, but the average rate in experimental class is greater than control class. It happened because students are not accustomed to interpretation and decision aspect. Students are more familiar with the investigation aspect which is often carried out by a teacher on direct learning method before this research conducted.

Teachers usually use lecturing method. It requires students to gain much information directly from the teacher. Receiving any information is regarding as investigation activity. Students in both experimental and control class are accustomed studying activity before this research, so that when we conducted this study, most of the students still use investigation activity.

4. Conclusion

Based on the study and discussion explained above, we can draw a conclusion that the implementation of Conceptual Understanding Procedures (CUPs) learning model in SMA N 1 Larangan gives positive effect to students' critical thinking skills in the material of Protista.

References

[1] Hattie J. 2003. Teachers Make a Difference. Australian Council for Educational Research: University of Auckland.
[2] Ruggiero VR. 2012. Beyond Feelings: A Guide to Critical Thinking. New York: McGraw-Hill Companies.
[3] Tamayo AM. 2013. *Asean Trade Integration 2015*.IAMURE International Journal. 7 (1): 66-84.
[4] Fisher A. 2008. Berpikir Kritis. Terjemahan Benyamin Adinata.Jakarta : Penerbit Erlangga.
[5] Eggen P & Donald K. 2012.Strategi dan Model Pembelajaran. Jakarta: Indeks.
[6] Mills D, Brian M, Pam M, & Susan F. 1999. CUP: cooperative learning that works. Physics Education. 34 (1): 11-15.
[7] Cakir M. 2008. Constructivist Approaches to Learning in Science and Their Implications for Science Pedagogy: A Literature Review. International Journal of Environmental & Science Education. 3 (4): 193-206.
[8] Ormond JE. 1998. Educational Psychology: Developing Learners, 2nd Edition. Upper Saddle River NJ: Merrill Prentice Hall.
[9] Suparno P. 2001. Teori Perkembangan Kognitif Jean Piaget. Yogyakarta: Kanisius.
[10] Isjoni. 2013. Cooperative Learning. BandungAlfabeta.
[11] Slavin RE. 2010. Cooperative Learning. Bandung: Nusa Media.
[12] Ahmad Z. & Mahmood N. 2010.Effects of Cooperative Learning vs. Traditional Instruction on Prospective Teachers' Learning Experience and Achievement.Journal of Faculty of Educational Sciences. 1 (43): 151-164.
[13] Sugiyono. 2010. Metode Penelitian Kuantitatif Kualitatif dan R&D. Bandung: Alfabeta.
[14] Gillies RM & Boyle M. 2010. Teachers' reflections on cooperative learning: Issues of implementation. Teaching and Teacher Education. 26 (1): 933-940.
[15] Ruggiero VR. 2012. Beyond Feelings: A Guide to Critical Thinking. New York: McGraw-Hill Companies.
[16] Djamarah SB. 2006. *Strategi Belajar Mengajar*. Jakarta: PT. Rineka Cipta

Citric acid compounds of tangerines peel extract (*Citrus reticulata*) as potential materials teeth whitening

F Pratiwi*[1], J K Tinata[2], A W Prakasa[3], Istiqomah[1], E Hartini[1], S Isworo[1]

[1]Faculty of Health, Dian Nuswantoro University, Semarang, Indonesia
[2]Faculty of Engineering, Dian Nuswantoro University, Semarang, Indonesia
[3]Faculty of Computer Science, Dian Nuswantoro University, Semarang, Indonesia

Corresponding author: fpratiwi28@yahoo.com

Abstract. Peel of citrus fruit (*Citrus reticulata*) has a variety of possible chemical compounds that may serve as a potential whitening teeth. This research is conducted on a laboratory scale; therefore, it needs to be developed on an application scale. A quasi-experimental was employed in this study. Citric acid extraction was carried out on the type of Sweet Orange (*Citrus Aurantium L*), Tangerine (*Citrus Reticulata Blanco* or *Citrus Nobilis*), Pomelo (*Citrus Maxima Merr, Citrus grandis Osbeck*), and Lemon (*Citrus Limon Linn*). Citric acid's ability test as teeth whitener was performed on premolar teeth with concentrations of 2.5%, 5%, and 10%. The experiments were replicated in 3 times, and teeth whiteness level was measured using *Shade Guide VITA Classical*. The result of this research showed that citric acid in every kind of orange peel with various concentration has different abilities on whitening teeth. The highest colour level obtained from Tangerine peel's citric acid concentration of 5%. Orange peel extract has the best teeth whitening abilities tested by the method of Gass Chromatography to know the active ingredients.

1. Introduction

Citrus is one of the horticultural plants which is produced in high quantity at Indonesian agriculture industry [1]. There are many ingredients inside the citrus fruit that beneficial to the body, according to Purdue University Horticulture and Landscape Architecture, in every 100 g of citrus contains 40 mg of calcium which beneficial to bones and teeth. However, with the exact amount, orange peel contained 161 mg calcium. Orange peel (*Citrus reticulata*) contain many compounds like *Tangeraxanthin, Tangeritin, Terpinen-4-ol, Terpinolene, Tetradecanal, Threonine, Thymol, Thyme- methyl-ether, Tryptophan, Tyrosine, Cis-3-hexenol, Cis-carveol, Citric-acid, Citronellal, Citronellic-acid, Citronellyl-acetate, Cystine, Decanal, Decanoic- acid, Decanol, Nobiletin* [2].

The whitening ability of citric acid extract was previously reported. The citric acid compound in citrus was known its ability in whitening the discoloured teeth, because it has OH group on the chemical structure [3]. The current research was aimed to formulate the composition of whitening compounds from the citric acid extract.

2. Methods

2.1. Materials

Sweet Orange (*Citrus auratium L*),Tangerine (*Citrus reticulata Blanco or Citrus nobilis*), Pomelo (*Citrus maxima Merr, Citrus grandis Osbeck*), and Lemon (*Citrus Limon Linn*). Citric acid's ability test as teeth whitener was performed on premolar teeth with concentrations of 2.5%, 5%, and 10%.

2.2. Procedure

2.2.1. Extraction of Orange Peels and Powder Preparation. Orange peels were separated from the fruit, and it was cut into little pieces. The orange peels were dried using the oven with 50 °C.

*2.2.2. The dr*ied orange peels were then ground into a fine powder using blender machine.

2.2.3. Teeth Colour Observation. Teeth colour was observed after discolouration process using orange peel extract. The citric acid was subjected for teeth soaking. Then, the Shade Guide Vita was applied as the guidance for teeth observation [4].

2.2.4. Teeth Discoloration. The discoloration of teeth was performed using 36 premolar teeth which were polished using nail polish on root part until CEJ part in 3 times. These teeth were then inserted to discoloration fluid that made of 10 g robusta coffee and 180 ml water (boil until 100°C). The solution was then changed every 24 hours during one week [5][6].

2.2.5. Orange Peels Citric Acid. The orange peel's powder of each 2 g, was mixed with 75 ml of water and 25 ml methanol. Then, it was stirred using a magnetic stirrer for 30 minutes with 7th speed, and it was kept in a volumetric flask. Next, the distillation phase for extracting orange peel citric acid was performed. The distillation step is to isolated citric acid compound from orange peel's powder. Sodium hydroxide (NaOH) is used to separate the citric acid, and the orange peel's powder distilled four times [7]. Then, after these processes, the distillation results were subjected to the GC-MS analysis.

2.2.6. Treatment Stage. The discoloured teeth were then employed into each flask with orange peel extract. The mixture was then placed in the incubator with 50 °C heating for 25 s, after that, it was then lowering the heat to 37 °C for 30 min. The process will produce white colour to the teeth.[6]

2.2.7. GC-MS Analysis. Gas chromatography analysis was performed by employing the samples that have been in the extraction was analysed by the method of gas chromatography FPD (flame. photometer Detector) Model Shimadzu GC-FID 2014. GC tool is equipped with AFC (Advanced Flow Controller). The temperature of the injector on the GC-MS was in maintenance at a temperature of 260 °C and pressure of the gas flow of Helium Ultra High Pure (UHP) was at 100 kPa and was utilised as a carrier Gas Chromatography – Mass Spectrometry (GC-MS).

2.2.8. Statistical Analysis. The data was analysed and interpreted using IBM SPSS 20 software. In this study, the analysis was conducted using two tests, there was a difference test of various types and concentrations of orange peel's extract to the ability to whiten teeth using One Way ANOVA, and the second test was Paired T-test that used to analyse level change between discoloured teeth and teeth after treatment.

3. Results and Discussion

3.1. Colour Changing After Testing

Colour measurement tests were applying the Shade Guide VITA Classical that have 16 category colours. The colour for shade guide starts from most white, bright, to most dark. The colour that has sorted as their order, and the order score is: B1=1, C1=2, A1=3, B2=4, D2=5, C2=6, B3=7, A2=8, B4=9, C3=10, A3=11, D3=12, D4=13, C4=14, A4=15, A35=16, and these the result of change discolored teeth that had soaked on extract acid citric orange peel in 30 min, with 4 kinds of orange peel and in various concentration.

Table 1. Result of ability test citric acid orange peel as teeth whitening at various concentrations

Citrus	Concentration	Discoloured	Score	Dental Score after Treatment			Mean
				I	II	III	
Sweet	2.5%			6	6	6	6
	5%	C4	14	4	4	4	4
	10%			2	2	2	2
Tangerine	2.5%			4	2	7	4.33
	5%	C4	14	2	1	1	1.33
	10%			7	7	7	7
Pomelo	2.5%			8	10	14	10.6
	5%	C4	14	10	10	7	9
	10%			7	10	10	9
Lemon	2.5%			7	7	4	6
	5%	C4	14	10	10	7	9
	10%			4	4	4	4

The data shown in the Table 1 indicates that citric acid on the tangerine peel with 5% concentration give the best result.

3.2. Data Analysis Result

From One Way Anova test, it is obtained the score Sig 0.000 (<0.05) which shows the different effect that significant between variation kinds and concentration orange peel (Table 2). On the other side, paired t-test also show sig 0.000 (<0.005) and prove that significant transformation between discoloured teeth and soaked teeth with orange peel extract (Table 3).

Table 2. Result of One Way ANOVA parametric statistical test

Source	df	Mean Square	F	Sig.
Intergroup	11	26.816	11.492	0.000

Table 3. Result of *Paired T-test* parametric statistical test

	Mean	T	df	Sig.
Pair 1 Discoloured score – Score after soaking on citrus peel	7.972	15.105	35	0.000

3.3. GC-MS Test Result

GC-MS tests carried out on samples that have the best ability to whiten teeth. Orange peel extract that has distilled to isolate the citric acid tested using GC-MS method for seeing active compound of citric acid as Table 4.

Table 4. GC-MS Test Result

Peak no.	Retention time	Area %	Name
1	43.771	21.46	methyl dihydro malvalate
2	47.782	20.17	Hexadecanoic acid, 1-(hydroxymethyl)-1,2-ethanediyl ester
3	51.298	33.38	9-Octadecenal, (Z)
4	67.940	16.69	Cholest-5-en-3-ol (3.beta.)-, carbonochloridate
5	67.969	8.30	Androstan-3-one, 17-hydroxy-1,17-dimethyl-, (1.alpha.,5.alpha.,17.beta.)

The table above telling us that orange peel's citric acid have several active compounds, like one that peaks no 2, there are *Hexadecanoic acid 1-(hydroxymethyl)-1,2-ethanediyl ester* that has the cluster as Figure 1.

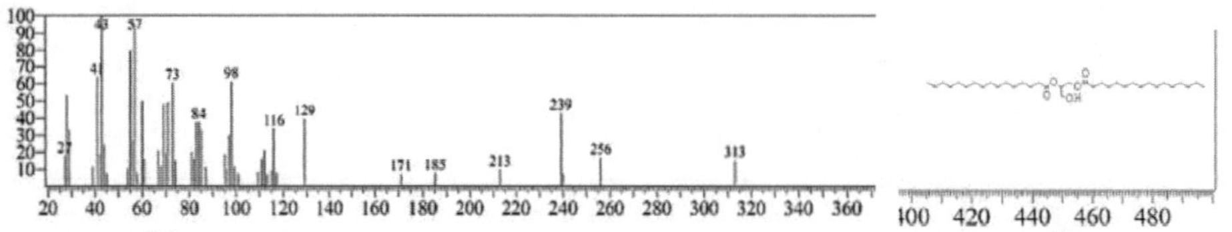

(a) (b)

Figure 1. (a) Mass Spectrum of Hexadecanoic acid and (b) Hexadecanoic acid cluster

From the chart above, Hexadecanoic acid has Hydroxide (OH) group that known can whiten teeth. Also Hexadecanoic acid has potent as anti-bacterial and anti-fungi. Then, as shown Figure 2 it has active compound Cholest-5-en-3-ol or cholesterol that can be as an emulsion stabiliser on toothpaste [8]., while as shown Figure 3, there are active compound Androstan which have OH group and can work as anti-inflammation on gum [9].

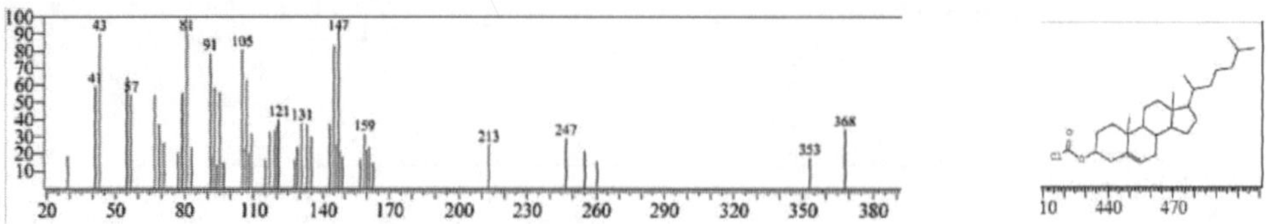

(a) (b)

Figure 2. (a) Mass Spectrum of Cholest and (b) Cholest cluster

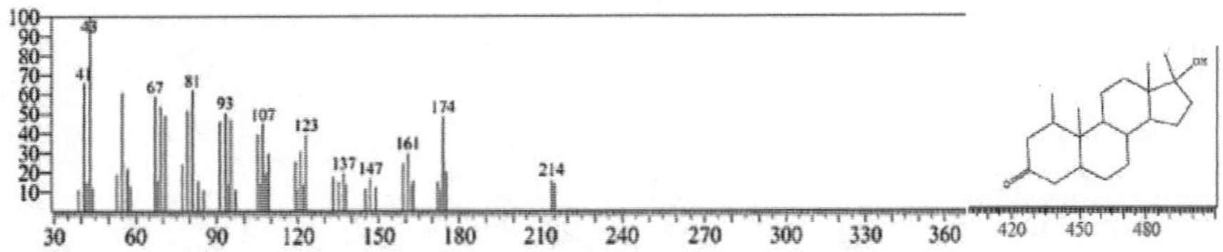

(a) (b)

Figure 3. (a) Mass Spectrum of Androstan and (b) Androstan cluster

4. Conclusion

Based on research which has been done, we get the result that citric acid containing on orange peel can whiten the teeth. Furthermore, these orange peel's citric acid have the active compound that functions as anti-bacterial, anti-fungi, emulsion stabiliser and anti-inflammation on gum. To obtain citric acid on an orange peel that can work as teeth whitener can be done with extraction. After testing on discoloured teeth, we have the result that tangerine peel with 5% concentration gives the best result on whitening teeth than other orange peels and concentrations.

References

[1] Penjor T et al. 2014 Characterization of limes (*Citrus aurantifolia*) grown in Bhutan and Indonesia using high-throughput sequencing *Scientific Reports (4) p 4853*

[2] Pan MH et al. 2002 Tangeretin Induced Cell Cycle G1 Arrest through Inhibiting Cyclin Dependent Kinases 2 and 4 Activities as well as Elevating Cdk Inhibitor p21 and p27 in Human Colorectal Carcinoma Cell *Carcinogenesis 23(10) p 1677-1684*

[3] Sundfeld RH et al. 2014 Microabrasion in tooth enamel discoloration defects: three cases with long-term follow-ups *Journal of Applied Oral Science 22(4) p 347-354*

[4] Zenthofer A et al. 2014 Selecting VITA classical shades with the VITA 3D-master shade guide *The International Journal of Prosthodontics 27(4) p 376-382*

[5] Rochmah N 2014 Lime (*Citrus aurantifolia*) Potential to The Whiten Discoloration Tooth Enamel *Insisiva Dental Journal 3(1) p 78-83*

[6] Kim JH et al. 2010 Tooth discoloration of immature permanent incisor associated with triple antibiotic therapy: a case report *Journal of Endodontics 36(6) p 1086-1091*

[7] Erukainure OL et al. 2016 Orange Peel Extracts: Chemical Characterization, Antioxidant, Antioxidative Burst, and Phytotoxic Activities *Journal of Dietary Supplements 13(5) p 585-594*

[8] Kamba EA et al. 2013 Utilization of Different Emulsifying Agents in the Preparation and Stabilization of Emulsions *International Journal of Materials and Chemistry 3(4) p 69-74*

[9] Khan NT et al. 2012 Synthesis of some potent immunomodulatory and anti-inflammatory metabolites by fungal transformation of anabolic steroid oxymetholone *Chemistry Central Journal 6 p 153*